BIOLOGY COLORING WORKBOOK

2nd Edition

Penguin
Random
House

BIOLOGY COLORING WORKBOOK

by I. Edward Alcamo, Ph.D.

Illustrations by John Bergdahl

Second Edition

Penguin
Random
House

The Princeton Review
555 W. 18th Street
New York, NY 10011
E-mail: editorialsupport@review.com

Published in the United States by Penguin Random House LLC, New York, and in Canada by Random House of Canada, a division of Penguin Random House Ltd., Toronto.

ISBN: 978-0-451-48778-0

Editor: Selena Coppock
Production Editors: Kathy G. Carter, Liz Rutzel
Production Artist: Debbie Silvestrini
Designer: Craig Patches

Printed in the United States of America on partially recycled paper.

Second Edition

ACKNOWLEDGMENTS

The Princeton Review would like to thank I. Edward Alcamo, PhD, for his outstanding work in building this book and John Bergdahl for illustrating every piece of art in that original text. Their hard work and imagination in creating this book made it the top seller that it continues to be.

The Princeton Review would also like to thank Deborah Silvestrini for her tireless work on the production side of this revision. Debbie is an outstanding designer and we are so lucky to have her on our team.

Last and very certainly not least, The Princeton Review would like to thank Jes Adams for her fantastic ideas, energy, inspiration, and thorough content review that comprise this second edition.

HOW TO USE THIS BOOK

This book contains 156 coloring plates, each consisting of text and directions and easy-to-follow diagrams. As the text proceeds, we take you to a concept or structure and ask that you locate and color it. As your senses become more involved in the diagram, you begin to see structures and relationships and you realize that biology can be understood and learned quite easily.

To get the most out of this book, you should have between 10 and 20 different colored pencils or felt-tipped pens. (Be careful of the pens, however, because they can be difficult to control and they may bleed through the pages). You should have a mix of light and dark colors to begin with, and can add to your collection as you go along. Try not to be intimidated by the diagram—just dive in and begin reading and coloring. You'll soon see the rewards of your work.

We've tried to simplify matters by leaving most choices up to you. We've omitted unusual symbols, and arrows or brackets to point out key parts or processes. The text will advise you of these as you move along. We also use capital or lowercase letters on occasion when there is a logical reason to do so. In the end, we hope you will use the coloring process to construct your unique image of the biological world.

TABLE OF CONTENTS

Register Your

1 Go to **PrincetonReview.com/cracking**

2 You'll see a welcome page where you can register your book using the following ISBN: 9780451487780

3 After placing this free order, you'll either be asked to log in or to answer a few simple questions in order to set up a new Princeton Review account.

4 Finally, click on the "Student Tools" tab located at the top of the screen. It may take an hour or two for your registration to go through, but after that, you're good to go.

If you have noticed potential content errors, please email EditorialSupport@review.com with the full title of the book, its ISBN number (located above), and the page number of the error.

Experiencing technical issues? Please e-mail TPRStudentTech@review.com with the following information:

- your full name
- e-mail address used to register the book
- full book title and ISBN
- your computer OS (Mac or PC) and Internet browser (Firefox, Safari, Chrome, etc.)
- description of technical issue

Book Online!

Once you've registered, you can…

- Download a printable Index of this book
- Access information about an assortment of Princeton Review test preparation programs
- Check to see if there have been any corrections or updates to this edition

Offline Resources

If you are looking for more biology review or medical school advice, please feel free to pick up these books in stores right now!

- *Cracking the AP Biology Exam*
- *Cracking the SAT Biology Subject Test*
- *The Best 167 Medical Schools*
- *The Princeton Review Complete MCAT*
- *Essential Anatomy Flashcards*

CHAPTER 1:

INTRODUCTION to BIOLOGY

CHARACTERISTICS OF LIVING THINGS

Biology is the scientific discipline that is concerned with the study of life. Strictly speaking, it is difficult to describe any one prototypical living thing, but in this plate we will describe some characteristics that all living things possess.

As you look over the plate, you will note that it consists of seven sections, each of which shows one characteristic of a living thing.

One of the characteristics of living things is internal organization. The cell itself is characterized by internal organization, and, in the case of multicellular organisms, the organism possesses an overall internal organization.

All living things must obtain materials and energy from the environment. These are used in metabolism, which is the sum of all of the chemical reactions that occur in an organism. In this plate, we see a **plant (A)** and an **animal (B).** Plants obtain nutrients from the soil, while animals obtain them through the ingestion of **food (C).** In both organisms, reactions are carried out and substances are synthesized and broken down. Energy is released from the organic substances that were obtained from the environment, and this energy fuels the organism.

All living things exhibit control (see the **bird (D)** in the figure), which refers to the coordination of their body activities, as well as the regulation of their internal and external processes. Control refers to the fact that the metabolic activities in living things occur in sequence and at specific rates. The flapping of the bird's wings, for example, requires that nerve impulses are coordinated as they travel to muscles and that muscles contract in sequence. Blood must be supplied to tissues, and waste products must be removed from them. Dynamic control is absent in nonliving things.

We have examined metabolism and control as two characteristics of living things. We now move to three additional characteristics as we continue our introduction. Continue your reading below as you color the plate.

Another essential aspect of life is reproduction. Here we see a **cell (E)** undergoing binary fission to produce two identical daughter cells. Reproduction is accompanied by an increase in size called growth. In the process of growth, living things increase the size of their structures. In this plate, we see growth in a **human (G).** Growth can also mean that cells increase in size, number, and specialization. As you can imagine, the fully formed human is much more complex than the fertilized egg cell from which it developed.

Living things also exhibit responsiveness, which means that they respond to changes to their environment. In the art, you can see a **responsive individual (F)** reacting to an injury. Hormones, nerve cells, and various other chemical substances bring about responses in humans.

We have now examined five characteristics that all living things possess. We will finish by examining adaptation and evolution, two more characteristics that are attributed to living organisms.

Adaptation refers to physical adjustments organisms make over relatively long periods of time to changes in their environments. In this plate, you can see an **adapted animal (H).** This is an animal with a coat that blends with the brown prairie grass among which it lives in the summertime. When snow falls, the animals loses its brown coat and takes on a white one in order to blend into its new environment.

Evolution refers to populations changing over time. Evolution is a slow change in the genetic makeup of a population (rather than an individual) that enables it to survive in its environment. We show the **evolving population (I)** as the evolution of the human species. By continuing to adapt to its changing environment, the human population survived the selective pressures of the environment over thousands of years. With evolution our population improved its ability to live in the environment, and avoided extinction. Other populations, which failed to make these adaptations, became extinct.

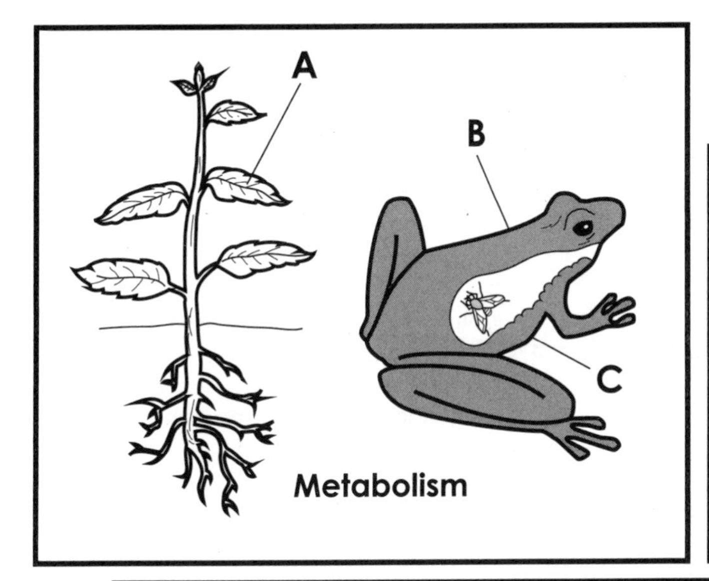

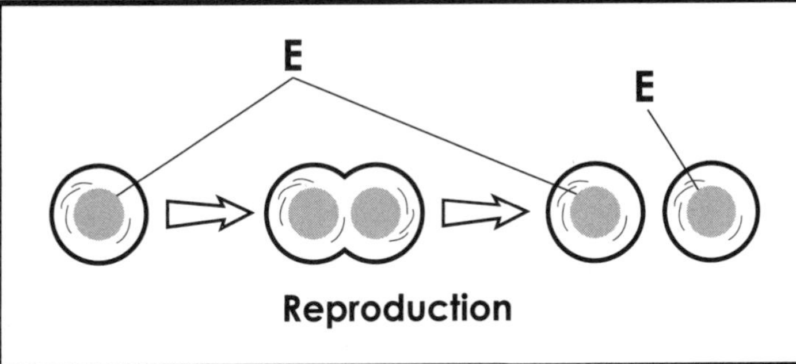

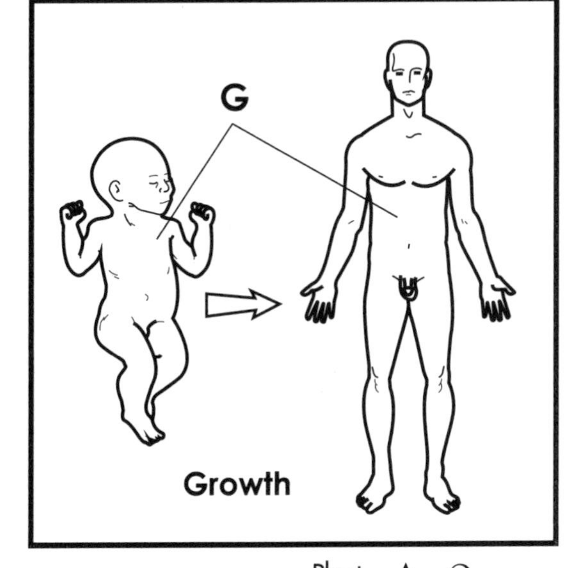

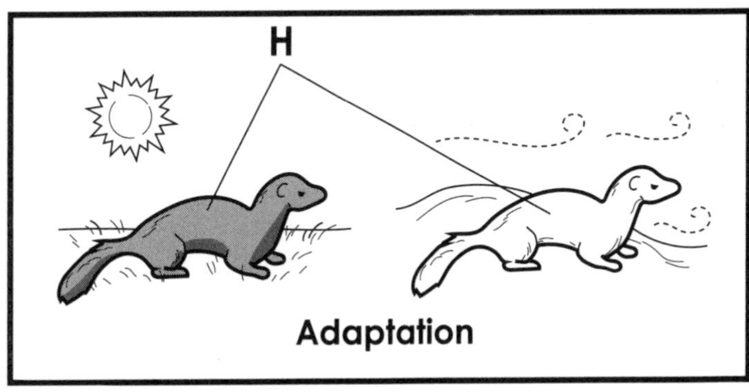

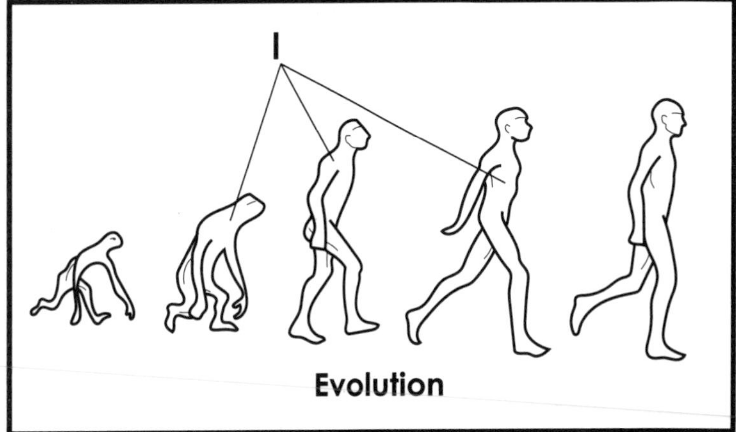

Plant	A	O
Animal	B	O
Food	C	O
Bird	D	O
Cell	E	O
Responsive Individual	F	O
Human	G	O
Adapted Animal	H	O
Evolving Population	I	O

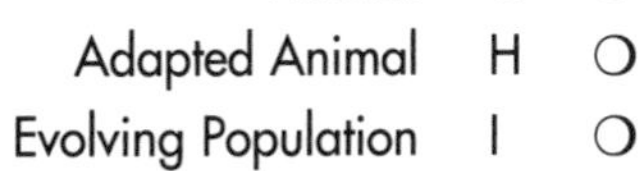

BIOLOGICAL ORGANIZATION

Organization in living things exists in a hierarchical structure. Organization exists within organisms, as we will discuss in this plate, and also within environments; this will be covered in the ecology chapter.

In this plate, we will show the hierarchy of organization that exists in a human. We will begin with the most basic level of organization at the bottom of the plate, and show the increasing complexity as we move up.

The most fundamental level of organization on Earth is the atom. All matter in the universe is composed of atoms. There are 92 naturally occurring elements, each of which is made up of a distinct atom. About 99% of the atoms in living things are carbon, hydrogen, oxygen, nitrogen, sulfur, or phosphorous.

At the bottom of the plate we show a typical **atom (A);** you should color the bracket in a bold color such as dark red or green. The atom consists of positively charged particles called **protons (A_1)** and uncharged particles, or **neutrons (A_2).** These particles are surrounded by **electrons (A_3),** which have a negative charge. Elements vary according to the number of protons present in the nuclei in their atoms. An atom of carbon, for example, has six protons, in its nucleus, while an atom of oxygen has eight protons.

Atoms join to form the **molecule (B).** A molecule consists of various kinds of atoms bonded to one another. In the plate, we show a **DNA molecule (B_1),** which is involved in heredity. It is composed of atoms of carbon, hydrogen, oxygen, nitrogen, and phosphorus. The atoms are linked and wound to form the familiar spiral staircase (double helix) of the DNA molecule. A light color should be used for it.

The most basic unit of life is the **cell (C).** In the plate, we show a muscle cell, which has both **cytoplasm (C_1)** and a **nucleus (C_2).** The cell is composed of molecules of protein, carbohydrate, lipid, and nucleic acid. It is long and slender in shape, and contracts to create movement. The human body consists of approximately 100 trillion cells.

Atoms and molecules are inert, but the cell is alive and has the properties of a living thing, which were discussed in the previous plate. We will now progress to the next level to see how cells are organized.

In some cases, cells operate independently. These are unicellular organisms, which include bacteria and protozoa. In humans, cells are generally grouped together to form tissues. In the plate, you can see one section of a typical **tissue (D)** outlined by the bracket. When muscle cells contract in unison, the entire muscle contracts. Other types of body tissues are nervous tissue, epithelial tissue, and connective tissue. They are discussed in a future plate, titled Cell Types and Tissues.

When various tissues join with one another, they can form an **organ (E).** In this plate, we show the heart as typical organ. The heart has two pumping chambers called **ventricles (E_1).** Carrying blood to and from the ventricles are the **major blood vessels (E_2).** Arteries take blood away from the heart of the body and veins bring blood back to it. The heart contains various types of tissues: muscle tissue that allows contraction, nerve tissue that controls the contract rate, and membranous tissue that enclose it.

Various organs in combination are called an **organ system (F).** An example of an organ system is the circulatory system. It consists of the **heart (F_1)** and numerous blood vessels. Certain of these blood vessels carry blood to and from the lungs and compose the **pulmonary circulation (F_2),** while other vessels carry blood to and from the other sections of the body and compose the **systemic circulation (F_3).** Another organ system is the digestive system. Organs in this system include the esophagus, stomach, pancreas, small and large intestines, and liver.

The highest level of biological organization is the **organism (G).** It is composed of numerous organ systems, including the digestive, nervous, circulatory, respiratory, excretory, and reproductive systems. The organ systems work closely together.

Biological organization can also be studied at the level of ecology, which we will discuss later in the book. A number of organisms grouped together constitute a population, and various populations form a community.

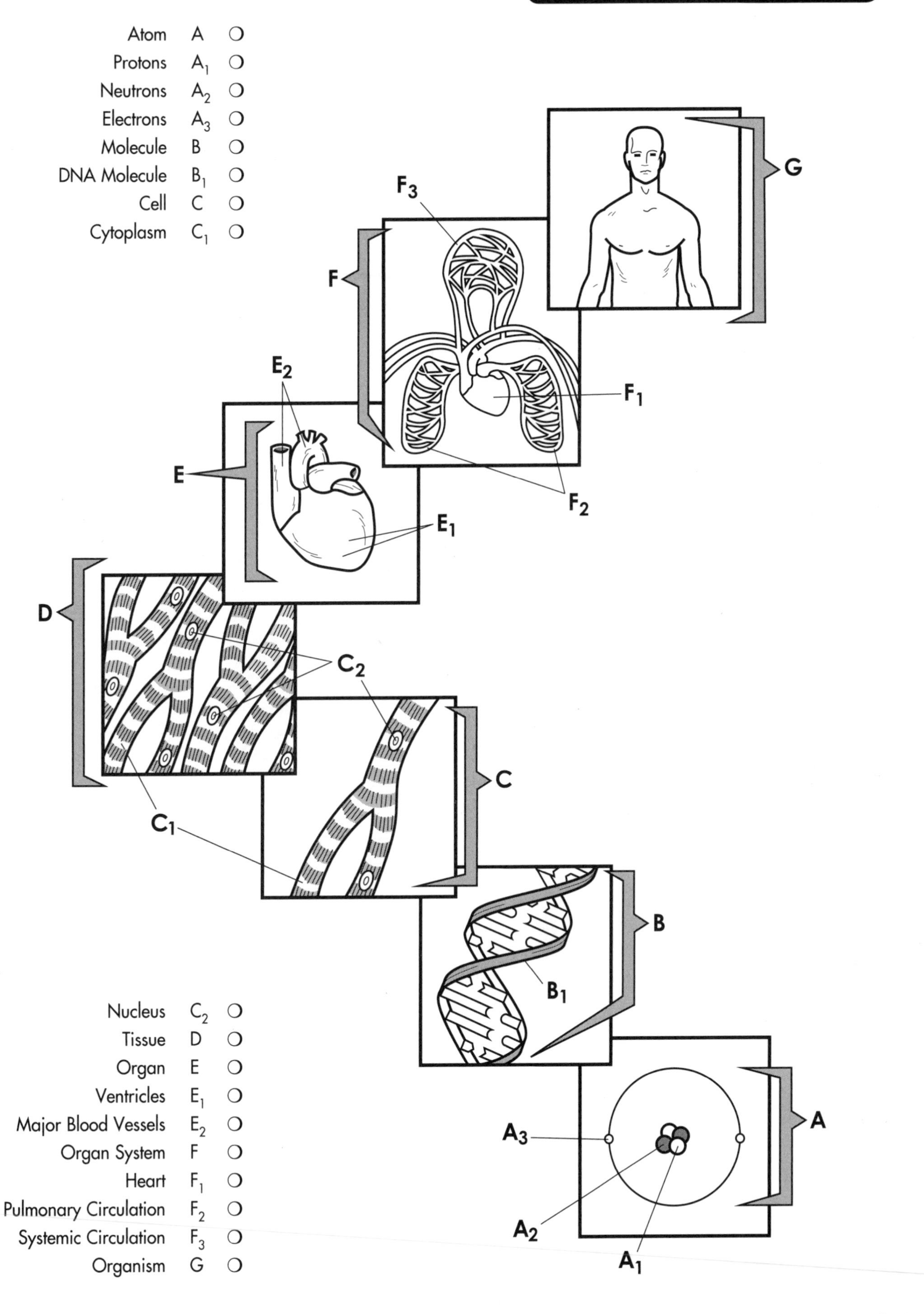
Atom A
Protons A_1
Neutrons A_2
Electrons A_3
Molecule B
DNA Molecule B_1
Cell C
Cytoplasm C_1
Nucleus C_2
Tissue D
Organ E
Ventricles E_1
Major Blood Vessels E_2
Organ System F
Heart F_1
Pulmonary Circulation F_2
Systemic Circulation F_3
Organism G
G
F_3
F
F_1
E_2
E
F_2
E_1
D
C_2
C
C_1
B
B_1
A
A_3
A_2
A_1

SIZE RELATIONSHIPS IN BIOLOGY

Organisms vary considerably in size. In this plate, we survey some members of the biological kingdom, and point out how they relate to one another in size.

This plate displays 13 different organisms, each of which is placed along the meter scale to show its size and relationship to other biological specimens. We will examine the specimens in three groups.

We will move from the smallest organism, which is on the left, to the largest. We begin with specimens that are too small to be viewed with any instrument, and those that can be viewed only with the **electron microscope (A).** The arrows that group the different size organisms should be colored in bold reds, greens, or blues.

Atoms cannot be seen under electron microscopes. An **atom (B)** measures approximately 0.1 nanometers (nm) in diameter, and a nanometer is a billionth of a meter. (So an atom is a ten-billionth of a meter in diameter.) The smallest specimen that can be viewed with an electron microscope is a **small molecule (C).** Such a molecule measures approximately 1 nm (one-billionth of a meter) in diameter. An amino acid and a glucose molecule are examples. Slightly larger specimens are large, **folded proteins (D).** Proteins can range in size from about 4 nm to about10 nm in diameter, as the bracket indicates.

Within the size range of 10 to 100 nm are the **viruses (E).** In the diagram, you can see a complex virus known as a bacteriophage. Its details are easily visible through the electron microscope. Viruses are inert particles that have the ability to replicate.

Thus far we have examined several specimens that are visible only though the electron microscope. We now turn to biological specimens that are visible with the light microscope. Continue your coloring as you read the paragraphs below.

The range of sizes of objects that is visible with the aid of the **light microscope (F)** is indicated by the arrow. The range starts at the objects that are approximately 100 nm. Included in this is the **chloroplast (G),** which is the photosynthetic organelle of the plant. Chloroplasts are approximately 1 micrometer (μm) in diameter, and a micrometer is one millionth of a meter.

As the diagram shows, **bacteria (H)** range in size from slightly less than 1 μm to about 10 μm. They are smaller than **plant and small cells (I),** which range from approximately 10 to 100 μm. Note that the electron microscope can be used to view these samples, and it is often used to view the submicroscopic structures in plant and animal cells.

Also within the realm of the light microscope is the **human egg cell (J).** This cell measures approximately 125 μm in diameter and is almost large enough to be seen by the naked, or unaided, eye.

We now move to those biological specimens that can be visualized with the unaided eye. Continue your reading below as you note the various size relationships among biological objects.

The range of the **unaided eye (K)** is indicated by the arrow, which should be colored a dark color. Within this realm is the **frog egg cell (L).** This specimen is slightly more than a millimeter (mm) in diameter. (A millimeter is a thousandth of a meter.)

The next specimen we will look at is the **insect (M).** Although the insect is larger than the frog egg cell, the size of insects varies considerably. The insect we display is approximately a centimeter (cm) in length. The next largest organism on our chart is the **rodent (N),** which is approximately 5 cm in length.

Now we examine the **human (O).** The average human is taller than a meter. A meter is equivalent to 39.37 inches, or slightly larger than three feet. The human we show is approximately five feet in height, or about 1.2 m.

The largest biological organism presented is the **whale (P).** Notice the enormous size of this animal relative to the other biological specimens. The animal is roughly 25 m in length, or approximately 75 feet. The whale is among the largest organisms known to inhabit Earth.

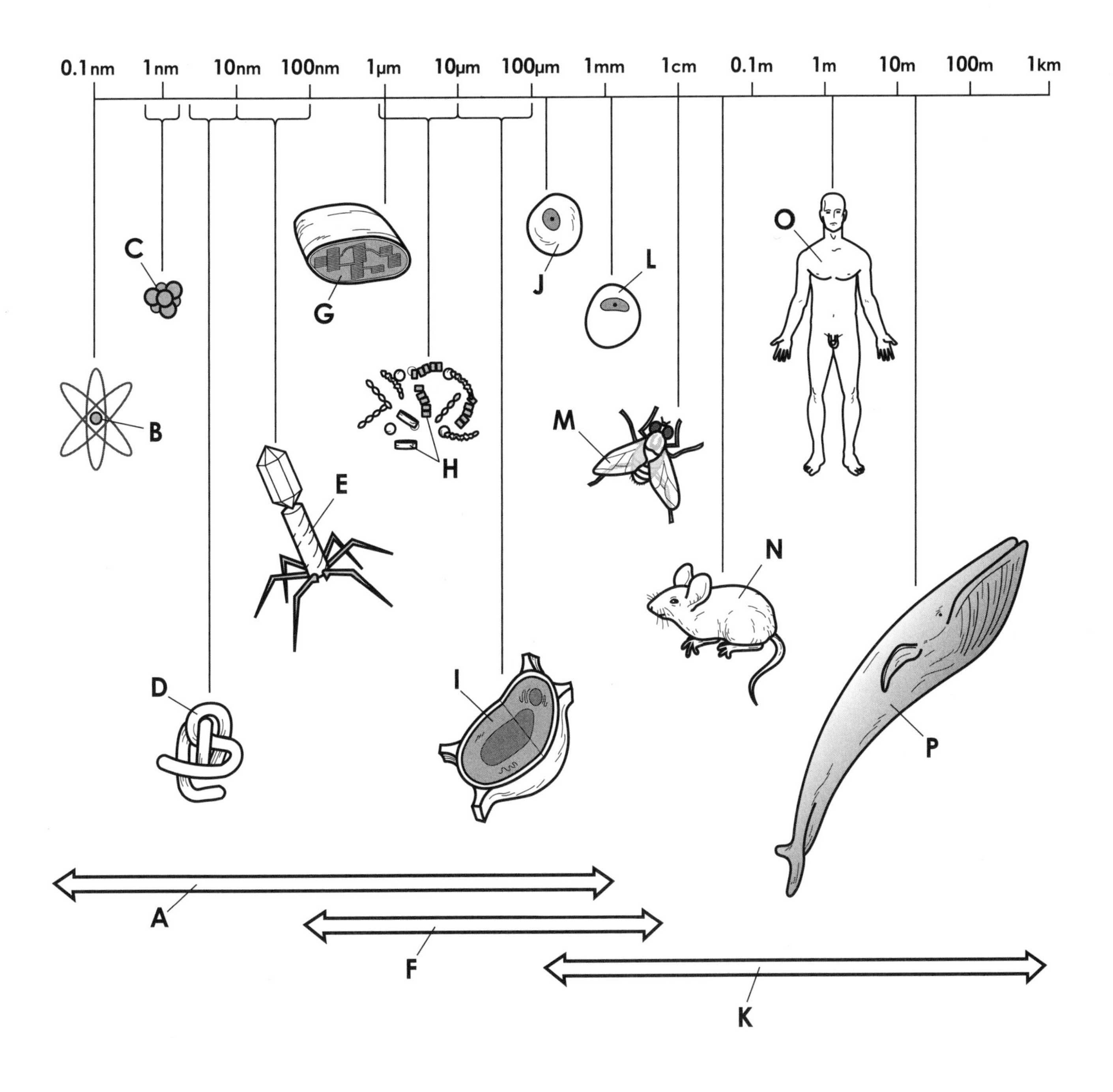

Electron Microscope Range A ❍
Atom B ❍
Small Molecule C ❍
Folded Protein D ❍
Virus E ❍

Light Microscope Range F ❍
Chloroplast G ❍
Bacteria H ❍
Plant/Animal Cell I ❍
Human Egg Cell J ❍

Unaided Eye Range K ❍
Frog Egg Cell L ❍
Insect M ❍
Rodent N ❍
Human O ❍
Whale P ❍

THE SCIENTIFIC METHOD

Science is not just a subject. Science is a process and scientists are truth-seekers. Science is the process of learning about the world around us. Everything we know about the world started with an observation and simple curiosity. In other words, somebody asked a question.

These types of curious questions are the foundation of science, and finding the correct answer(s) is the magic of the scientific process. Thinking about something carefully and thoroughly is called critical thinking. It requires using facts, information, and observations to judge information before believing it.

Finding the truth can be a difficult process. It must begin with an open mind. Questions can only be answered by using evidence. Evidence is what separates a scientific conclusion from a guess. Since evidence is crucial to the decision-making process, it needs to be indisputable information.

The process of science should be pursued carefully. Scientists must decide which choices are best for science and best for society. They must also avoid personal bias. If scientists do not have open minds, their bias could influence their critical thinking. Personal bias could cause scientists to twist observations to support the conclusion that they are rooting for, but this is unethical! Personal desires should always stay out of science. A tiny twist here or there completely destroys the neutral scientific process.

Answering a question is important, but the way it is answered is also important. Some questions are difficult to answer because doing so may cause harm. For example, a car company would benefit from knowing exactly what happens to a human passenger in a collision test. Yet, this is not done because it is unethical to intentionally crash a car with humans in it. This is why crash test dummies were invented. Scientists avoid harmful experiments.

When experiments are used to answer a question, the **Scientific Method** is a handy process used to work through them. It has 5 steps:

1. Start with a Question: Scientific questions should be specific and measurable.

It is important to start with the right question. Sometimes science has to solve big questions, but first scientists break them down into smaller questions. Each question must be something that tests for a specific **quantitative trait.** Quantitative is a fancy way of saying that something can be counted or measured. Traits that are not measurable are called **qualitative traits.**

Here are some examples of quantitative traits:

- quantity
- size (height, weight, length, width)
- speed
- pH
- temperature
- pressure
- absorbance wavelength

Quantitative traits are reported in numbers and are not swayed by the scientist's opinion.

Here are some examples of qualitative traits:

- better/worse
- happiness
- beauty
- humor
- health
- texture
- color

Qualitative traits are described in words (instead of numbers) and are opinion-based. This means that each person can experience and describe them differently.

Scientific questions should never ask about things that vary with opinion. Evidence must be something that can be measured. Now is a good time to start coloring the plate that follows this text. Start with diagram 1, where you can see a scientist asking a question.

2. Come up with a Hypothesis: What do you expect to happen?

Hypothesizing is a brainstorming thought process (color diagram 2). It requires you to think about possible answers to your question, and about the things that will factor into the experiment. It will help you design an experiment that can prove or disprove some of your hypotheses.

It is sometimes helpful to think about your experiment as a test of your favorite hypothesis. Pretend that one of your hypotheses is right. Then, think about how you can prove it is right and/or how you can prove that the other hypotheses are wrong. A perfect experiment eliminates all possible hypotheses except one. Later, you can use the hypothesis as a double check to make sure your experiment is designed correctly.

3. Experiment: Set up a situation to answer your question and gather evidence.

Experiments allow scientists to set up the perfect situation so they can gather the evidence they need to answer a question (color diagram 3). These 6 steps will help you design an experiment:

Step 1—Identify the different scenarios you want to compare. Think about your original question and your hypothesis. What are you testing?

Step 2—Make groups that represent each of the scenarios. To test something, there must be groups that share a common trait, and groups that either don't have that trait OR have something else instead. The difference between your groups is called the **independent variable.**

Every experiment needs different scenarios to compare. Sometimes, you will be comparing situations in which you are adding or changing something. In these cases, the normal group (or no treatment group) is called the **control group.** Control groups are important because they give you a baseline, or a measure of background effects. The groups that are not the control group are called **experimental groups.**

Sometimes it can be tricky to figure out what you need to use as the control group. Just remember, you need to compare your experimental group to the control group, so it should basically be the same as your experimental groups, except it has a background or baseline version of the independent variable.

Step 3—Make sure everything else is the same between the groups.
To know for sure that the difference between groups is caused by the independent variable, everything else must be kept the same. Things that are the same are called **constants.** Only the factor that is being compared should be different between groups.

Step 4—Use multiple samples or repeat the experiment. Once is never enough.
Think about what would happen if you took two measurements and they were very different. You wouldn't know which one was true and which was the fluke. The best way to make sure your evidence is reliable is to see if it happens again. Generally, repeating things three or more times is best.

Step 5—Think carefully about how you will collect data.
Be sure to think through the whole process before you start! Make sure that the length of time for the experiment makes sense. For longer and more detailed experiments, it is often necessary to take more measurements. You could take a measurement every week, or every day, or even every hour. Just make sure you think out your plan ahead of time.

Step 6—Do it.
Remember, each group must be treated exactly the same. The only difference should be the independent variable. Data collection must also be conducted in the same way for each group. Another name for the data you gather is the **dependent variable.** It is a variable because it should vary between the different groups. It is dependent because it depends on the independent variable.

4. Analysis: Summarize your data.

Once the experiment is over, it may seem like the time to relax, but this is when the brainwork really begins. Your experiment should have provided data. If it didn't, you need to rethink the purpose and design of your experiment. The data is called your **results.**

When you summarize your data, your goal is to describe what the results show. For example:

- Are there any patterns or trends?
- Does something increase or decrease? If so, why?
- If two groups gave the same results, what does this mean?

Displaying your data in a graph usually makes it easier to understand. Bar graphs and line graphs are common ways to show data (color diagram 4). Different types of graphs are helpful for different types of data.

5. Conclusion and Future Steps: Answer your question and figure out what's next.

Once you have poured over the numbers, it is time to answer the question you asked at the beginning of this process. Think about your hypothesis. If it is true, would you get these results? Sometimes thinking about possible results is helpful to do before you even look at the real results. A lot of numbers can be confusing, so knowing what to look for ahead of time can be helpful.

Try to summarize the answer in a single sentence. Does it prove or disprove your hypothesis? Now is a good time to color the last part on this plate, diagram 5.

Finding a conclusion just means that a particular question has been answered, but science itself is an ongoing process. New questions are raised all the time, especially if the hypothesis was incorrect (which happens all the time). In fact, an answer might open the door to a hundred new questions. Questions that seemed simple might have complex answers.

As new information is discovered, the truth can grow and change. Scientists know this and so they are hesitant to use the word *truth*. Instead, if something is very well known, they call it a theory. Have you heard of the Theory of Relativity or the Theory of Evolution? These have been shown to be true many, many times, so scientists basically consider them to be true. Unless, of course, someone proves them to be incorrect.

An Example of the Scientific Method

Now that you've learned about the Scientific Method, let's go through an example so you can see exactly how it works in a real situation.

1. The Question: Pretend you have a new plant and you need to decide where it should live. Don't ask these questions:

- Does this plant grow better in the front or the back window?
    - What does "better" mean? Taller? Faster? Greener? Number of flowers?
- Where does this plant grow fastest?
    - There are too many options here; you can't test every possible location. Be specific.

Instead, you should ask this question:

- Will this plant grow faster in the front window or the back window?

2. The Hypothesis: We have three possible outcomes:

- Front window causes more growth
- Back window causes more growth
- Both windows cause the same growth

Now, if you had to make an educated guess, maybe you would pick the window that looks the sunniest. Let's pretend the sunniest window is the back window. Here is your hypothesis: This plant will grow faster in the back window than in the front window.

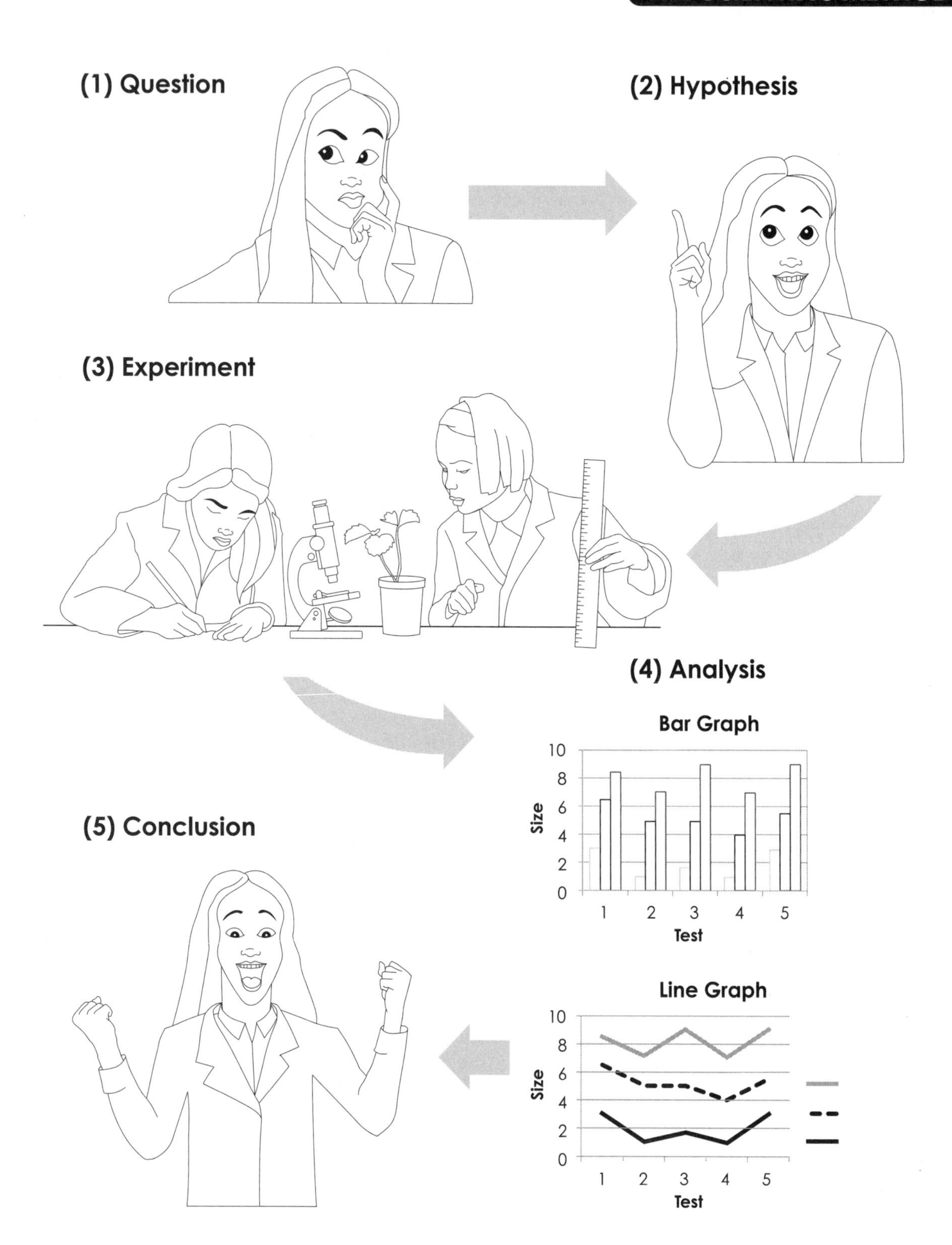
(1) Question
(2) Hypothesis
(3) Experiment
(4) Analysis
Bar Graph
10
8
6
4
2
0
Size
1
2
3
4
5
Test
(5) Conclusion
Line Graph
10
8
6
4
2
0
Size
1
2
3
4
5
Test

3. The Experiment

Step 1—The Scenarios

- We want to compare plant growth in a front window vs. plant growth in a back window.

Step 2—The Groups

- One group should have plants grown in a front window and one group should have plants grown in a back window. Location is the independent variable.

Step 3—Keep everything else the same

- Only the location should be different. The plant, pot, soil, amount of water, and length of the experiment should be identical between the two groups. These are the constants.

Step 4—Use multiple samples

- Use more than one plant in each window (multiple samples). It is best to use at least three.

Step 5—Think about how to collect data

- We asked which location caused the plant to grow the fastest. We will therefore be measuring plant growth (this is the dependent variable). However, we cannot just measure the final height of the plant. This does not tell us how much the plant grew.
- We must measure the initial height of the plant and then the height at the end of the experiment. If you forget to measure the initial height of the plant, then you will not be able to calculate the amount of growth over the course of the experiment.
- We will run our plant experiment over 3 weeks and take one measurement at the beginning and one at the end.
- The two groups of plants should be measured. Then they should be placed in their respective windows and their height should be measured with a ruler after 3 weeks. Any watering or treatment of the plants should be the same for both groups for the entire three weeks.

Step 6—Do It!

We expect the plant growth to vary from the front window to the back window.

- Independent variable: plant location
- Dependent variable: plant growth

4. The Analysis

Here is the data collected:

Plant	Initial height (cm)	Final height (cm)	Total growth (cm)
Front-1	6	14	8
Front-2	6.5	14	7.5
Front-3	8	17	9
Back-1	6	9	3
Back-2	7	9.5	2.5
Back-3	6.5	9.5	3

By subtracting the initial height from the final height, we can calculate the amount of growth that occurred during the three weeks of the experiment. The final column shows the total growth of the plants.

You could summarize your data using a bar graph:

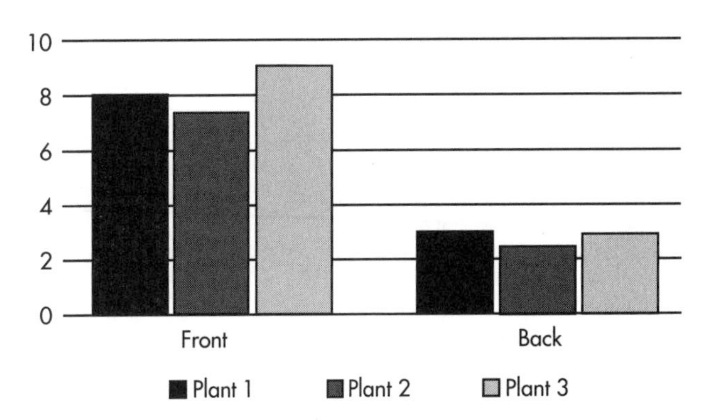

The bars are drawn to be the same height as the numbers in the final column in the table. So, the front plants have bars that are 8, 7.5, and 9 centimeters. The bars are taller for the front window, which represents more plant growth from the front window. Plants grew taller in the front window than in the back window.

5. The Conclusion

- Original question: Will this plant grow faster in the front window or the back window?
- Hypothesis: The plant will grow faster in the back window.
- Conclusion: The plant grew faster in the front window. Our hypothesis was wrong.

Back to the Drawing Board

If the window we thought was the sunniest (the back window) did not cause faster plant growth, then what about the front window was so special? The brainstorming process begins again! You could ask a new question: Why do plants in the back window grow slower than those in the front window? There are lots of possible hypotheses:

- Your fat cat sits in the sunny back window and blocks the sunlight, so the plants in the front window actually get more sunlight because they don't have the feline eclipse every afternoon.
- The back window might be sunnier, but it is drafty and colder because the back door opens more often.
- The back window is closer to a heat vent and the soil gets dried out quickly from the blowing air.

Summary of the Scientific Method

Here is a summary chart of how to use the Scientific Method well:

Parts of Scientific Method	**Strong Experiment Characteristic**	**Weak Experiment Characteristic**
Question	• Specific • Measurable	• Broad • Subject to opinion
Hypotheses	• Specific • Mention a quantitative trait	• Vague • Mention a qualitative trait
Experiments	• Different groups to compare • Groups are same except one trait • Repeats • Control Group	• Groups have many variables • Don't collect all required data • No repeats • Forget to use control group
Analysis	• Organize data • Identify a trend or pattern	• Some data is ignored • Patterns or trends are not identified
Conclusion	• Refers to Question/Hypothesis • Based on data • Specific	• Does not answer question • Not based on data/evidence • Too general

NOTES

CARBOHYDRATES

A thorough understanding of biology requires that one have an understanding of the fundamentals of chemistry, because life is basically a chemical process.

Many of the chemical substances associated with living things are referred to as organic substances; organic substances are substances that contain carbon. (All other substances are called inorganic.) Four classes of organic compounds are studied in depth in this book: carbohydrates, lipids, proteins, and nucleic acids. This plate centers on the first group, the carbohydrates. Carbohydrates are primarily used as sources of energy in living things.

Looking over this plate, you will notice that we discuss two types of carbohydrates: the simple sugars and polysaccharides. We will note what these substances are composed of and what some of their functions are in living things. Continue to read as you color the plate.

All plants, animals, and microorganisms use carbohydrates as sources of energy. Carbohydrates are also used as structural building blocks. For instance, the cellulose of plant cell walls is composed of carbohydrates. Carbohydrates are made up of carbon, hydrogen, and oxygen atoms. In the upper portion of the plate, we show a molecule of **glucose (A).** This is a basic carbohydrate known as a monosaccharide, or simple sugar. It consists of six carbon atoms located at the positions that are designated 1 through 6. Carbon 1 and carbon 5 are connected by an oxygen molecule, as the diagram shows.

Fructose (B) is also a monosaccharide. This molecule also consists of six carbon atoms, but they are arranged differently than the carbons in glucose. Notice that the oxygen atom joins carbons 2 and 5, and that fructose is a five-membered ring, while glucose's ring contains six members.

Extending from both glucose and fructose molecules are a number of –OH groups, or **hydroxyl groups (C).** When the molecules combine, an –OH group leaves the glucose molecule and an –H atom leaves the fructose. They unite to form a **water molecule (E).** The two monosaccharides then bond with one another to form a molecule of **sucrose (D),** and this double sugar is called a disaccharide.

Other disaccharides include maltose (two glucose units combined) and lactose (one glucose unit and one galactose unit). When disaccharides are used for energy, they are digested by enzymes that break them up into monosaccharides, and the monosaccharides are metabolized in the process of cellular respiration (as shown in the plate on Glycolysis). The process of the digestion of the disaccharide lactose (milk) requires the enzyme lactase. Lactose intolerant persons cannot produce lactase and therefore experience gastric difficulties from the ingestion of milk.

We now turn to complex carbohydrates known as polysaccharides. Continue your coloring as before, using the same colors as used previously. Two polysaccharides will be discussed in the paragraphs below. Each is essential for the maintenance and structure of biological organisms.

Polysaccharides are molecules that can consist of hundreds or thousands of monosaccharide units. The first polysaccharide molecule we will look at (in the bracket) is **starch (F),** which is found in plants such as corn and wheat.

Starch molecules represent a storage form of glucose; as the diagram shows, a starch molecule is composed of many glucose molecules (A) (we left off the hydroxyl groups of glucose here for simplicity). As you can see, the glucose molecules are bound to one another through an oxygen atom that unites carbons 1 and 4. The starch molecule is extensively twisted to form the final molecule.

When starch from food is taken into the body, the enzyme amylase (present in the mouth and small intestine) breaks it down into smaller units. After further digestion, glucose units are liberated, and these glucose units are absorbed into the bloodstream, and taken up by the cells. Cells use them in the metabolic process of respiration, which yields high-energy molecules necessary for the life of the cell.

The second polysaccharide we will mention is **glycogen (G),** indicated by the bracket at the bottom of the page. Glycogen is often referred to as animal starch. It is stored in the liver and muscles when the body has to store excess glucose molecules. Note that the glycogen molecule is composed of glucose units, but that glycogen is highly branched—this distinguishes glycogen from starch.

As we mentioned previously, one example of a structural polysaccharide is cellulose. Cellulose is very similar to starch except that the units are combined such that they cannot be digested by any of our digestive enzymes. For this reason, the cellulose in plant cells remains undigested and acts as roughage—which is essential to our diet.

A polysaccharide is an example of a macromolecule, also called a polymer. Polysaccharides may be branched or they may be long and linear. The types of linkages in polysaccharides determine which chemical reactions they take part in.

Simple Sugars

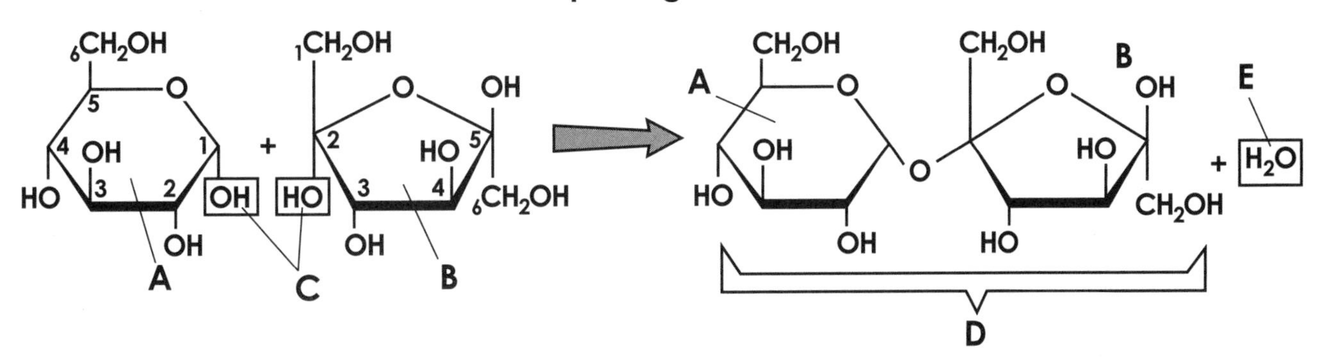

Polysaccharides

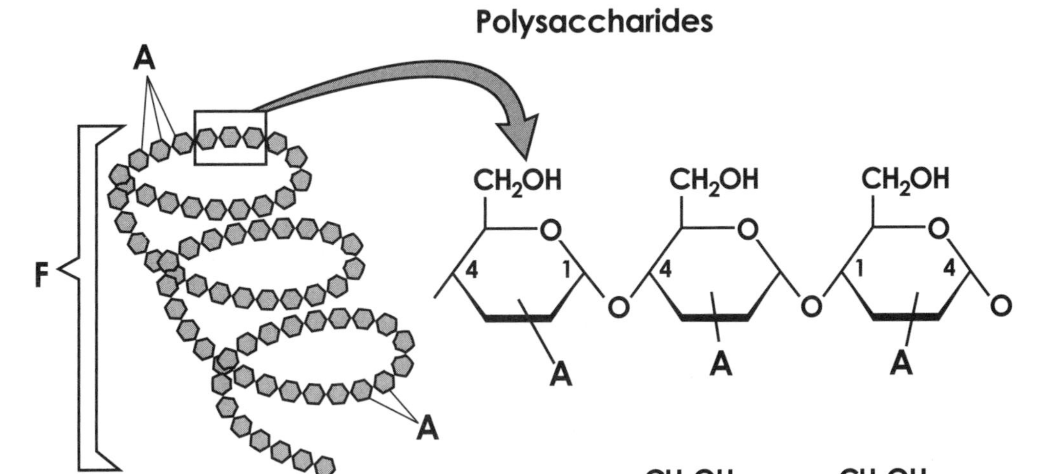

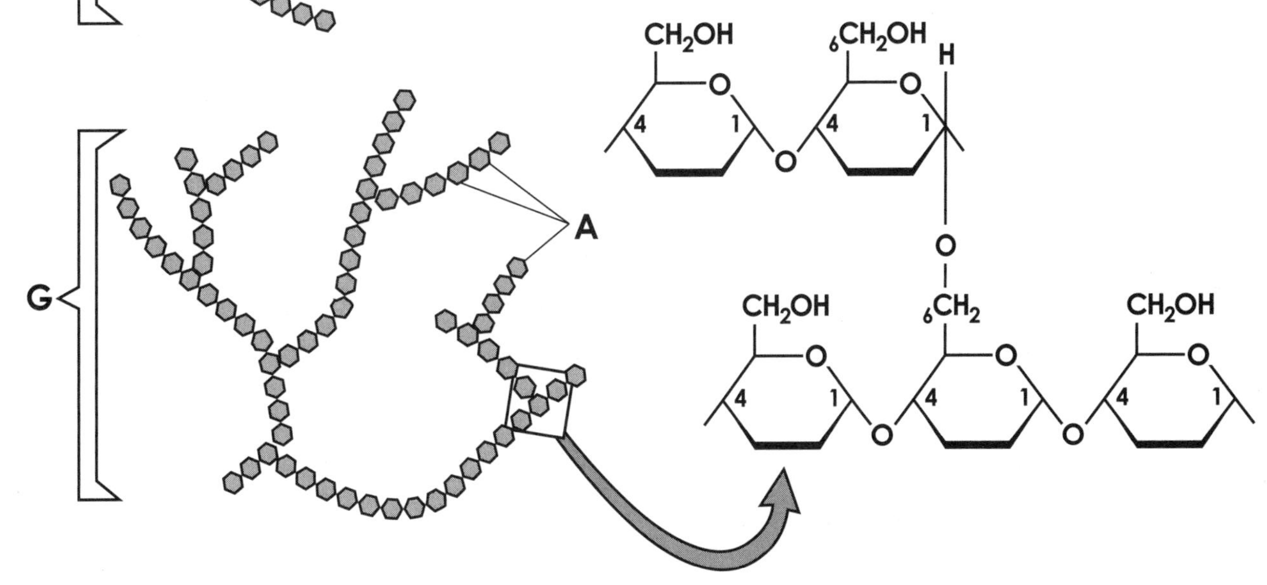

Glucose Molecule A ○
Fructose Molecule B ○
Hydroxyl Group C ○
Sucrose Molecule D ○
Water Molecule E ○
Starch Molecule F ○
Glycogen Molecule G ○

LIPIDS

The four types of organic molecules that are associated with life are carbohydrates, lipids, proteins, and nucleic acids.

In the preceding plate, carbohydrates were considered, and in this plate, we will focus our attention on lipids. Lipids are a group of organic molecules that dissolve in oils, but not water.

In this plate, we will discuss fats, phospholipids, and cholesterol, all important lipids. They consist solely of carbon, hydrogen, and oxygen. Focus on the top portion of the plate, where we consider fats.

Fats are very efficient energy-storage molecules that yield about twice the amount of chemical energy per gram as do carbohydrates. Fats are important in the construction of plasma membranes, and they also provide physical and thermal insulation to animals.

In the upper portion of the plate, we consider two types of fats: a saturated fat and an unsaturated fat. The saturated fat is built from two types of subunits: a **glycerol molecule (A),** which contains three atoms of carbon and numerous oxygen and hydrogen atoms. The box containing the molecule should be shaded in a pale color.

In the saturated fat, three **saturated fatty acid chains (B)** are chemically bound to the glycerol subunit. A fatty acid is a long hydrocarbon chain, as the diagram shows. A saturated fatty acid contains its maximum number of hydrogen atoms; single bonds are represented by straight lines in the diagram. When three saturated fatty acids are connected to glycerol as shown, the result is a saturated fat, or triglyceride.

In the unsaturated fatty acid, there are fewer than the maximum number of hydrogen atoms in the **unsaturated fatty acid chains (C);** there are some double bonds, which are represented by two parallel lines. Three unsaturated fatty acid chains bonded to glycerol form an unsaturated fat, or triglyceride.

Saturated and unsaturated fats are extremely important to the metabolism of organisms. Fats are broken down into two-carbon units, and these units are used in the Krebs cycle (discussed later). They undergo a series of conversions and release their energy in the form of ATP molecules. Fats serve as a supplemental energy source when carbohydrate stores are exhausted.

Having discussed fats, we turn to another important lipid, the phospholipid. Read about this organic compound as you color the diagram.

One of the key uses of phospholipids is in the formation of the cell (plasma) membrane. The cell membrane consists of two layers of phospholipids with associated proteins. We will discuss membranes in depth in a later plate.

Phospholipids are basically made up of a glycerol molecule (A) with a **phosphate group (D)** attached to the first carbon at the extreme right. This phosphate group consists of a phosphorus atom and four oxygen atoms, and you should use a light color for it. The second and third carbons of the glycerol molecule bear fatty acid chains. Note that the fatty acid on the left is a saturated fatty acid (B), while the fatty acid on the right is unsaturated (C).

In the phospholipid bilayer of the cell membrane, the phosphate groups point toward the outside of the cell and the fatty acid chains extend toward each other. The phosphate end is the **polar end (E)** because it bears a negative charge. The opposite end of the molecule is the **nonpolar end (F),** and this section of the molecule lacks an electrical charge. In the construction of the cell membrane, millions of these molecules stand next to one another to form a structure similar to a picket fence. Molecules entering or leaving the cell must pass through this double layer of lipids, and the membrane acts as a selective barrier for the cell.

We now turn our attention to the final type of lipid, cholesterol. This molecule should be colored as you read about it below.

Steroids comprise an important group of lipids that are insoluble in water and consist of carbon, hydrogen, and oxygen atoms arranged in rings. Estrogen and androgens, the sex hormones in humans, are steroid hormones.

One familiar steroid is cholesterol. In the diagram, we show this complex molecule that has a **sterol ring (G).** In humans, cholesterol is used as a precursor to sex hormones, but an excess of cholesterol is a problem because it can clog arteries and veins, which result in restricted blood flow. The liver synthesizes cholesterol for the body, but the diet also provides some, and if this intake is high, excessive cholesterol accumulates.

A

B

Saturated Fat

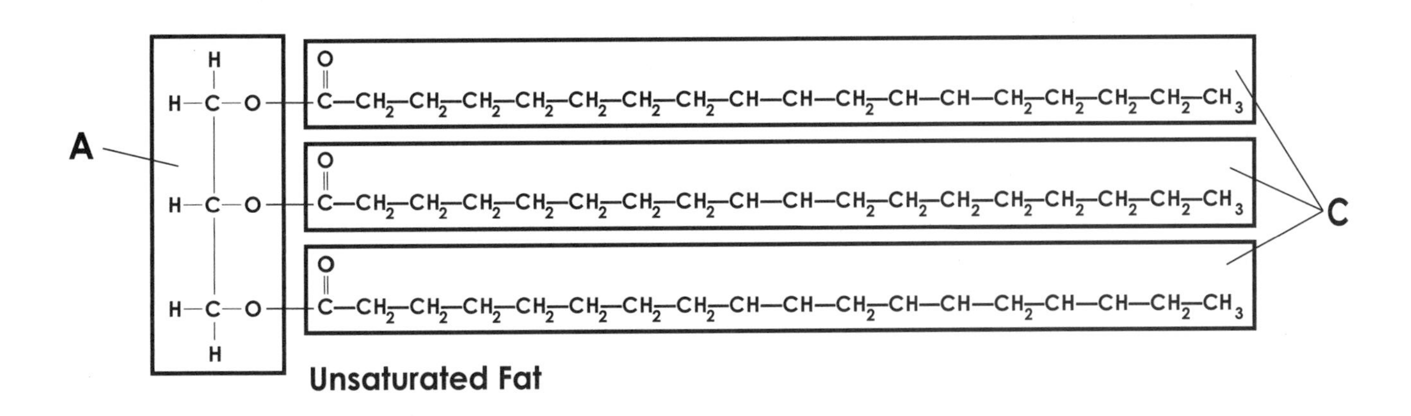

Unsaturated Fat

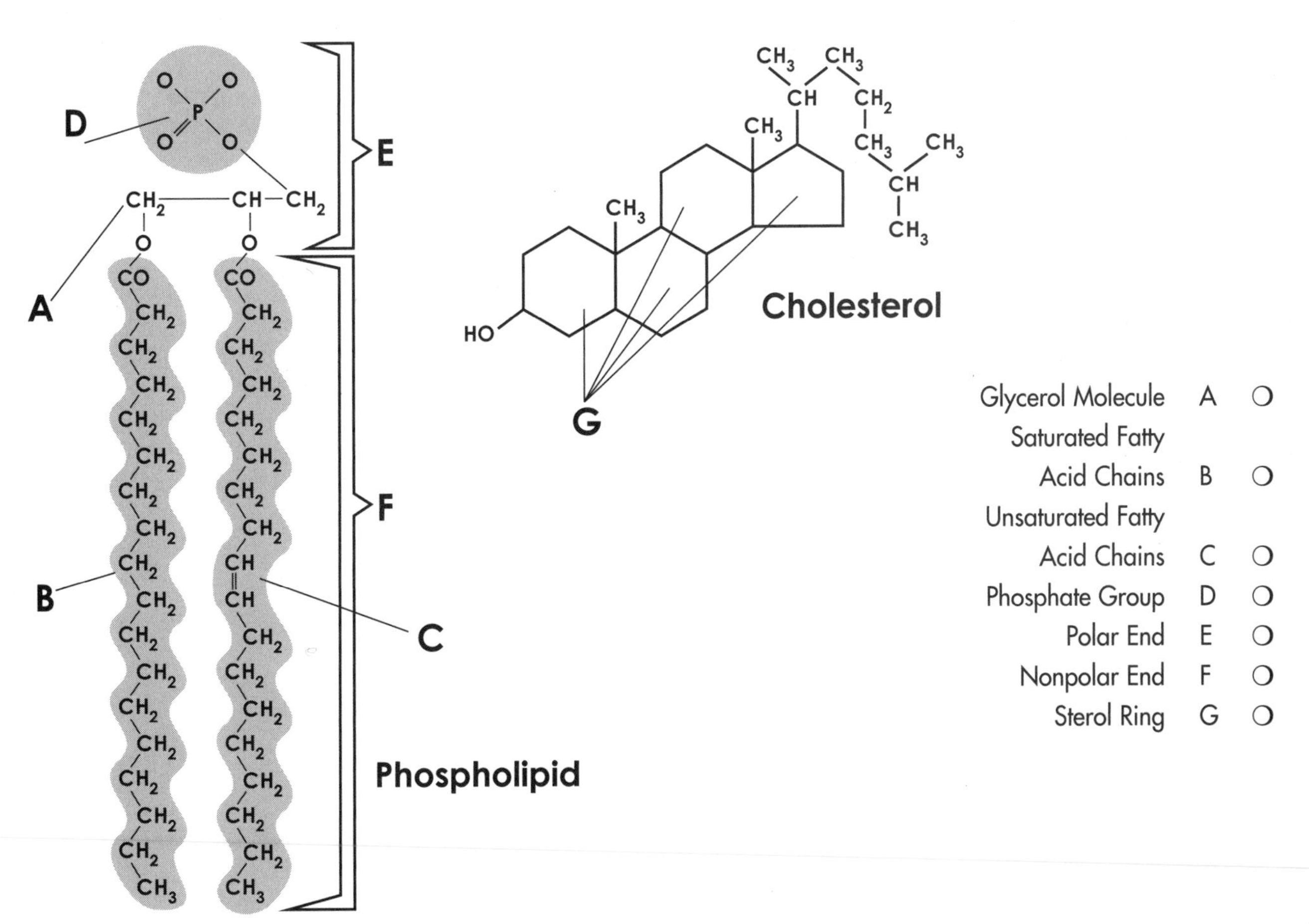

Glycerol Molecule A ❍
Saturated Fatty Acid Chains B ❍
Unsaturated Fatty Acid Chains C ❍
Phosphate Group D ❍
Polar End E ❍
Nonpolar End F ❍
Sterol Ring G ❍

PROTEINS

Organic molecules provide the body with structural materials to form cells, tissues, and organs; regulatory substances to direct and govern the interactions of molecules; and energy to fuel the chemical operations of cells.

In previous plates, we discussed the structure and function of carbohydrates and lipids, and here we will examine proteins. Proteins are vital to the formation and function of many cellular structures and processes. They are also among the most diverse organic molecules in the living organism.

This plate shows how amino acid subunits are joined to form protein molecules. Once formed, proteins can assume a number of shapes, as the remainder of the plate will illustrate.

Proteins are molecules that are formed from units called amino acids. A protein may contain as few as ten amino acids, or it may contain thousands. The sequence of amino acids in proteins gives them unique functional characteristics.

In this plate, we show how amino acids combine with one another to form a linear protein called a peptide. The upper portion of the plate shows two **amino acids (A)** outlined by brackets. Each amino acid contains a **carboxyl group (B)** (shaded on the left amino acid and outlined on the right one). Amino acids also contain **amino groups (C).** In the left amino acid, the amino group is outlined, and in the right one, it has been shaded. Light colors are recommended to color these groups. The amino group contains nitrogen and hydrogen atoms. The –R stands for a general alkyl group, and each of the 20 amino acids bears a distinct alkyl side chain.

In the synthesis of a peptide, the –OH group of the carboxyl group of one amino acid and the –H of the amino group of the next amino acid are enzymatically removed. The nitrogen from one amino acid bonds with the carboxyl carbon of the adjacent amino acid, and this bond is called the **peptide bond (D).** We have now formed a dipeptide, the smallest protein. In the course of the formation of the peptide bond, a molecule of water (H_2O) is given off.

Notice the far right of the dipeptide, where it could join with the carboxyl group of an adjacent amino acid, and the far left of the dipeptide, where it could also be extended. Additional amino acids are added to the growing chain of peptides, and when the final amino acid has been added, a polypeptide results.

The order and number of amino acids in the peptide chain is determined by the cell's genes; this is discussed in a future plate. Once the peptide has formed, additional modifications occur, as the following diagrams illustrate. Continue your coloring as you read below.

As amino acids are added to the growing peptide, a polypeptide results, and when the polypeptide is modified to its working structure, it is called a protein.

The linear sequence of amino acids in a protein is referred to as a protein's primary structure. In the diagram entitled Primary Structure, we show six amino acids (A) linked together by peptide bonds (D). You may choose to color the six amino acids different colors. The amino acids in this peptide are valine (val), leucine (leu), lysine (lys), tyrosine (tyr), and histidine (his).

The secondary structure of proteins refers to the folding and twisting patterns of the protein chain. One pattern that proteins can assume is a **helix (E)**, indicated by the bracket. The individual amino acids in the helix can be seen.

Certain polypeptide chains fold back on themselves to form **folded proteins (F).** These proteins are globular proteins, and their three-dimensional shape is shown. The protein in the figure is one of the four polypeptides in a hemoglobin molecule of the human blood cell, and the **heme group (G)** is found within the folded protein.

The final structure we show is the quaternary structure, in which **multiple polypeptides (H)** are organized together. The **four chains** (**H_1, H_2, H_3, H_4**) are different, but one is identical to the folded protein of the tertiary structure shown earlier. This is the final structure of the hemoglobin protein found in red blood cells. It is important for the transport of oxygen throughout the body.

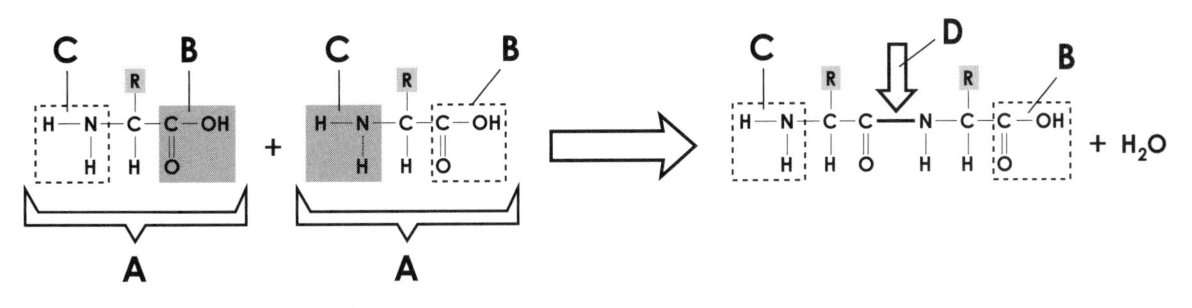

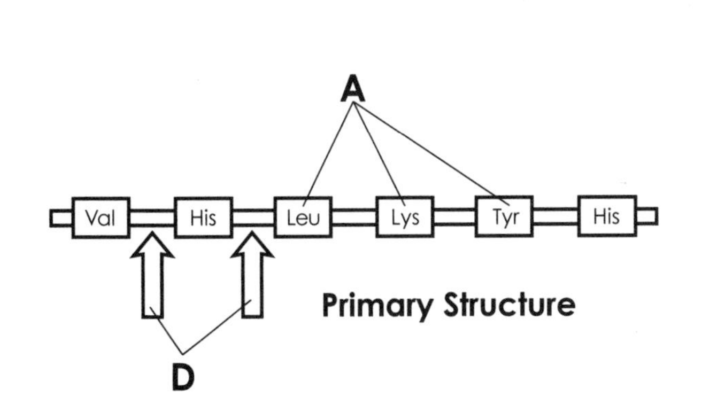

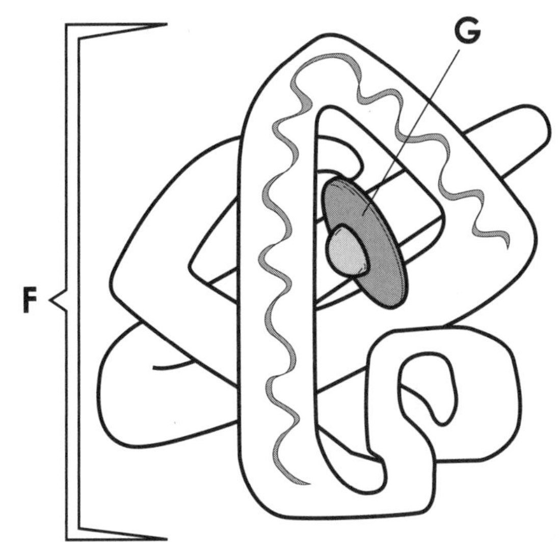

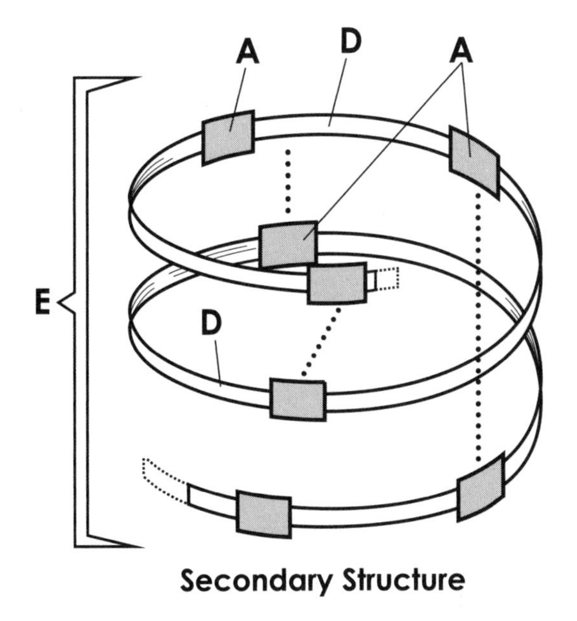

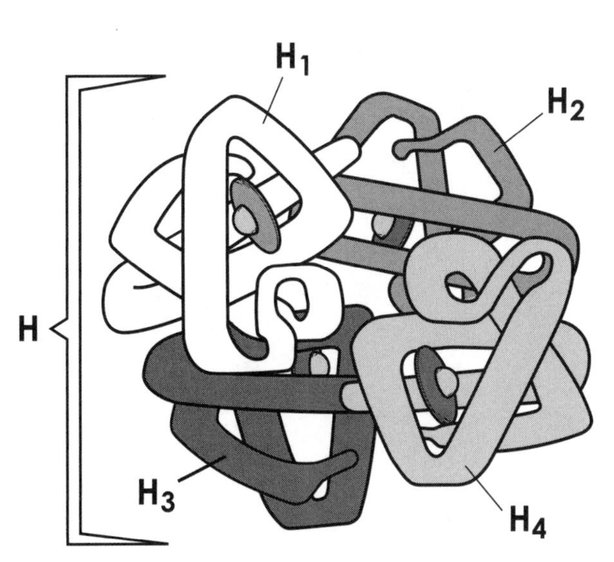

Amino Acids	A	❍	Helix	E	❍	Chain #1	H_1	❍
Carboxyl Group	B	❍	Folded Protein	F	❍	Chain #2	H_2	❍
Amino Groups	C	❍	Heme Group	G	❍	Chain #3	H_3	❍
Peptide Bond	D	❍	Multiple Polypeptides	H	❍	Chain #4	H_4	❍

CHAPTER 2:

BIOLOGY of the CELL

THE ANIMAL CELL

The basic unit of all living systems is the cell. In this plate, we will describe some of the features of animal cells. We will study the plant cell in a few plates.

> This plate consists of a diagram of a section of an animal cell. Under a light microscope, the animal cell seems relatively simple, but the electron microscope reveals a wealth of structures that contribute to its activities. As you read about structures in the following paragraphs, color them in the plate. Light colors are recommended, because the structures are small.

It is impossible to locate a typical animal cell in nature because none exists; here we present a composite cell.

The cell is enclosed by a **cell (plasma) membrane (A),** which is composed of phospholipids and proteins. Various biochemical mechanisms permit small nutrients to pass through the membrane to the cell's interior, and will be discussed in a future plate.

Within the cell membrane is the cytosol, which is also known as the **cytoplasm (B).** This fluid portion of the cell suspends organelles, and enzymes and other proteins are produced within the cytosol.

The cytosol contains an internal protein framework called the **cytoskeleton (C).** Tracing the fibers with a dark color will help highlight their presence. Microfilaments within the cytoskeleton provide the mechanism for contraction in muscle cells, and other cytoskeleton fibers called microtubules participate in cell reproduction.

Extending out from the cell membrane are projections called **microvilli (D).** These fingerlike projections are found in cells of the digestive tract, where absorption takes place. Longer hairlike extensions called **cilia (E)** are found on cells of the respiratory tract, where they trap dust particles in mucus in order to prevent them from entering the lungs.

> We now move to some of the submicroscopic structures within the cell, and relate them to cell functions. Continue your coloring as above. Light colors are recommended to keep you from obscuring the details in the plate.

The first internal cell structure we will study is the **centrosome (F).** The centrosome contains two bodies called **centrioles (F_1).** As the plate indicates, centrioles are situated at right angles to one another and are composed of microtubules; they are involved in mitosis.

Ribosomes (G) are seen at numerous locations within the cell. These ultramicroscopic bodies are the "workbenches" of the cells; they are the sites of protein synthesis from amino acid subunits. Ribosomes are especially numerous in cells that synthesize proteins, such as pancreatic cells, muscle cells, and epidermal cells.

An important organelle of the cytoplasm is the **mitochondrion (H).** The mitochondrion is a double-membrane enclosed organelle that produces ATP, which is the energy currency of the cell. Cells that require a large amount of energy such as muscle cells and sperm cells contain many mitochondria, while fewer exist in less active cells.

The center of genetic activity in the cell is the **nucleus (I).** With the exception of red blood cells and gametes (sex cells), all human cells have 46 chromosomes in their nucleus. A body of RNA called the **nucleolus (I_1)** is suspended in the fluid-like **nucleoplasm (I_2)** in the nucleus. The genes within the nucleus are specific nucleotide sequences of DNA that contain the biochemical instructions for the synthesis of particular proteins.

> We complete the plate by examining the last few cellular structures important to the activity of the cells. Some of these structures are involved in protein synthesis. Continue using light colors, since these structures are relatively small.

The **endoplasmic reticulum (J),** also called the ER, is a system of interconnected membrane channels in the cytosol. These membranes may or may not have ribosomes associated with them. If ribosomes are associated with the ER, it is referred to as **rough ER (J_1).** Rough ER predominates in cells that are actively synthesizing protein for export. Where the endoplasmic reticulum has few or no ribosomes, it is known as **smooth ER (J_2).** The smooth ER is not actively involved in protein processing and usually has tissue-specific functions. For example, it can contain enzymes involved in steroid hormone biosynthesis (such as in the gonads), or in the degradation of environmental toxins (such as in the liver). After proteins have been manufactured, they are generally stored in a series of flattened membranes called the **Golgi body (K).** The Golgi body sorts and packages proteins for secretion from the cell.

The cell stores digestive enzymes in organelles known as **lysosomes (L).** Enzymes in lysosomes help break down organic molecules into components that are useful to the cell in protein synthesis and energy metabolism. Enzymes are also stored in **peroxisomes (M).** This is the site in which toxic compounds are neutralized. For this reason, peroxisomes are abundant in liver cells where they participate in the breakdown of alcohol, among other toxins.

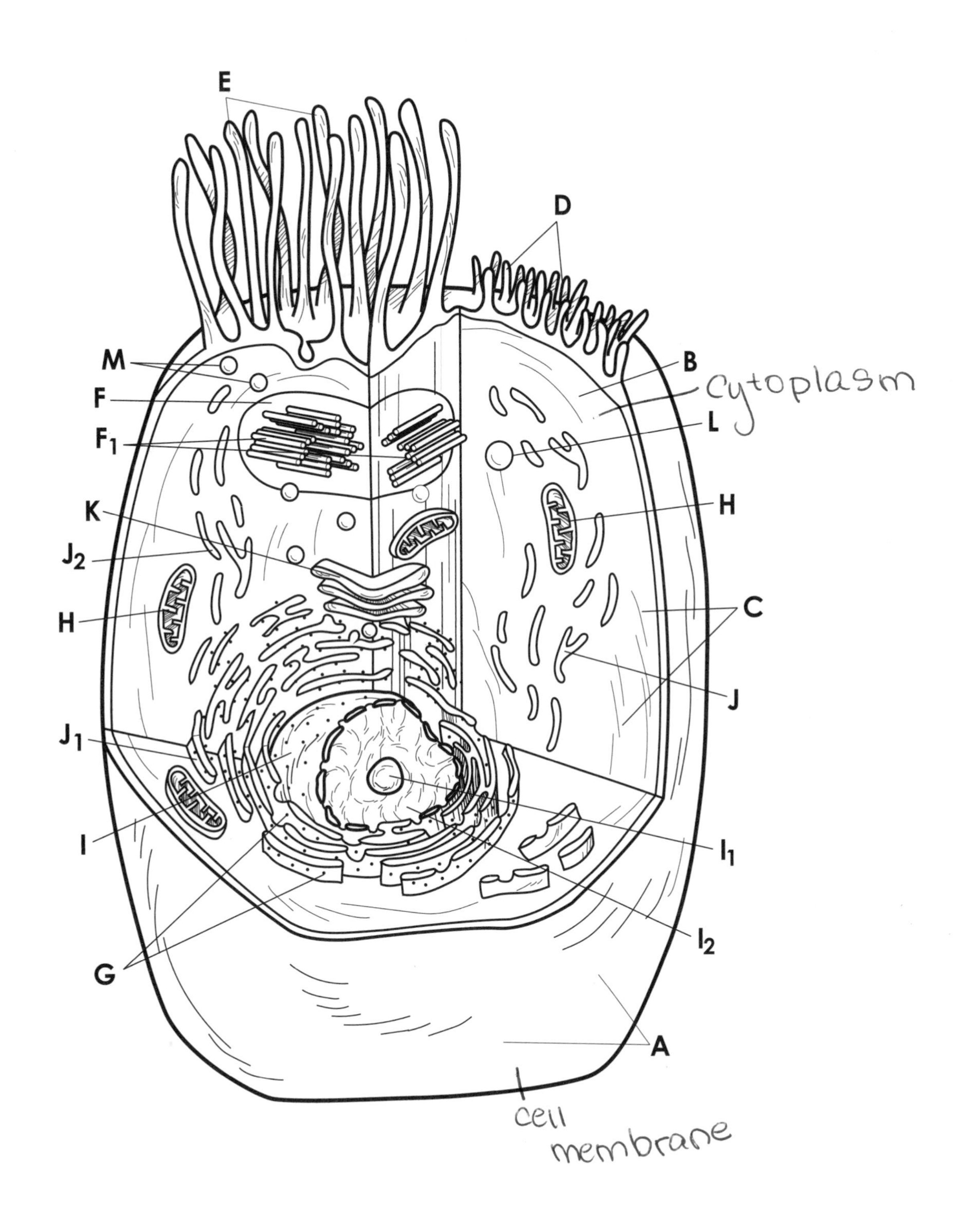

Structure	Label	
Cell Membrane	A	❍
Cytosol (Cytoplasm)	B	❍
Cytoskeleton	C	❍
Microvilli	D	❍
Cilia	E	❍
Centrosome	F	❍
Centrioles	F_1	❍
Ribosomes	G	❍
Mitochondrion	H	❍
Nucleus	I	❍
Nucleolus	I_1	❍
Nucleoplasm	I_2	❍
Endoplasmic Reticulum	J	❍
Rough ER	J_1	❍
Smooth ER	J_2	❍
Golgi Body	K	❍
Lysosome	L	❍
Peroxisome	M	❍

CELL TYPES AND TISSUES

There are hundreds of different types of cells in plants, animals, and microscopic organisms, and each type of cell has a specific function. In multicellular organisms, cells usually function in organized groups called tissues. In this plate, we will examine the various types of cells and tissues found in the human body.

This plate shows some of the numerous types of cells and tissues that exist in the human body. Start your work by looking over the plate, and then continue your reading below.

There are four major types of tissues in the human body. The first type that we'll discuss is **epithelial tissue (A),** seen here enclosed by a box that you should color. Epithelial tissue is located at the body's surface and where the body meets with agents of the external environment, such as lining of the digestive tract. There are three common shapes of epithelial cells: squamous, columnar, and cuboidal.

The epithelial tissue found on the body's surface is made up of flat cells called **squamous cells (A_1),** seen here lying on what's called the basement membrane. These cells provide the body with protection from dehydration and some mechanical injury. Squamous cells also make up the blood vessels and the air sacs in the lungs. The center drawing depicts **columnar cells (A_2),** which line the digestive tract, absorbing nutrients from food. Microvilli on the surface of these cells provide them with massive surface areas relative to their size, which increases the rate of absorption of nutrients. The third type of epithelial cell is the **cuboidal cell (A_3).** These cube-shaped cells make up the epithelia of kidney tubules and many glands.

We now move to the next two types of tissue: muscle and nerve tissue. Continue your reading as you color the appropriate portions of the diagram.

A second type of tissue in the human body is **muscle tissue (B),** which is outlined by a box in the art. This type of tissue consists of long, fibrous **muscle cells (B_1),** in which you can see dark nuclei. The muscle shown is skeletal muscle; skeletal muscle can be found, for instance, in the arms and legs. Muscle tissue permits movement of the appendages and lines the hollow structures of the body such as blood vessels.

A third type of tissue is **nerve tissue (C).** The **nerve cells (C_1)** that make up this tissue are long and have numerous extensions through which nerve impulses travel on their route throughout the body. Nerve cells are often called neurons. An accumulation of neurons and supporting tissue exists in the brain.

We conclude the plate by examining the final type of tissue, connective tissue. We will mention six types of connective tissue.

Connective tissue (D) supports and binds the other tissues of the body. The first type of connective tissue we'll mention is **cartilage (D_2).** Cartilage is made up of **chondroblasts (D_1)** embedded in a rubbery substance known as chondrin. It maintains the shape of organs such as the outer ear.

On the opposite side of the plate is a second type of connective tissue, **bone (D_4).** Bone tissue contains cells called **osteoblasts (D_3),** shown as dark dashes within the rings. You should use spots of color for them. Bone material exists in concentric rings; the osteoblasts are confined to spaces called lacunae and secrete the calcium phosphate and collagen that make bone hard. Bones provide support for the body.

Two other kinds of connective tissue are **tendons (D_6)** and **ligaments (D_7).** Tendons are made up of fibrous connective tissue, which in turn is made up of strands of protein produced by **tendon-forming cells (D_5).** Tendons connect muscle to bone, as you can see in the diagram. **Ligaments (D_7)** are similar to tendons, but they connect bone to bone.

At the bottom right of the plate, we see another type of connective tissue called fat (adipose). **Fat (D_9)** is also called adipose. It is an organic substance made of **fat or adipose cells (D_8).** The nuclei of the fat cells can be seen as spots along the perimeters of the cells, but the main portion of the cell is composed of fat droplets. Fat provides insulation for the body and stores fuel for metabolic activity.

The final connective tissue we will examine is blood. Blood contains **red blood cells (D_{10}),** which are shaped like discs and are sometimes stacked. Red blood cells carry oxygen to cells and pick up carbon dioxide from them for expulsion from the body. Blood also contains **white blood cells (D_{11}),** which act as the body's defense system, phagocytosing bacteria and other harmful organisms and producing substances used in the immune system. The final type of blood cell is the **platelet (D_{12}).** Platelets are cell fragments involved in blood clotting. A future plate will discuss blood cells in more detail.

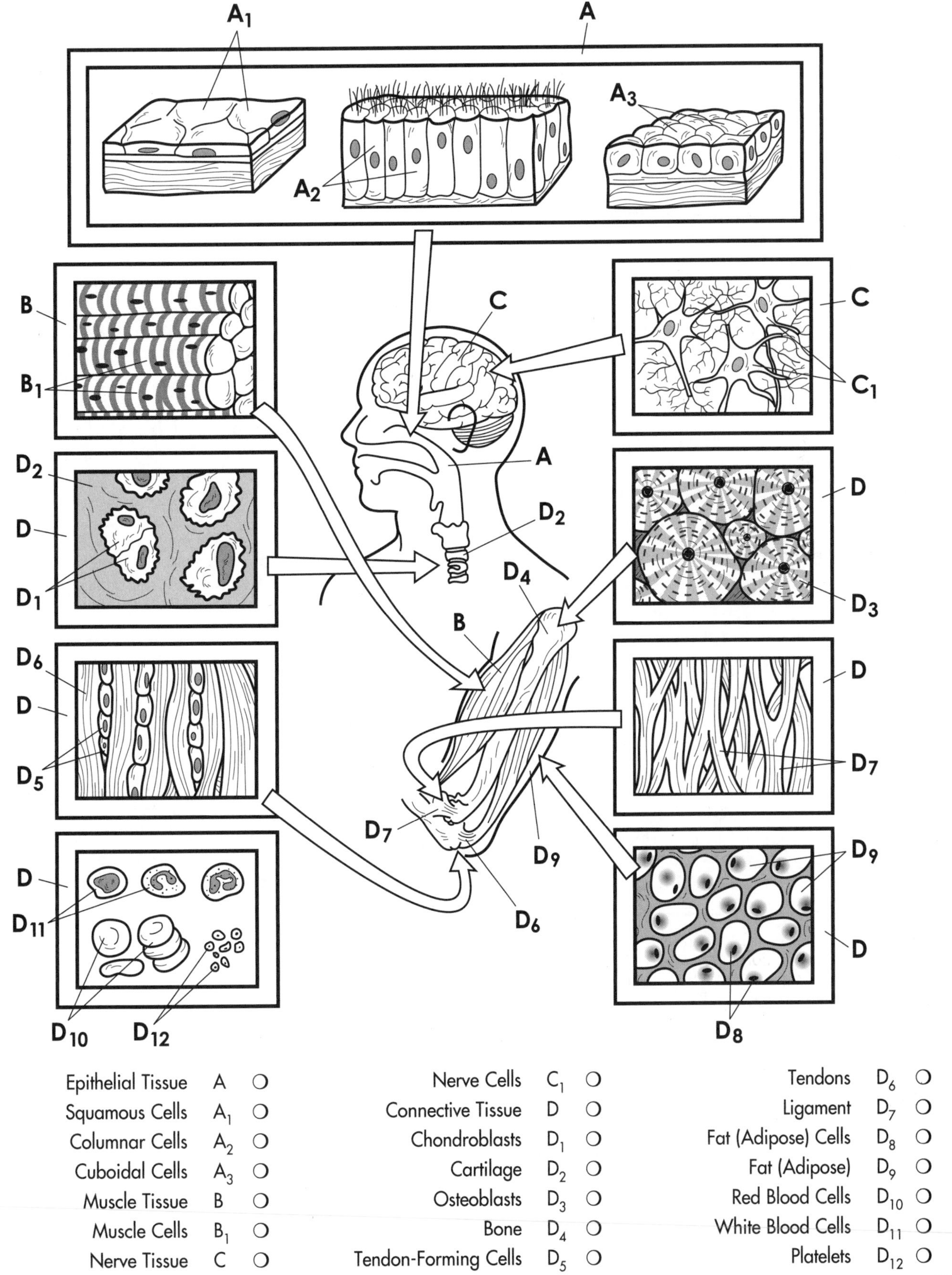
A_1
A
A_3
A_2
B
C
C
B_1
C_1
D_2
A
D
D
D_2
D_1
D_4
B
D_3
D_6
D
D
D_5
D_7
D_7
D_9
D_9
D
D
D_{11}
D_6
D_{10}
D_{12}
D_8
Epithelial Tissue A
Squamous Cells A_1
Columnar Cells A_2
Cuboidal Cells A_3
Muscle Tissue B
Muscle Cells B_1
Nerve Tissue C
Nerve Cells C_1
Connective Tissue D
Chondroblasts D_1
Cartilage D_2
Osteoblasts D_3
Bone D_4
Tendon-Forming Cells D_5
Tendons D_6
Ligament D_7
Fat (Adipose) Cells D_8
Fat (Adipose) D_9
Red Blood Cells D_{10}
White Blood Cells D_{11}
Platelets D_{12}

THE PLANT CELL

The tissues of all organisms are composed of cells, and it is within cells that the metabolic functions of living things take place. Most cells are microscopic, but some, such as the frog egg cell, can be seen with the naked eye.

In this plate, we examine the features of a plant cell. Wherever possible, you should compare the plant cell to the animal cell discussed in the first plate of this chapter.

Looking over the plate, notice that we are showing a cross section of a typical plant cell. Just for our study, we have combined the features of most plant cells in this hypothetical one. As you read about the parts of the cell, color them in the plate.

We will begin our study of the plant cell at its surface with the **cell wall (A).** This structure protects and supports the plant cell and renders it somewhat rigid. The polysaccharide cellulose is the main component of the plant cell wall. Animal cells do not have a cell wall.

Inside the cell wall is the flexible **cell membrane (B),** also known as the plasma membrane. The cell membrane regulates the movement of materials into and out of the cell, and allows for communication between cells. Phospholipids and proteins make up this membrane, and it is similar in structure and function to the plasma membrane of an animal cell.

The main component of the interior of the cell is the **cytoplasm (C),** or cytosol. You should use a light color for this liquid mass, which is the site of many metabolic activities. Three types of protein fibers compose a **cytoskeleton (D)** within the cytoplasm, which provides a framework for many cellular activities.

We have examined the outer surface and the main body of the cell, and we will now focus on the cell organelles. Spots of darker colors are recommended for smaller organelles, and light colors for the larger ones. Continue your coloring as you read the paragraphs below.

Located within the cytoplasm are ultramicroscopic bodies called **ribosomes (E),** which should be indicated with spots of black. These are the "workbenches" of the cell; they are the sites of protein synthesis.

Ribosomes are commonly located along a folded membrane called the ER, or **endoplasmic reticulum (F),** which forms an internal network within the cytoplasm. Membrane components and lipids are synthesized at the ER. When ribosomes are located along the ER, it is referred to as **rough ER (F_1),** and if there are no associated ribosomes, it is **smooth ER (F_2).** The arrows that point to the two types should be colored dark colors.

Similar to the animal cell, the centers of energy metabolism in the plant cell are the **mitochondria (G).** Along the inner membranes of these organelles, the energy from carbohydrates is released and used to produce ATP molecules.

Other key organelles in the plant cell cytoplasm are **chloroplasts (H),** which are not present in animal cells. Chloroplasts are the site of photosynthesis, in which the Sun's energy is converted into chemical energy in the form of glucose and other carbohydrates. Photosynthesis is an essential metabolic process that we will explore in depth in future plates.

Still another important organelle is the **plastid (I),** which is a structure that stores nutrients and pigment molecules. The center of the plant cell contains a large space known as the **vacuole (J),** which contains water, sugars, ions, pigments, and other substances. It also applies pressure to the cell membrane, causing it to expand and stick close to the cell wall.

Two other cytoplasmic organelles also deserve attention; the first is the **lysosome (L).** This body contains digestive enzymes that break down compounds. Also in the cytoplasm is the **Golgi body (M),** which is made up of a series of about 10 to 20 flattened membranes. This organelle modifies, packages, and secretes proteins after they are synthesized at the endoplasmic reticulum. Plant cells may also contain enzyme-filled **peroxisomes,** which are not shown in this plate.

Now that we have discussed the many components of the cytoplasm, we will focus on another prominent structure, the nucleus.

The membrane-bound body in which the chromosomes are located is the **nucleus (N).** The nucleus is surrounded by a nuclear membrane, which is characterized by shallow depressions called **nuclear pores (N_1).** RNA passes out of the nucleus through these pores and travels to ribosomes for translation. Small proteins and nucleotides also pass into the nucleus through these pores. The fluid substance inside the nucleus is called **nucleoplasm (N_2).** The final structure we will look at is the **nucleolus (O).** This is a region inside the nucleus made of proteins, DNA, and RNA. It is the site of ribosome synthesis and assembly.

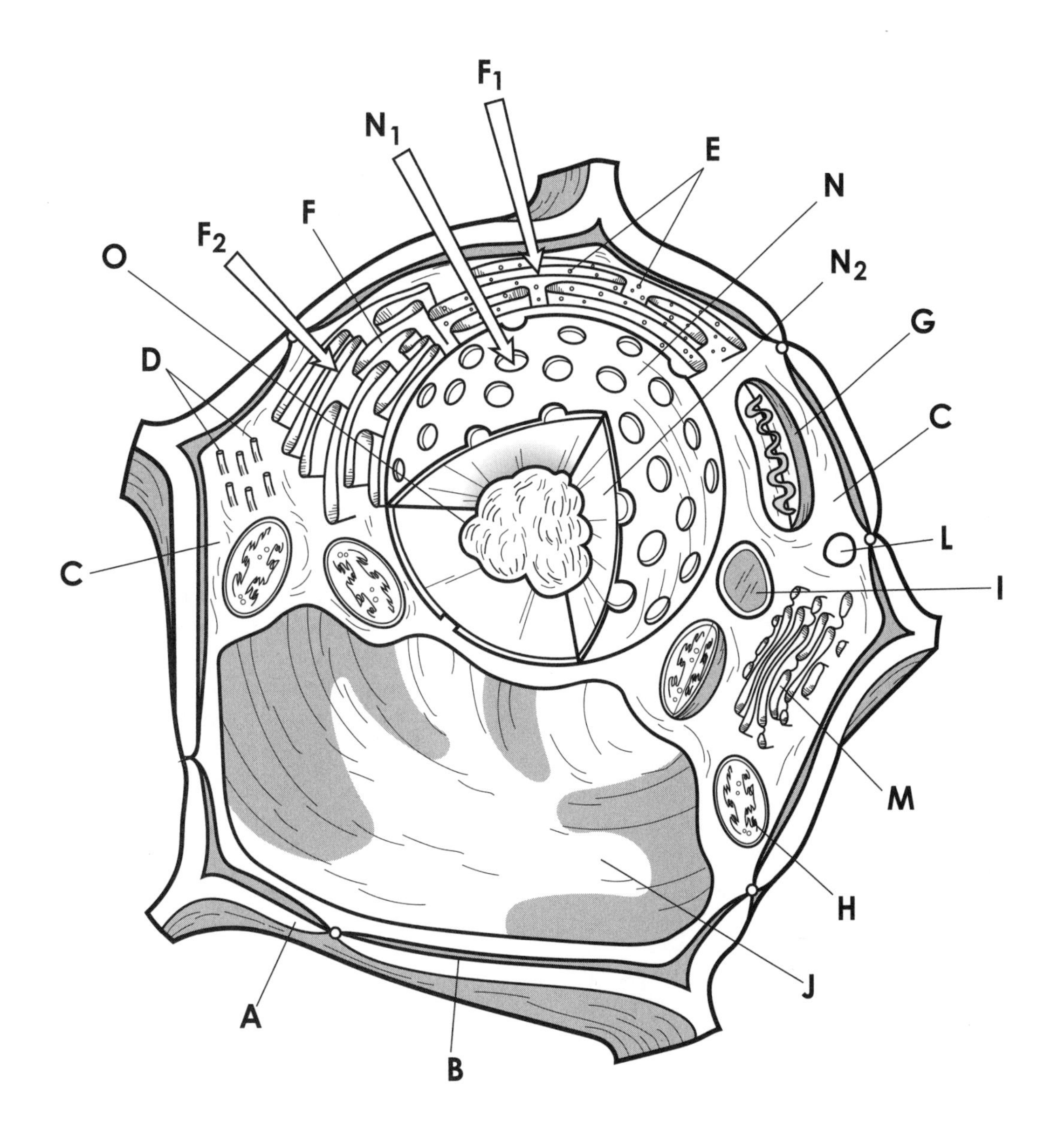

Cell Wall	A	❍
Cell Membrane	B	❍
Cytoplasm	C	❍
Cytoskeleton	D	❍
Ribosomes	E	❍
Endoplasmic Reticulum	F	❍
Rough ER	F_1	❍
Smooth ER	F_2	❍
Mitochondrion	G	❍
Chloroplast	H	❍
Plastid	I	❍
Vacuole	J	❍
Lysosome	L	❍
Golgi Body	M	❍
Nucleus	N	❍
Nuclear Pore	N_1	❍
Nucleoplasm	N_2	❍
Nucleolus	O	❍

THE CELL MEMBRANE

The cell membrane is also known as the plasma membrane, and we use the terms interchangeably in this book. The cell membrane is responsible for bringing essential materials into the cell and excreting metabolic waste products. In this plate, we will examine some of the components of the membrane. You should keep in mind that other membranes such as the endoplasmic reticulum and the nuclear membrane are similar to the cell membrane. These organelles were discussed in a previous plate.

This plate presents an enlarged view of the cell membrane. We will identify the various structures that make up the membrane and mention their activities. Begin your reading below.

The cell membrane is made up of proteins and carbohydrates as well as a phospholipid bilayer. It is an extremely thin structure that measures about 5 to 10 nanometers (nm) in thickness, and it can be seen clearly only through an electron microscope. The currently accepted hypothesis of membrane structure is referred to as the fluid mosaic model, and was proposed by Singer and Nicholson in 1976.

The most prominent element of the cell membrane is a fluid bilayer of lipids, in which a number of proteins are embedded. In the plate, the bracket outlines the **lipid bilayer (A),** in which you can see individual **phospholipids (B).** As the detailed diagram at the bottom of the plate indicates, a phospholipid consists of a somewhat circular phosphate group head, and two long, fatty acid chain tails. The head region is said to be **hydrophilic and polar (C)** because it is water-soluble, while the tail portions are **hydrophobic and nonpolar (D)** because they are not water-soluble. Notice that the hydrophilic heads of the lipid bilayer point toward the cell's exterior and interior, while the hydrophobic tails point inward. The brackets pointing out the hydrophilic heads and hydrophobic tails should be colored in bold colors, but the phospholipid (B) itself should be a single pale color.

Another type of lipid within the lipid bilayer is the **cholesterol molecule (E).** Cholesterol helps to maintain the fluid condition of the bilayer by breaking up the closely associated phospholipids. The detailed view shows several cholesterol molecules, which are types of steroid lipids.

We have discussed the basic structure of the cell membrane, and now we will focus on the proteins and carbohydrates associated with it.

Proteins that are embedded in the cell membrane carry out various cellular functions such as nutrient and energy transport and message transmission. One type of embedded protein is the **integral protein (F),** which spans the entire width of the lipid bilayer and protrudes at both sides. These proteins function as channels through which ions and molecules can travel into and out of the cell.

Other membrane proteins include **peripheral proteins (G),** which are not embedded in the lipid bilayer, but sit on the outside and are bound to exposed regions of integral proteins. Peripheral proteins are often connected to **cytoskeleton filaments (H)** on the cell's interior. Several filaments are shown in the plate.

Another type of protein in the membrane is the **alpha helix protein (I),** which is wound like a coil. It extends through the membrane, as the plate indicates, and acts as a channel for nutrients entering the cytoplasm.

Having mentioned the lipids and proteins involved in the cell membrane, we will now focus on carbohydrates. Continue your reading below and complete the coloring of the plate.

Glycoproteins (J) consist of a protein with an attached **carbohydrate (K).** In the diagram, we show a string of hexagonal molecules that represent the glucose molecules in a polysaccharide. The carbohydrate molecules are involved in cell recognition as receptors, and they also aid in the cell's adhesion to other cells. For example, hormones attach to the carbohydrates on the membranes of target cells. Research on these membrane carbohydrates is ongoing.

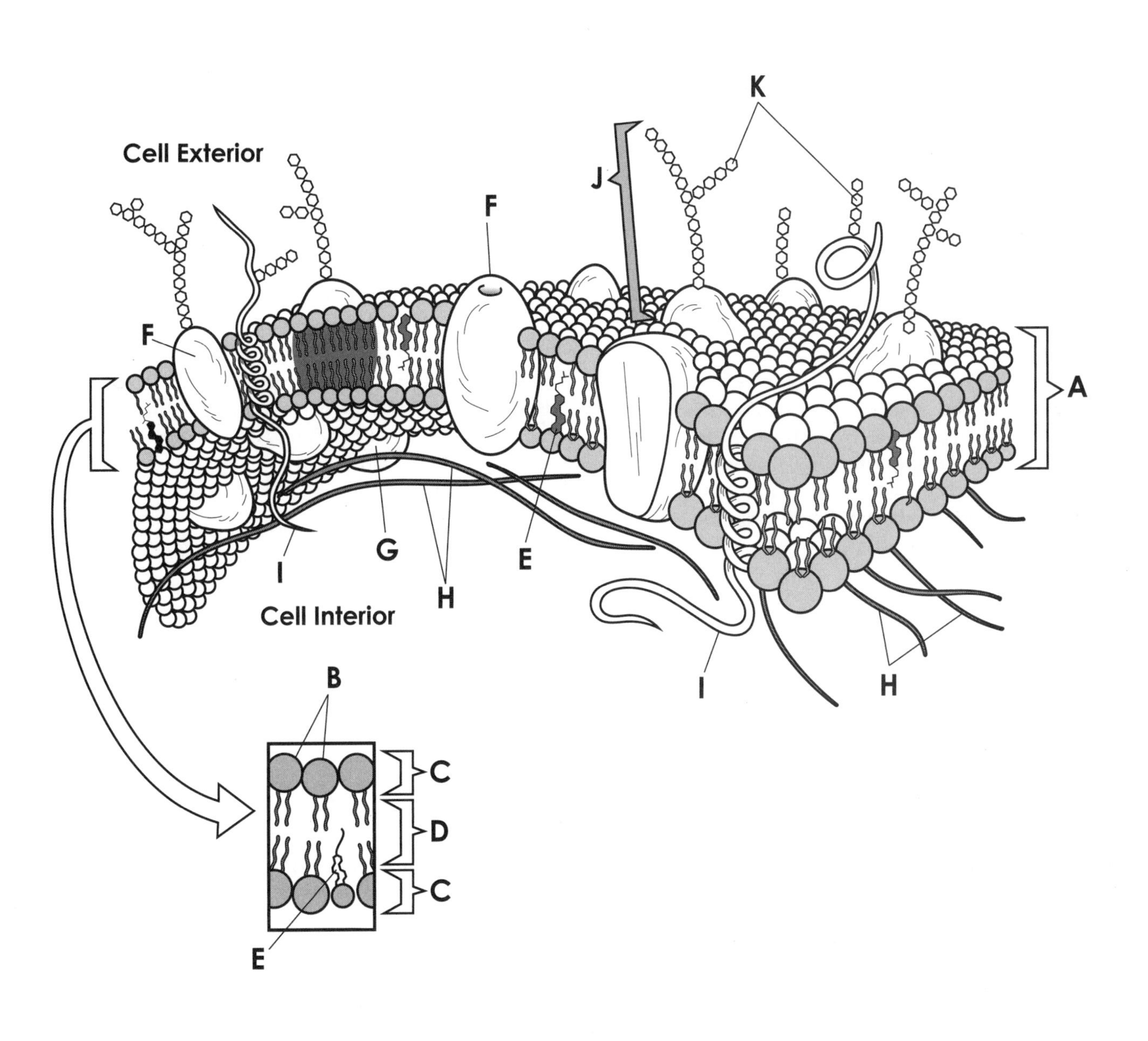

Lipid Bilayer	A	❍	Cholesterol Molecule	E	❍	Alpha Helix Protein	I	❍
Phospholipids	B	❍	Integral Protein	F	❍	Glycoprotein	J	❍
Hydrophilic Polar Head	C	❍	Peripheral Protein	G	❍	Carbohydrate	K	❍
Hydrophobic Nonpolar Tail	D	❍	Cytoskeleton Filaments	H	❍			

PASSIVE TRANSPORT

Despite their differences in size and shape, all cells are enclosed by a cell membrane that consists of a double layer of phospholipids interspersed with proteins. Its unique structure permits some substances to cross the plasma membrane rapidly, while others are unable to cross it, or cross it slowly. Thus, the plasma membrane regulates the substances entering and leaving the cell. In this plate we explore three methods for the passive transport of molecules through the plasma membrane. Passive transport processes are ones that do not require cellular energy to proceed.

Looking over the plate, notice that it is composed of three diagrams, each of which depicts a form of transport.

A plasma membrane that permits the passage of certain substances is said to be semipermeable. For example, a semipermeable cell membrane might not be permeable to certain large molecules, but might be permeable to oxygen and carbon dioxide (this means that they pass freely across the membrane). The force that propels oxygen and carbon dioxide across the membrane is called diffusion. Diffusion is the net movement of molecules from a region of high concentration to one of low concentration.

In the first diagram, we illustrate the process of diffusion in the absence of a membrane. A beaker contains both **water molecules (A)** and a **crystal (B)** of colored material. Within the triangle are a number of **crystal molecules (C).** You should use contrasting colors for the crystal molecules and water.

Molecules diffuse, or move, from areas where they are highly concentrated to areas where their concentration is lower. In the second beaker, you can see the **movement of crystal molecules (C_1)** away from the area of the triangular area, and a **movement of water molecules (A_1)** into the triangle area where the crystals started.

In the third beaker, equilibrium has been reached, and you can see completely **mixed water and crystal molecules (D).** Diffusion has taken place and no more net movement of the crystal molecules will occur.

We now turn to a special kind of diffusion called osmosis. This time we will use a cell and show movement across its membrane. The movement is passive, meaning that it occurs without the input of any energy. Continue coloring as you read the paragraphs below.

The net movement of water molecules across a membrane is a special kind of diffusion called osmosis. In the first view, we see a **red blood cell (E),** which should be colored in a light color. The concentrations of salt and **water (F)** inside and outside the cell are identical, so that the **movement of water (G)** occurs at equal rates, into and out of the cell. The inside and outside environments are said to be isotonic.

In the second view, the red blood cell (E) is placed in a very salty solution. Water molecules begin to flow out of the cell, as shown by the movement of water (G), and the cell will shrink. The outside environment is said to be hypertonic relative to the interior, which is hypotonic.

Now examine the third view, in which the red blood cell has been placed in a solution that contains no salt. There is a higher salt concentration inside the cell, and water flows through the membrane into the cell to dilute the salt; this causes the cell to swell. The inside environment is said to be hypertonic relative to the exterior cell environment, which is hypotonic.

While some water molecules can simply diffuse across a plasma membrane without the help of a protein, most cross the plasma membrane through aquaporin proteins. This is a type of passive transport called **facilitated diffusion,** because a protein helps a small molecule cross into or out of the cell. The molecule crossing the membrane is still traveling down its gradient (from a region of high concentration to an area of lower concentration), so no energy is required.

Carbohydrate molecules can also cross the plasma membrane via facilitated diffusion. Continue your reading below as you color the third diagram in the plate.

Facilitated diffusion is the movement of molecules across the membrane with the aid of a membrane protein. This diagram shows a high concentration of **carbohydrates (H)** at the cell exterior; this is shown by the **concentration gradient (I).** Facilitated diffusion takes place through **transport proteins (J)** embedded in the **cell membrane (K),** and diffusion occurs with the concentration gradient.

In the second view, diffusion has begun. The carbohydrate molecules are moving from the area of high concentration to an area of low concentration inside the cell. One carbohydrate molecule is presently being transported. In the third view, the carbohydrate molecule is added to the one already in the cell. The remaining carbohydrate molecules will follow it until the number of carbohydrates on either side of the membrane is equal.

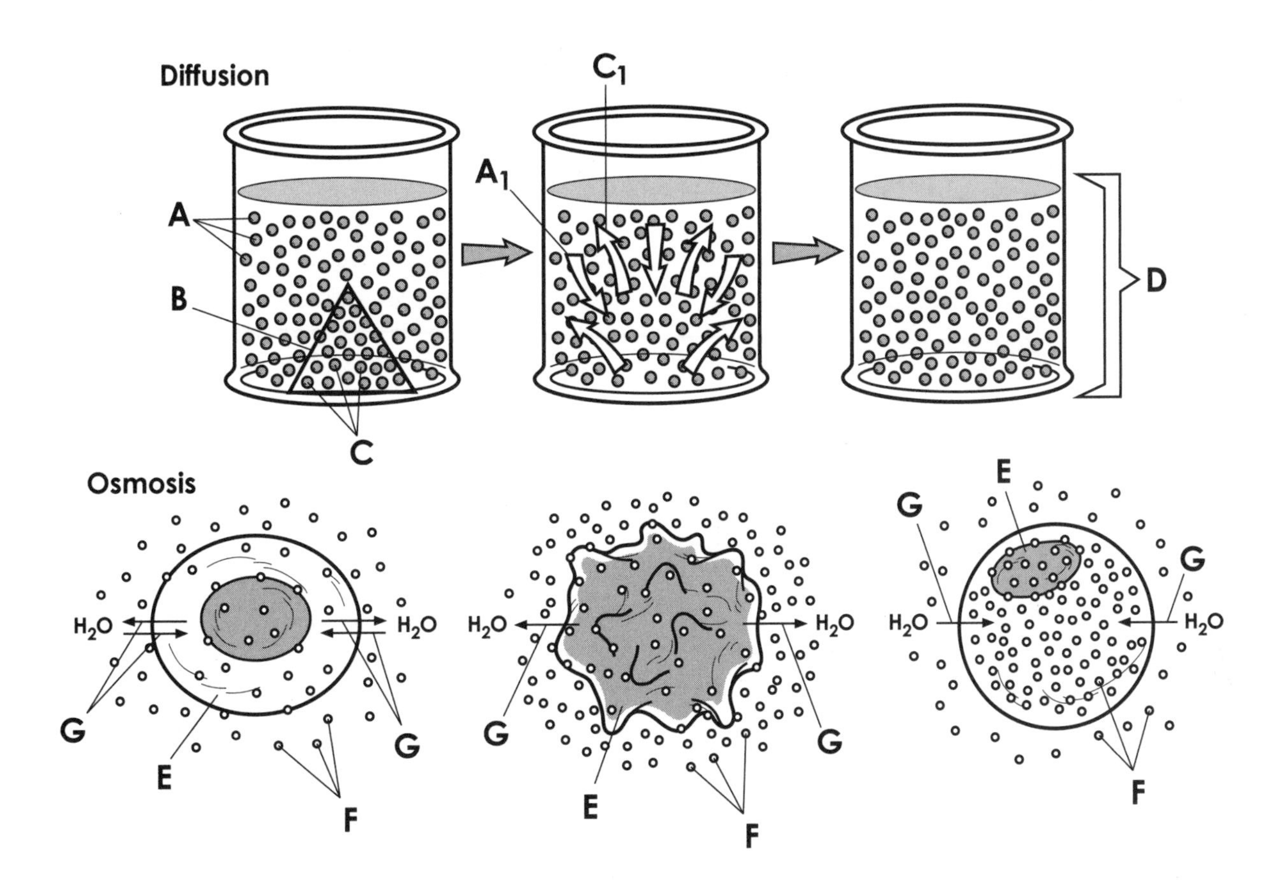

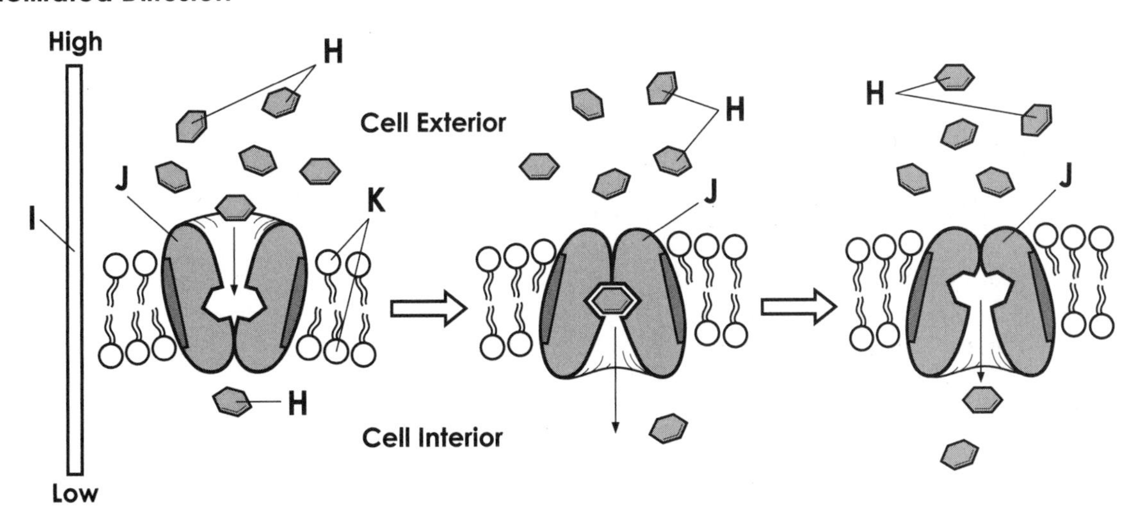

Water Molecules A ❍
Movement of Water Molecules A_1 ❍
Crystal B ❍
Crystal Molecules C ❍
Movement of Crystal Molecules C_1 ❍
Mixed Water and Crystal Molecules D ❍
Red Blood Cell E ❍
Water F ❍
Movement of Water G ❍
Carbohydrates H ❍
Concentration Gradient I ❍
Transport Protein J ❍
Cell Membrane K ❍

ACTIVE TRANSPORT AND ENDOCYTOSIS

The cell membrane encloses the cell, forming a barrier that separates the interior and exterior environments. The membrane may be relatively permeable or impermeable, prohibiting the passage of most molecules. It can also be selectively permeable, allowing certain substances to pass, but not others. In the previous plate, we discussed passive methods of transport and in this plate, we discuss two methods of active transport. Both of these active transport methods require the input of energy by the cell.

We will discuss two types of active transport in this plate. The upper portion of the plate shows a process of active transport that involves a transmembrane protein, while the lower portion shows the process of endocytosis.

The process of active transport requires energy because the molecules being moved are traveling from a region of low concentration to a region of high concentration. That is, they are being transported against their concentration gradient.

The first process we will discuss is referred to merely as active transport. In the diagram, we see some **amino acids (A)** at the cell's exterior. Within the cell, the number of amino acid molecules is higher than on the outside, as you can see in the diagram. (Notice that the bar representing the **concentration gradient (B)** shows an increase in amino acid concentration from exterior to interior.) The amino acids must be moved against the concentration gradient if they are to enter the cell. Active transport involves special proteins called **transport proteins (C),** in the **cell membrane (D).** Light colors should be used for the cell membrane.

In the second view, you can see that active transport has begun, and an amino acid is enclosed within the transport protein (C). An **ATP molecule (E)** supplies energy and is consumed during this transportation, and its breakdown results in an **ADP molecule (F)** and **phosphate ion (G).** Moving to the third view, we see that active transport is complete, and the amino acid is in the cell's interior; this is how amino acids are absorbed after the digestion of protein by cells that line the digestive tract.

We will now look at a second form of active transport, endocytosis. Endocytosis refers to the movement of particulate matter into cells, as the diagrams in the lower half of the plate illustrate. Continue your coloring as you read about endocytosis.

Certain molecules such as polypeptides, polysaccharides, and DNA are too large to be transported into the cell by carrier proteins, so they must be endocytosed.

In endocytosis, the **particles (H)** that are to be taken into the cell are represented by dots that you should color in a bright color. They are suspended in the **extracellular fluid (I),** which should be left white or shaded in a pale color. Inside the membrane (D), in the cell's interior, you can see the **cell cytoplasm (J).**

In the first view, you can see that the membrane is beginning to fold inward, and in the following drawings it continues to invaginate, forming a bubble that eventually pinches off. In the third view, the membrane has pinched off from the surface membrane and is now a **vesicle (K).** The structure of the vesicle is identical to that of the plasma membrane, and as you can see, it contains the particles.

The materials in this vesicle will soon be broken down by enzymes that are derived from a cellular body known as the lysosome. Once digested, the cell will use the products in cellular processes.

Biologists recognize two types of endocytosis. The first, called phagocytosis, occurs when a cell takes in particulate matter for digestion. For example, white blood cells are responsible for engulfing bacteria and destroying them. The second form of endocytosis is pinocytosis, in which nutrients are taken into the cell. The root cells of plants use this method for obtaining dissolved nutrients from the soil.

Active Transport

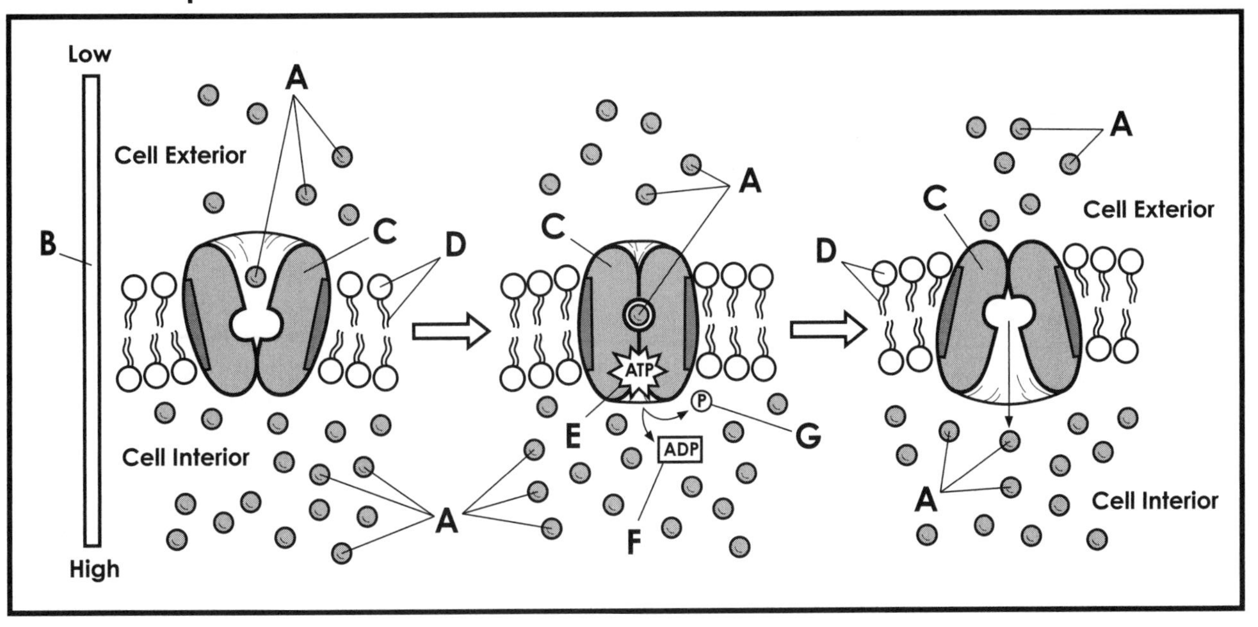

Endocytosis

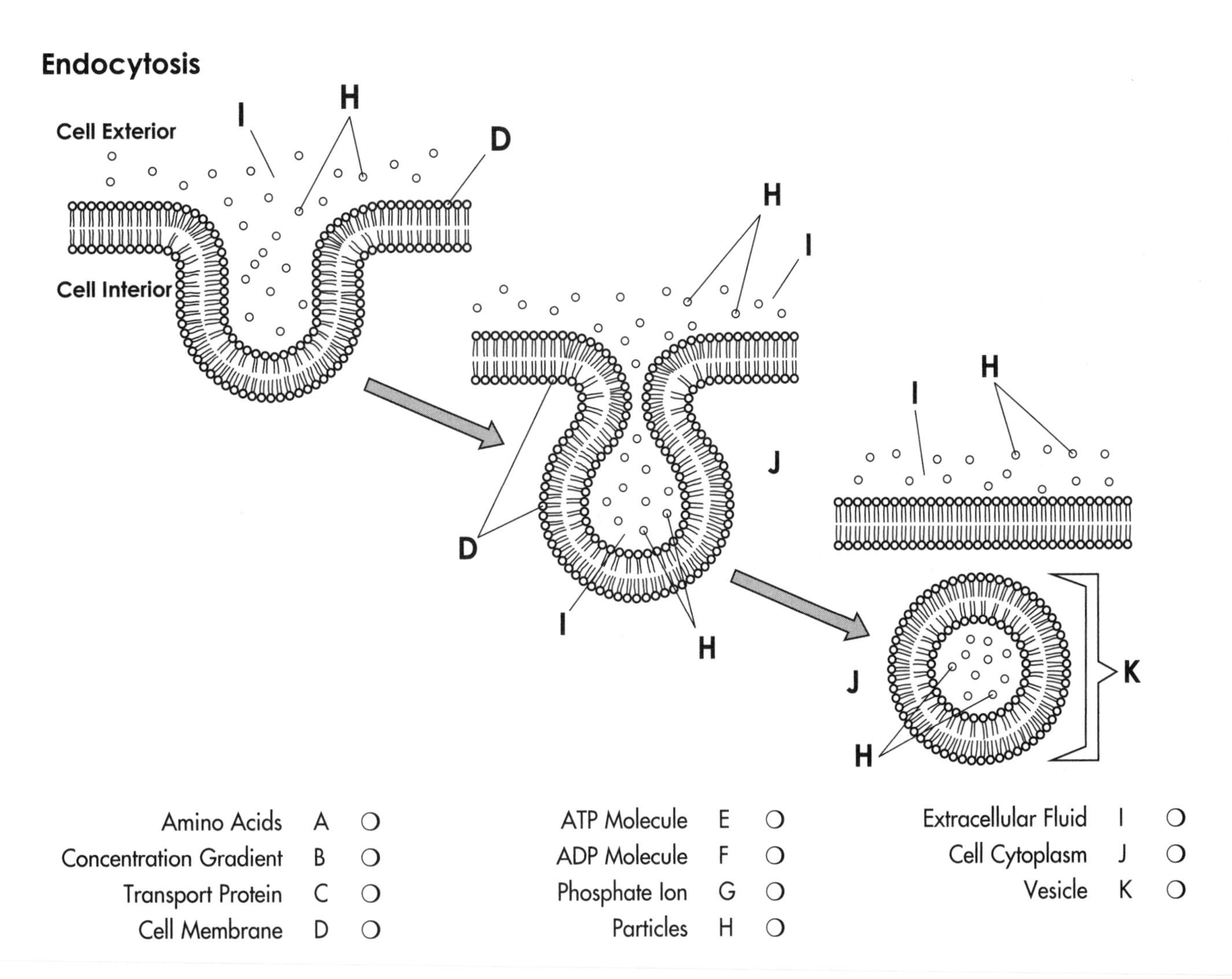

Amino Acids	A	❍	ATP Molecule	E	❍	Extracellular Fluid	I	❍
Concentration Gradient	B	❍	ADP Molecule	F	❍	Cell Cytoplasm	J	❍
Transport Protein	C	❍	Phosphate Ion	G	❍	Vesicle	K	❍
Cell Membrane	D	❍	Particles	H	❍			

ENZYMES

The chemical reactions that occur in the cells of living things take place in part because of the actions of biological catalysts called enzymes. Each type of enzyme catalyzes only one type of reaction, and since thousands of different kinds of reactions occur in cells, thousands of different enzymes exist. Almost all enzymes are proteins, so they are synthesized through the mechanism of protein synthesis discussed in Chapter 4. There are some RNA molecules with enzymatic activity; these are called **ribozymes.** The ribosome contains some ribozymes that help with peptide bond formation, and the process of gene splicing (where exons are kept and introns are deleted) also uses ribozymes. Both these processes will be reviewed in the plates on Protein Synthesis.

> In this plate we will examine the function of enzymes in chemical reactions. Our focus will initially be on how enzymes affect rates of reaction, and then we will turn to a study of enzyme activity. Take a look at the text below and color the appropriate structures.

Energy must be supplied in order for most cellular chemical reactions to proceed. The initial input of energy required to start a reaction is called the energy of activation. As is shown in the diagram entitled Enzyme Function, chemical **reactants (A)** can be converted into **products (B)** only after a substantial input of energy. Bold reds, greens, or blues should be used to color the arrows. Enzymes act to lower the activation energies of chemical reactions; in enzyme-catalyzed reactions, **reactants (C)** react more quickly to form **products (D)** because the activation energy of the reaction is far less. Dark colors should be used for these arrows.

> We will now discuss how enzymes lower the activation energy of chemical reactions.

An **enzyme (E)** is a biological catalyst that speeds up a chemical reaction without itself being consumed or altered. Enzymes' names can be easily recognized because they usually end in –ase. For example, the enzyme protease acts on protein, and lactase acts on lactose. Some enzymes are named specifically according to the type of reaction they catalyze. One example of this is synthetase, which catalyzes synthetic reactions; another is hydrolase, which catalyzes hydrolysis reactions.

The key portion of the enzyme is the **active site (E_1).** The active site is the region in which reactants bind and, essentially, the region where the chemical reaction takes place. We will now describe one way that enzymes operate—enzymes can operate in several different ways to speed reactions. In this process, the reactant that binds loosely to the active site at the start of the reaction is called the **substrate (F).** In reaction 1, we see the enzyme and substrate combine at the active site. This reaction forms what is known as an enzyme-substrate complex. In reaction 2, a chemical change takes place at the active site and the result is the formation of two **products (G).** In reaction 3, the products are released from the active site and the enzyme is recycled for use in another reaction. For reactions in which there are two or more substrates, both bind loosely to the enzyme, which brings them close together. This allows them to react more quickly than they would otherwise, after which they are released.

> One way to remove an enzyme from a chemical system is to inhibit it. How inhibition takes place is the subject of the third diagram of the plate. Color the appropriate structures as you complete the plate.

There are numerous ways in which enzymes can be inhibited. For example, heat can denature an enzyme, or change its structure, making it unable to bind to a substrate. Enzymes can also be denatured by acidic environments.

Many chemical substances interfere with the activity of enzymes by binding to them, and these are referred to as inhibitors. Inhibitors are classified by the way in which they bind to an enzyme. Notice that the enzyme (E) can react with the substrate (F) as well as with an **inhibitor (H).** In these cases, there is competition for the enzyme's active site. When large amounts of the inhibitor are present, it is more likely that the inhibitor, and not the substrate, will bind at the active site, as shown in chemical reaction 2. Once the inhibitor is blocking the active site, the substrate molecule cannot bind, as you can see in reaction 3, and no chemical reaction will occur.

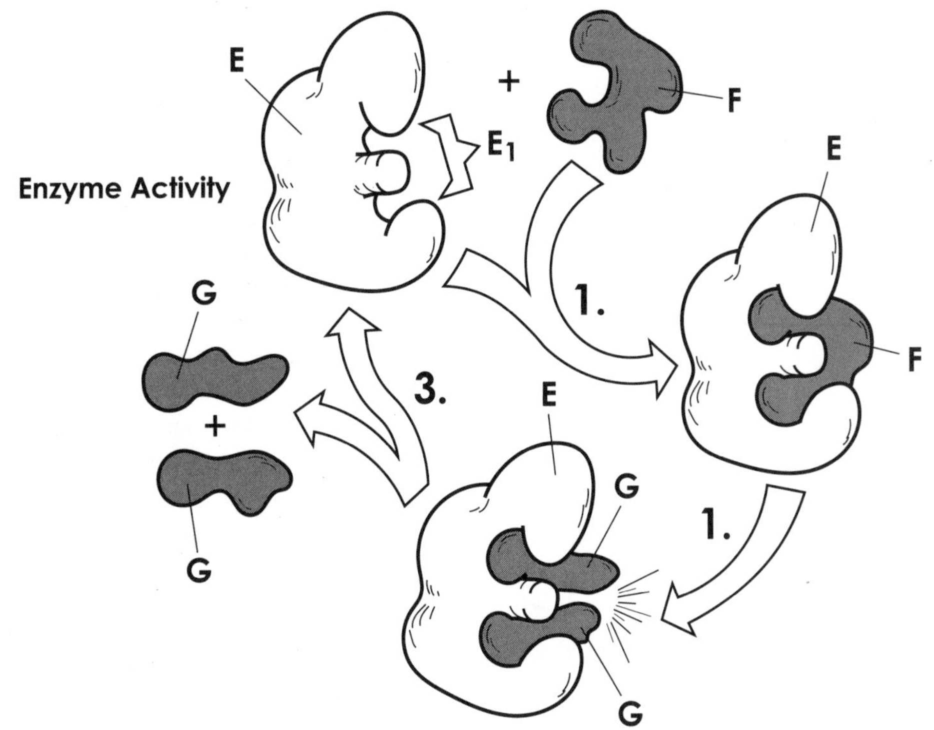

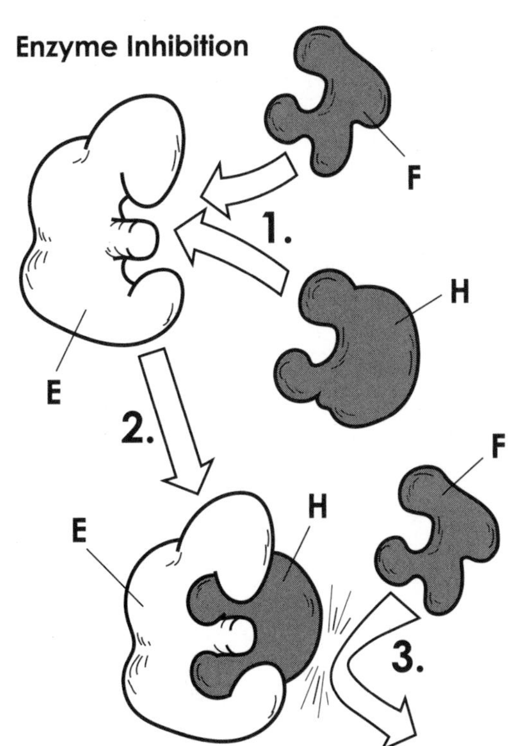

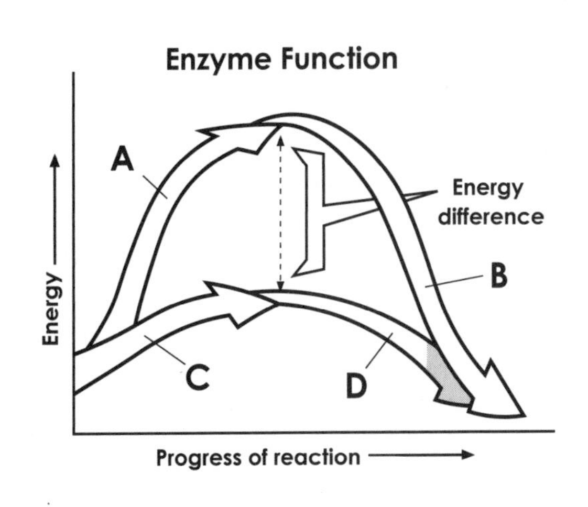

Reaction Without Enzyme
Reactants A ❍
Products B ❍

Reaction With Enzyme
Reactants C ❍
Products D ❍

Enzyme E ❍
Active Site E_1 ❍
Substrate F ❍
Products G ❍
Inhibitor H ❍

ENERGY FLOW IN LIVING THINGS

The total amount of energy that exists in the universe remains constant, but energy can change from one form to another. For example, the chemical energy in gasoline can be released and transformed into heat energy and the energy of motion.

This type of transformation of energy occurs in many of the processes that take place in living things. In this plate, we will examine the flow of energy through living things and identify the molecule that serves as the main energy source in all life processes.

This plate shows how energy exists in different forms at different times in living things. As you encounter the terms, color the appropriate structures in the diagram.

All of the energy on the Earth comes from the **Sun (A);** the **Sun's energy (A_1)** is what drives chemical reactions and the processes of life. This solar energy is trapped in a photosynthesizing organelle of the plant called the **chloroplast (B)**; we discuss this organelle in detail later in the book. It is found in all plant cells and some protists.

A number of chemical reactions take place in the chloroplast to transform solar energy into chemical energy. **Carbon dioxide (C)** and **water (D)** are necessary for the process of **photosynthesis (E),** and the products of photosynthesis include **carbohydrates (F),** which are represented by a candy bar, and molecular **oxygen (G).** The bonds of the carbohydrates now contain some of the Sun's energy; photosynthesis has transformed the Sun's energy into the chemical energy of the carbohydrate. Oxygen is given off as a waste product of photosynthesis, and it is expelled from the plant cell into the atmosphere.

Having explained how the Sun's energy is converted to the chemical energy found in carbohydrates, we will now discuss another transformation of energy. Continue your reading below, and focus on the right side of the diagram as we continue to study energy flow in living things.

Plants, humans, and many other living organisms use carbohydrates as their essential source of energy. Carbohydrates are transported to an organelle called the **mitochondrion (H),** where they are combined with oxygen molecules in the process of **respiration (I),** illustrated by the arrow. During chemical reactions in the mitochondrion, the energy from carbohydrates is released and used to form the energy-rich molecule **adenosine triphosphate (J).** (Adenosine triphosphate is commonly abbreviated as ATP.) Carbon dioxide and water are byproducts of respiration; notice that both are essential for photosynthesis. To summarize, the energy of the Sun is first transformed into the energy of carbohydrates and then into the energy in the ATP molecule.

We will conclude with a brief examination of the ATP molecule. Recall that the energy of the ATP molecule comes from the Sun. As you read, color the appropriate structures in the diagram.

The adenosine triphosphate (ATP) molecule (J) is shown at the bottom of the plate. You should use a light shade to color the interior of the box, and darker colors should be used for the components of ATP. These components include an **adenine molecule (J_1)** and a **ribose molecule (J_2).** Adenine is one of the four nitrogenous bases found in DNA and RNA, and ribose is a five-carbon carbohydrate. Attached to the ribose molecule are three **phosphate groups (J_3).**

Living things use energy in the form of ATP, breaking it down into **adenosine diphosphate (K)** and an inorganic phosphate group. Adenosine diphosphate (ADP) contains adenine (J_1) and a ribose molecule (J_2), but only two phosphate groups (J_3). During this breakdown, seven kilocalories of energy are given off for use by the cell.

In the following plates, we will study the processes by which ATP is created, such as glycolysis, the Krebs cycle, electron transport, and chemiosmosis.

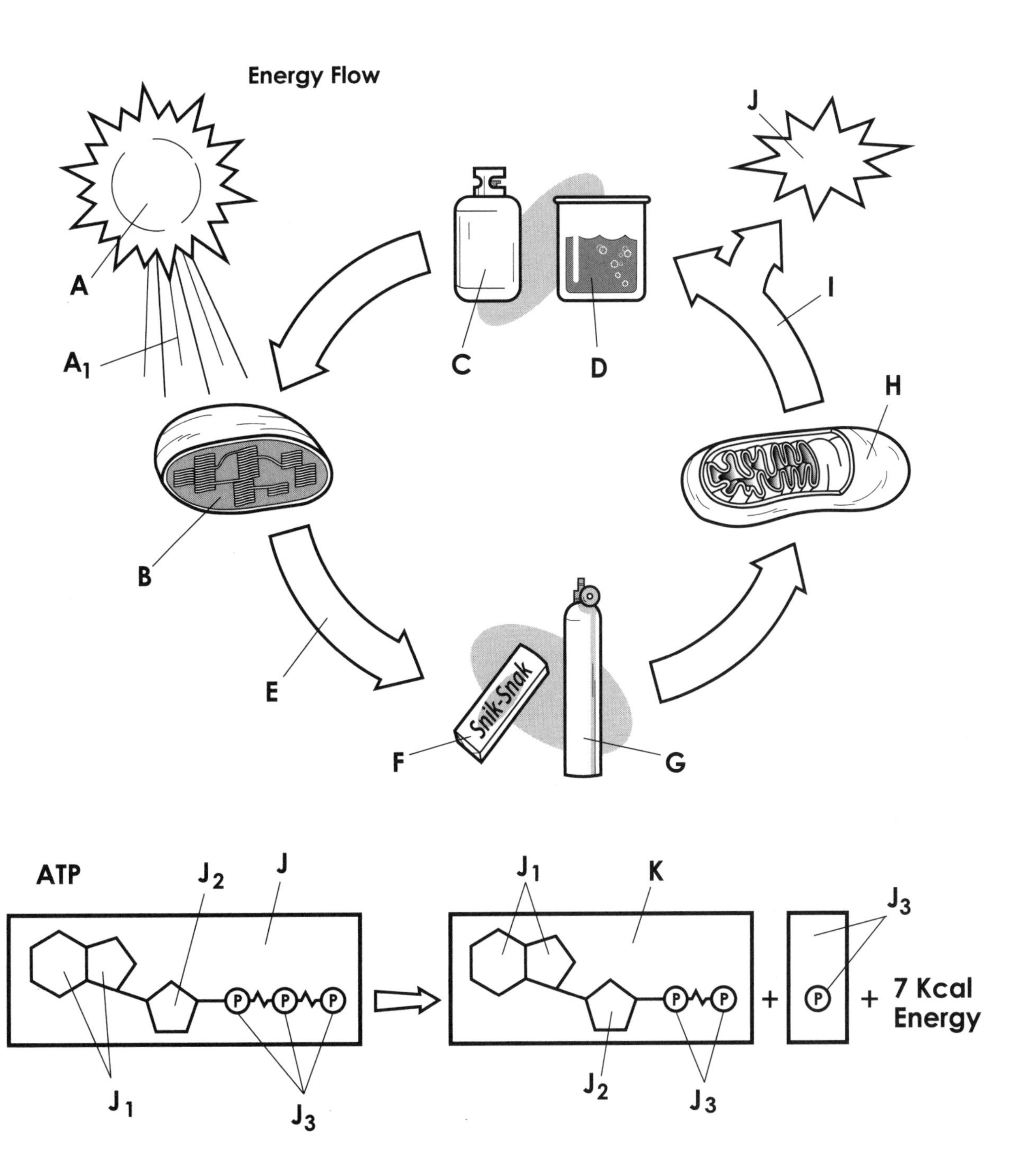

Sun	A	❍	Photosynthesis	E	❍	Adenosine Triphosphate	J	❍
Sun's Energy	A_1	❍	Carbohydrates	F	❍	Adenine	J_1	❍
Chloroplast	B	❍	Oxygen	G	❍	Ribose	J_2	❍
Carbon Dioxide	C	❍	Mitochondrion	H	❍	Phosphate Groups	J_3	❍
Water	D	❍	Respiration	I	❍	Adenosine Diphosphate	K	❍

STRUCTURES OF PHOTOSYNTHESIS

Photosynthesis is the biochemical process through which plants convert the Sun's energy into a usable chemical form. During photosynthesis, a plant produces carbohydrates that provide energy for the plant and are modified in numerous ways to serve as important cellular components.

Photosynthesis is also essential to animals, including humans, who obtain all their food either directly or indirectly from plants. In addition, photosynthesis replenishes the atmospheric oxygen used in animal metabolism.

The reactions of photosynthesis take place within the chloroplasts of plant cells and in the cytoplasm of cyanobacteria. This plate focuses on chloroplasts and describes their structure and function in photosynthesis.

In this plate, we present a series of diagrams starting with the leaf and progressing to the submicroscopic structures involved in photosynthesis.

We will begin with a survey of the main photosynthetic structure of the plant, the **leaf (A).** Although the leaf is considered the center of photosynthesis, this process also occurs in cells of the plant stem.

In diagram 2, we show a cross section of the leaf. The surface of the leaf is covered by a thin waxy layer called the **cuticle (B),** under which lie the cells of the **epidermis (C).** Beneath the epidermis are several layers of cells called **mesophyll cells (D).** Some of these cells are tall and stacked against each other; these make up the palisade layer of mesophyll cells. Other cells are more cubical and loosely packed; these comprise the spongy layer of the mesophyll. Mesophyll cells contain the main structures that carry out photosynthesis. At the lower portion of diagram 2 are stomates, where carbon dioxide necessary for photosynthesis enters the leaf.

We have begun our survey of photosynthetic structures by focusing on the leaf and some of its details. We will now take a single cell of the leaf and display its photosynthetic structures. Continue your reading as you color the plate.

Take a look at the single plant cell in diagram 3. This cell is rectangular compared to an animal cell, because plant cells have cell walls that maintain their box-like rigidity.

In diagram 3, we show a single mesophyll cell (D) and some of its major features. For example, the **nucleus (E)** is situated along the edge of the cell because the large central vacuole has pushed it to the side, and within the cytoplasm are a number of **chloroplasts (F).** These bodies can be seen with a light microscope, but the smaller structures we will mention in this plate can only be seen with an electron microscope. The mesophyll cell contains numerous chloroplasts, which are where the photosynthetic structures are found.

The next view is of a single chloroplast (F) in diagram 4. The fluid-filled space within the chloroplast is known as **stroma (G),** which is a matrix that holds the functional components of photosynthesis. We now move to diagram 5, in which the chloroplast has been further magnified. You can see stacks of membranous, sac-like vesicles called **thylakoids (I).** Thylakoids are disc-shaped, and a stack of them composes what is called a **granum (H).**

We complete the plate with a study of diagram 6, in which a granum (H) is enclosed by a bracket. The region between the thylakoid membranes is the **thylakoid space (J),** and this space is also sometimes called the lumen. The space around the thylakoids is the stroma of the chloroplast.

Embedded in the thylakoid membranes themselves we see a number of **photosynthetic pigments (K).** These pigments, which include chlorophyll, are the biochemical substances involved in photosynthesis. Also embedded in the membrane is a chemical complex called **ATP synthase (L),** at which energy from the Sun is converted to the energy of ATP molecules. We will explain how this takes place in the next plate.

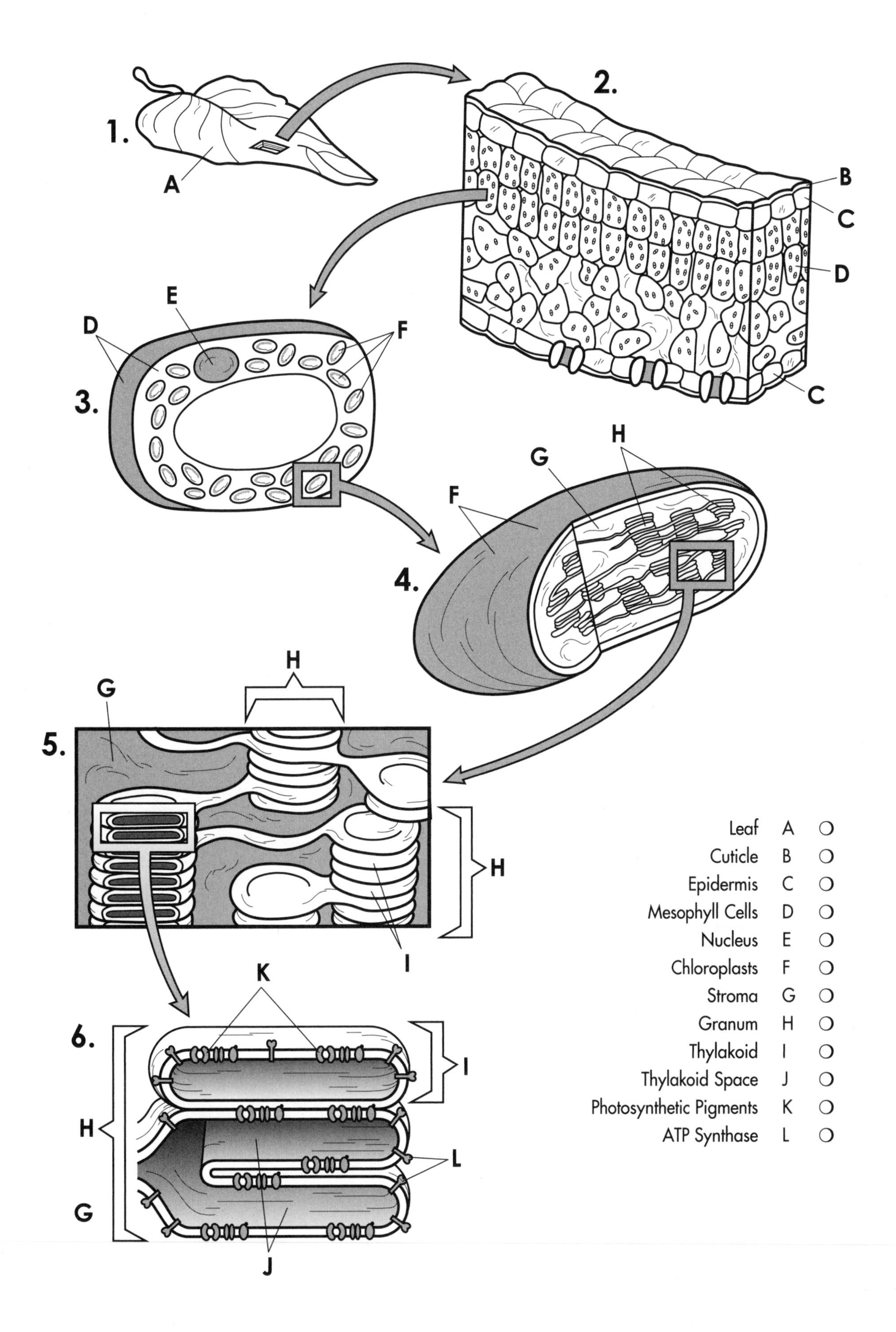
1.
A
2.
B
C
D
C
3.
D
E
F
4.
F
G
H
5.
G
H
H
I
6.
K
I
H
L
G
J
Leaf A
Cuticle B
Epidermis C
Mesophyll Cells D
Nucleus E
Chloroplasts F
Stroma G
Granum H
Thylakoid I
Thylakoid Space J
Photosynthetic Pigments K
ATP Synthase L

PHOTOSYNTHESIS—THE LIGHT REACTIONS

Photosynthesis is the biochemical process by which sunlight, oxygen, and water are converted into energy contained in the chemical bonds of carbohydrates. Photosynthetic organisms include green plants, algae, and certain species of bacteria. These organisms are key players in the cycles of life on Earth, since all atmospheric oxygen and a large quantity of food come from photosynthesis.

The two main processes of photosynthesis involve a series of energy-fixing (light) reactions and a series of carbon-fixing (dark) reactions. In the light reactions, energy from sunlight is trapped in the chemical bonds of ATP, while in the second process, this ATP is used to form carbohydrate molecules. The dark reactions are the subject of the next plate.

> This plate contains three diagrams that depict the energy-fixing reactions of photosynthesis. The biochemistry of these reactions can be difficult to comprehend, so go through the reading slowly.

Photosynthesis takes place inside chloroplasts in specialized membranes called thylakoids (which were mentioned in the last plate). In the main diagram of this plate, we show a large leaf. If you want to color it, use a very pale color.

The process of photosynthesis begins with the Sun's **light energy (A).** This energy enters leaf cells and is absorbed and transferred to a series of chlorophyll molecules within a complex cluster called a photosystem. The first photosystem involved in this transfer is **photosystem II (B),** and this photosystem contains chlorophyll molecules that absorb light that has a wavelength of 680 nanometers (nm).

When the complex in photosystem II is activated by light energy, it gives up **electrons (C),** which are then absorbed by an **electron acceptor (D).** This electron acceptor is part of what is called an energy transfer system, through which the electrons move until they reach **photosystem I (F).** During these transfers, hydrogen ions are pumped from the stroma into the interior of the thylakoid. As hydrogen ions leak back across the membrane through special carrier proteins called ATP synthases, **ADP (E_1)** is phosphorylated, forming **ATP (E).**

Now photosynthesis continues. Light energy (A) is absorbed by photosystem I (F), whose chlorophyll pigments absorb light energy that measures 700 nm. Once again, energy is transmitted by the chlorophyll in the complex, and an electron is given off to an electron acceptor, ferredoxin.

The electron acceptor then transfers the electron to a molecule of nicotinamide adenine dinucleotide phosphate, or **$NADP^+$ (G_1),** which takes on a **hydrogen ion (H)** to become **NADPH (G).** NADPH is used in carbon fixation in the next plate.

We complete the process of photosynthesis by referring back to the original P680 molecule, which has lost an electron that must be replaced. A **water molecule (I)** breaks down, forming free hydrogen ions and **diatomic oxygen (J),** and contributing an electron to the P680 complex. The oxygen is released to the atmosphere.

> We have completed the discussion of the energy-fixing reactions of photosynthesis, and will now spend a moment on an alternative one called the cyclic reactions.

An alternative process of photosynthesis occurs in certain types of bacteria and is used to produce ATP, but does not produce NADPH, nor does it involve water or oxygen. In this cyclic reaction, light energy stimulates photosystem I (F) to emit an electron (C). An electron acceptor picks up the electron and passes it on through a series of molecules until it eventually returns to the photosystem. During this process, ADP (E_1) combines with a phosphate molecules to produce ATP (E). This reaction is cyclic because the electron moves out of the photosystem and then back into it.

> We conclude by looking at the key elements of the light stages of photosynthesis, in which ATP is produced. This reaction is entitled Chemiosmosis. Read about the process as you focus on the final diagram of the plate.

Chemiosmosis is the mechanism through which ATP is produced in the chloroplast of the plant cell. Light energy (A) enters a specific **chlorophyll molecule (B_1)** of photosystem II, and then we see **energy flow (L)** as this energy moves into photosystem I (F). All of this takes place within the **thylakoid membrane (K).** During this transferal of energy, there is **hydrogen ion flow (M)** across the membrane, from the **stroma (P)** into the **thylakoid space (O).**

As hydrogen ions move back into the stroma, they travel through an enzyme complex called **ATP synthase (N).** The hydrogen ion flow coincides with the formation of molecules of ATP (E) from ADP. The ATP formed in chemiosmosis is an essential factor in the carbon-fixing reactions of photosynthesis, which are discussed in the next plate.

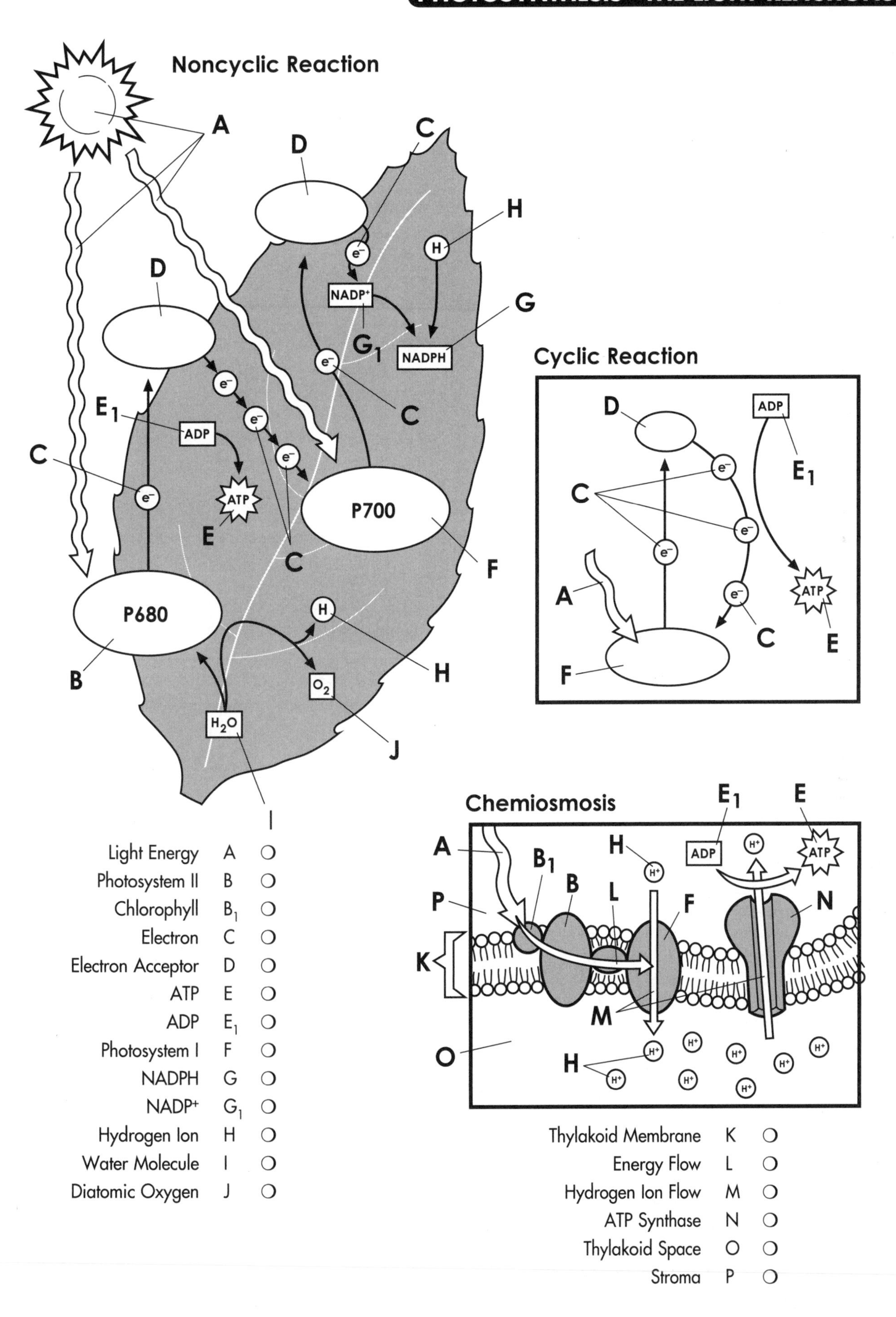
Noncyclic Reaction
P680
P700
ADP
ATP
NADP+
NADPH
H2O
O2
Cyclic Reaction
Chemiosmosis
Light Energy A
Photosystem II B
Chlorophyll B1
Electron C
Electron Acceptor D
ATP E
ADP E1
Photosystem I F
NADPH G
NADP+ G1
Hydrogen Ion H
Water Molecule I
Diatomic Oxygen J
Thylakoid Membrane K
Energy Flow L
Hydrogen Ion Flow M
ATP Synthase N
Thylakoid Space O
Stroma P

PHOTOSYNTHESIS—THE DARK REACTIONS

The term *photosynthesis* refers to two major processes that are made up of complex series of biochemical events. The first process (which is made up of the energy-fixing reactions) was discussed in the last plate. ATP and NADPH are generated in those reactions, and are used in the second process, the carbon-fixing reactions, which we will now discuss. The carbon-fixing reactions are sometimes referred to as the dark reactions, since they do not require light. They can occur in both its presence and absence, since their energy source is ATP.

In the carbon-fixing reactions, organic molecules are formed. Carbon dioxide provides the carbon backbone for the new molecules, and fuels the reactions. The process is often called the Calvin Cycle, after Melvin Calvin, who discovered some of the key reactions.

In this plate, we present a cycle in which carbon dioxide provides the carbon necessary for the synthesis of organic compounds such as carbohydrates. The series of reactions is cyclic, meaning that it follows a circular biochemical pathway.

In the second major process of photosynthesis, energy from sunlight is converted to the chemical energy contained in the bonds of carbohydrates. These reactions occur at the outer surface of the thylakoid membrane, and the resulting carbohydrates are stored in plant cells to be used in cellular respiration, or distributed to other cells.

We begin this process with three molecules of **carbon dioxide (A),** which you can see at the top of the cycle. In reaction 1, carbon dioxide combines with a molecule called **ribulose bisphosphate (B),** abbreviated as RuBP. This molecule contains five carbon atoms and two phosphate groups, which are indicated by the P's. When the two molecules combine, a six-carbon molecule is formed. This molecule is split immediately into molecules of **3-phosphoglyceric acid (PGA) (C).** Note that each PGA molecules has three carbon atoms. This completes the first step of the carbon-fixing reactions.

We have seen how carbon dioxide from the atmosphere is incorporated into the cycle of carbon-fixing reactions. Essentially, a five-carbon molecule combines with the carbon dioxide molecule to yield a six-carbon molecule that immediately splits into two three-carbon molecules. We will now look at reaction 2.

In the second reaction of the process, the six PGA molecules are converted to six molecules of **1,3-diphosphoglyceric acid,** or **DPGA (D).** An **ATP molecule (E)** provides the energy needed to make this reaction proceed, and in the process is converted to **ADP (F).** This ATP was formed in the energy-fixing reactions described in the previous plate, so you can see that the energy-fixing reactions are essential for the carbon-fixing reactions. Only two DPGA molecules are shown.

We now move to reaction 3. Here the DPGA molecules convert to molecules of **phosphoglyceraldehyde, or PGAL (G).** The **NADPH (H)** formed in the energy-fixing reactions is used up in this conversion and becomes **$NADP^+$ (I).** Note also that one phosphate group has been lost from the DPGA in the formation of PGAL.

In reaction 5 we see that some of the PGAL molecules are used to form **carbohydrates (J),** including glucose, lactose, cellulose, and others. The interconversions that lead to these carbohydrates are extremely complex, but it is at this point that the carbon of carbon dioxide is incorporated into carbohydrate molecules, which all plants and animals use as energy. In other words, this is where the energy from the Sun is incorporated into the energy of carbohydrates.

We have now passed through most of the reactions of the carbon-fixing portion of photosynthesis. We have seen the point at which ATP and NADP are consumed and where carbohydrates are formed. We complete the cycle by showing how RuBP is regenerated. Continue to read below as you finish coloring the plate.

In order for the cycle to continue incorporating carbon dioxide, RuBP must be regenerated. Notice that in reaction 6, five PGAL molecules are used to form **ribulose phosphate (K),** also known as RP. This is a five-carbon compound that's somewhat similar to the ribose in RNA. It is formed through a complex interaction of organic compounds that we will not discuss. You should note that PGAL is used for two important purposes in photosynthesis.

The cycle is completed as RP is converted to RuBP. Note that ATP is used once more in reaction 7; it provides the phosphate group that attaches to the carbon skeleton in RuBP. Once the RuBP is formed, it is available to unite with CO_2 molecules from the atmosphere and enter another turn of the cycle.

The carbon-fixing reactions of photosynthesis are quite complex. Light energy is converted first to ATP energy, and then ATP energy is used in the formation of carbohydrates.

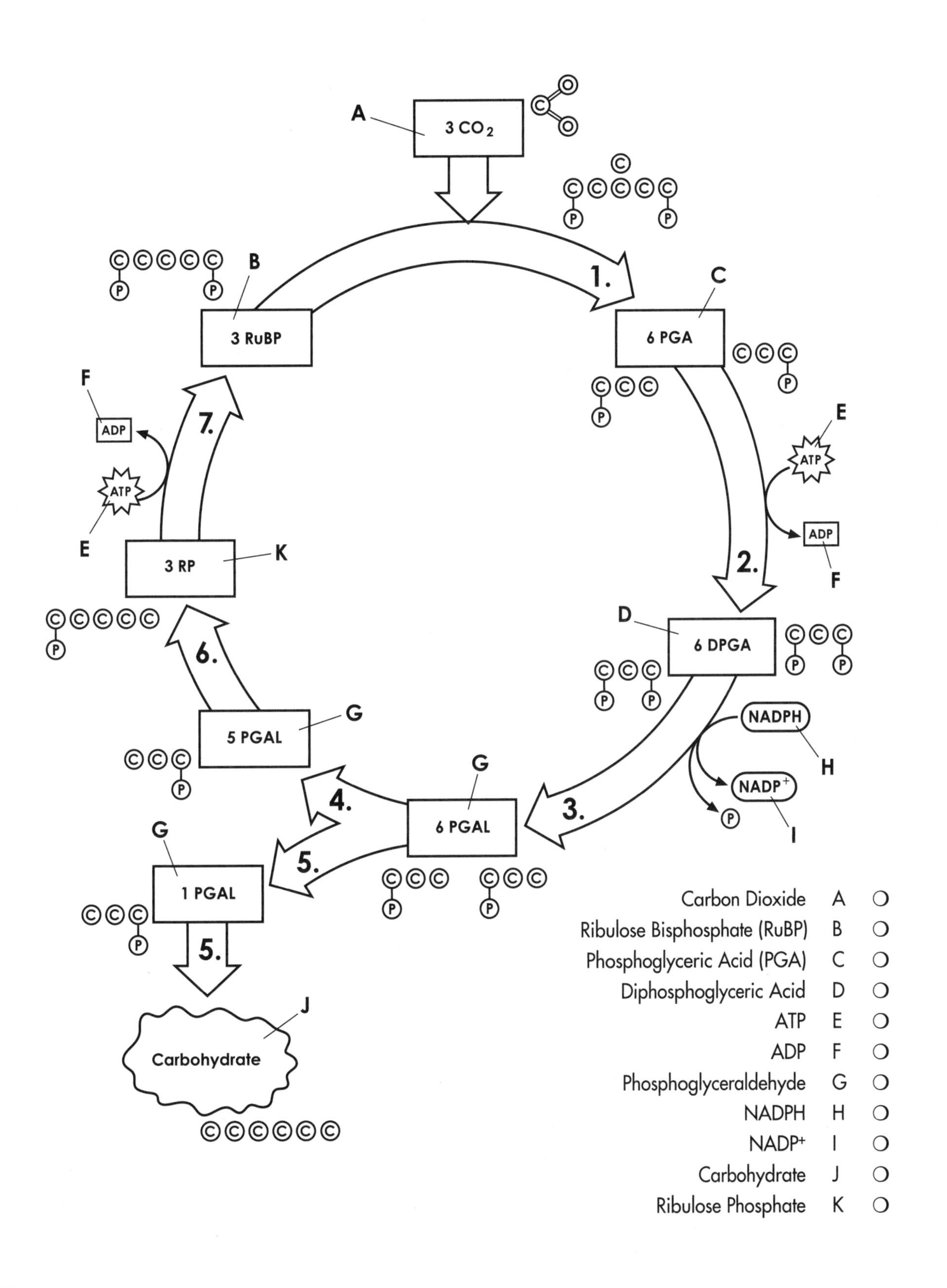

Carbon Dioxide A ❍
Ribulose Bisphosphate (RuBP) B ❍
Phosphoglyceric Acid (PGA) C ❍
Diphosphoglyceric Acid D ❍
ATP E ❍
ADP F ❍
Phosphoglyceraldehyde G ❍
NADPH H ❍
$NADP^+$ I ❍
Carbohydrate J ❍
Ribulose Phosphate K ❍

GLYCOLYSIS

Plants trap the energy from sunlight and convert it to the chemical energy contained in the bonds of carbohydrates. They accomplish this through photosynthesis, in which the principal carbohydrate formed is glucose. Other organisms, including humans, use the energy stored in glucose to create high energy molecules of ATP through the process of cellular respiration. ATP is an energy source that powers the metabolic and anabolic processes of living things, as explored in an earlier plate.

Cellular respiration involves four major steps: glycolysis, which is explored in this plate; the Krebs cycle; electron transport; and chemiosmosis, all of which are discussed in following plates. Bear in mind that the overall object of cellular respiration is to transfer the energy of glucose molecules into ATP.

This plate contains a series of cellular reactions numbered 1–9. A molecule of glucose is at the top of the series and two molecules of pyruvic acid are at the bottom. This plate will highlight some of the main conversions that occur along the way. You should color the carbon, oxygen, and phosphates. This multistep metabolic pathway takes place in the cytoplasm of cells. The reactions are catalyzed by many different enzymes. Remember that general information on enzymes was discussed in an earlier plate.

Glycolysis begins with a molecule of glucose, which, as you may remember, is a six-carbon carbohydrate that contains **carbon (C)** and **oxygen (O).** Its atoms should be colored with bold colors. The first reaction in glycolysis is indicated by the number 1. Its arrow should be colored with a dark color. Note that a molecule of adenosine triphosphate, or ATP, is consumed in the reaction to supply energy. This yields a molecule of ADP. The **phosphate group (P)** of the ATP molecule attaches to the glucose molecule and is seen in the second molecule. This reaction is catalyzed by an enzyme called hexokinase, or HK.

The second molecule is glucose-6-phosphate. In reaction 2 of the process, an enzyme converts it into fructose-6-phosphate. In reaction 3, another ATP molecule is consumed, and we see that a second phosphate group appears in the newly formed molecule, which is called fructose-1, 6-bisphosphate. This step is catalyzed by an enzyme called phosphofructokinase, or PFK.

Thus far in glycolysis, a molecule of glucose has been converted to a molecule of fructose-1, 6-bisphosphate. Enzymes are responsible for all of these conversions and two ATP molecules have been consumed in the process.

Returning to the plate, note that in reaction 4, the molecule splits. The six-carbon fructose molecule becomes two three-carbon molecules. Each of the three carbon molecules should have its carbon atoms and phosphate groups colored. No carbon has been lost in the reaction.

In reaction 5, the three-carbon molecules are converted to 1, 3-diphosphoglyceric acid, another three-carbon molecule. A phosphate group is supplied from the cytoplasm. Also, in this reaction, a molecule of nicotinamide adenine dinucleotide, or NAD^+, forms a molecule of NADH, which is extremely important in electron transport. Two molecules of NADH are formed at this point because the reaction occurs twice per six-carbon glucose.

In reaction 6, energy is given off to form a molecule of ATP. Since the reaction happens on both the left and right sides, two ATPs are formed, and the product of this reaction is 3-phosphoglyceric acid.

In reaction 7, the phosphate group is relocated and 2-phosphoglyceric acid is produced, and in reaction 8 a double bond is formed, and the result is phosphoenolpyruvic acid, or PEP. No ATP is produced during these reactions, nor is any used.

In reaction 9, the phosphate group from PEP is transferred to ADP, and two ATPs are produced. The net result of glycolysis is two pyruvic acid molecules. The conversion of pyruvic acid into acetyl CoA is catalyzed by the pyruvate dehydrogenase complex (or PDC), a multi-enzyme complex.

We have come to the end of glycolysis, and we close the plate by briefly reviewing what has taken place.

Reviewing glycolysis, we see that a molecule of glucose has been converted to two molecules of pyruvic acid. The original glucose molecule contained six carbons, and the two pyruvic acid molecules contain a total of six carbon atoms, so no carbon has been lost in the process. Two molecules of ATP were "invested" at reactions 1 and 3, but two ATP molecules were obtained in reactions 6 and 9. And since these reactions happen twice, we end with a total of four ATP molecules. Therefore, the total gain in glycolysis is two ATP molecules. In addition, reaction 5 has given us two molecules of NADH, which will be used in electron transport, where more ATP will be produced. The two molecules of pyruvic acid are used in the Krebs cycle, as explored in the next plate.

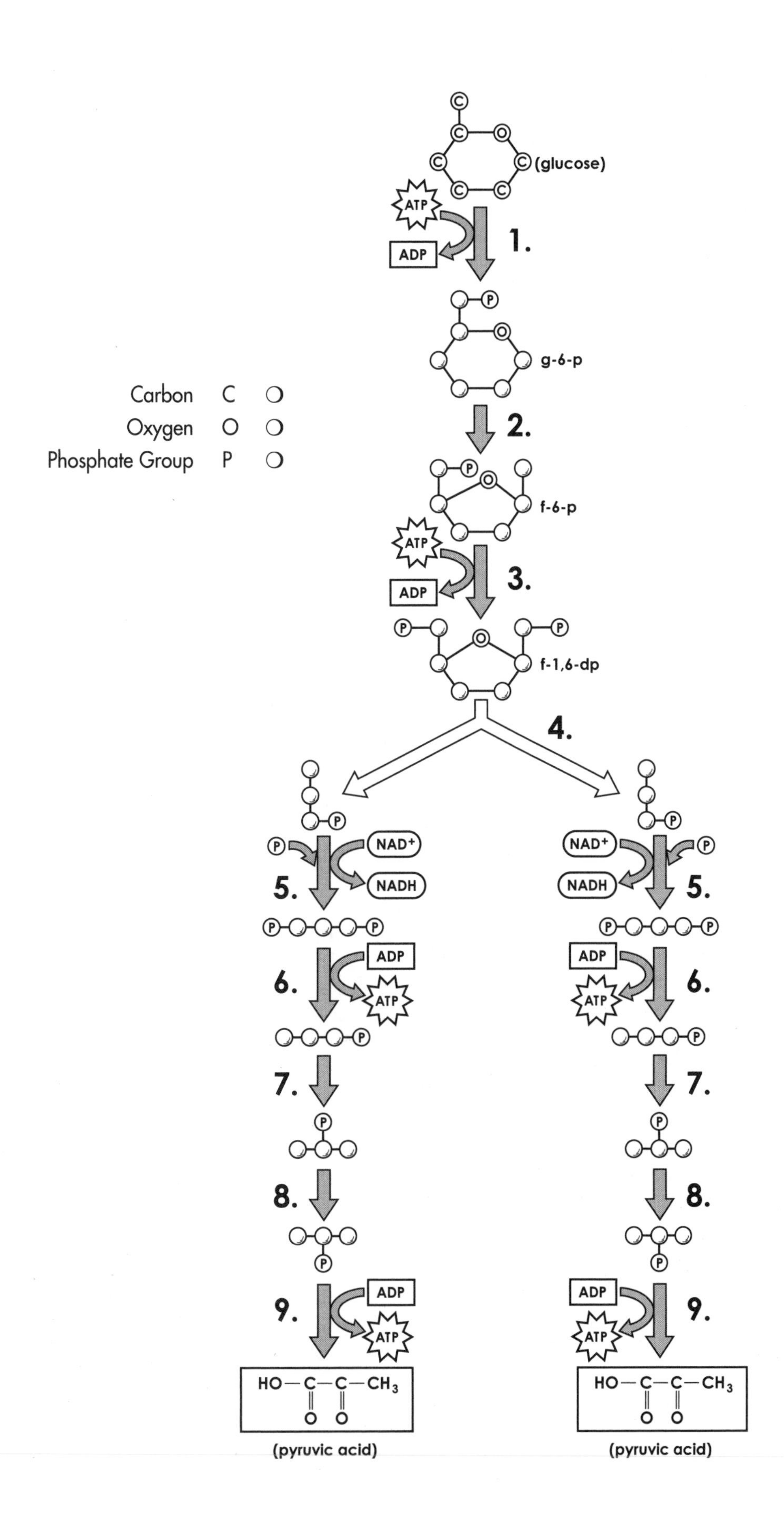
(glucose)
ATP
ADP
1.
g-6-p
Carbon C
Oxygen O
Phosphate Group P
2.
f-6-p
3.
f-1,6-dp
4.
NAD+
NADH
5.
6.
7.
8.
9.
HO—C—C—CH3
O O
(pyruvic acid)

THE KREBS CYCLE

The second important process in cellular respiration is the Krebs cycle (also called the Citric Acid Cycle). The Krebs cycle is an extension of the process of glycolysis. It takes place within the mitochondria of plant and animal cells and along the cell membranes of bacteria. In the Krebs cycle, a molecule of pyruvic acid is metabolized and its three carbon atoms are released as molecules of carbon dioxide. Along the way, several molecules of NADH are produced, and these enter the electron transport chain.

> As you look over the plate, you will note that it contains a cyclic series of events involving eight chemical reactions. As we did with glycolysis, we have simplified the structures of the molecules by drawing only their carbon backbones. We recommend using the same colors you did for glycolysis.

The Krebs cycle consumed the pyruvic acid produced during glycolysis; first color the **carbon atoms (C)** of pyruvic acid. In the first reaction, pyruvate is converted into a molecule of **acetyl CoA.** CoA stands for coenzyme A, and is the substance that activates pyruvate. During the conversion, a molecule of carbon dioxide, designated CO_2, is formed. The remaining two carbon atoms of the pyruvate combine with the CoA molecule to form acetyl CoA. NAD^+ is also involved in reaction 1 of the Krebs cycle; it is protonated and becomes NADH, which is shuttled off to the respiratory chain (discussed later). The conversion of pyruvic acid into acetyl CoA is catalyzed by the pyruvate dehydrogenase complex (or PDC), a multi-enzyme complex.

Now acetyl CoA is ready to enter the Krebs cycle. In reaction 2, an enzyme catalyzes a reaction between acetyl CoA and a four-carbon molecule called oxaloacetate. The result is citric acid, which contains six carbon atoms. The arrow for reaction 2 should be colored in a dark color. During this reaction, coenzyme A is released back into the cellular environment.

> At this point, pyruvate is in the Krebs cycle. A carbon dioxide molecule has been released, and acetyl CoA has fused with an oxaloacetic acid molecule to form citric acid, giving off CoA. We now continue with the Krebs cycle, focusing on reaction 3.

During reaction 3 of the Krebs cycle, citric acid is converted to isocitrate, and no carbon is lost. But in the next reaction, reaction 4, a carbon atom is lost as a molecule of CO_2. (Notice that the molecule that results from reaction 4, α-ketoglutarate, has only five carbon atoms.) Another molecule of NADH also forms during the course of reaction 4, and this high-energy molecule is shuttled to the respiratory chain to aid in the production of ATP.

We now move to reaction 5. Here, carbon dioxide is given off as five-carbon α-ketoglutarate is converted to the four-carbon molecule succinyl-CoA. During this process, another NADH molecule is produced. In the next reaction, succinyl-CoA is converted to succinate with an accompanying protonation of NAD^+. Also at this point, guanosine diphosphate, or GDP, combines with a **phosphate (P)** group to form GTP.

> We now continue and complete the Krebs cycle by examining reactions 7, 8, and 9.

In reaction 7, the four-carbon succinate converts to four-carbon fumarate, and this process involves a substance called flavin adenine dinucleotide, or FAD. This molecule is activated to form a high-energy molecule known as $FADH_2$, which is involved in electron transport.

In reaction 8, fumarate converts to a four-carbon molecule of malic acid, and in the next reaction, reaction 9, an enzyme converts maleate to oxaloacetate. During this process, another NAD^+ molecule is protonated.

The compound that results from reaction 9 is oxaloacetic acid, and you may notice that it was this molecule that combined with acetyl CoA at the beginning of the Krebs cycle. Oxaloacetate now combines with acetyl CoA that came from another pyruvic acid molecule that was formed in glycolysis.

For every molecule of acetyal CoA that enters the Krebs cycle, three NADH molecules, one $FADH_2$ molecule, one GTP molecule, and two CO_2 molecules are produced. The NADH and the $FADH_2$ molecules will be used in the electron transport mechanism and the chemiosmosis (which are discussed in the next plate) and the GTP is available for immediate use. GTP contains high energy bonds and like ATP, can be used to power costly biological events (such as replicating the DNA genome or building proteins). The carbon dioxide molecules are expelled into the environment when we exhale.

Carbon C ❍
Phosphate P ❍

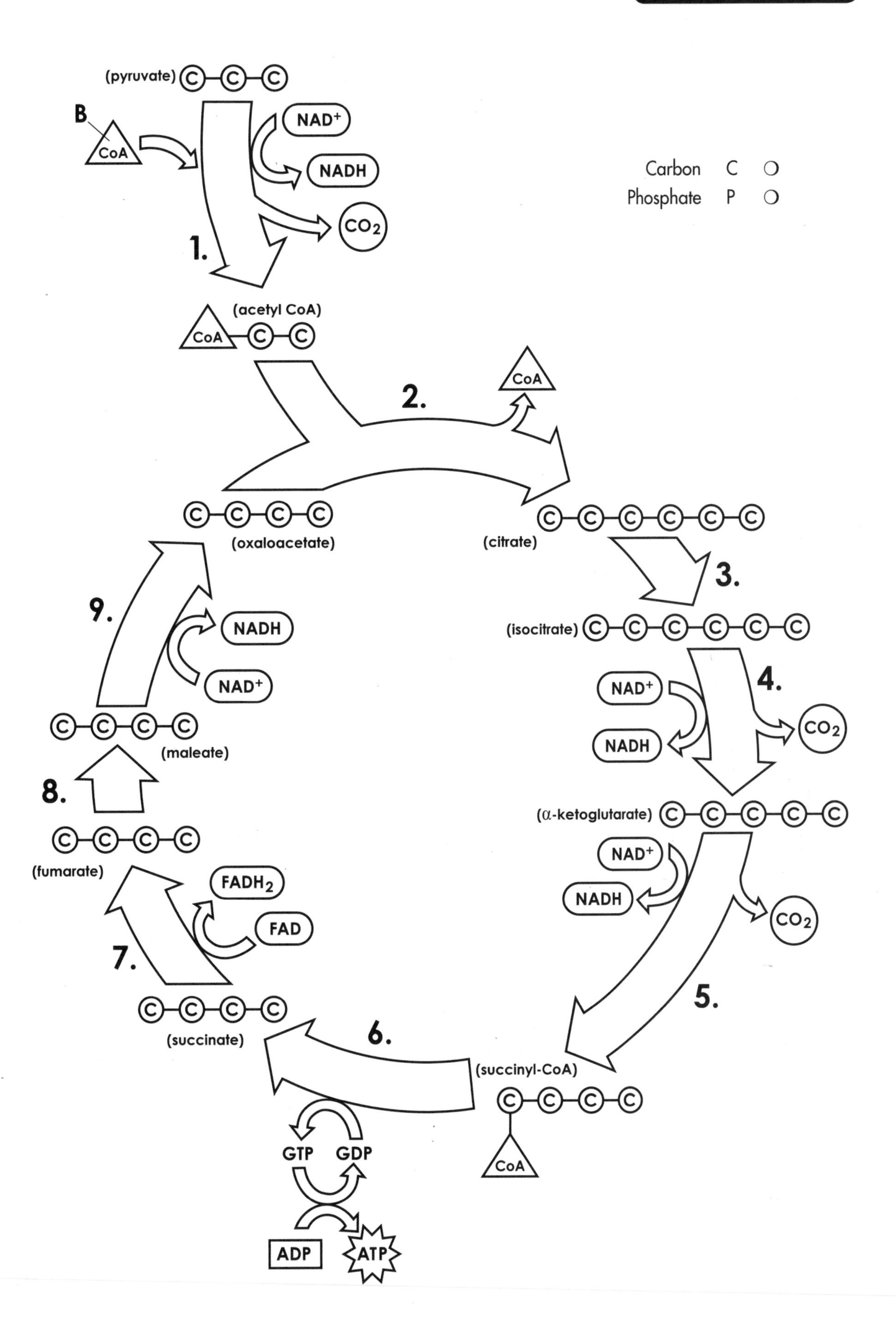

ELECTRON TRANSPORT AND CHEMIOSMOSIS

In the process of cellular respiration, glucose molecules are shuttled through glycolysis to yield pyruvic acid. The pyruvic acid then participates in the Krebs cycle, which is covered by the previous plate. The reactions of the Krebs cycle are important because they yield molecules of NADH and $FADH_2$. These molecules are then involved in the processes of electron transport and chemiosmosis, and their energy is used to produce ATP.

This plate shows a view of the mitochondrion and the reactions that take place along its membrane. The biochemistry is complex, and is an essential factor in this life of cells. Begin your reading below.

The process of cellular respiration begins in the cytoplasm of the cell with glycolysis. It continues along the inner membrane of the mitochondrion, where the Krebs cycle and electron transport occur. These foldings of the inner mitochondrial membrane are called **cristae (A),** the space between the cristae and outer membrane is the **intermembrane space (B),** and the area enclosed by the cristae is the **matrix (C).** You should use very light colors for these spaces.

In the plate, we have taken a section of the cristae and enlarged it to show the details of the reactions involved. Notice that this crista, like the cell membrane, consists of a phospholipid bilayer with embedded proteins.

We will now begin to discuss the process of electron transport. Follow along carefully as you read, since the biochemistry is complex.

We begin the biochemistry of electron transport with **NADH (D),** which was produced in the Krebs cycle and glycolysis. An NADH molecule loses two **electrons (F)** (indicated by the arrow), which pass through an enzyme called **NADH dehydrogenase (G).** Simultaneously, four **hydrogen ions (E)** are pumped through the enzyme and into the intermembrane space. As a result of this reaction, NADH is oxidized to **NAD^+ (D_1)** and can once again enter the Krebs cycle. NADH dehydrogenase is also known as coenzyme Q reductase. Enzymes in the electron transport chain are named for their redox roles in this process.

After moving through NADH dehydrogenase, the electrons are passed to a molecule called **coenzyme Q, or ubiquinone (H). $FADH_2$ (I)** also feeds electrons to coenzyme Q, as the arrow indicates, and in the process is oxidized to become **FAD (I_1).** Just like NAD^+, FAD is recycled back to the Krebs cycle. Notice that coenzyme Q and NADH dehydrogenase are both embedded in the crista.

Cytochrome C reductase (J) receives electrons from coenzyme Q, and as these electrons pass through, another four **hydrogen ions (E)** are pumped into the intermembrane space. The electrons continue on their way, and are next taken up by **cytochrome C (K),** which is located at the intermembrane-space surface of the crista.

Cytochrome C now passes the electrons to **cytochrome C oxidase (L),** and as this occurs, another two hydrogen ions are pumped through the membrane into the intermembrane space. The electrons then leave the electron transport chain and combine with **oxygen (M).** Each oxygen atom takes on two hydrogen ions to become **water (N).** Because of this role, oxygen is called the terminal electron acceptor, and this important function is why your cells require oxygen to survive.

We have just described the electron transport system, in which electrons move from one molecule to another, and their energy is used to pump hydrogen ions into the intermembrane space. In total, ten protons were pumped per NADH molecule. Each $FADH_2$ causes six protons to be pumped across the inner mitochondrial membrane.

We now move to the final part of this process, chemiosmosis, in which ATP molecules are produced. Continue to read as you color the final portion of this plate.

Up to this point, hydrogen ions have accumulated in the intermembrane space of the mitochondrion. Now, in chemiosmosis, they pass back across the membrane into the matrix, down their concentration gradient, through **ATP synthase (O).** As four hydrogen ions pass through ATP synthase, their energy is used to combine a molecule of **ADP (P)** and a phosphate group, to create a molecule of **ATP (Q).**

The process you learned about on this page is often called oxidative phosphorylation, because the oxidation of high-energy electron carriers (NADH and $FADH_2$) is coupled to the phosphorylation of ADP, to produce ATP.

Although this plate shows the synthesis of a single ATP molecule, eukaryotes can generate a total of 30 ATP molecules from processing one molecule of glucose through glycolysis, the Krebs cycle, electron transport, and chemiosmosis. The ATP molecules produced within the matrix of the mitochondrion now flow through a **channel protein (R),** exiting the mitochondrion for use anywhere else in the cell. The mitochondrion is commonly known as the powerhouse of the cell because it is the center for ATP production.

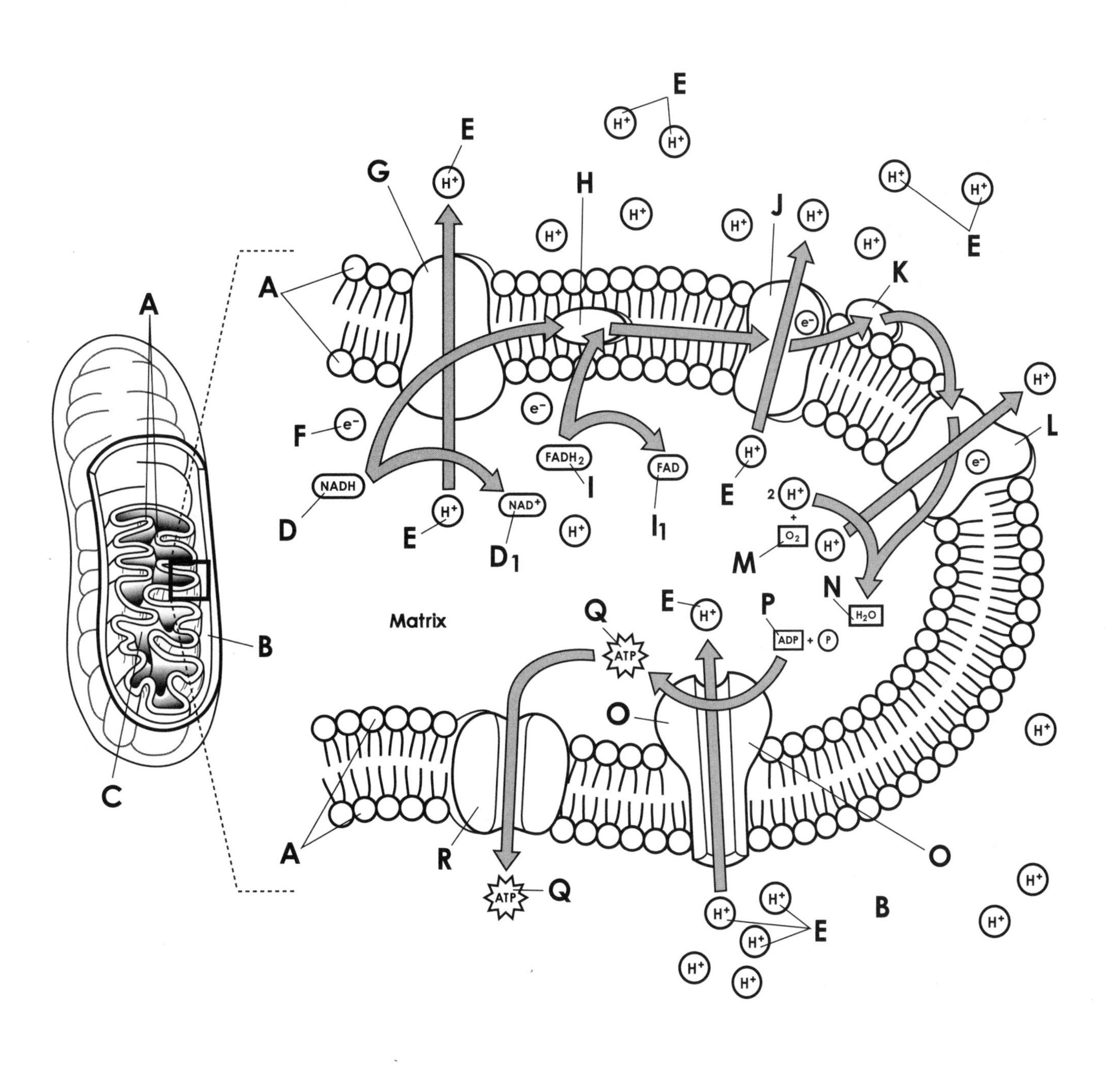

Cristae A ❍	NADH Dehydrogenase G ❍	Oxygen M ❍
Intermembrane Space B ❍	Coenzyme Q H ❍	Water N ❍
Matrix C ❍	$FADH_2$ I ❍	ATP Synthase O ❍
NADH D ❍	FAD I_1 ❍	ADP P ❍
NAD^+ D_1 ❍	Cytochrome C Reductase J ❍	ATP Q ❍
Hydrogen Ion E ❍	Cytochrome C K ❍	Channel Protein R ❍
Electrons F ❍	Cytochrome C Oxidase L ❍	

CELL SIGNALING

Receptors are an important class of integral membrane proteins that transmit signals from the extracellular space into the cytoplasm. Each receptor binds a particular molecule in a highly specific interaction. The molecule that serves as the key for a given receptor is termed the ligand. Hormones and neurotransmitters are examples of ligands.

Cell signaling pathways have some common characteristics:

- An extracellular signal is transduced into the cell, and causes intracellular effects; this is called signal transduction.
- The signal is amplified as it is transduced throughout the cell.
- Signaling cascades use second messenger molecules (such as PIP_3 or cAMP); these are usually small molecules, and they can diffuse throughout the cell. They are made and destroyed quickly.

There are three main types of signal-transducing cell-surface receptors: ligand-gated ion channels, catalytic receptors, and G-protein-linked receptors. In each of the diagrams on the facing page, color the **extracellular region (A)** the same color, the **plasma membranes (B)** the same color, and the **intracellular regions (C)** the same color. Use three of your lightest colors for these regions. Color each of the **cell surface receptors (D)** a bright color (such as blue) and the **ligands (E)** a different bright color, like orange.

1. Ligand-gated ion channels (D_1) in the plasma membrane open an ion channel upon binding a particular ligand (E), usually a neurotransmitter. Without ligand binding (top diagram), the **ion gate is closed (F),** and **ions (G)** cannot enter the cell. Color the ions in this diagram green. Upon ligand binding (middle), the **ion gate opens (H)** and ions can cross the plasma membrane. This causes a cellular response. The system resets (bottom) when the ligand dissociates, **closing the ion gate once again (I).** Ligand-gated sodium channels on the surface of muscle cells are an example of this type of receptor. When the neurotransmitter acetylcholine binds to this receptor, the receptor undergoes a conformational change and becomes an open Na^+ channel. This results in a massive influx of sodium, which depolarizes the muscle cell and signals it to contract.

2. Catalytic receptors (D_2) have an enzymatic active site on the cytoplasmic side of the membrane, meaning the receptor itself is an enzyme. Enzyme activity is initiated when a ligand binds at the extracellular surface. Most of these receptors are protein kinases, which means they attach phosphate groups to proteins. The presence of phosphate groups can regulate the activity of a protein.

The growth hormone receptor (D_2) is an example of a catalytic receptor. It activates several proteins near the plasma membrane, including Ras and PI3K. This causes accumulation of PIP_3 (a second messenger) and activation of downstream proteins such as PDK1 and Akt, resulting in many changes in cell biology. Be sure to color the arrows that end in an arrowhead in green (these are activation events) and the lines that end in perpendicular lines in red (these are inhibition events). Color the proteins in this diagram (Grb2, Ras, Sos, PI3K, PTEN, GEF, GAP, PDK1, Akt) variations of pink or purple, and the cell events (growth, translation, apoptosis, autophagy, cell cycle, metabolism, cytoskeleton rearrangements, and intracellular trafficking) shades of yellow or brown.

3. G-protein-linked receptors (D_3) do not directly transduce signals, but transmit them into the cell with the aid of a second messenger, such as cAMP. In this example, the ligand (E) epinephrine (also known as adrenaline) binds a G-protein linked receptor (D_3). G proteins bind guanine nucleotides, such as GTP or GDP. Ligand binding activates many **G-proteins (J),** by substituting a bound GDP with a GTP. Each G-protein activates many **adenylyl cyclase enzymes (K),** each adenylyl cyclase makes lots of **cAMP (L)** from ATP, and each cAMP activates many **cAMP-dependent kinases (M).** This can lead to many different reactions, one of which is shown on this plate (N). Here, a cAMP-dependent kinase is adding a **phosphorylate (P)** group to an enzyme. cAMP-dependent kinases can do this to many enzymes in the cell. Some of these enzymes will be activated, and others inactivated by phosphorylation, with the end result that the entire cell harmoniously works toward the same goal: energy mobilization.

Ligand-Gated Ion Channel

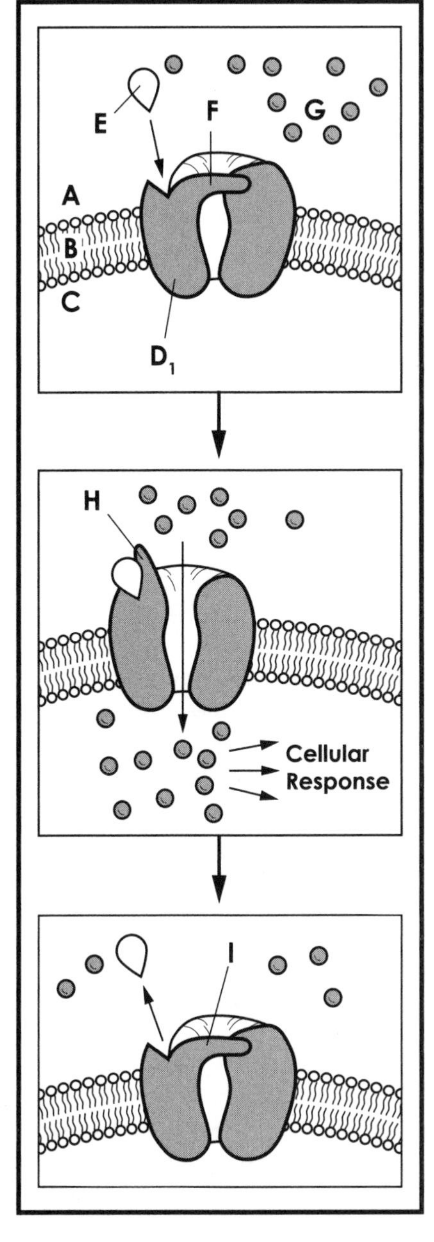

Catalytic Receptor

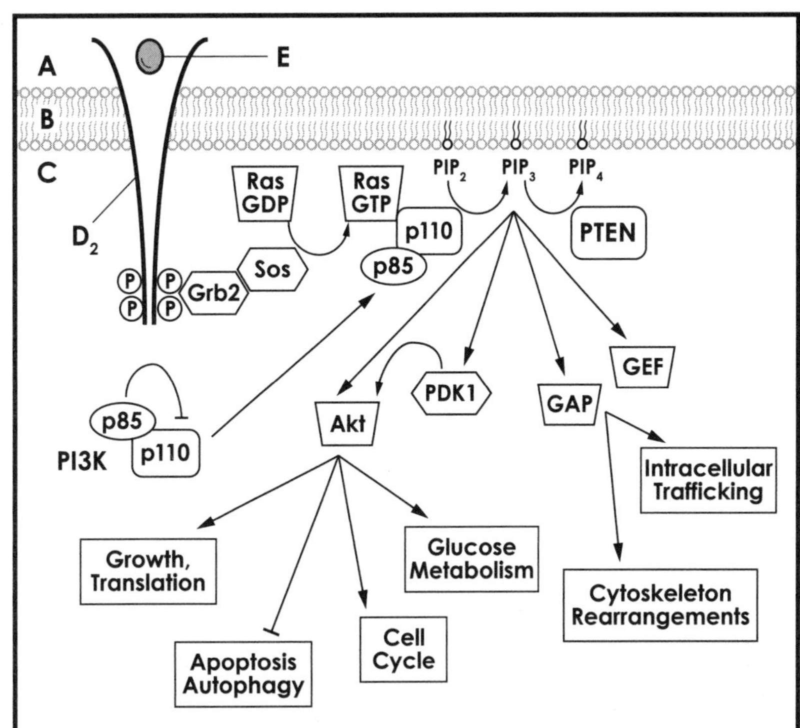

G-Protein-Linked Receptor

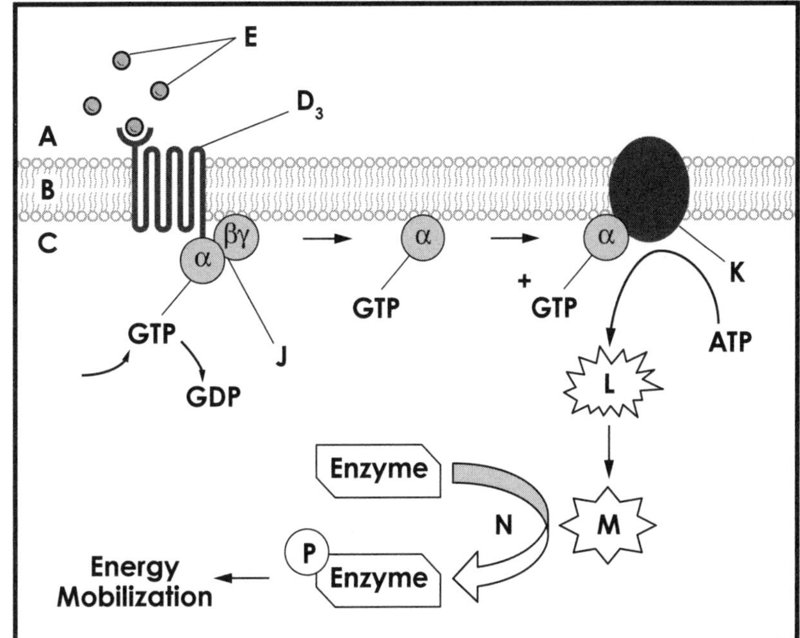

Extracellular Region	A	❍	G-protein-linked Receptor	D_3	❍	Adenylyl Cyclase	K	❍
Plasma Membrane	B	❍	Ligand	E	❍	cAMP	L	❍
Intracellular Region	C	❍	Closed Ion Gate	F	❍	cAMP-dependent Kinase	M	❍
Cell Surface Receptors:	D	❍	Ions	G	❍	Enzyme Phosphorylation	N	❍
Ligand-gated Ion Channel Receptor	D_1	❍	Open Ion Gate	H	❍	Phosphate Group	P	❍
Catalytic Receptor	D_2	❍	Closed Ion Gate	I	❍			
			G-protein	J	❍			

THE CELL CYCLE

The cells of all living things grow and multiply through a cycle that's made up of four phases. During three of these phases, the cell is growing and is metabolically active. During the fourth phase, it is undergoing division, which yields two new cells. In this plate, we examine the four phases of the cell cycle and note the important characteristics and subdivisions of each phase. A subsequent plate explores the phases of mitosis in detail.

As you look over this plate, note that it contains numerous subdivisions that represent phases of the cell cycle. Bold colors may be used for these phases since there is little overlap and there are no fine details to obscure.

The **cell cycle (A)** takes place over different periods of time in different types of cells, and as you know, different types of cells coexist in many organisms. For example, in human fibroblast cells, the cell cycle may encompass about fifteen hours, while in brain cells, the cycle may take many years to complete.

The two major periods of the cell cycle are **interphase (B)** and the M phase (also known as the phase of **cell division (C)**). As the plate indicates, interphase encompasses three smaller periods and is the period of time between cell division. The same bold color may be used for all three portions of interphase, and a different color should be used for the M phase. Reds, blues, greens, or purples are suggested.

During interphase, the cell is extremely active and carries on routine cellular and physiological activities. For example, cells of the pancreas are actively producing insulin, which facilitates the passage of glucose molecules into cells. During the M phase of cell division, the rate of metabolism is reduced and the cell undergoes division to form two cells.

We now focus on the three phases of interphase during the cell cycle. As before, bold colors should be used.

Three shorter phases make up the interphase period of the cell cycle. The first phase is known as the **G_1 phase (D).** During this time period, metabolism is occurring at a high rate, many proteins are synthesized, and cell growth is vigorous. The cell's organelles also increase in number and size. The G in G_1 stands for "gap," so this phase is also referred to as Gap 1 phase. This name stems from the fact that the processes occurring in G_1 aren't easily seen with a light microscope, and so initially looked like a gap in activity to biologists. We now know that dozens of crucial cell processes are happening in G_1.

The second phase of interphase is the **S phase (E).** In the S phase, some activities related to cell division take place (S stands for synthesis). The cell's DNA replicates, ensuring that future cells obtain similar copies of its hereditary material, and proteins associated with the DNA are produced during this phase.

The cell prepares to reproduce during the **G_2 phase (F).** More of the proteins that are essential for cell division are produced during this phase, and these proteins move to appropriate sites. The centrioles used for cell division complete their replication during this phase. In addition to these activities, the cell continues its growth and many of its physiological processes.

We complete the plate by focusing on the process of cell division that takes place during the M phase. More detailed descriptions of cell division are given in the next plate; a brief overview is given here.

At the conclusion of the G_2 phase, the cell enters its M (mitosis) phase of cell division. This phase consists of two main processes: the first is **mitosis (H),** in which the chromosomes separate and segregate themselves on opposite sides of the cell, and the second is **cytokinesis (G),** in which the cell actually splits. The results of cytokinesis and mitosis are shown in the plate.

Mitosis occurs as a series of events that are separated into four phases, and the process is continuous through these four phases. During **prophase (H_1),** distinct and compact chromosomes appear. These structures are made of tightly packed DNA and protein, and are formed so the genetic material can be easily moved around the cell. During **metaphase (H_2),** the chromosomes line up along the equator. During **anaphase (H_3),** the chromosomes separate, and one member of each pair moves to opposite poles of the cell. Lastly, during **telophase (H_4),** the chromosomes arrive at the opposite poles of the cell and two distinct nuclei begin to form.

The processes that take place during the M phase of the cell cycle lead to new cells that are referred to as daughter cells. A single **mother cell (a)** has passed through the G_1, S, and G_2 phases and enters cell division to produce two **daughter cells (b).** Each of the two new cells will now enter interphase and the cycle will be repeated.

Some cells are able to exit the cell cycle during G_1, meaning they don't divide. These cells have a few options. They can undergo programmed cell death, or **apoptosis (I)** (discussed in a future plate). Cells can also remain alive and metabolically active, but not actively dividing. For example, **senescence (J)** is irreversible; these cells cannot re-enter the cell cycle. Many red blood cells are senescent, as are most of the cells on the surface of your skin. Some cells leave G_1 and enter a phase called **G_0,** or **quiescence (K).** This is a reversible process, so quiescent cells are able to re-enter the cell cycle at G_1. Many nerves and heart muscle cells spend most of their lives in quiescence.

Finally, the cell cycle is a tightly regulated process. The G_1/S checkpoint is a complex singling pathway that makes sure the cell is ready to proceed to the S phase, or DNA replication. The G_2/M checkpoint makes sure the DNA genome is replicated and repaired before mitosis. In either case, the cell cycle can be paused if problems are found. This gives the cell a chance to repair problems before cell division.

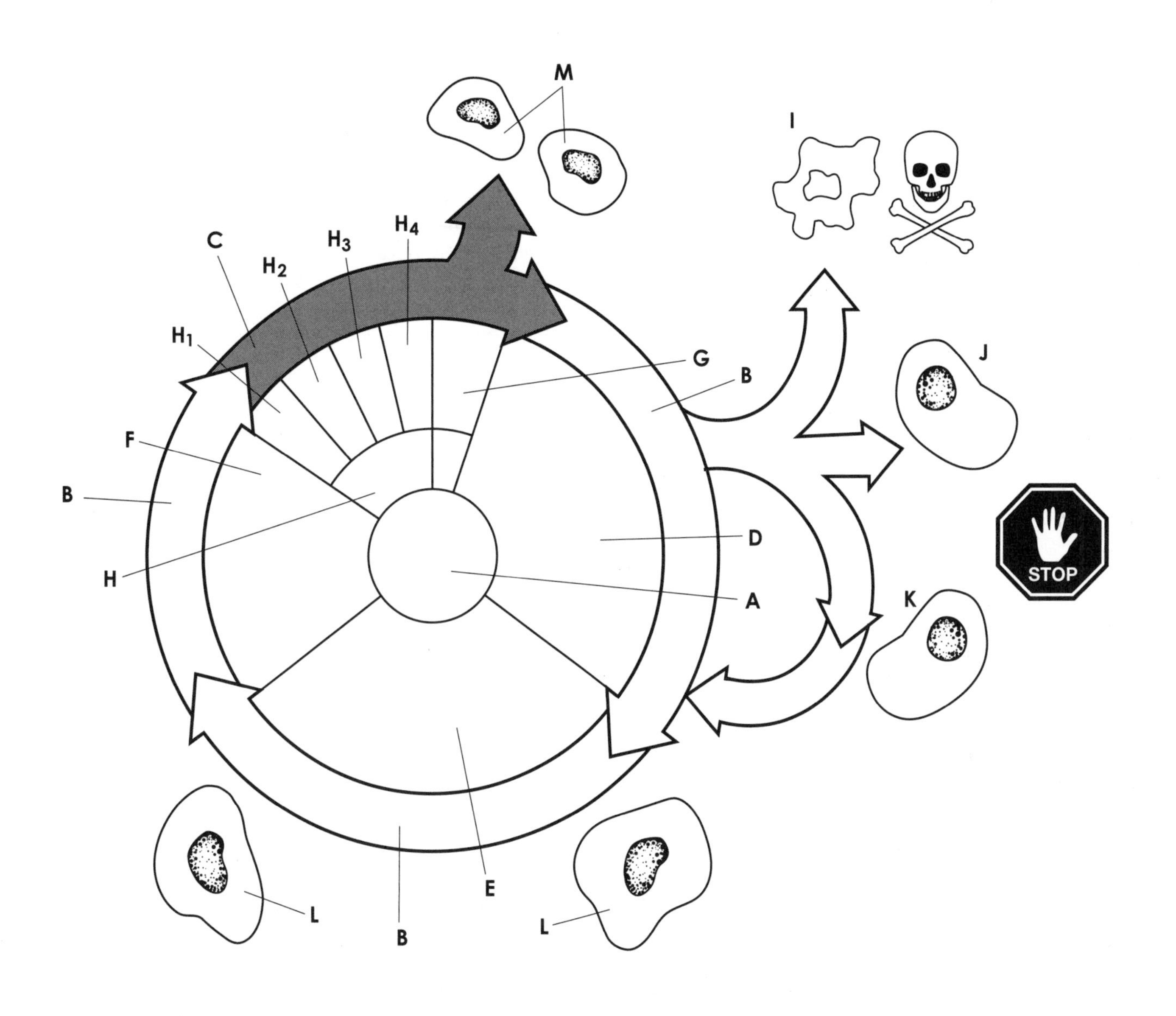

The Cell Cycle	A	❍	Cytokinesis	G	❍	Telophase	H_4	❍
Interphase	B	❍	Mitosis	H	❍	Apoptosis	I	❍
M Phase (Cell Division)	C	❍	Phases of Mitosis			Senescence	J	❍
G_1 Phase	D	❍	Prophase	H_1	❍	Quiescence	K	❍
S Phase	E	❍	Metaphase	H_2	❍	Mother Cell	L	❍
G_2 Phase	F	❍	Anaphase	H_3	❍	Daughter Cells	M	❍

MITOSIS

During the cell division phase of the cell cycle, the cell undergoes mitosis and then cytokinesis. Mitosis is the process in which the duplicated chromosomal pairs separate and, in cytokinesis, the cell splits to form two new cells. This plate will explore the process of mitosis.

This plate shows six diagrams that portray the phases of a cell undergoing mitosis. This process is a continuous one, and is generally described in terms of four phases: interphase, prophase, anaphase, and telophase. Use the same colors throughout the six diagrams. These diagrams tend to be fairly detailed, so light colors are recommended.

As you may remember, the DNA in the nucleus of the cell replicated during the S phase of the cell cycle, but is not distinguished as distinct chromosomes during the first phase of mitosis, interphase. The **nucleus (B)** contains the DNA in a diffuse mass called chromatin. The **nucleolus (C)** is seen clearly in the interphase cell, and the **nuclear membrane (D)** encloses the nucleus. Color the **cytoplasm (A)** a light color.

Two submicroscopic bodies (also duplicated prior to mitosis) that participate in mitosis are the centrosomes. Each of the centrosomes contains two cylindrical structures that are arranged at right angles to each other, called **centrioles (E),** which are involved in the organization of microtubules during cell division.

We now begin the process of mitosis; our cell is in early prophase. The same colors that were used for the interphase cell should be used here.

Prophase is the longest phase of mitosis. It begins when the chromatin of the cell nucleus condenses to form distinct chromosomes. Because DNA replication has taken place during interphase, each chromosome is composed of two identical strands, known as **chromatids (G_1).** Notice that, in early prophase, the centrioles (E) are surrounded by a series of microtubules that radiate outward; these are called **asters (F).**

In late prophase, the centrioles (E) have moved to opposite poles of the cell and the asters (F) are still visible. **Spindle fibers (H)** can be seen extending between the centrioles and should be traced with a light color such as yellow. Spindle fibers are composed of microtubules and associated proteins. Notice that the chromatids (G_1) have continued to compact, becoming shorter and thicker. The nuclear membrane begins to break apart and disappear as the cell proceeds through late prophase.

As mitosis continues, the chromatids separate into chromosomes. Continue your reading and color the structures as you encounter them in the plate.

The next phase of mitosis is metaphase. Here the chromatid pairs align themselves along the equator of the cell, at an area called the metaphase, or equatorial plate. Each chromosome contains two sister chromatids (G_1), which are connected at the centromere of the chromosome. The centromere of a chromosome is the center of the "X" shape, where all the chromosome arms meet. Surrounding the centromere are **kinetochore proteins (I);** this is where the mitotic spindle attaches to the chromosome. At this stage, the spindle fibers (H) are distinct, and they extend out from the centrioles. The remainder of the cytoplasm (A) should be colored in a light color.

In anaphase, sister chromatids are pulled apart. Each chromatid is now a **chromosome (G_2).** Four chromosomes are seen moving to the bottom of the diagram, and four to the top of the diagram. The chromosomes resemble "V's" because the spindle fibers lead them by their centrioles. An equal number of chromosomes move to the opposite poles of the cell. In a human cell, for example, 46 chromosomes move to one pole and 46 chromosomes move to the opposite pole.

We will now examine a cell in telophase, the phase that signals the end of mitosis and immediately precedes cytokinesis.

As the dividing cell enters telophase, you can see that the chromosomes (G_2) arrive at opposite ends of the cell, where they become thinner and less distinct. The spindle fibers (H) begin to break down in this phase, the nuclear membrane (D) begins to form around the chromosomal material, and the nucleolus (C) reappears.

As telophase comes to an end, the cytoplasm (A) is divided between the two new daughter cells. At the center of the cell in animal cells, a **cleavage furrow (J)** begins to form as the membrane pinches in from both sides. The appearance of the cleavage furrow signals the end of telophase and the beginning of cytokinesis. The furrow pushes inward from opposite sides of the cell until two cells are created. These cells are referred to as the daughter cells.

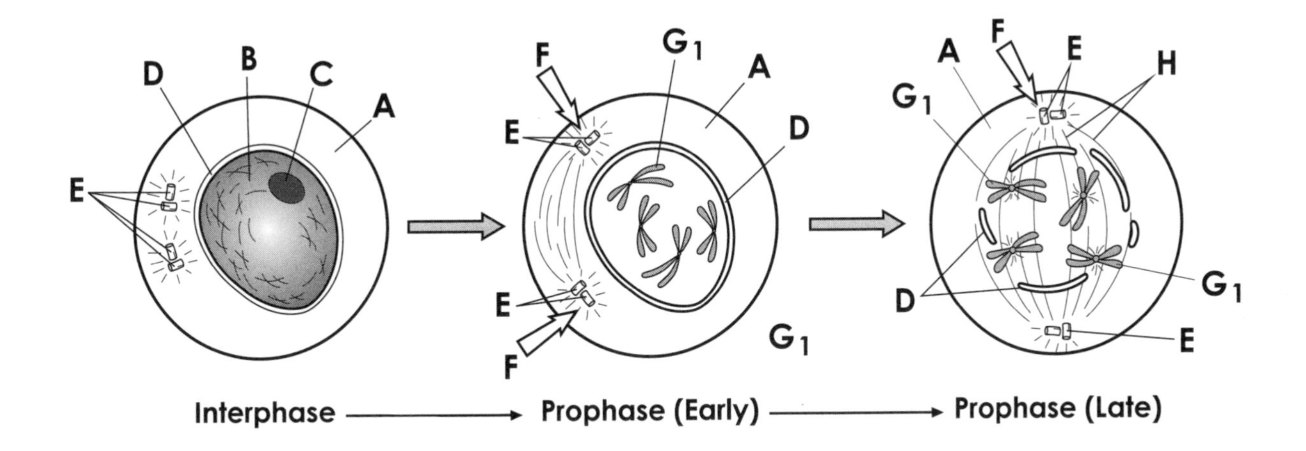

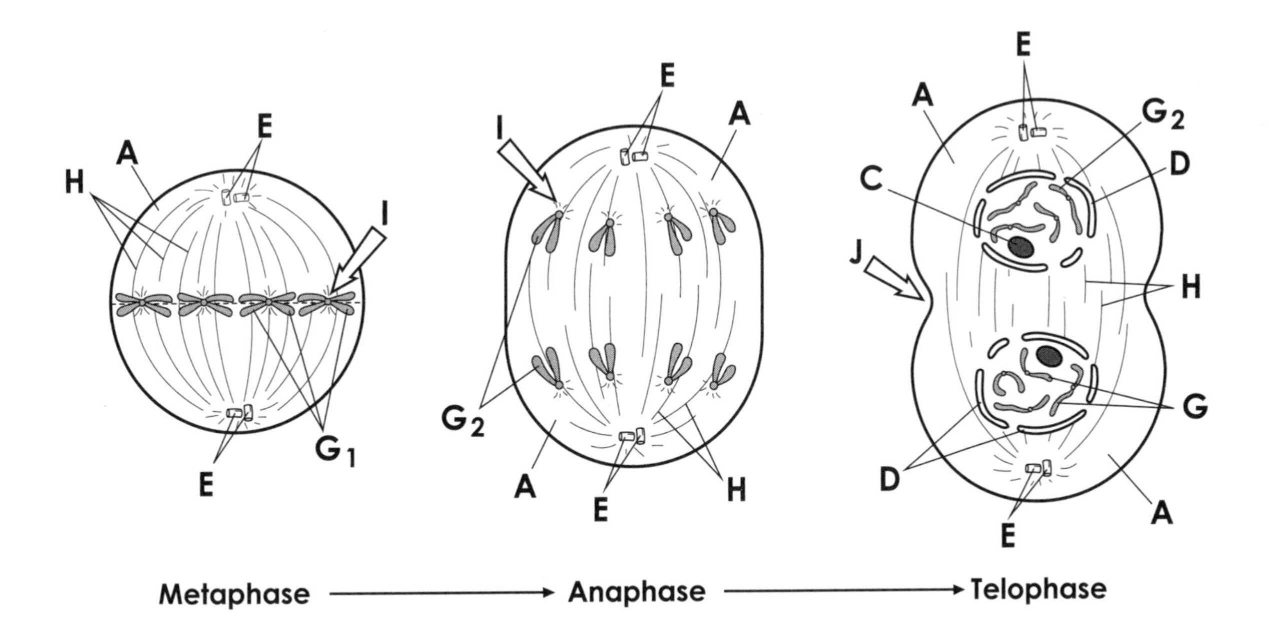

Cytoplasm A ○
Nucleus (Chromatin) B ○
Nucleolus C ○
Nuclear Membrane D ○

Centrioles E ○
Asters F ○
Chromatids G_1 ○
Chromosomes G_2 ○

Spindle Fibers H ○
Kinetochore I ○
Cleavage Furrow J ○

APOPTOSIS

Apoptosis, or programmed cell death, allows a cell to shrink and die while simultaneously minimizing damage to neighboring cells and limiting the exposure of other cells to its cytosolic contents. It is a normal part of development and health maintenance in multicellular organisms.

Cell death can be induced two different ways:

1. The **extrinsic pathway (A):** A stressor outside the cell (such as nitric oxide, a toxin, or cytokines) activates a cell receptor to signal cell death from outside the cell.
2. The **intrinsic pathway (B):** Extensive cell stress or damage induces cell death from inside the cell.

Both pathways activate a signaling pathway that results in **cytochrome C (C)** being released from the inner mitochondrial membrane of the **mitochondria (D).** This protein is normally found embedded in the inner mitochondrial membrane and is an important member of the electron transport chain.

Cytochrome C release from the mitochondria causes assembly of the **apoptosome,** a large protein complex. An important part of this is a **reaction (E)** that forms **active caspase enzymes (G)** from their inactive precursors (called **procaspases, (F)**). There are two different types of caspases: initiator caspases are activated first and effector caspases are activated later. Initiator caspases respond to extra- or intracellular death signals by clustering together; this clustering allows them to activate each other. The activation of initiators leads to the activation of the effector caspases in a cascade of activation. Effector caspases then cleave a variety of cellular proteins to trigger apoptosis. Once the cell has started apoptosis, it begins to shrink and form **blebs (K).**

Once active, caspases (G) function as proteases, which means they degrade proteins. Caspase enzymes function to dismantle the cell and break down cellular components. For example, the genome in the nucleus is fragmented to approximately 200 **base pair fragments (H).** Membrane-bound organelles are fragmented and proteins in the cell are degraded. The **cytoskeleton (I)** is broken down **(J),** which dismantles the cellular infrastructure.

Finally, the cell retracts and detaches from the extracellular matrix and also from its neighbors **(M).** The cell breaks into several smaller pieces containing cell components and destroyed nucleus; these pieces are called **apoptotic bodies (N).**

A cell undergoing apoptosis emits signals to attract professional phagocytes, like **macrophages (L).** They help clean up the mess left behind. Professional phagocytes that are called over to the dying cell engulf all the cellular material that is left behind **(O).** They degrade this material using their lysosomes, and thus function to clear away the debris.

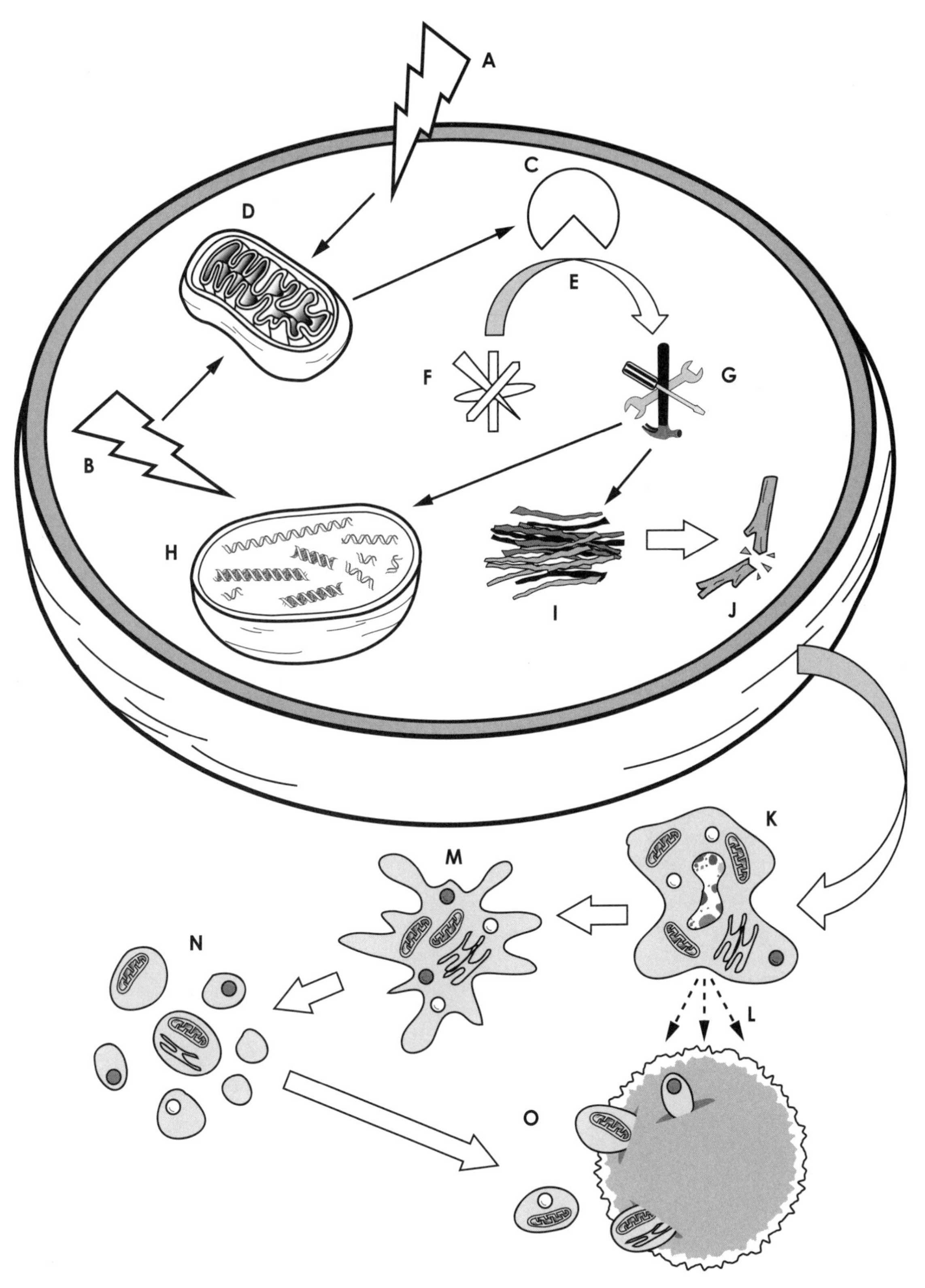

Extrinsic Pathway	A	❍
Intrinsic Pathway	B	❍
Cytochrome C	C	❍
Mitochondria	D	❍
Caspase Activation	E	❍
Procaspase	F	❍
Active Caspase	G	❍
Genome Fragmentation	H	❍
Cytoskeleton	I	❍
Degraded Cytoskeleton	J	❍
Cell Blebbing	K	❍
Attraction Signals to Phagocytes	L	❍
Cell Retraction	M	❍
Apoptotic Bodies	N	❍
Phagocytosis	O	❍

CANCER

There are several mechanisms that keep cell growth in check. For example, **normal cells (A)** can arrest their growth for short or long periods of time **(B).** These cells aren't dead and continue to undergo many normal cell processes (like metabolism), but don't divide via mitosis. Stressed or damaged cells can undergo **apoptosis (C),** which you learned about in the previous plate.

Despite these (and lots of other) protection mechanisms, sometimes cells start growing out of control **(D).** This can have disastrous consequences and is often driven by an accumulation of mutations in the genome. These mutations can be caused by any of the following factors:

- Light, such as ionizing radiation (X-rays, α particles, γ rays) or UV
- Mutagens, chemicals, or drugs
- Errors in cell processes, such as DNA replication, DNA repair, recombination or chromosome segregation during mitosis or meiosis
- Biological agents, such as viruses

There are two effects of these mutations:

1. Oncogenes are induced, via amplifying the gene, over-expressing the gene, or mutating the gene so it codes for an overly active form of the protein. The proteins coded by oncogenes normally promote cell growth and proliferation. An increase in oncogene function induces cell growth.

2. Tumor suppressor genes are inhibited, by deleting them from the genome or inactivating them. Tumor suppressor genes normally inhibit cell growth and proliferation. A decrease in tumor suppressor gene function also induces cell growth.

These changes in oncogene and tumor suppressor gene function have lots of effects:

- **Aberrant cell growth (D, F)**
- Even more mutations accumulate in the DNA genome.
- The cells resist death, or avoid apoptosis.

This can result in the formation of a **benign tumor (F).** Benign tumors stay in the same location and the cells remain in the tumor. The presence of a benign tumor does not mean the same thing as cancer, and is usually not a major problem. However, eventually even more mutations can accumulate and this can allow the tumor to induce blood vessel formation; this is called **angiogenesis (E).** Some of the tumor cells can then become mobile and leave the tumor mass **(G).** These mobile cells can travel in the blood or lymphatic system to other parts of the body **(H)** and if the conditions are right, they can settle into a new organ and grow into a secondary tumor **(I).** This process of tumor spreading is called **metastasis** and is one of the hallmarks of cancer. **Cancer** means "crab," as in the zodiac sign. The name is derived from the observation that malignant tumors grow into the surrounding tissue, embedding themselves like clawed crabs. Keep in mind, however, that not all cancers present as malignant solid tumors. Some cancers, such as leukemia, are due to blood cells transforming.

Cancer is not one disease, but a collection of hundreds of diseases. Each of these diseases is caused by different mutations and genetic changes. Cancer treatments are thus specific to certain cancer types, which is why some cancers have a good prognosis and others do not yet. Many types of cancers have been effectively cured by the development of really good therapeutics. Other types of cancers are still stumping cancer researchers and discovery is ongoing.

Better screening tests have been developed for many types of cancers. This includes tests like mammograms, CT scans, MRI scans, fecal sample tests and colonoscopies, genetic testing, blood tests, and prostate exams.

Incidence rates of certain types of cancer have decreased in recent years because some types of cancers are preventable. For example, increased sunscreen use can decrease the risk of developing skin cancer, and not smoking can help prevent lung cancer. A few cancers can be prevented by vaccines (such as HPV-induced cervical cancer).

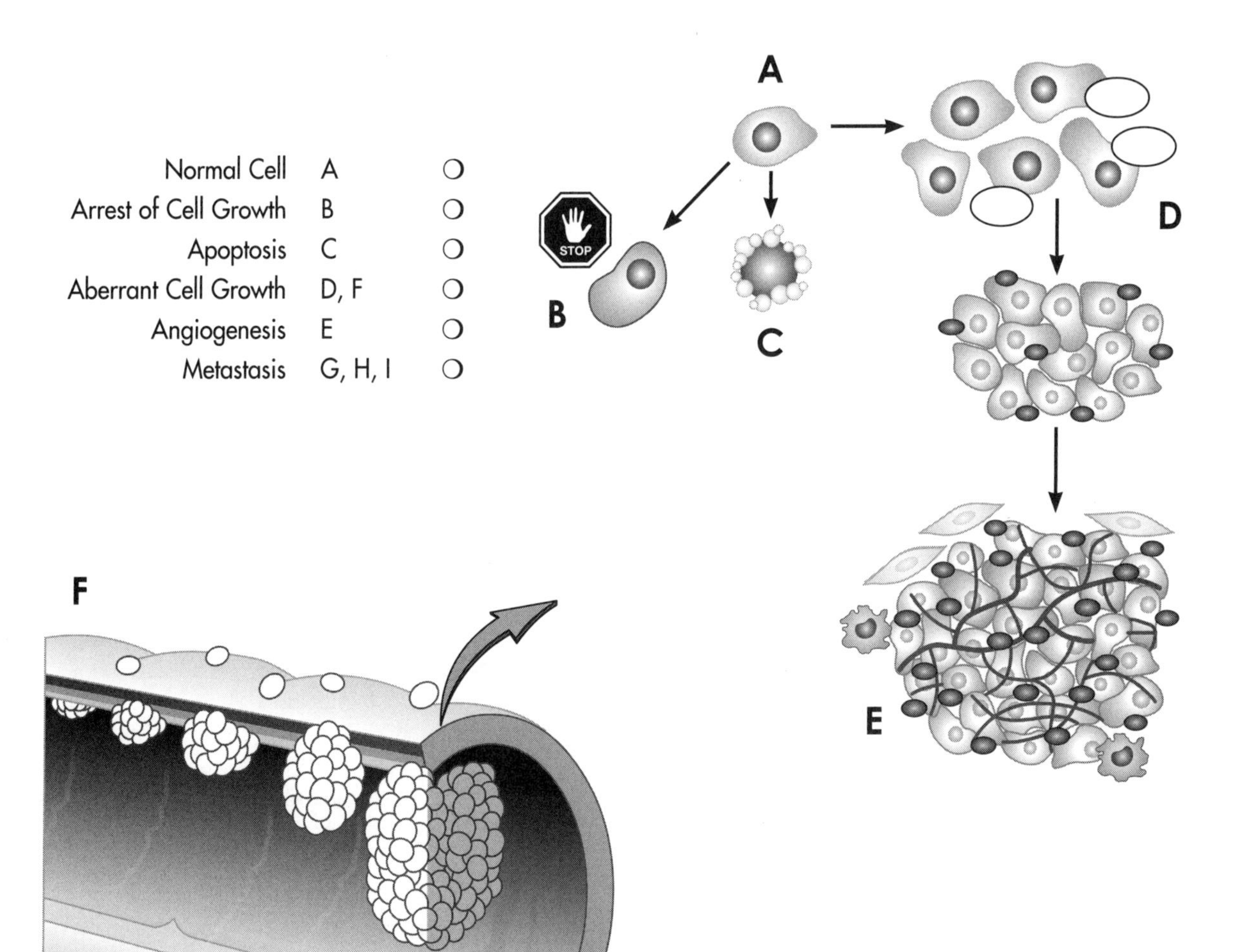

Normal Cell A
Arrest of Cell Growth B
Apoptosis C
Aberrant Cell Growth D, F
Angiogenesis E
Metastasis G, H, I
A
B
STOP
C
D
E
F

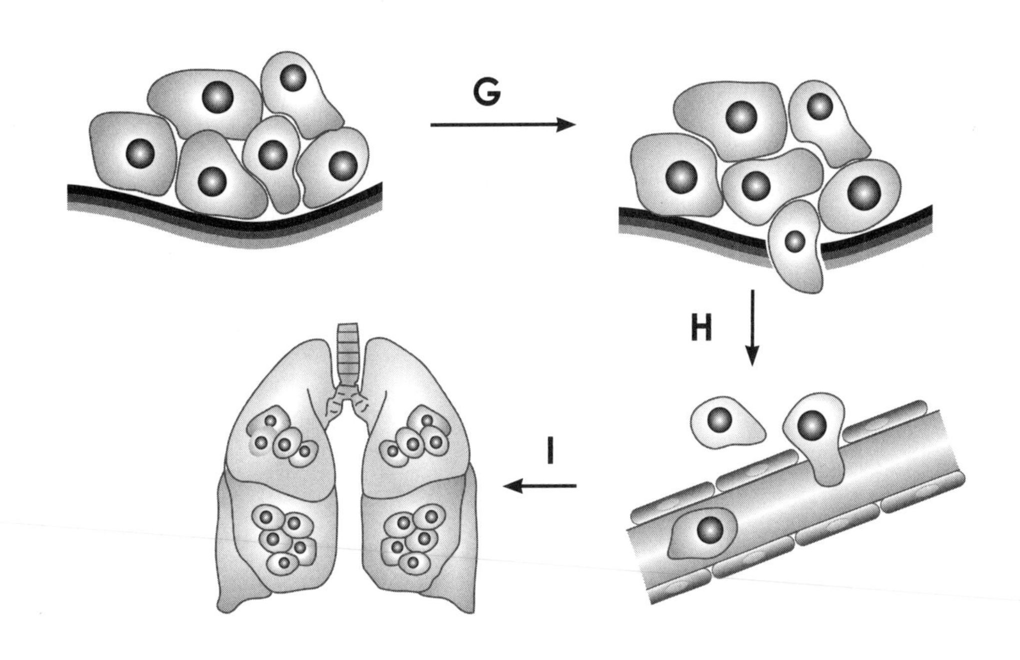

G
H
I

MEIOSIS

The process of mitosis, discussed in a previous plate, occurs in cells that are reproducing during growth and wound healing, and the replacement of dead cells. The two cells that arise from mitosis are genetically identical to their parent cell. Certain cells undergo another form of cell division known as meiosis. In this process, a single parent cell produces up to four cells, each of which has half the number of parental chromosomes. The parent has two sets of chromosomes and is said to be diploid (2n), while each of the cells that result from meiosis has a single set of chromosomes and is said to be haploid (n). In humans, n = 23. This means a typical human somatic cell contains 46 chromosomes and is diploid. A somatic cell is any cell of a living organism other than the reproductive cells.

Meiosis takes place in the reproductive organs and results in cells that are used during reproduction. These cells, which are sperm and egg cells, are called gametes. At fertilization, the fusion of two haploid gametes forms a single cell, called the zygote, which is diploid.

In this plate, we trace the two main phases of meiosis. Many of these processes are similar to those of mitosis, and you should refer to the previous plate whenever necessary. We will follow a single pair of chromosomes through the process of meiosis and will note how they are distributed to four cells.

The process of meiosis involves two rounds of cell division, known as **meiosis I (A)** and **meiosis II (B).** The bars that indicate these two rounds should be colored. The first round results in daughter cells that are haploid and thus have half the number of chromosomes. In the second round, these chromosomes split in two, as sister chromatids move to either end of the cell and are distributed to the gametes. Each round of meiosis contains a prophase, metaphase, anaphase, and telophase, as is the case in mitosis.

We will begin with meiosis I. Here we see a parent cell with a distinctive **nucleus (C)** and **nucleolus (D).** The **cytoplasm (E)** should be colored in a pale color. The **centriole (F)** functions in meiosis as it does in mitosis. The phase designated 1a represents prophase.

Prophase continues in view 1b. Here a single pair of chromosomes is considered (remember that humans have 23 pairs of chromosomes per cell). We see **homologous chromosome 1 (G_1)** and **homologous chromosome 2 (G_2).** The DNA in each chromosome has replicated. Here, the chromosomes have come together, and crossing over (Chapter 3) may take place.

View 1c represents metaphase. The homologous chromosomes line up along the equator of the cell, and we see that each consists of sister chromatids. Chromosome 1 has **sister chromatids 1 (H_1),** and chromosome 2 consists of **sister chromatids 2 (H_2).**

Anaphase is shown in view 1d. Sister chromatids 1 move to the left pole of the cell while sister chromatids 2 move to the right. In telophase, sister chromatids 1 are contained in the left daughter cell, and sister chromatids 2 are in the right daughter cell. This means each daughter cell receives one chromosome (and so is haploid), but each chromosome contains two sister chromatids (because it is still replicated).

This marks the end of meiosis I.

At the end of meiosis I, the chromosome pair has separated and a chromosome that consists of two sister chromatids has moved to each daughter cell. The sister chromatids are held together at the centromere. Each of the two daughter cells will now enter meiosis II.

The two daughter cells now enter meiosis II, shown at the top of the second column. View 2a shows prophase. Again, we see the centrioles (F) and the cytoplasm (E), which should be colored in a pale color. Sister chromatids 1 (H_1) are in the left cell, and sister chromatids 2 (H_2) are in the right cell. In view 2b, the sister chromatids line up along the equator of each cell. The mitotic spindle binds kinetochore proteins around the centromeres to separate sister chromatids.

Now, in view 2c, anaphase is in process, and the sister chromatids are considered chromosomes. In the left cell, **chromosome 1 (I_1)** moves to one side of the cell, while **chromosome 2 (I_2)** moves to the other. **Chromosome 3 (I_3)** and **chromosome 4 (I_4)** separate in the second cell. As telophase commences, in view 2d, the chromosomes are situated at the poles, and the nuclei are taking shape once again. Cell division (cytokinesis) begins.

In the final view, 2e, we see the four cells that result from cytokinesis. Each cell is haploid, meaning that it contains a single chromosome from the original chromosome pair. Recall that we began with two chromosomes. Now in the final view, each cell has one chromosome from that original pair. In the human male, these cells will undergo further development to become sperm cells, and in the human female, one of these cells will become an egg cell.

Meiosis is linked to sexual reproduction in plants and animals because haploid cells join to form a fertilized diploid cell. In animals, the haploid stage is very brief, but in simple plants, the haploid stage predominates over the diploid stage, as you will see in the plates on plant biology.

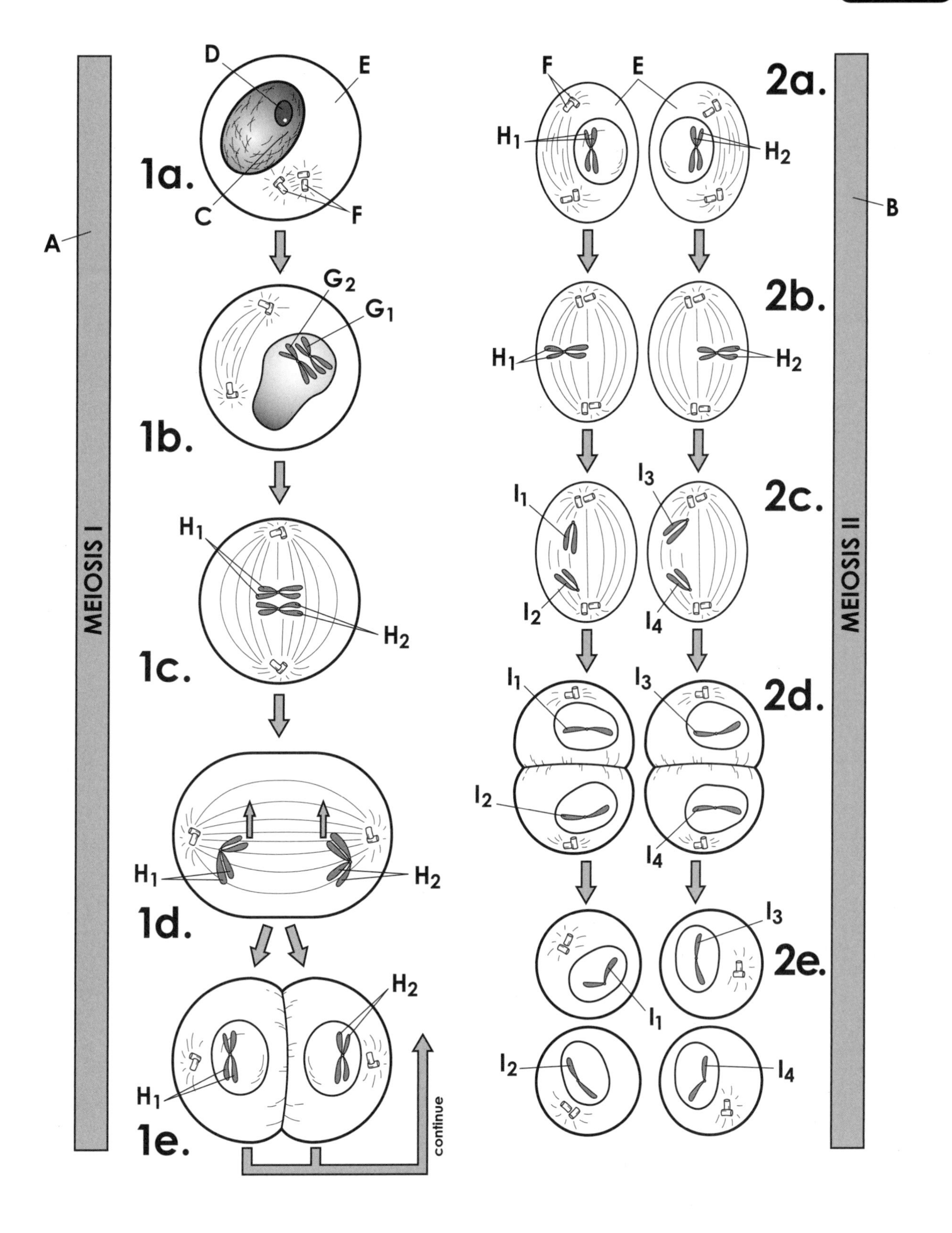

Meiosis I	A	❍	Centrioles	F	❍	Chromosome 1	I_1	❍
Meiosis II	B	❍	Homologous Chromosome 1	G_1	❍	Chromosome 2	I_2	❍
Nucleus	C	❍	Homologous Chromosome 2	G_2	❍	Chromosome 3	I_3	❍
Nucleolus	D	❍	Sister Chromatids 1	H	❍	Chromosome 4	I_4	❍
Cytoplasm	E	❍	Sister Chromatids 2	I	❍			

CHAPTER 3:

PRINCIPLES of GENETICS

MENDELIAN GENETICS

The principles of genetics were established in the 1860s by Gregor Mendel. Mendel's two main principles describe how genes pass through generations, and we will describe the fundamentals of these principles in this plate.

This is the first of several plates that cover the science of genetics. Here, we will discuss Gregor Mendel's seminal experiments on pea plants. We will introduce some of the terminology of genetics and see how Mendel arrived at two essential principles of this discipline of biology.

Gregor Mendel performed a series of experiments using the common garden pea plant. Pea plants were readily available to Mendel, were fairly easy to grow, and other scientists had studied their breeding behaviors.

Mendel began by studying two types of seeds (peas); **yellow seeds (A)** and **green seeds (B).** When he planted a yellow seed, it grew into a **yellow-seeded pea plant (C),** and when he planted a green seed, it developed into a **green-seeded pea plant (D).** The first generation of yellow-seeded and green-seeded plants is called the parent generation, or P_1 generation.

At this point, Mendel had two types of pea plants, one that produced yellow seeds (peas), and one that produced green seeds. Continue your reading below as you color the plate.

In the next experiment Mendel performed, he bred, or "crossed" a yellow-seeded plant and a green-seeded plant. This type of cross is called a monohybrid cross. In a monohybrid cross, the parent plants differ by only a single trait; in this case, that trait is seed color.

The offspring of the parental generation is referred to as the **F_1 generation (E);** you can see our F generation in the middle drawing. We will call the plant the **F_1 generation plant (F).** All of the F_1 generation plant's seeds were yellow, which meant that the green trait had somehow been hidden. You should color the peas in the F_1 generation plant yellow (A).

Mendel deduced that, since the seeds in the F_1 generation were not a blend of the colors of the two parents, and since all of the seeds were yellow, the yellow trait was the apparent, or dominant trait, and the green trait was hidden, or recessive. At this point, Mendel was not sure whether the factor that produced the green trait had disappeared completely.

Mendel's F_1 generation plants all had yellow seeds. He assumed that yellowness dominated over greenness. To determine what had happened Mendel proceeded as explained below. Continue your coloring as you continue to read about his experiments.

In his next experiment, Mendel permitted the pea plants from the F_1 generation to interbreed; the result was an **F_2 generation (G).** The bracket that outlines this generation should be colored. Mendel observed something startling in the **F_2 generation plant (H):** Most of the seeds (peas) were yellow (A), but a small number were green (B). The green trait had reappeared!

Mendel took a careful count of the number of yellow and green peas in the F_2 generation plant and discovered that about three-quarters of the seeds were yellow, while about one-quarter were green. In other words, there was a color ratio of 3:1, yellow to green.

From hundreds of experiments like these, Mendel concluded that each hereditary trait is determined by two factors, which we now call genes. Mendel suggested that each parent contributed one gene for a trait. These alternative forms of genes are called alleles. In a diploid organism, there are two copies of every gene, so there can be two different alleles. If the two alleles are the same, the individual is said to be homozygous for a trait, and if the individual has two different alleles, then the individual is said to be heterozygous for a trait, or hybrid.

Mendel further proposed in his first law (the Law of Segregation) that alleles separate, or segregate, during the formation of sperm and egg cells in plants and animals. He also determined that each pair of alleles separates independently of other pairs of alleles. This second law of Mendel's is called the Law of Independent Assortment.

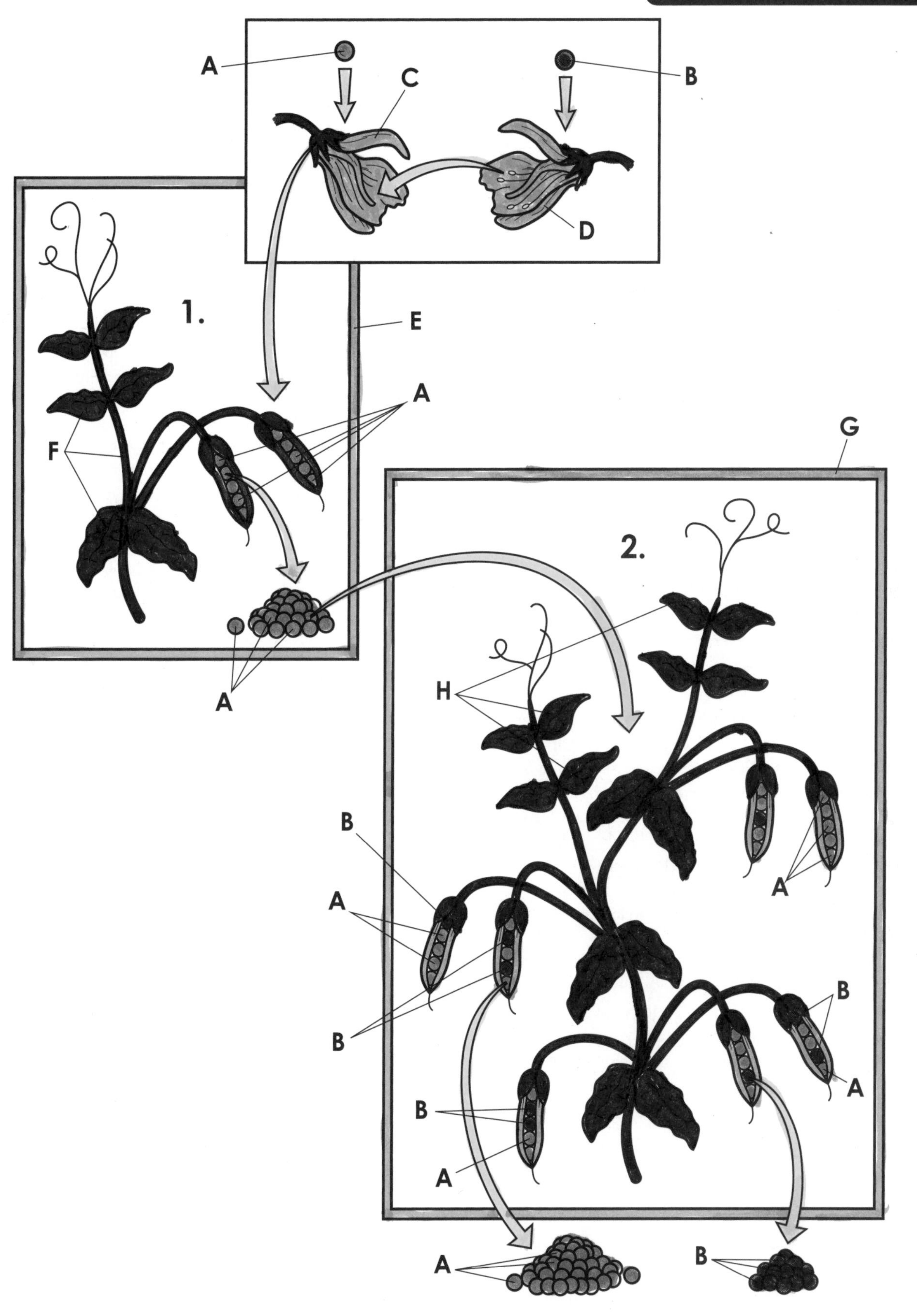

Yellow Seed A
Green Seed B
Yellow-seeded Pea Plant C
Green-seeded Pea Plant D
F_1 Generation E
F_1 Generation Plant F
F_2 Generation G
F_2 Generation Plant H

THE MONOHYBRID CROSS

The observations made by Gregor Mendel in the 1860s laid the foundation for the science of genetics. Mendel proposed that in a gene pair, one allele dominates over the other, recessive, allele. He theorized that alleles segregate (separate) independently during the formation of sperm and eggs and that they come together again in the new individual. In this plate, we see how this works in pea plants and we demonstrate an abbreviated way of expressing the crosses that take place.

This plate shows two ways in which crosses are expressed in genetics. In the first view, we see the Law of Segregation in action as the alleles separate and then come together in the offspring. The second view shows the shorthand way that geneticists keep track of alleles in crosses. You should use the same yellow and green colors for the pea plants that you did in the previous plate.

Gregor Mendel worked with pea plants that were either yellow or green. He began with P_1 generation of plants that produce either yellow seeds (peas) or green seeds (peas). The true-breeding yellow-seed bearing plant is the **homozygous dominant individual (A).** This individual has two alleles for yellow seed color. The alleles are represented by two capital Y's.

The second parent in Mendel's cross had green seeds (peas). Since the green allele is masked in the offspring, this individual is the **homozygous recessive individual (B).** Note that we represent the homozygous recessive by two small y's.

Gametes (sex cells) are formed during the process of meiosis. Each gamete has a member of the allele pair. Therefore, when the homozygous dominant individual produces gametes, all have the **dominant allele (C).** We show only one gamete, but all of these gametes are identical. In the same way, when the homozygous recessive individual produces gametes, they possess the recessive allele. Thus, we show a **gamete with the recessive allele (D).**

When these gametes come together to form a fertilized egg, the alleles are united and a new individual results. This individual has one allele for the yellow color (which is dominant) and one allele for the green color (which is recessive). Since the yellow allele dominates, all individuals will produce yellow seeds. These individuals are all said to be heterozygous, or **hybrids (E).**

We have now seen the basis for the Law of Segregation in which the alleles separate as gametes are formed and then come together in the new individual. We will explore this principle further as we continue with the next generation.

The first generation produced by individuals (P_1) is called the F_1 generation. When Mendel interbred these F_1 individuals, he crossed hybrids with hybrids (E). Once again, the alleles separate. In the brackets, we see gametes with the yellow allele (C) and gametes with the green allele (D).

When gametes come together, the F_2 generation forms. In one case, the offspring has yellow seeds because its alleles are YY. Geneticists say that the phenotype is yellow and the genotype is YY. When two individuals have a dominant Y gene and a recessive y gene, they are a hybrid. Their phenotypes are yellow, and their genotypes are Yy. To the right, the gametes unite to form an individual with green seeds (B). Here the genotype is yy and the phenotype is green. Of the four possible combinations, three plants have yellow seeds and one has green seeds. This is the 3:1 ratio observed by Mendel. The ratio of genotypes is 1:2:1 [one homozygous dominant individual, two heterozygous individuals (hybrids), and one homozygous recessive individual].

We have seen how the Law of Segregation works in an actual monohybrid cross. To simplify matters, geneticists use the Punnett square.

The Punnett square is a shorthand way of showing how the gametes behave in a monohybrid cross. It is set up in the following way: The alleles are separated, and one is placed at the head of each column. This is for the male. For the female, the gametes are separated and placed along the left side of the square. Then, by simply crossing the gametes, we can see which four individuals will form. In this case, each individual formed is a hybrid (E), or is heterozygous for this particular trait.

Next, we cross the F_1 generation. Here, one pair of alleles is placed at the top of the square (Yy) and the other is to the left (Yy). Now we simply cross the gametes. We again produce the homozygous dominant individual (A), the homozygous recessive individual (B), and the two heterozygous individuals (E). This corresponds to the 1:2:1 ratio seen on the left side of the plate. The Punnett square tells us the genotype of the offspring and, by examining the genotype, we can determine the phenotype.

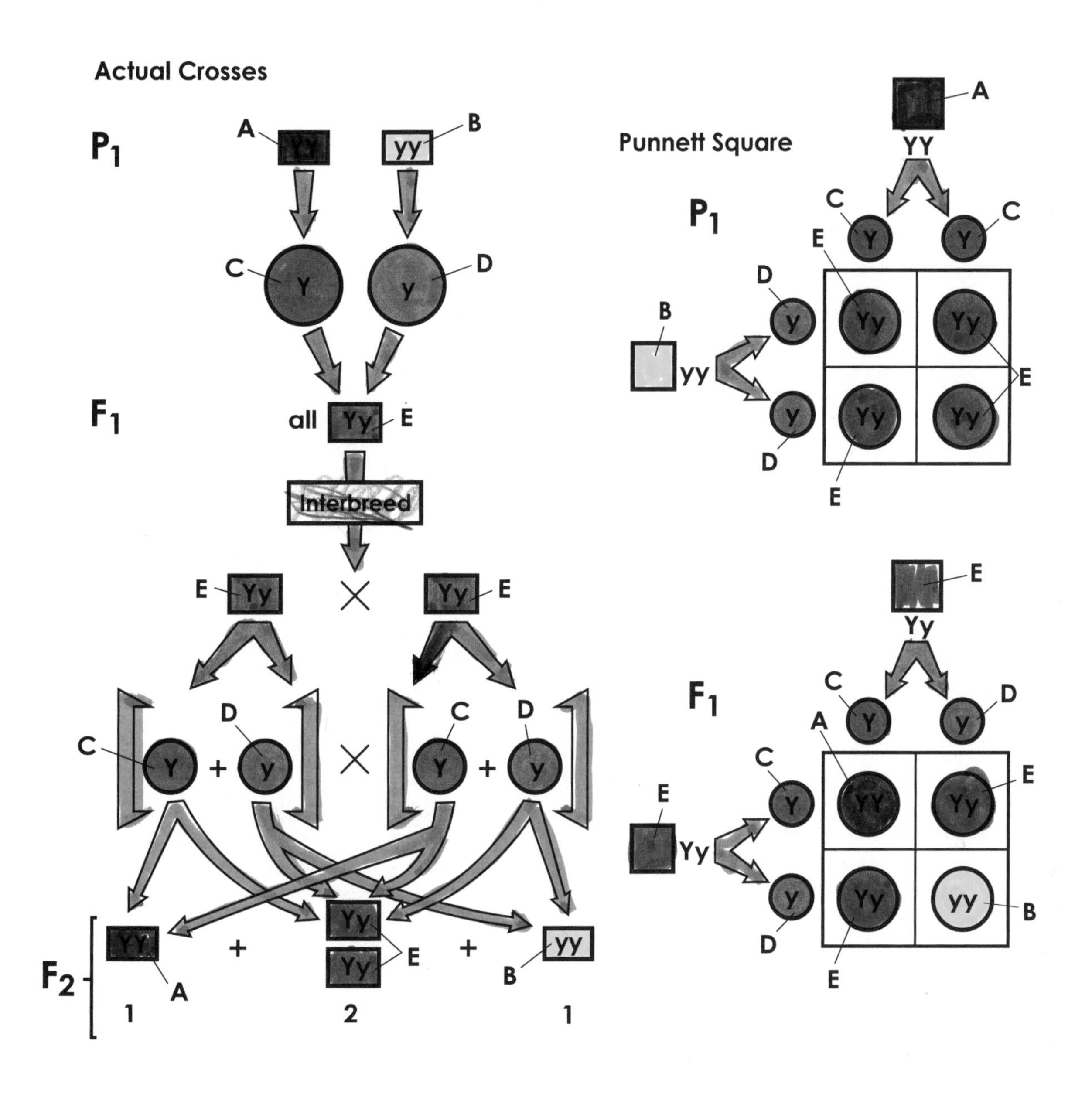

Homozygous Dominant A

Homozygous Recessive B

Gamete with Dominant Allele C

Gamete with Recessive Allele D

Heterozygous (Hybrid) Individual E

THE DIHYBRID CROSS

Mendel's first crosses were monohybrid crosses, which are used to examine the distribution of any one set of alleles in offspring. Later he also utilized the dihybrid cross, which is a cross with plants that differ in two ways. Mendel's observations of the results of dihybrid crosses led to his second law, the Law of Independent Assortment. This law states that the inheritance of one pair of alleles occurs independently of the inheritances of another pair.

Looking over this plate, you may note that it is very similar to the plate in which monohybrid crosses were explained. The difference is that here we are dealing with two characteristics. We follow two pairs of alleles through three generations and see how they assort independently.

Mendel used a dihybrid cross to study the inheritance pattern of two characteristics of the seeds (peas) of pea plants: roundness (R) or wrinkles (r); and yellow (Y) or green (y).

Mendel began with true-breeding plants. The seeds of one plant were round and yellow. This plant contained two alleles for the roundness trait (RR), and two alleles for the yellowness trait (YY). Thus, it was a **homozygous dominant individual (A).** Color this individual yellow. The second individual had seeds that were wrinkled (rr) and green (yy). This parent was a **homozygous recessive individual (B).** Color these seeds green.

When Mendel crossed the homozygous dominant individuals with the homozygous recessive individuals, he obtained a number of **heterozygous individuals (C),** also known as hybrids. All of the individuals of the F_1 generation had round, yellow seeds because each inherited one allele for roundness and one allele for yellowness from the parents, both of which are dominant; the recessive alleles were masked in this cross. The genotype of all of the individuals is RrYy; the phenotype is round and yellow.

We have seen how individuals heterozygous for two traits were obtained. We now see what happens when a dihybrid is crossed with another dihybrid. Continue your reading as you focus on the next portion of the plate.

Mendel next performed a dihybrid cross between a male and female plant, both with genotype RrYy. The dihybrid shown to the right produced a set of **hybrid gametes (D)** identical to the gametes produced by the other dihybrid. During meiosis, the alleles assort independently and produce four different kinds of gametes. These gametes are equivalent to a sperm cell and an egg cell **gamete with dominant alleles (E);** two **gametes with mixed alleles (F);** and a **gamete with two recessive alleles (G).** These reproductive cells will come together to form the new individual.

Now we determine the alleles by making a Punnett square. Sixteen combinations can occur in the offspring. Note, for example, that when the gamete (RY) unites with the gamete (RY), the result is an individual with the genotype RRYY; and as the diagram shows, the seed is round and yellow.

Now look at the second row of individuals, and look at the last individual in the column. This individual has inherited the alleles Rryy, and so has a round seed. However, it has inherited two alleles (yy) for greenness, so it is round and green. The genotype of the seed is Rryy, and its phenotype is round and green.

Examine the first individual in the second column. Its genotype is RRYy. This individual bears round seeds (as shown) and is yellow. You should color it yellow.

The genotypes of all the other individuals of the F_2 generation are shown in the plate. Your job is to determine their phenotypes and add the correct color: are they yellow or green?

At the bottom of the diagram is the ratio of phenotypes that result from the dihybrid cross. Count the number of offspring that have round, yellow seeds; the number with round, green seeds; the number with wrinkled yellow seeds; and the number with wrinkled green seeds. Place the numbers in the space to determine their ratio. Your numbers should be in a ratio of 9:3:3:1. If these are not your numbers, then review the colors you have selected for the phenotypes in the F_2 generation. The dihybrid cross shows that alleles separate and pass to offspring independent of other pairs of alleles.

THE DIHYBRID CROSS

Homozygous Dominant Individual A
Homozygous Recessive Individual B
Heterozygous (Hybrid) Individual C
Hybrid Gametes E
Gamete with Dominant Alleles E
Gamete with Mixed Alleles F
Gamete with Recessive Alleles G

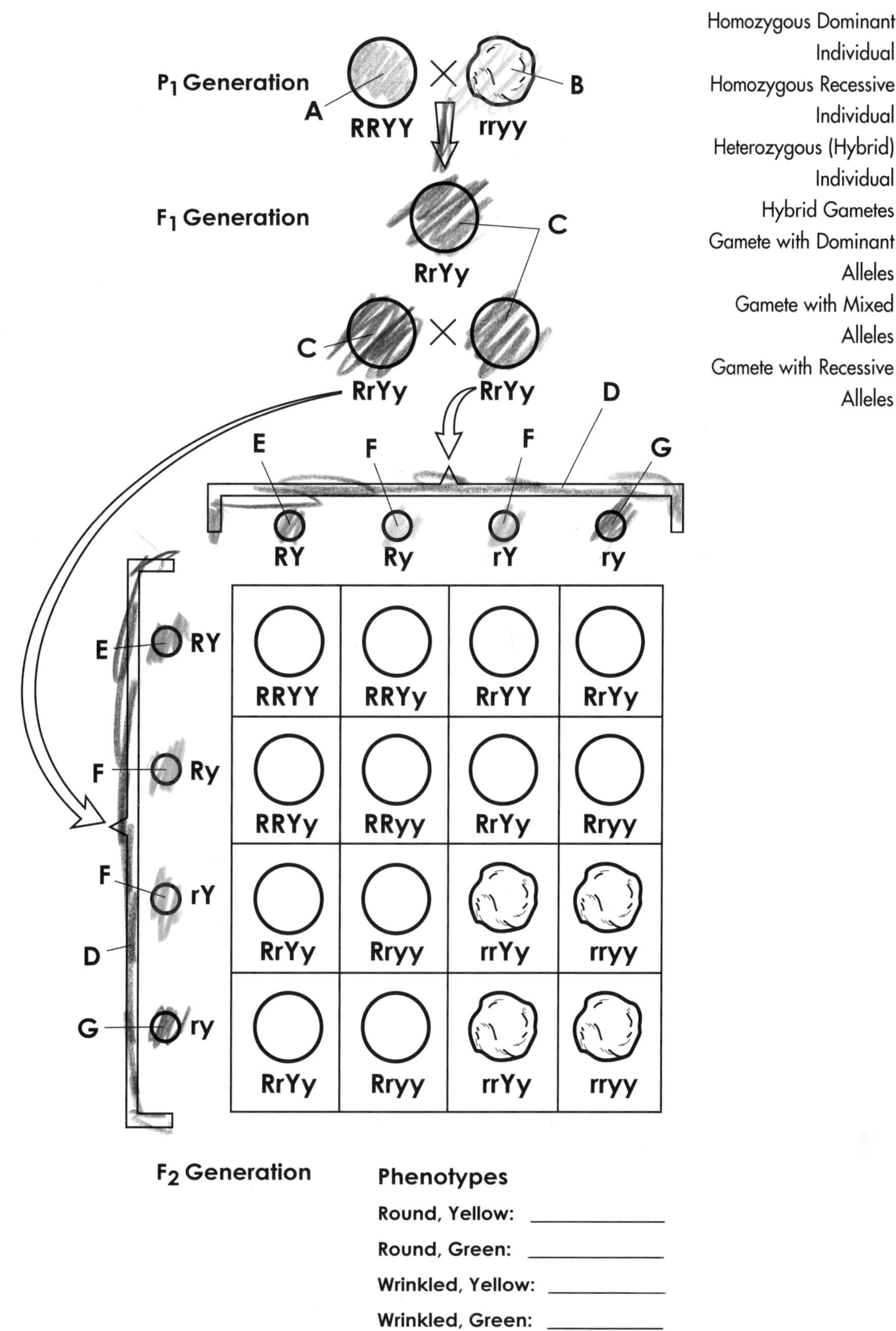

Phenotypes

Round, Yellow: ____________

Round, Green: ____________

Wrinkled, Yellow: ____________

Wrinkled, Green: ____________

THE TESTCROSS

In order to perform genetic experiments on individuals, it is important to know their genotypes, their genetic makeup. A gene can have many alleles, which are variations of the gene. A diploid organism has two copies of each gene, and these two alleles can be the same or different. When the alleles are identical, the individual is homozygous; when the alleles are different, the individual is heterozygous.

To determine the genotype of an individual, Mendel developed a procedure called a testcross. In this procedure, an individual of unknown genotype is crossed with an individual that is homozygous recessive for that trait. We will see how this process works in this plate.

This plate displays a group of peas whose genotype for roundness or wrinkledness is unknown. The objective is to show how their genotype is determined via a testcross.

As the plate shows, we begin with a series of round peas. These are the **individuals of unknown genotype (A).** Their phenotype is round, but they may be **homozygous (B)** individuals that have the genotype RR, or they may be **heterozygous individuals (C)** that have the genotype Rr. We do not know from the phenotype whether the genotype is RR or Rr.

The testcross is performed by taking the individuals of unknown genotype and crossing them with homozygous recessive individuals. Now we focus on possibility number one. Let us assume that the unknown individuals are homozygous and have the genotype RR (B). During meiosis, they form a set of **gametes (E),** which are either sperm or egg cells. Each is a **gamete with a dominant allele (F).** To the right is the **homozygous recessive individual (D)** used in the testcross. It also produces gametes (E), denoted by the bracket. Each is a **gamete with a recessive allele (G).**

Now fertilization takes place. Within the Punnett square, all individuals formed from this cross have both a dominant allele (R) and a recessive allele (r). All four individuals are heterozygous and all have round seeds. You may use any color you wish for these four, but the same color should be used for all four seeds. Thus, the only possible result of this cross is plants with round seeds. We may therefore conclude that if the unknown individual had a genotype of RR, all offspring in the F_1 generation would have round seeds.

We now focus on the second possible genotype (Rr). Again we perform a testcross, but this time we will see different results. Continue your reading below as you continue to study and color the diagram.

For the second possibility, we assume that the unknown individual has the genotype Rr. This individual (Rr) is crossed with the homozygous recessive individual (rr).

In the Punnett square, note that the unknown individual forms gametes (E). The gametes may have the dominant allele (F) or the recessive allele (G). At the left side of the Punnett square, the homozygous recessive individual forms gametes, both of which contain recessive alleles (G).

Now we perform the crosses. Two of the resulting individuals have the genotype Rr, and two have the genotype rr (two individuals have round peas and two have wrinkled peas). Thus, 50% of the offspring have round peas and 50% have wrinkled peas. You may use any color you wish for these four, but the same color should be used for round seeds, and you should use a different color for wrinkled seeds. If we reason backward, then the only possible parent that could have given us 50% round and 50% wrinkled peas is a parent with the genotype Rr. Thus, we can deduce the genotype by observing the phenotypes of the F_1 generation.

The testcross is a way of determining the genotype of the individual parent by observing the phenotype of the offspring. We saw in these two possibilities that if the unknown parent genotype was RR, then all the offspring of a cross with an rr parent would have round peas. Alternately, if the unknown parent genotype was Rr, then 50% of the offspring of a cross between that parent and an rr individual would have round peas and 50% would have wrinkled peas. As you can see, the testcross is a convenient way of determining what genotype was present in the parent individuals.

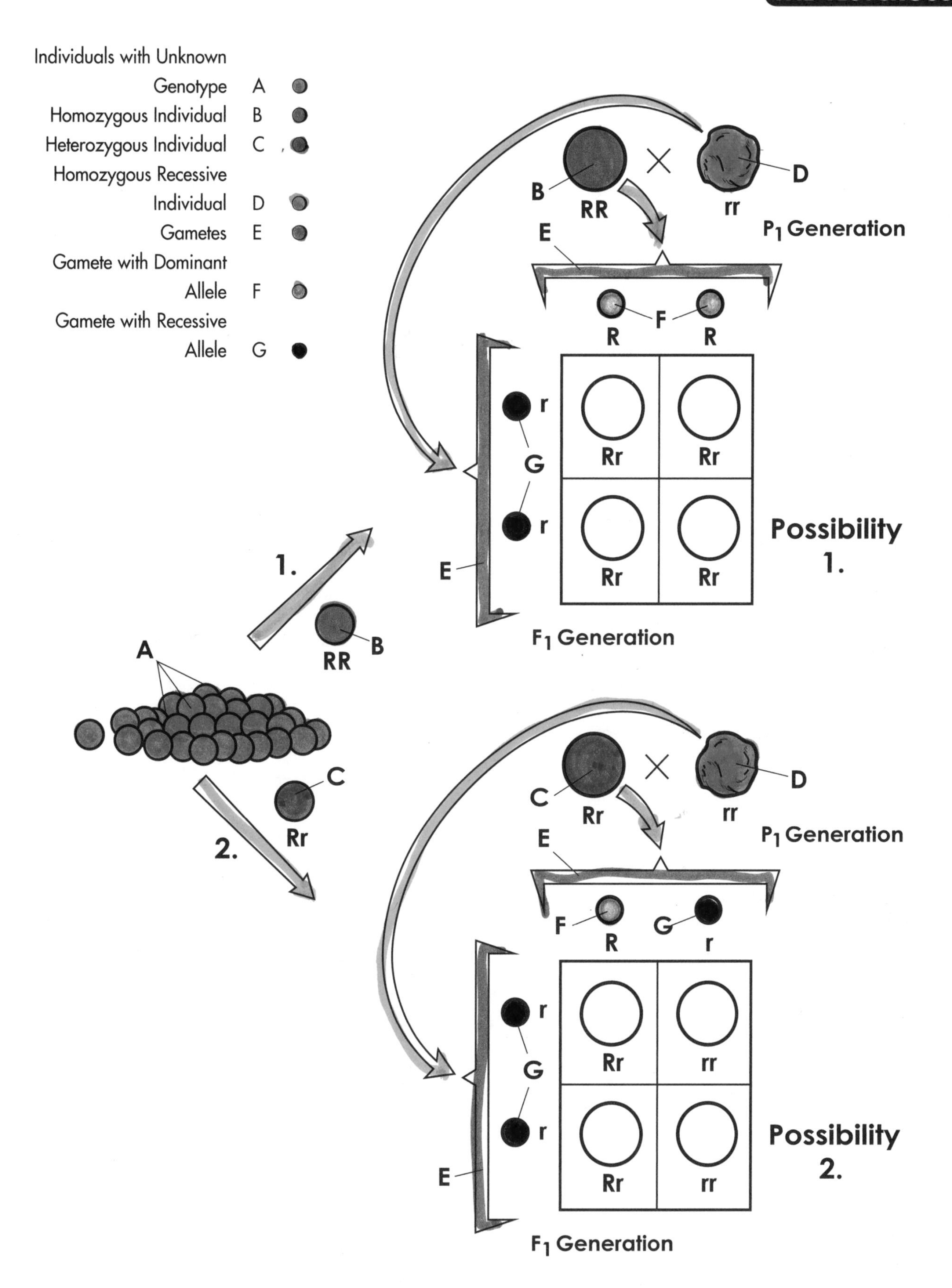

Individuals with Unknown Genotype A
Homozygous Individual B
Heterozygous Individual C
Homozygous Recessive Individual D
Gametes E
Gamete with Dominant Allele F
Gamete with Recessive Allele G
B
RR
D
rr
E
P1 Generation
F
R
R
r
G
r
Rr
Rr
Rr
Rr
E
Possibility 1.
F1 Generation
1.
B
RR
A
C
Rr
2.
C
Rr
D
rr
E
P1 Generation
F
R
G
r
r
G
r
Rr
rr
Rr
rr
E
Possibility 2.
F1 Generation

INCOMPLETE DOMINANCE

In Mendel's experiments with pea plants, the dominant allele masked the expression of the recessive allele; the heterozygous offspring had phenotypes that were dictated by the dominant allele. We saw, for instance, that yellow dominates over green in seed color, and that round dominated over wrinkled in seed shape. In heterozygous individuals, the dominant round and yellow alleles are expressed if they are present, and the recessive wrinkled or green alleles are hidden.

Sometimes, two alleles are both expressed so that the phenotype of the hybrid is intermediate between the two phenotypes of the parent individuals. This pattern of inheritances is called incomplete dominance. It causes an apparent blending of the phenotypes, but it is not a true blending because the alleles for the trait have not been altered.

In this plate, we will study the phenomenon of incomplete dominance in flowers called snapdragons. Begin your work by focusing on the first portion of the plate, and read the paragraphs below.

Snapdragons have two true-breeding strains. The **red snapdragon (A)** bears red flowers and the **white snapdragon (B)** bears white flowers. You should color the **stem (C)** green. We will start by assuming that the genotype of the red snapdragon is RR, and that the genotype of the white snapdragon is WW.

One example of incomplete dominance can be seen when white snapdragons are crossed with red ones. All the members of the resulting F_1 generation are **pink snapdragons (D);** the heterozygous individuals have phenotypes that are intermediate between the phenotypes of the two parent plants. Keep in mind that the alleles for flower color have not changed even though the phenotype is blended. The genotype of the offspring is RW, so this heterozygous plant contains a gene from each of the parent strains.

As we have noted above, the alleles are not changed in the offspring in cases of incomplete dominance. This will become apparent when we interbreed the hybrids of the F_1 generation. Continue your reading below, and focus on the lower portion of the plate.

When two pink snapdragons are crossed, you can predict the phenotype pattern in the F_2 generation if you know Mendel's Law of Segregation. We cross two hybrid **pink snapdragons (D),** both members of the F_1 generation. The snapdragon on the left produces sperm cells by meiosis and we will put these **gametes (E)** vertically on the left. (They are enclosed in a bracket). One is a **gamete with the red allele (F)** and the other is a **gamete with the white allele (G).** At the top of the Punnett square are the egg cells that were produced by the second snapdragon, and again there is one gamete with the red allele (F) and one with the white allele (G).

Now we are ready to do the crosses. The first individual that results from this cross will have the genotype RR, and will be a red snapdragon (A). Two individuals of the cross will have the genotypes RW, just as their parents did in the F_1 generation. They are pink snapdragons (D). One individual in the F_2 generation will be a white snapdragon (B), and will have the genotype WW.

The results of this cross show that the individual alleles were not changed, despite the fact that neither of them was expressed strongly in the F_1 generation. We saw a ratio of genotypes of 1:2:1, which translates into 1RR:2RW:1WW. Interestingly, the ratio of phenotypes is also 1:2:1, that is 1 red:2 pink:1 white. Thus, the genotype and phenotype ratios are identical. (Recall that in the monohybrid cross, the genotype ratio in the F_2 generation was 1:2:1, while the phenotype ratio was 3:1.)

One example of incomplete dominance in humans can be seen in Tay-Sachs disease. An individual with two recessive alleles exhibits the symptoms of this genetic disorder. However, a heterozygous individual shows slight symptoms, indicating that the presence of the dominant allele does not completely overshadow the recessive allele.

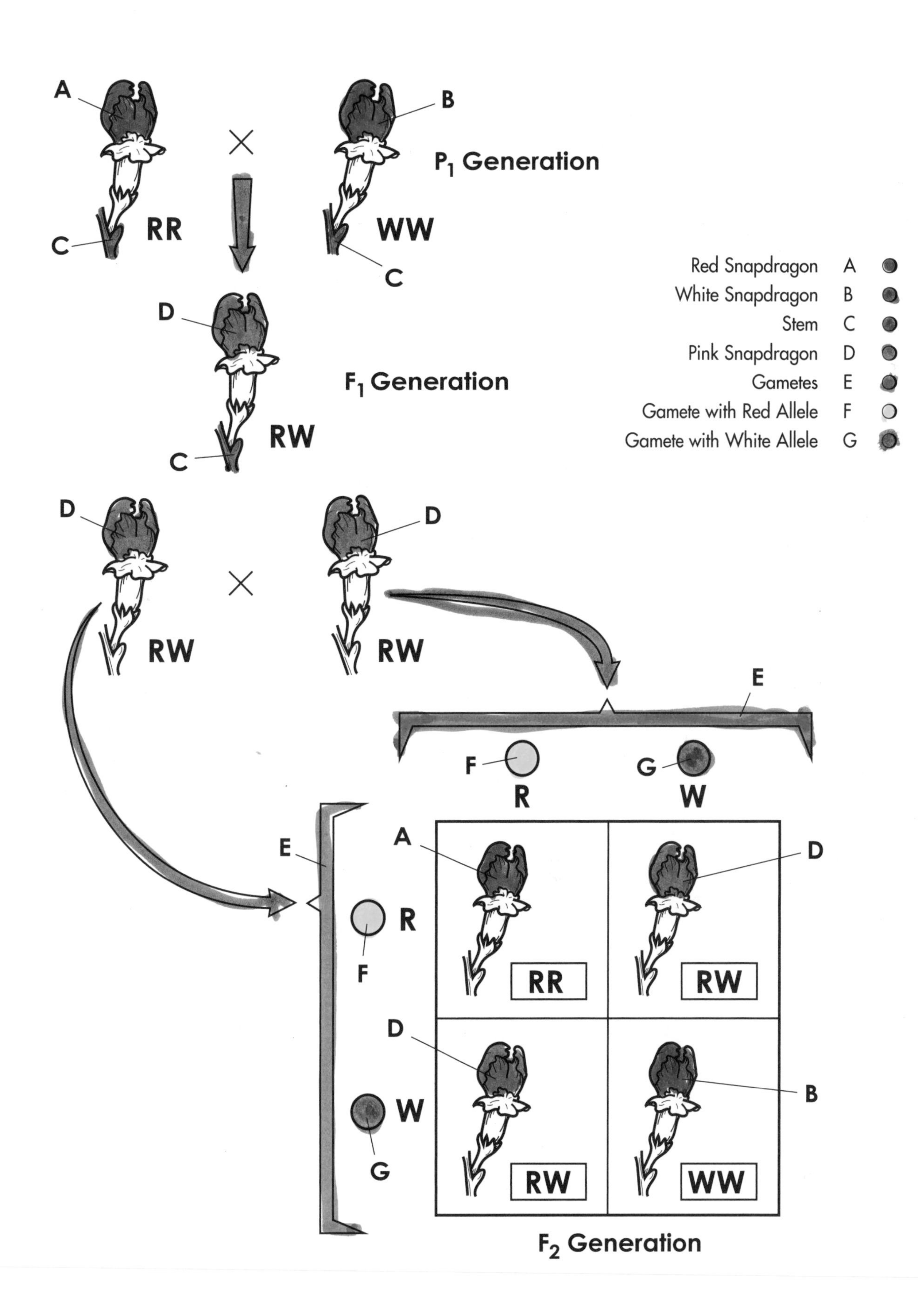
A
B
RR
WW
C
C
P₁ Generation
D
F₁ Generation
RW
C
D
D
RW
RW
E
F
G
R
W
E
A
D
R
F
RR
RW
D
B
W
G
RW
WW
F₂ Generation
Red Snapdragon A
White Snapdragon B
Stem C
Pink Snapdragon D
Gametes E
Gamete with Red Allele F
Gamete with White Allele G

MULTIPLE ALLELES AND CODOMINANCE

A single individual can have only two alleles for a given gene, but many alleles for a trait may be present in a population. Blood type in humans is a good example of multiple alleles. The blood types A, B, AB, and O result from the pairings of three different alleles of a single gene. These alleles are designated A, B, and O. Alleles A and B direct the synthesis of glycoproteins A and B, which are round on the surface of red blood cells, while the O allele does not direct the synthesis of any proteins.

In this plate, we will show how multiple alleles contribute to creating different blood types. Start reading below as you color the plate.

As we mentioned above, there are three alleles that contribute to blood type in humans, but only two alleles can exist in a particular individual. The type O allele is recessive to the alleles A and B, and A and B are co-dominant. This means that both will express themselves if they are present. Individual humans therefore have one of six possible genotypes: AA, BB, AB, AO, BO, or OO. However, there are only four possible phenotypes (A, B, AB, and O) because genotypes AA and AO yield blood type A,while genotypes BB and BO yield blood type B.

Now we will see how these alleles determine the phenotypes of the offspring of two individuals. We will begin with a male whose blood type is AB; he has an **A allele (A)** and a **B allele (B).** You should use two dark colors to distinguish the alleles. The female in this cross has a phenotype of blood type B, and her genotype is BO, which means that she possesses a B allele (B) and an **O allele (O).**

During meiosis, the male and female produce **gametes (C).** The female produces **egg cells (D),** and each egg cell has either a B allele or an O allele. The male produces **sperm cells (E),** and these sperm cells have either the A allele or the B allele.

Now we are ready to perform the crosses. If the egg cell that contains the B allele unites with the sperm cell that contains the A allele, then the offspring will have both A and B alleles, and this person's blood type will be AB. If the sperm and egg cells that have the B allele join, the offspring will be homozygous and have blood type B. If the egg cell that has the O allele joins the sperm with allele A, the blood type will be A. If the same egg were to join with the sperm carrying the B allele, the blood type would be B.

Overall then, the B allele is dominant to the recessive O allele, and the A allele is also dominant to the recessive O allele. However, when an individual has both the A and B alleles, both are expressed. One doesn't mask the other one. This type of inheritance is called codominance, because an individual with a heterozygote genotype expresses both alleles together.

Now that we have seen how blood type is inherited, we will turn to two possible crosses between individuals who have different blood types. As you proceed, you will be asked to determine possible blood types that result from mating pairs.

Take a look at the second diagram in the plate. Here you can see a male that has blood type B (genotype BB), and a female that has blood type A (genotype AO). We have put the alleles from the egg cell across the top of the Punnett square. You should put the alleles of the sperm cells produced by the male in the left-hand column.

Now make the appropriate crosses between the egg and sperm cell and figure out the four possible genotypes of the offspring; then determine the phenotypes. Is it possible that this couple could have a child whose blood type is type O? Is it possible that this couple could have a child whose blood type is AB?

Having seen how multiple alleles function in a cross, we will now look at a third example to see how blood typing can be used in paternity suits. Continue reading below as you focus on the third diagram in the plate.

Put the possible egg cells produced by the female that has blood type AB at the top of the Punnett square in the third diagram. Place the possible alleles in the sperm cells of the male in the left-hand column. Now perform the crosses in the squares and determine the genotypes of all the possible offspring from this mating pair. Determine which phenotypes are possible in the children of this pair. Could a child with blood type O result from this mating pair? Could a child of blood type AB result from this mating pair? If either a type O or type AB individual claimed to be the offspring, would this be a legitimate claim?

1.

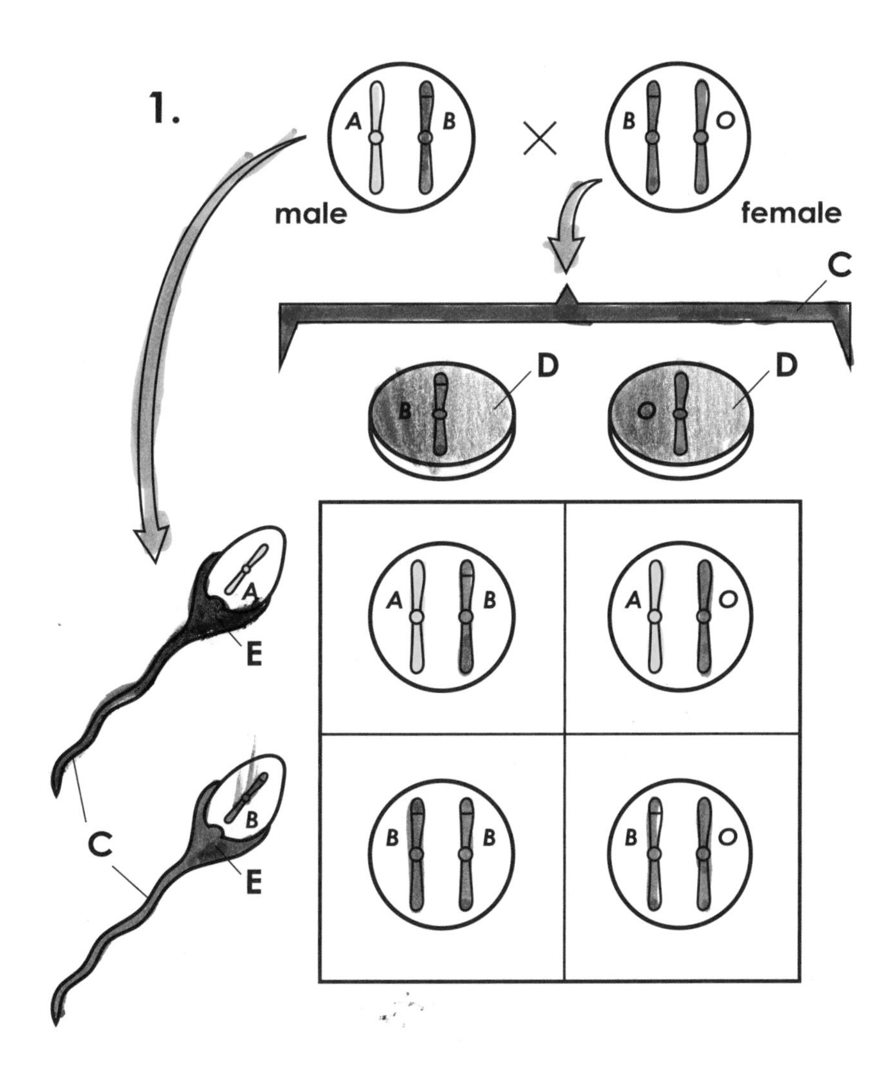

Allele A A
Allele B B
Allele O O
Gametes C
Egg Cells D
Sperm Cells E

2.

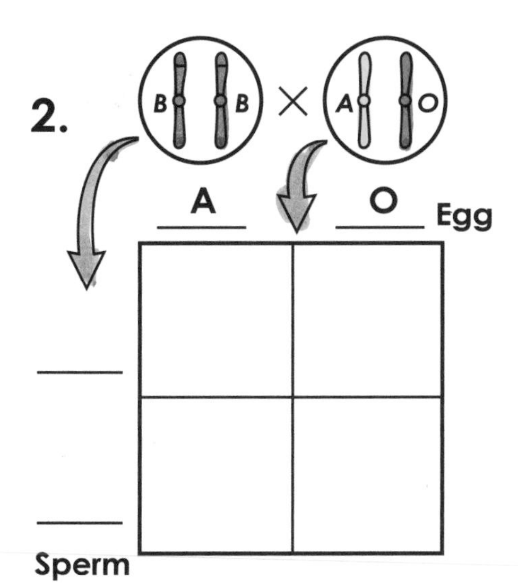

3.

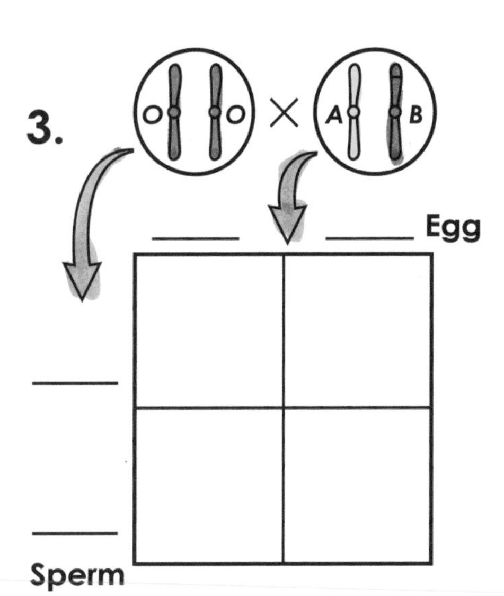

POLYGENIC INHERITANCE

In many instances, a trait is governed by several genes found on different pairs of chromosomes that interact with each other. This phenomenon is called polygenic inheritance, and the genes involved are known as polygenes.

One example of polygenic inheritance in humans is skin color. Scientists believe that skin pigmentation in humans is controlled by three separately inherited genes. Thus, a variety of skin colors is possible, depending on how many dark-skin alleles are present in the individual. In this plate, we will examine the genetic basis for variations in skin color as an example of polygenic inheritance.

Looking over the plate, you will notice that we show three generations of individuals and their genotypes. We will follow the genetic patterns of very light and very dark individuals, and show that their offspring can have at least 7 phenotypes and 27 genotypes.

We will assume that human skin color is governed by at least three allelic pairs of genes, and that each pair is located on a different chromosome. We will designate the three genes A, B, and C. Each gene consists of two possible alleles, one for maximum pigmentation (very dark) and one for no pigmentation (very light). By convention, the dominant gene for very dark pigmentation is represented by a capital letter. The recessive gene for no pigmentation is represented by a lowercase letter.

In this plate, we will start with the phenotype of an extremely dark individual who is **homozygous dominant (A).** You should use a very dark color to color this box. This person mates with an individual with an extremely light complexion, who is **homozygous recessive (B).** You should shade this box in a very light color.

When these individuals produce offspring, each contributes alleles for skin pigmentation. The homozygous dominant parent contributes alleles ABC, while the homozygous recessive individual contributes alleles abc. The offspring is a **heterozygous individual (C)** that has three alleles for pigmentation and three for no pigmentation, and its skin color is intermediate between that of its parents.

We will now consider the next generation, in which two heterozygous individuals mate. We will see how a wide range of skin color exists in the F_2 generation. Continue your reading below as you color.

Now we will cross two heterozygous individuals that both have the genotype AaBbCc. The offspring that result from this union are shown in the F_2 generation. We have set up these results in order, from darkest phenotype to lightest.

One of the possible results of the mating is an **individual with six dominant alleles (D).** You should color this extremely dark individual the same color as its homozygous dominant grandparent.

The next column contains a series of **individuals that have five alleles for darkness (E).** Notice that the first individual in the column has the alleles AaBBCC. This individual is slightly less dark than the individual to the left (D). Now take a look at the column that contains **individuals with four dominant alleles (F).** Notice that each individual's genotype contains four capital letters. These individuals have a slightly lighter skin color than the previous group, so a lighter color should be used to color them.

Now look at the group of **individuals with three dominant alleles (G),** which is the largest group of the same phenotype. These individuals have the same skin color as their parents. (Notice that the third individual down from the top also has the same genotype as its parents.) A still lighter color should be used to color them.

As we continue our survey, we see the **individuals with two dominant alleles (H).** These individuals are lighter than both of the parents, because each has only two alleles for dark skin color. Continuing, we encounter **individuals with one dominant allele for skin color (I).** These individuals have very light skin pigmentation, since they have only one dominant allele. The final case that we see is an **individual with no dominant alleles (G)** and a very light complexion.

Counting the number of columns, you will notice that there are seven possible phenotypes produced by two people who are heterozygous for the three genes, and the number of boxes shows there are 27 possible genotypes in the F_2 offspring.

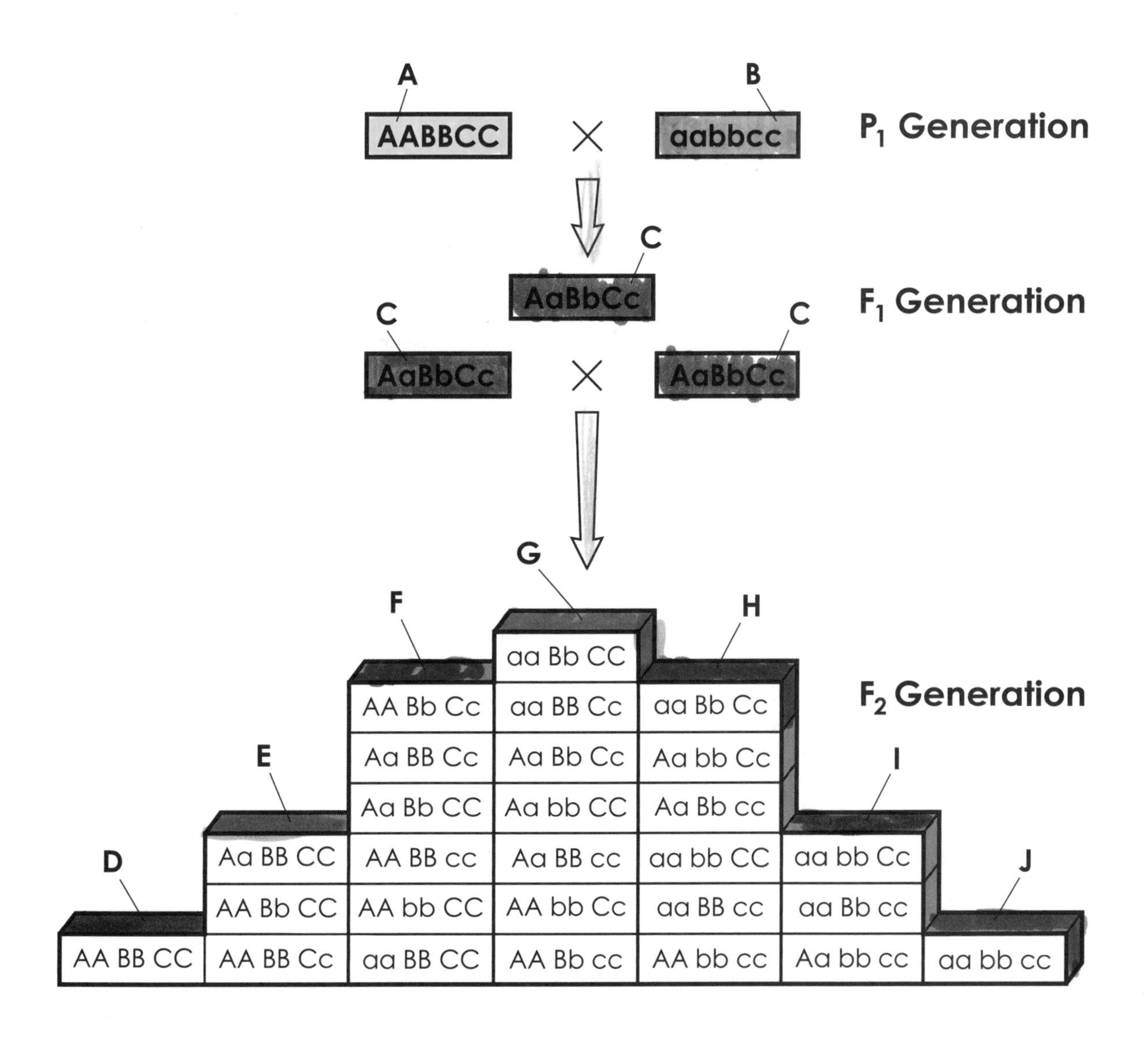

Homozygous Dominant Individual A
Homozygous Recessive Individual B
Heterozygous Individual C
Individual with Six Dominant Alleles D
Individual with Five Dominant Alleles E
Individual with Four Dominant Alleles F
Individual with Three Dominant Alleles G
Individual with Two Dominant Alleles H
Individual with One Dominant Allele I
Individual with No Dominant Alleles J

SEX DETERMINATION

Chromosomes can be distinguished from one another by their appearance; they differ both in size and in the position of their centromere. Human cells have 46 chromosomes in total, but these chromosomes can be matched in pairs: There are two of each type, and these pairs are called homologous chromosomes. For this reason, we can say that there are 23 pairs of homologous chromosomes in each cell. The exception to this is the two sex chromosomes, designed X and Y. Human females have a pair of homologous X chromosomes, but human males have one X chromosome and one Y chromosome, and the Y is significantly smaller than any of the other chromosomes. The X and Y chromosomes are called the sex chromosomes, while the others are referred to as autosomes.

In this plate we will see how the laws of genetics determine which sex chromosomes the offspring receive; this determines whether they will be male or female.

This plate presents a simple monohybrid cross between a male and female. We will follow one pair of chromosomes, the sex chromosomes. Start your work by looking over the plate as you read about sex determination.

In this plate, we show a **male (A)** and a **female (B).** A cell from each of the individuals is shown, and the **male cell (C)** possesses two different chromosomes. One is the **X chromosome (E)** and the other is the **Y chromosome (F).** Use red and green to distinguish these chromosomes, and notice that the Y chromosome is smaller than the X. In the **female cell (D),** we see two identical X chromosomes.

Now we have seen how the cells of males and females differ in the sex chromosomes they contain. We will now follow the pattern of inheritance of sex chromosomes. Continue your reading as you color the appropriate parts of the plate.

During the reproductive process, the female cell undergoes meiosis in the ovary and produces a set of **gametes (G),** enclosed by a bracket. The bracket should be colored in a dark color. Notice that both **egg cells (H)** are identical, in that both carry an X chromosome. Since a gamete receives one chromosome of a chromosome pair, all of the mother's egg cells will have an X chromosome.

Now look at the left side of the plate. The father's cell undergoes meiosis in the testes to produce two types of **sperm cells (I):** one with the X chromosome (E), and one with the Y chromosome (F). Either of the two types of sperm cells may fertilize the egg cell.

We now will see the possible crosses that can result. Continue your reading below as you color the appropriate diagrams in the plate.

Because the male produces equal numbers of sperm cells that have X and Y chromosomes, there is a 50% chance that an X-bearing sperm cell will fertilize the egg cell in the ovary. There is also a 50% chance than a Y-bearing sperm cell will fertilize the cell. In the Punnett square, the first row shows the offspring when an X-bearing sperm fertilizes the egg; both offspring are female (B). Now look at the bottom row of the square; when a Y-bearing sperm cell has fertilized the egg cell, both offspring are male (A).

In effect, the sperm cell determines the sex of the offspring individual, since the woman's sex cells are identical for each pregnancy. The numbers of chromosomes vary in different animals, but all animals have one set of sex chromosomes.

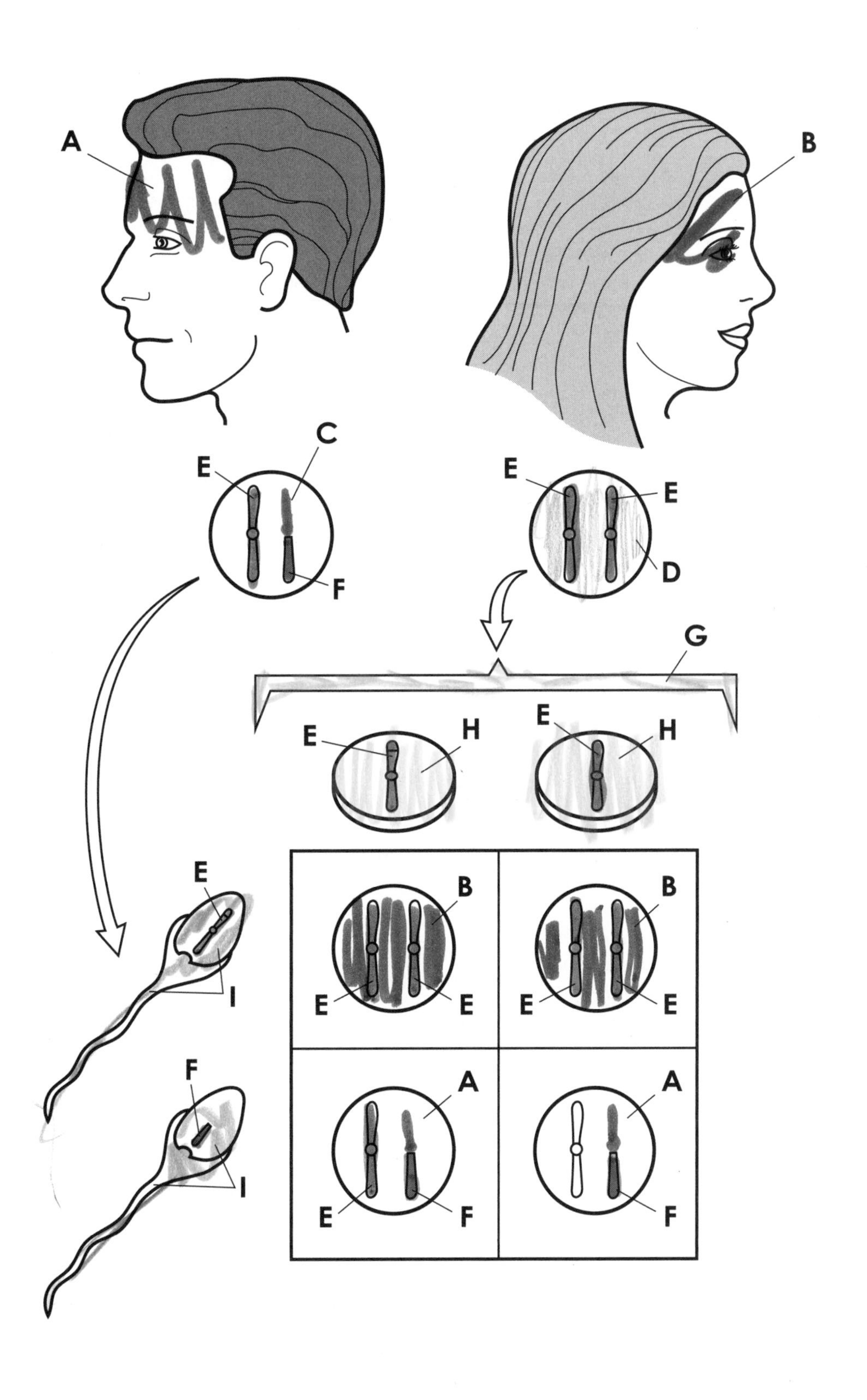

Male A
Female B
Male Cell C
Female Cell D
X Chromosome E
Y Chromosome F
Gametes G
Egg Cell H
Sperm Cell I

SEX-LINKED TRAITS: COLOR BLINDNESS

In humans, certain genes are found on one sex chromosome and never on the other; these genes are called sex-linked genes. For instances, the human X chromosome contains the genes for color vision as well as for blood clotting, and these genes do not exist on the Y chromosome. Females have two X chromosomes, so that any allele that's missing or defective on one X chromosome may exist on the other. Males, by contrast, have to express whatever allele appears on their X chromosome, because they don't have another X chromosome allele that can "mask" it. In this plate, we will show how sex-linked traits work.

Looking over the plate, you will notice that we show a cross between a male and female. Both the male and female have normal color vision, but the female is a carrier for the trait of color blindness. Begin your reading below.

Color blindness is a much-studied sex-linked trait that occurs in humans. The allele for normal color vision is dominant, and the allele for color blindness is recessive. The allele for color blindness renders individuals unable to distinguish shades of red or green—these colors appear gray. About 8% of American males are colorblind, but only 0.6% of females can't see colors.

In this plate, we cross a **male (A)** with a **female (B).** Notice that we show a **male cell (C)** and a **female cell (D),** and that the male cell contains an **X chromosome (E)** and a smaller **Y chromosome (F).**

Examining the female cell (D), we see that one of the X chromosomes carries the **color blindness allele (G),** which is recessive. This part of the allele should be colored in darkly to distinguish it. Since the woman has an allele for normal vision on her other X chromosome, the normal allele will mask the defective one, and she will see red and green normally. The male has no color blindness allele.

We have seen that the male does not have the color blindness trait, and that the woman is a carrier of the color blindness trait, although she sees colors normally. We will not examine the offspring of this pair to see how the color blindness trait expresses itself.

When the female in our study produces gametes (in meiosis), two types of **egg cells (H)** result. One of these egg cells has the color blindness trait (G), while the other cell has the **allele for normal vision (J).**

When the male forms **sperm cells (I),** two different types are possible. During meiosis, the alleles separate, and one sperm cell ends up with the X chromosome (E), while the other has the Y chromosome (F).

We will now perform the cross. Take a look at the Punnett square; in the offspring in the upper left corner, one X chromosome has come from the female and one has been contributed by the male. The result is a female (B) that has one colorblind allele and one normal allele; she will have normal vision. Now look at the upper right portion of the Punnett square—this individual is also female, with two X chromosomes (E), neither of which has the color blindness trait. She has normal vision.

Next examine the bottom row of offspring. In the bottom left, we see a male (A) who has acquired an X chromosome from the mother and a Y chromosome from the father—he has acquired the color blindness trait (G). Because there is neither another X chromosome with a normal allele that will mask the colorblindness trait, nor an offsetting normal trait on the Y chromosome, this male will be colorblind. Finally, examine the individual in the lower right. This male (A) has acquired an X chromosome from the mother and a Y chromosome from the father. The X chromosome carries the normal vision trait, so this individual will be able to see red and green.

To summarize the results of this cross, we can say that the possible four results of this mating are a normal-vision carrier female, a normal-vision homozygous female, a colorblind male, and a normal-vision male. Thus, there are two chances in four that the offspring will be male, and if it's a male, there is a 50% chance that he will be colorblind.

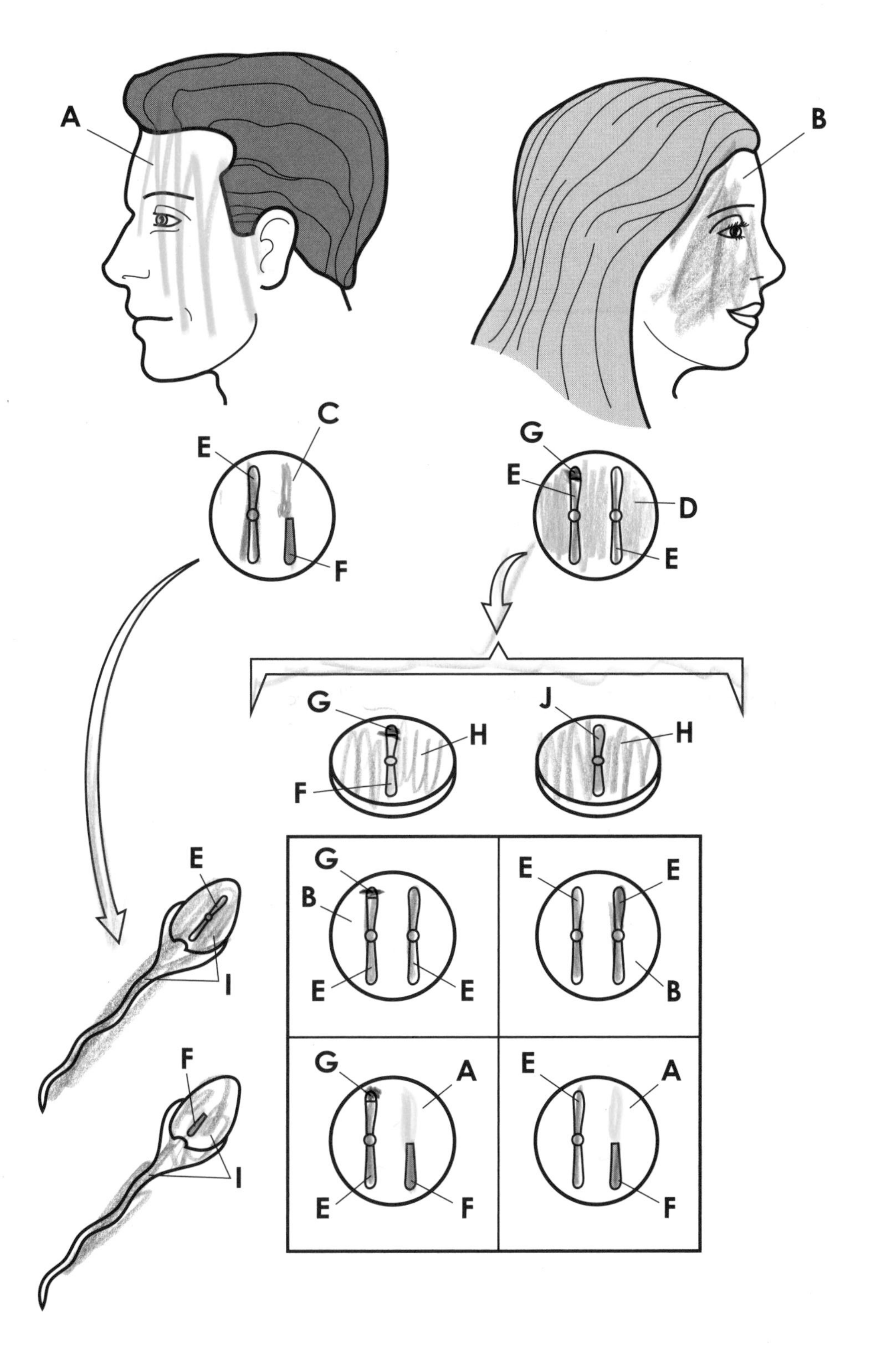

Male A
Female B
Male Cell C
Female Cell D
X Chromosome E
Y Chromosome F
Color Blindness Allele G
Egg Cell H
Sperm Cell I
Allele for Normal Vision J

CHROMOSOMAL ALTERATIONS

Permanent changes in chromosomes known as mutations may be passed to the offspring of a mating pair if they exist in cells that produce sperm or egg cells.

One kind of mutation affects only a single gene, while other types of mutation involve the rearrangement of several of them. For instance, pieces of chromosomes may be lost or exchanged between nonhomologous chromosomes. When altered chromosomes are passed to offspring, variation increases.

This plate displays four different types of alterations that can occur in chromosomes. Focus on the first two alterations, entitled Deletion and Inversion.

The first chromosomal alteration we will discuss is deletion, which is illustrated in the upper left portion of the plate. You should begin by coloring the normal chromosome with genes **A** to **G** using seven distinctive colors. When gene deletion occurs, a portion of the chromosome is lost, usually from the end. In our diagram, the chromosome that has undergone deletion is missing gene A. The remainder of the chromosome should be colored with the same colors that were used in the normal chromosome. A deletion sometimes results in the loss of an important gene, with severe consequences to the organism.

We will now turn to the second chromosomal alteration, called Inversion. You should continue to use the same colors for the genes. We will now show how inversion differs from deletion.

We will start again by looking at the normal chromosome, which contains genes A through G. When an inversion takes place, a segment of chromosome turns around 180°. Notice that genes C and D have inverted, so that the sequence of genes in the altered chromosomes is different.

At first glance, it may seem that the chromosome is not affected because the same genes are present, but the position of a gene in a chromosome is very important. For example, a gene may be separated from its nearby regulatory region as a result of inversion, so its rate of expression may be altered, or it may cease to be expressed at all. Scientists believe that chromosomal inversion may be a factor in developing cancer cells.

We will now focus on a third type of alteration called Translocation, in which two chromosomes are involved. Continue your coloring as you read below.

A translocation involves the movement of a chromosomal segment from one chromosome to another. The two chromosomes involved are nonhomologous, which means that they are chromosomes from different chromosomal pairs. Begin by coloring genes A through G with the colors you used before, and then color the genes of the second chromosome, **genes H to N,** with different colors. Now take a look at the point at which translocation has taken place. Genes F and G from the first chromosome have moved to the second chromosome, and genes M and N have moved from the second chromosome to the first. Chromosomes 1 and 2 are now considerably different than they were originally. Certain translocations have been linked to cancer, and abnormal gametes can result from this alteration.

The final type of chromosomal alteration that we will consider is Duplication. Focus on the lower right portion of the plate. The same colors that you used for the genes before should be used here.

In duplication, a chromosomal segment doubles itself. For instance, here we see the normal chromosome with genes A to G on the left, but genes D and E appear twice after duplication occurs in the abnormal chromosome.

Duplication occurs when a broken segment of one chromosome attaches to its homologous chromosome. One effect of the repeating genes may be duplicate proteins in an individual. For example, there are two alpha chains in hemoglobin molecules in human red blood cells. The two molecules may result from a single gene that duplicated in an ancient ancestor so that the modern descendent now produces two proteins instead of one. Therefore, duplication may be a factor in evolution.

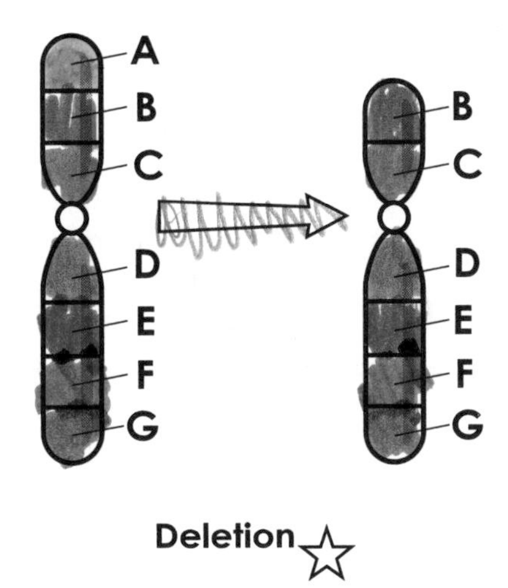

Deletion ☆

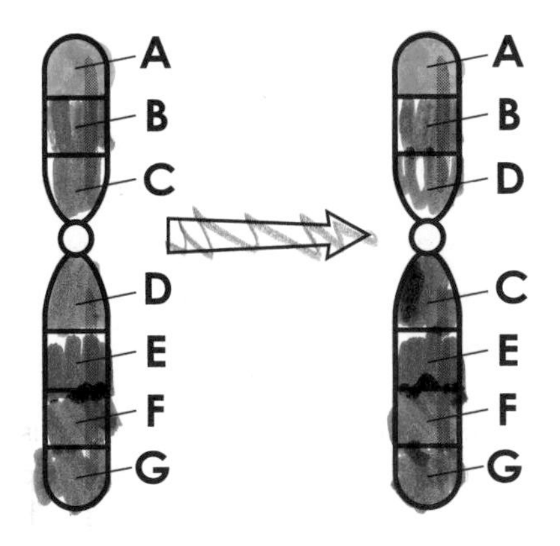

Inversion ☆

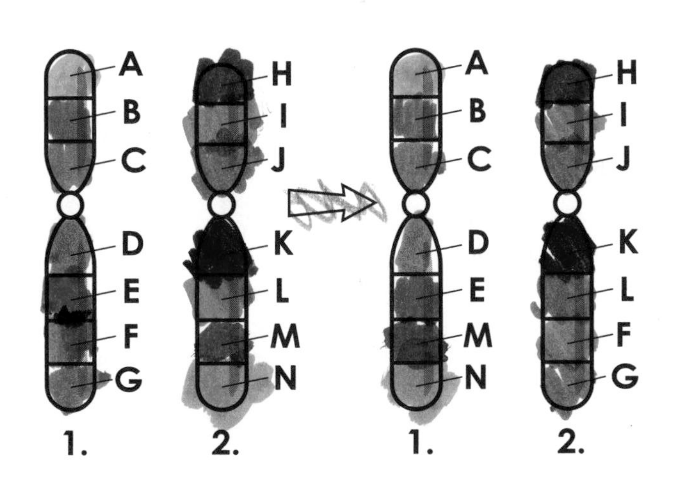

Translocation ☆

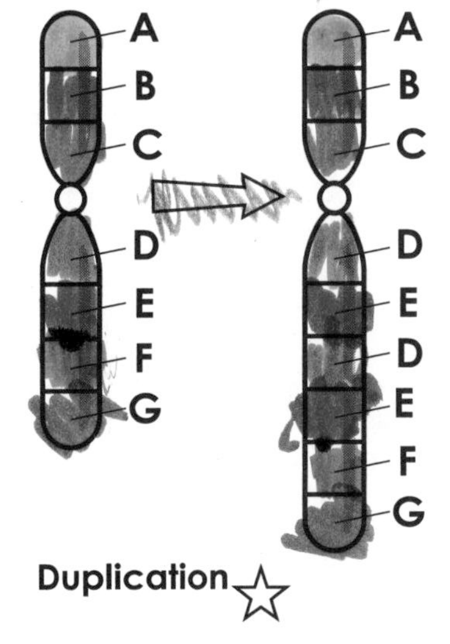

Duplication ☆

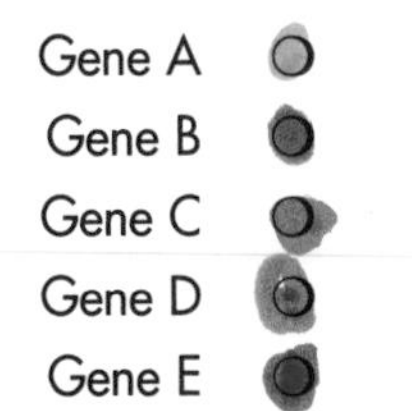

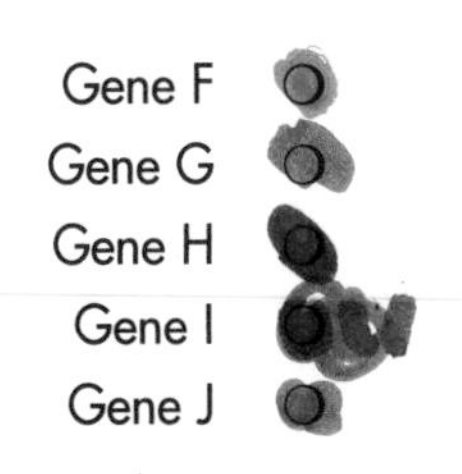

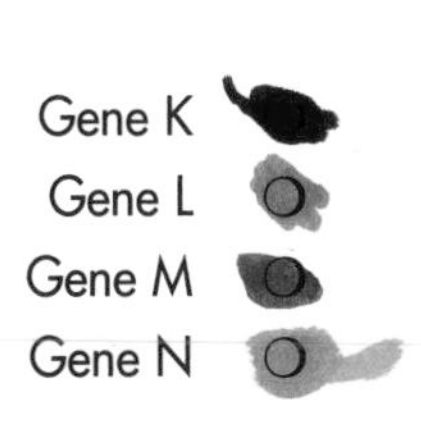

CROSSING OVER

The process of crossing over may cause permanent changes in the genetic makeup of chromosomes. Crossing over occurs during prophase I of meiosis and takes place between paired homologous chromosomes.

In this plate, we will follow a homologous pair of chromosomes and see what happens when crossing over occurs, and when it fails to occur. We will point out that only two types of gametes are possible without crossing over, but four different types of gametes are possible when crossing over does occur, making it an essential source of variability in species.

This plate describes two processes. In the first process, we follow a homologous pair of chromosomes (four sister chromatids) in the absence of crossing over.

In the first process, variability is not introduced into a species because crossing over does not take place. We begin by looking at the pair of chromosomes in diagram 1, at the top. This is a homologous pair, meaning that the DNA has duplicated just before the start of meiosis. The result of this DNA duplication was a pair of sister chromatids for each homologous chromosome. The first homologous chromosomes now consist of **sister chromatids A (A)** and **sister chromatids B (B).** Light colors should be used for them to keep from obscuring the genes.

In this example, we are going to look at two genes on the same chromosome. As we see in the first diagram, sister chromatids A have **gene 1 allele L (C)** and **gene 2 allele R (D).** Both sister chromatids have the same genes and the same alleles because they are copies of one another and are identical. Sister chromatids B are also the same as each other. Both have **gene 1 allele M (E)** and **gene 2 allele S (F);** four contrasting dark colors should be used for these two genes, each of which has two different alleles (1L, 1M, 2R, 2S). We are not concerned with dominance or recession here, so don't worry about how these genes are expressed or inherited.

In diagram 2, the sister chromatids appear as they did in diagram 1. Diagram 2 represents the point in prophase I at which the homologous chromosomes come together and sister chromatids form a tetrad. The chromatids stand side by side, and crossing over has not yet occurred.

In diagram 3, the chromatids have separated during anaphase I, and you can see four **chromosomes (G);** a light color should be used for all of them. Now, in telophase, we note that each chromosome has a copy of gene 1 and a copy of gene 2. In addition, the combination of alleles is the same as what we started with. Two chromosomes have 1L and 2R; these came from sister chromatids A. Two chromosomes have 1M and 2S and these came from sister chromatids B.

We now consider the second part of the diagram, where crossing over takes place. We will see the effects of crossing over—variability in the chromosomes.

Once again, we focus on sister chromatids A (A) and B (B). This diagram is similar to the one above it, but as we move to diagram 2, we see a difference. Here the sister chromatids exist in the tetrad formation, but crossing over occurs. Notice that the second and third chromatids cross one another; an exchange of genes is taking place.

As we move to diagram 3, we see the effects of crossing over. Chromatid 2 in sister chromatids A now has the M allele of gene 1 (E) and its original R allele of gene 2 (D). In sister chromatids B, the third chromatid now has the L allele of gene 1 (C) plus its original S allele of gene 2 (F). Thus the second and third chromatids are different because of crossing over. Only the first and fourth chromosomes remained unchanged.

We now move to the fourth diagram, in which the chromatids have separated during anaphase. Now they exist as chromosomes.

When we examine the genetic composition of the chromosomes, we see a dramatic difference because of the crossing over. Chromosome 1 has gene 1 allele L (C) and gene 2 allele R (D) as expected, but chromosome 2 has gene 1 allele M (E) and gene 2 allele R (D). The third chromosome has gene 1 allele L (C) and gene 2 allele S (F), and the fourth chromosome has gene 1 allele M (E) and gene 2 allele S (F). They are now four different chromosomes, and when these chromosomes are distributed to sperm or egg cells, four different cells can result. Without crossing over, only two different cells could result.

Crossing over is tremendously significant in the evolutionary process. For example, a chromosome might acquire an advantageous allele that joins an already advantageous allele. At fertilization, the offspring might receive this allele pair and have a genetic advantage. Thus, the offspring would be favored by natural selection to reach reproductive age and pass the advantageous alleles to its offspring. This is one of the ways in which crossing over contributes to evolution.

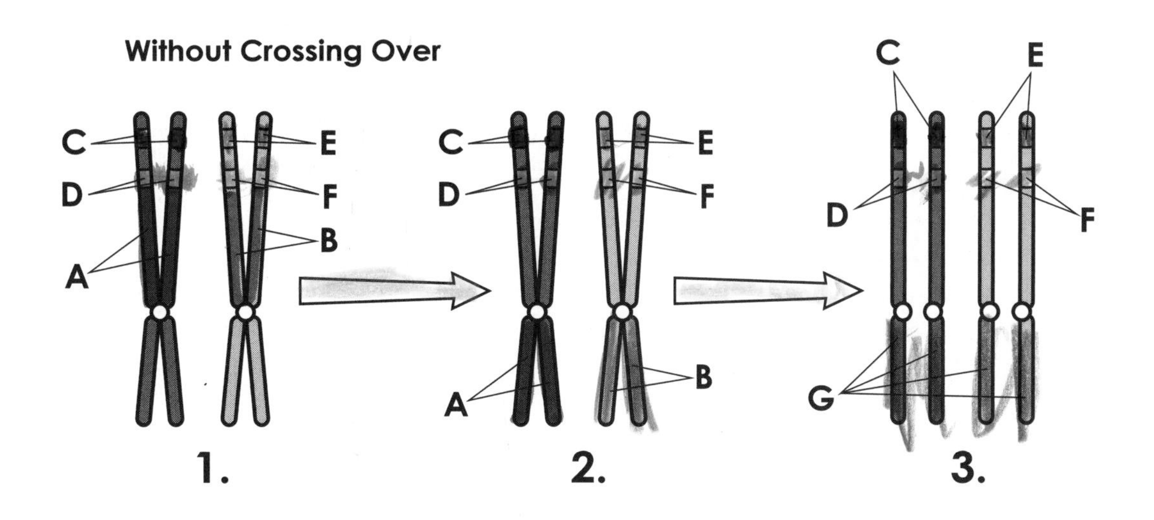

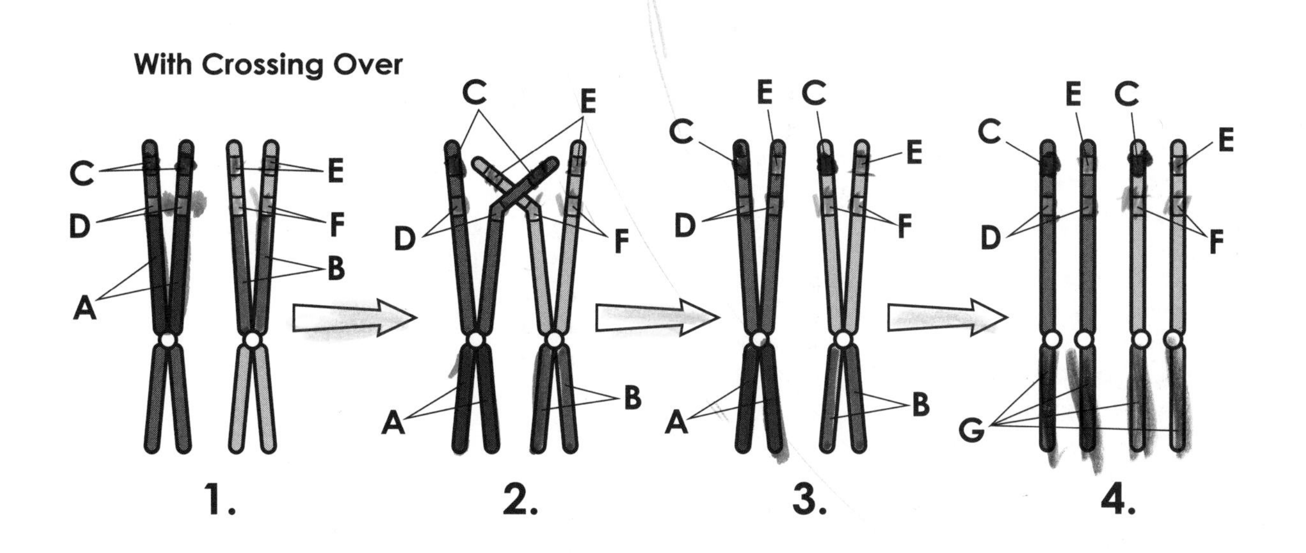

Sister Chromatids A A
Sister Chromatids B B
Gene 1 Allele L C
Gene 2 Allele R D
Gene 1 Allele M E
Gene 2 Allele S F
Chromosomes G

NONDISJUNCTION

Changes in chromosome number may be passed to offspring if the changes occur during the formation of gametes (sex cells). For example, the cells of a human normally contain 46 chromosomes, but it is possible for them to contain 47 or 45. These abnormal chromosome numbers are called aneuploidies, and they can have various effects on the individual. People with Down's Syndrome, for example, have an extra chromosome #21.

Changes in chromosome number can occur as a result of nondisjunction, which is the failure of chromosomes to separate properly during meiosis. The result of nondisjunction is that one gamete contains extra chromosomes, while another contains too few. We will see how nondisjunction occurs in this plate, and what effects it can have on individuals.

We will follow the activity of two pairs of homologous chromosomes and see what happens when they fail to separate properly. Start by taking a look at the upper part of the plate.

Nondisjunction is one of the most common errors that occurs during meiosis; it can occur in meiosis I if homologous chromosomes fail to separate, and during meiosis II if chromatids fail to separate properly and travel to the same daughter cell.

The first part of the plate, entitled Process of Nondisjunction, shows a simplified version of the process during meiosis II. We first show a **normal cell (A),** which produces sperm cells in the testes. In this cell, we will show only two pairs of chromatids for simplicity, and in this case, the **first chromatids (B)** are larger than the **second chromatids (C).** We recommend that you use contrasting colors for the two pairs of chromatids.

The arrows that lead from the cell represent the process of meiosis II, during which the chromosomes separate and sperm cells are formed. Only two of the four **sperm cells (D)** are shown here. (We are showing only the cell that experiences nondisjunction, and not the other one produced during meiosis I that undergoes normal division to produce regular gametes.) We see that the first chromatid pair (B) separates as normal, and one of its chromatids is found in each sperm cell, but nondisjunction occurs in the second chromatid pair, so that both chromatids pass to one sperm cell. The sperm cell to the left receives no chromatid from the second chromatid pair, and it has a single chromosome from the first.

Having shown how abnormal sperm cells are formed, we will now show what happens if these sperm cells unite with egg cells. It should be remembered that millions of sperm cells are produced during meiosis, so that the sperm cells that underwent nondisjunction may not be involved. Continue your coloring as you consider the defects caused by nondisjunction.

We will now see what happens if these abnormal gametes unite with normal egg cells. In diagram 1 we have an **egg cell (E),** which contains one member of each of two chromatid pairs from the female. You can see the **first egg cell chromatid (F)** and the **second egg cell chromatid (G).** When the normal egg cell unites with a sperm cell (D) that has only a single chromosome (B), the result is a fertilized egg cell with three, instead of four chromatids. This is a **cell with a missing chromosome (H).**

Now let's consider a second situation, shown in diagram 2. The normal egg cell unites with the sperm cell (D) that has one extra chromosome. There is the normal chromosome from the first chromatid pair (B), plus both chromosomes from the second chromatid pair (C). When the union takes place, the fertilized egg cell is a **cell with an extra chromosome (I).**

As we mentioned above, children that have three copies of chromosome 21 develop a disorder called Down's Syndrome, which is characterized by intellectual disabilities and physical abnormalities. Situations in which an extra chromosome is present are known as trisomy.

When nondisjunction occurs in the sex chromosomes, an extra sex chromosome may be present, or one may be lacking. For example, an individual with XXY chromosome combinations has what's call Klinefelter Syndrome. This individual is male, but has enhanced female characteristics. When an individual has only one X, she has Turner Syndrome and is female, but has enhanced male characteristics.

Process of Nondisjunction ☆

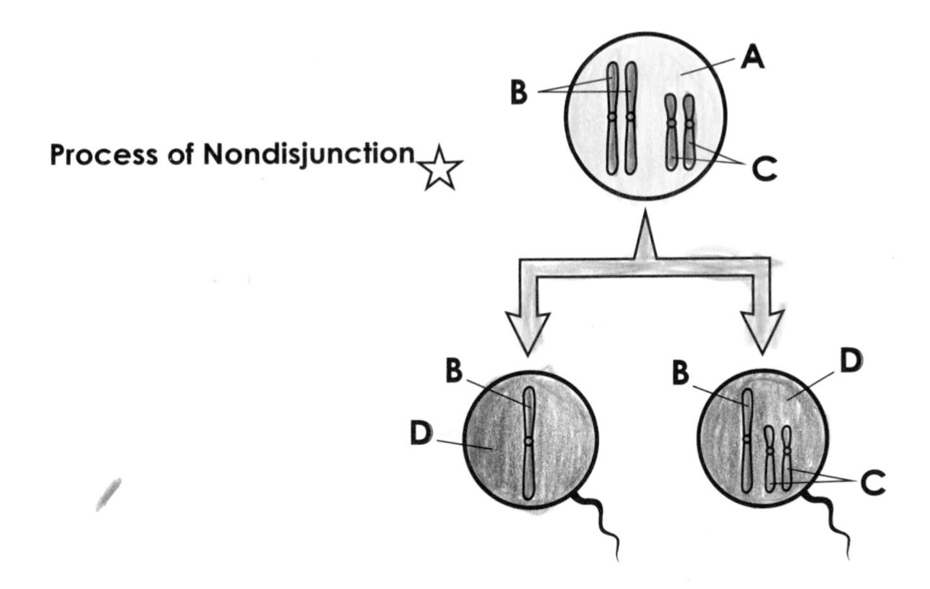

Normal Cell A	Egg Cell E	Cell with Missing Chromosome H
First Chromatids B	First Egg Cell Chromatid F	Cell with Extra Chromosome I
Second Chromatids C	Second Egg Cell Chromatid G	
Sperm Cell D		

Effects of Nondisjunction ☆

1.

2.

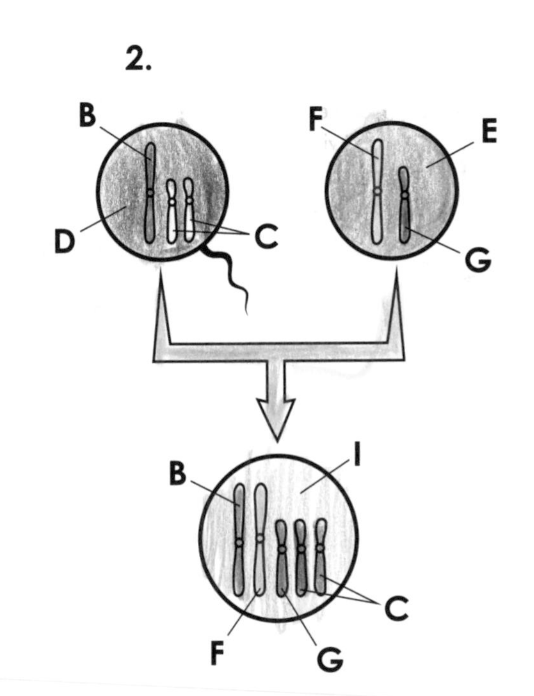

CHAPTER 4:

DNA and GENE EXPRESSION

STRUCTURE OF DNA

Two types of nucleic acids exist: deoxyribonucleic acid (DNA) and ribonucleic acid (RNA). DNA is the genetic material of organisms, while RNA is used during the construction of proteins. This plate will examine the structure of DNA. RNA's structure is studied in a succeeding plate in this chapter.

This plate illustrates the components of a molecule of DNA. Letters have been correlated with the names of some of the components; most textbooks use these letter abbreviations. Light colors such as grays and yellows should be used for the first part of the plate.

DNA exists in the chromosome of the living eukaryotic cell, and in the cytoplasm of prokaryotic cells. DNA is composed of repeating units known as nucleotides. Each nucleotide has three components: a molecule of carbohydrate deoxyribose, a phosphate group, and a nitrogenous base. At the upper portion of the plate, two nucleotides are shown. At the left is a nucleotide composed of a **phosphate group (P),** a **deoxyribose molecule (D),** and a nitrogenous base called **adenine (A).** the three components should be lightly shaded to avoid obscuring their individual atoms.

The deoxyribose molecule contains a five-carbon carbohydrate ring bound to the phosphate group at its $-CH_2$ group. On its opposite side, the deoxyribose molecule is bonded to the adenine molecule. The adenine contains five nitrogen atoms, which is why it is called a nitrogenous base.

A second nucleotide is shown at the right. It consists of a nitrogenous base called **thymine (T)**, bonded to a deoxyribose molecule (D), which is inverted here. The deoxyribose is in turn bonded to a phosphate group (P). As before, light shading should be used to denote the three portions of the nucleotide.

Adenine and thymine nucleotides are held to one another by two **hydrogen bonds (H),** one of which is indicated by an arrow, which should be colored boldly. Hydrogen bonds are weak chemical bonds formed between hydrogen and nearby electronegative atoms. In DNA, two hydrogen bonds exist between A and T, and three exist between G and C.

We will now examine how the nucleotides bind to one another to form DNA. Continue your coloring as you read, and use the same colors in the DNA molecule that you used for the nucleotides.

The four nitrogenous bases that make up DNA are thymine, adenine, **cytosine (C),** and **guanine (G).** The bases G and A are derived from a precursor called purine, so they are referred to as the purines. In contrast, C, T, and U (which you will learn about later) are the pyrimidines. Let's take a look at the DNA double helix.

Begin at the top of the molecule, and note that the first nucleotide contains adenine (A), and that it is attached to deoxyribose (D). The deoxyribose is connected to a phosphate group (P), which in turn is connected to another deoxyribose molecule. The latter is connected to a cytosine molecule (C), as well as another phosphate group (P). A deoxyribose molecule (D) follows, which is connected to an adenine (A). This pattern continues with alternating deoxyribose molecules and phosphate groups as the ribbon-like strand continues and curves. Each deoxyribose molecule is connected to one of the four nitrogenous bases.

Now move to the right side of the molecule and follow the ribbon, beginning at the upper right. As you follow it, note that it contains deoxyribose molecules that alternate with phosphate groups, and that again, connected to each deoxyribose molecule is one of the four nitrogenous bases. The second strand of DNA is very similar to the first strand.

We will complete the plate by noting how the two strands of DNA unite to form the double-stranded DNA molecule. If you have not yet completed your coloring of all the parts of the two strands, do so at this point. Then read below.

In the complete DNA molecule, two single strands oppose one another in a ladder-like arrangement, in which the nitrogenous bases line up opposite one another according to the principle of complementary base pairing. Adenine always lines up opposite thymine, and cytosine always lines up opposite guanine. As we mentioned earlier, hydrogen bonds then hold the bases together. The nitrogenous bases thus form rungs of a ladder.

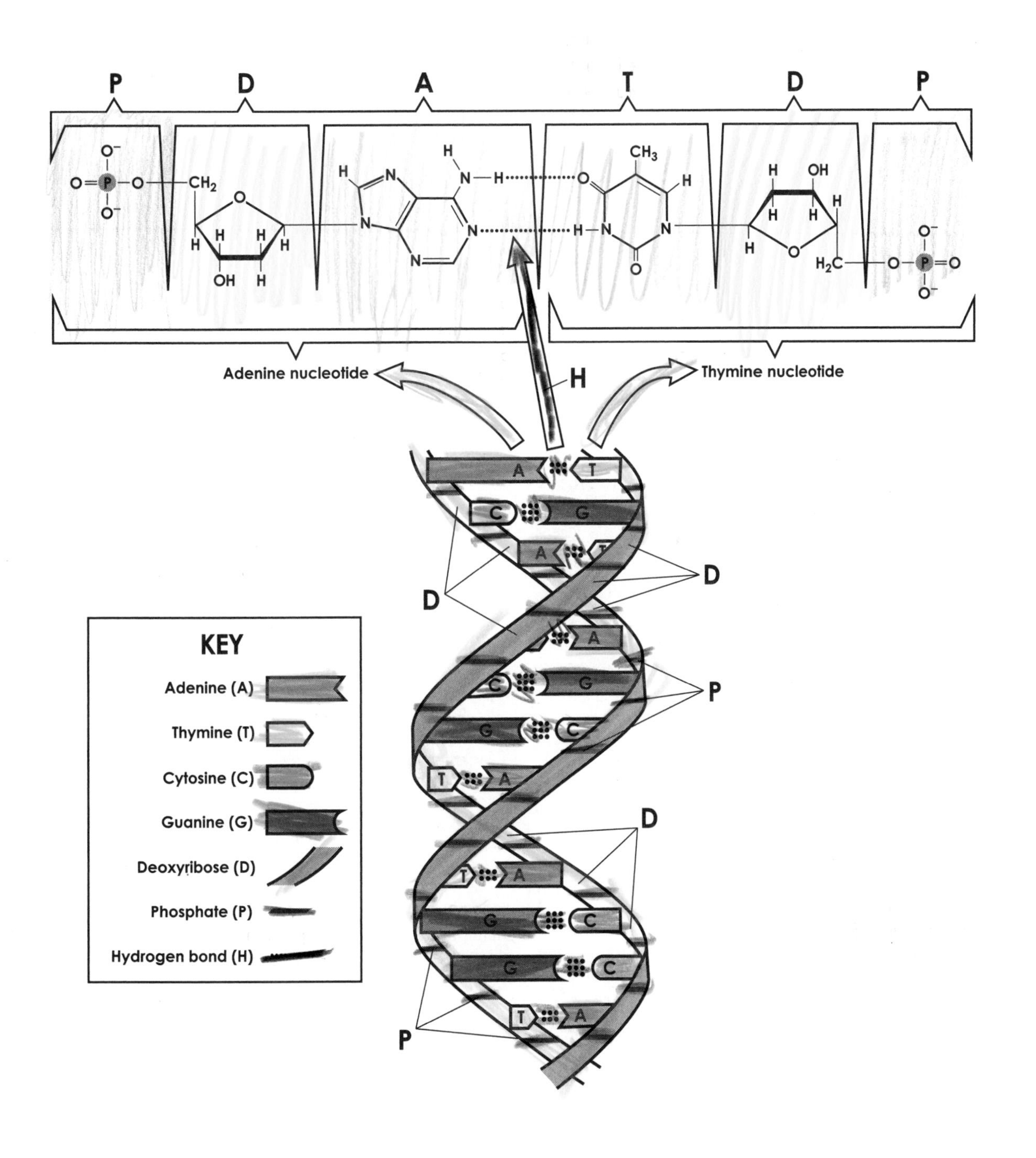

Phosphate Group P
Deoxyribose D
Adenine A

Thymine T
Hydrogen Bond H

Cytosine C
Guanine G

REPLICATION OF DNA

The result of the cell cycle is the division of the mother cell into two daughter cells. This cell division takes place in almost all types of plant, animal, and microbial cells. Before a cell divides, the DNA in its nucleus replicates to ensure that identical copies of its genes are passed to each of the daughter cells. This replication occurs during the S phase of the cell cycle and has already been completed when mitosis begins. This plate explores general mechanisms of DNA replication.

> As you look over the plate, note that we are presenting a single illustration of a double-stranded DNA molecule undergoing replication. Many of the colors that were used in the previous plate should be used here. As you read the paragraphs below, color the appropriate structures in the plate.

The DNA molecule is a double helix composed of two strands of DNA. Each strand is made up of alternating deoxyribose molecules and phosphate groups. Forming the rungs on the DNA ladder are the four nitrogenous bases, which are connected to the deoxyribose backbone. In the plate, you should use a medium color on the **old deoxyribose-phosphate backbone (O):** There are two of these backbones, and both should be colored.

After you color the deoxyribose-phosphate backbones of the old DNA molecule, you should select four different colors with which to color the four different nitrogenous bases of the old DNA molecule. These bases are **adenine (A), thymine (T), cytosine (C),** and **guanine (G).**

> We will now begin the construction of the two new strands of DNA. As you read about the process of their construction, color the appropriate portions of the plate.

The replication process begins with an uncoiling of the original double-stranded DNA molecule. A specific enzyme untwists the double helix and separates the DNA molecule into its two complementary strands. This forms a **replication fork (D),** and an arrow points to it.

Each strand of the DNA molecule now serves as a model, or template, for the construction of a complementary DNA molecule. To construct this complementary DNA molecule, **new deoxyribose-phosphate molecules (N)** will be needed, as well as new base pairs. On the left, notice that the new backbone is being synthesized as the molecule untwists, while on the right side, it is being constructed in the opposite direction, with new molecules being added in fragments, starting at the bottom. The strand on the left is therefore called the leading strand, and the one on the right side is the lagging strand.

> We will complete the construction of the two new DNA molecules by mentioning the principle of complementary base pairing. Continue using the colors for the nitrogenous bases that you used above.

Once the deoxyribose-phosphate backbones have been constructed, the nucleotides will join with one another so that base pairing takes place in a specific pattern. Thymine (T) molecules will always pair with adenine (A) molecules, and cytosine (C) molecules will always pair with guanine (G) molecules. Thus, the order of nitrogenous bases on the original template strand determines the order of the bases on the new strands. For example, we see on the left side the base sequence: C-G-T-T-A-G-A-G-G-T. These code for a new strand with the base sequence: G-C-A-A-T-C-T-C-C-A. As you can see, this is what ensures that each new DNA molecule is identical to its parent strand.

Hydrogen bonding holds the bases of the two strands together, and a double helix forms, so that each new DNA double helix consists of an old strand and a new strand. This method of DNA replication is referred to as semiconservative replication.

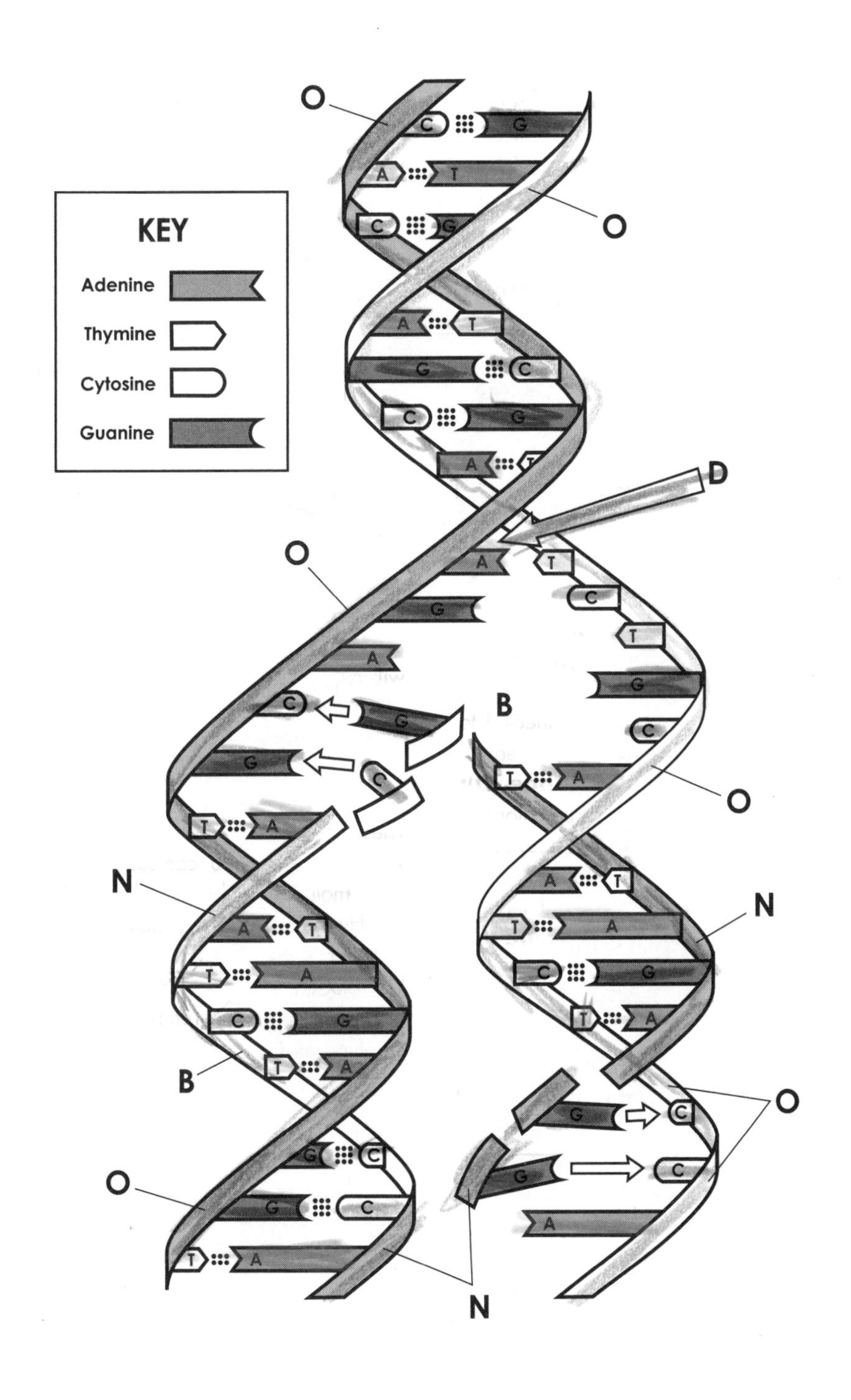

Deoxyribose-Phosphate Backbone (old) O
Deoxyribose-Phosphate Backbone (New) N
Adenine A
Thymine T
Cytosine C
Guanine G
Replication Fork D

PROKARYOTIC DNA REPLICATION

As we explained in the last plate, the replication of DNA involves the unwinding of the parent strands and complementary base pairing between the two new strands so that each new DNA molecule contains one old and one new strand. This is the semiconservative model of DNA replication.

The process of DNA replication differs in prokaryotic and eukaryotic cells. Eukaryotic DNA replication will be explained in a succeeding plate, and this plate explains prokaryote DNA replication.

This plate contains four views of a typical prokaryote bacterium, in which DNA is undergoing replication. The first two views are at the top of the page; the second two are at the bottom.

Unlike the DNA of eukaryotic cells, the genetic material of bacteria exists as a single circular molecule of DNA.

The bacterium displayed is a relatively simple cell. The **bacterial cell wall (A)** lies outside the **bacterial cell membrane (B),** and it possesses several **flagella (C).** Dark colors can be used for these structures. The **cytoplasm (D)** should be colored in a light color.

The single circular DNA molecule is within the cytoplasm of the bacterial cell. You should use a light color to highlight this molecule. The nitrogenous bases are the short lines that link the outer and inner strands. The entire circular strand of **original DNA (E)** is shown.

The replication of DNA in the prokaryotic chromosome begins at a point called the **origin of replication (O),** which is indicated by an arrow. A bold color should be used to color the arrow. DNA replication begins when an enzyme breaks the hydrogen bonds between the two strands of DNA at the origin of replication; this break establishes the replication fork.

At the origin of replication, the paired bases are separated and the strands of the double helix begin to separate. Their unwinding is facilitated by an enzyme that is part of a replication complex. Next, an **enzyme (F)** called DNA polymerase begins the synthesis of new DNA molecules. It does so by adding nucleotides to one another.

In diagram 2, the production of a new DNA strand begins. We see the **original DNA (E)** as well as some of the **new DNA (G).** The enzyme (F) is synthesizing new DNA at the point indicated as it proceeds around the original strand of DNA. You should use different colors for the original DNA (E) and the new DNA (G).

Prokaryotes have only one circular chromosome, and this one chromosome has only one origin of replication. Because of this, as replication proceeds, the partially duplicated genome begins to look like the Greek letter θ (theta). You can see this in diagram 2. Hence the replication of prokaryotes is referred to as theta replication.

Continue your reading below as the replication process continues in diagram 3.

We now focus on diagram 3. Here the replication proceeds and we see the original DNA (E) as well as much of the **new DNA (G).** The replication forks have been traveling in opposite directions and are close to meeting at opposite sides of the circle, and the new strands of DNA appear as ever-enlarging double loops.

In diagram 4, the loops have separated. Each new DNA molecule is double stranded, and each consists of one strand of old DNA (E) and one of new DNA (G). Once the two DNA molecules have formed, the cell is ready to undergo binary fission and split in two. The single chromosome is attached to the plasma membrane of the bacteria, and after replication has taken place, the two copies separate from one another just before binary fission takes place. The plate entitled Bacteria discusses this multiplication process.

The process of prokaryotic DNA replication also occurs in the cytoplasm, mitochondria, and chloroplasts of eukaryotic cells. This provides evidence that bacteria were the source of mitochondria chloroplasts in eukaryotic cells, as you will see in the plate entitled The FIrst Eukaryotic Cells.

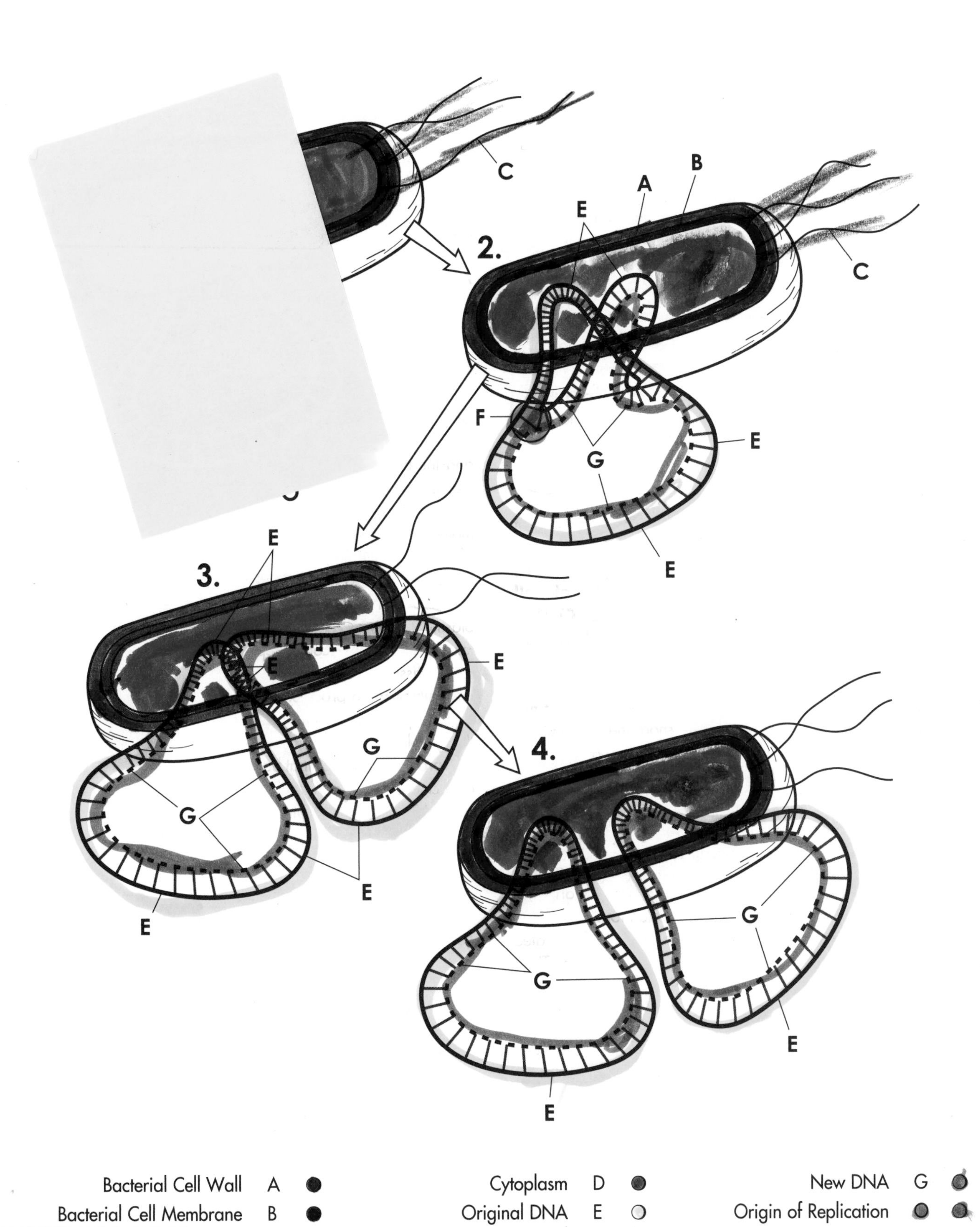

Bacterial Cell Wall A
Bacterial Cell Membrane B
Flagellum C
Cytoplasm D
Original DNA E
Enzyme F
New DNA G
Origin of Replication

EUKARYOTIC DNA REPLICATION

The volume of DNA in eukaryotic cells is immense. For example, the human genome is made of DNA, contains 3.2 billion base pairs, and codes for about 21,000 genes. Because there is such a huge amount of DNA, replication in eukaryotic cells would take an extremely long time if it occurred in the same way as it does in prokaryotic cells. But there are processes that are unique to eukaryotic cells, and we will study them in this plate.

This plate displays a single strand of DNA in a eukaryotic cell. We point out various areas of interest along the long DNA molecule to help situate you as you learn about eukaryotic DNA replication.

In this plate, we show a simplified eukaryotic cell. It is enclosed by a **eukaryotic cell membrane (A),** and the main portion of the cell is the **eukaryotic cell cytoplasm (B).** A major organelle within the cytoplasm is the **nucleus (C).**

Within its nucleus, the eukaryotic cell contains a number of chromosomes, which are made up of DNA. In all human cells except the sex cells and red blood cells, there are 46 chromosomes. Approximately three billion base pairs are in all the 46 chromosomes. In the plate, a single chromosome has been stretched out of the cell for study; this **chromosome (D)** represents any of the 46 human chromosomes. As the chromosome extends from the nucleus, we see the **original DNA (E).** We will study this DNA in detail in a moment.

We have begun work in the plate by pulling out a single DNA molecule for study. Our purpose is to explain unique situations that occur in the course of the replication of eukaryotic DNA. As you proceed, you should color carefully to distinguish the features of this diagram.

In the replication of DNA in bacterial cells (studied in a previous plate), a single origin of replication develops in the chromosome. But if a eukaryotic DNA molecule were to replicate with a single origin, an immense amount of time would be necessary for replication. For this reason, several **origins of replication** exist in the eukaryotic DNA molecule. One such origin, designated by an arrow, is shown in the plate **(O).**

In numerous places along the DNA molecule, we see bubbles. In these areas, **enzymes (F)** have unzipped the strands of DNA at the origin of replication and have begun to transcribe in opposite directions on each side of the DNA molecule. The enzymes are seen at the left and right side of each of these bubbles. A pale or gray color should be used to color them.

An important part of the replication enzyme complex is the enzyme DNA polymerase. As DNA polymerases proceed in opposite directions from the origin of replication, they synthesize two new strands of DNA, and a number of **nucleotides (H)** are used in this synthesis. Therefore, we see in the plate that **new DNA (G)** is forming within the confines of the bubbles. As we go from the first to the fourth largest bubble, we see more details of the new DNA. Nucleotides continue to be added to the growing DNA molecule.

The process will continue until all the replication forks meet and the replication process is complete. We can see that, at the end of the DNA molecule, for example, two **DNA double helices (I)** have emerged. The bracket that surrounds these double helices should be colored in a dark color. Eventually the two helices will continue to separate until they reach the first large bubble.

In eukaryotic DNA, hundreds of origins are formed along the length of a DNA molecule. Multiple replication complexes work more slowly than those used in prokaryotic cells. For example, bacterial replication proceeds at a rate of about one million base pairs per minute, but eukaryotic cells have a replication rate that ranges from 500 to 5,000 base pairs per minute. Since the newly synthesized DNA strands grow in bidirectional patterns, the replication of eukaryotic cells averages a few hours. Once DNA replication is complete, the eukaryotic cell is ready to begin the process of mitosis.

EUKARYOTIC DNA REPLICATION

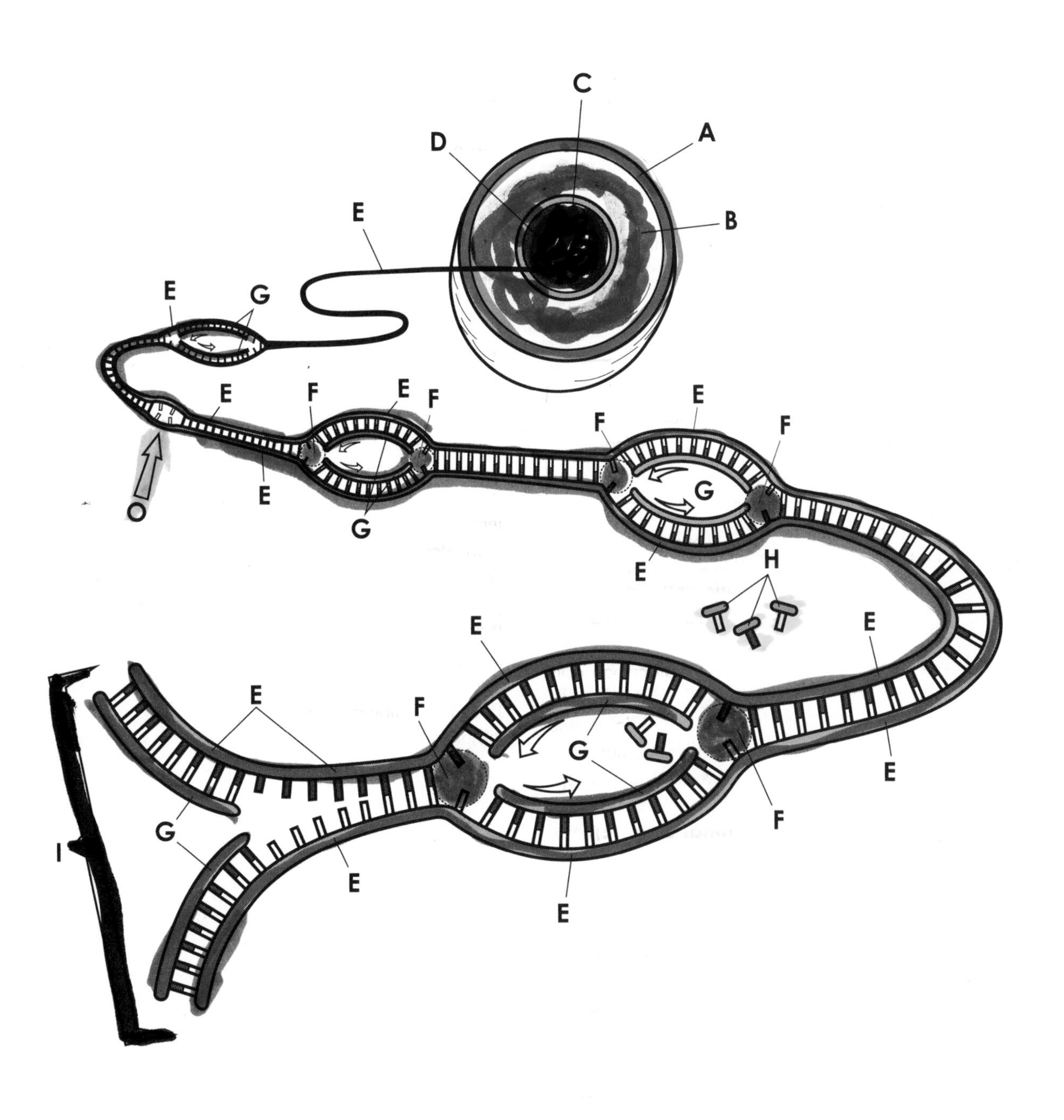

Eukaryotic Cell Membrane A
Eukaryotic Cell Cytoplasm B
Nucleus C
Chromosome D

Original DNA E
Enzyme F
New DNA G

Origin of Replication O
Nucleotides H
DNA Double Helices I

MOLECULAR MECHANISMS OF DNA REPLICATION

In the last few plates, you learned about general mechanisms of DNA replication, and how both prokaryotic and eukaryotic cells perform this important process. This lesson builds on this knowledge, and will show you the molecular mechanisms of DNA replication. Previous plates hinted at some of these concepts and enzymes, but here we will explore the details.

When it is not being replicated, DNA is tightly coiled. You will learn more about DNA packing on the next plate. Replication cannot begin unless the double helix is unwound and separated into two single strands. To do this, the **origin recognition complex (A_1)** scans along the chromosome (like a train on a track) until it finds an origin of replication **(A_2),** which is a specific sequence of nucleotides in the DNA. Once it finds one, the complex calls in many other proteins. First comes an enzyme called **helicase (B).** Helicase uses the energy of ATP hydrolysis to break hydrogen bonds holding the two template **DNA strands (C)** together. This unwinds the double-stranded DNA (starting at the origin of replication) and forms a **replication bubble (D).** Each replication bubble has two **replication forks (E)**—one on either side. As DNA replication proceeds, helicases widen the replication bubble and the forks move apart.

When helicase unwinds the DNA helix at the origin of replication, the helix just outside the bubble gets wound more tightly. Enzymes called **topoisomerases (F)** cut one or both of the DNA strands to release this extra tension. Another potential problem is that single-stranded DNA is much less stable than double-stranded DNA. **Single-strand binding proteins (G)** protect single-stranded DNA and keeps it separated. Replication may now begin.

Several **RNA primers (H_1)** must be synthesized for each template strand. This is accomplished by an RNA polymerase called **primase (I).** Primer synthesis is important because the next enzyme, DNA polymerase, cannot start a new DNA chain. It can only add nucleotides to an existing nucleotide chain. The RNA primer is usually 8–12 nucleotides long, and is later replaced by DNA.

Daughter DNA is created as a growing polymer. **DNA polymerase (J)** catalyzes the elongation of the daughter strand using the **parental template DNA (C),** and elongates the primer by adding nucleotides (or dNTPs) to its 3′ end. This means the new strand of DNA is built from 5′ to 3′, and the template DNA is read from 3′ to 5′. DNA polymerase can also proofread its work and correct any base pairing mistakes it may make. Rapid elongation of the daughter strands follows. Since the two template strands are antiparallel, the two primers will elongate toward opposite ends of the replication bubble, away from the origin of replication. These new DNA strands are called **leading strands (K),** because their synthesis leads into the replication forks. In this plate we have drawn the new DNA strands schematically, but remember that this is a specific and tightly regulated process of base pairing. DNA polymerase reads the DNA code of the template DNA and makes a new DNA strand by adding complementary base pairs: Adenine pairs with thymine via two hydrogen bonds, and cytosine pairs with guanine via three hydrogen bonds.

You can see that right now you've colored only half the DNA that needs to be replicated. We have yet to talk about the top right and bottom left parts of the replication bubble. These sections are made differently because their synthesis doesn't lead into the replication fork, but away from it. Remember, DNA can be synthesized only from 5′ to 3′! To replicate these sections, a **primer (H_2)** is laid down and DNA polymerase builds a short section of new DNA called a **lagging strand (L).** DNA is thus replicated in small fragments called **Okazaki fragments (M).** As the replication bubble widens, more template is made available for new Okazaki fragments to be made. Eventually, DNA polymerase replaces RNA primers with DNA, and **DNA ligase (N)** connects Okazaki fragments together, and also connects the leading and lagging strands. This completes DNA replication.

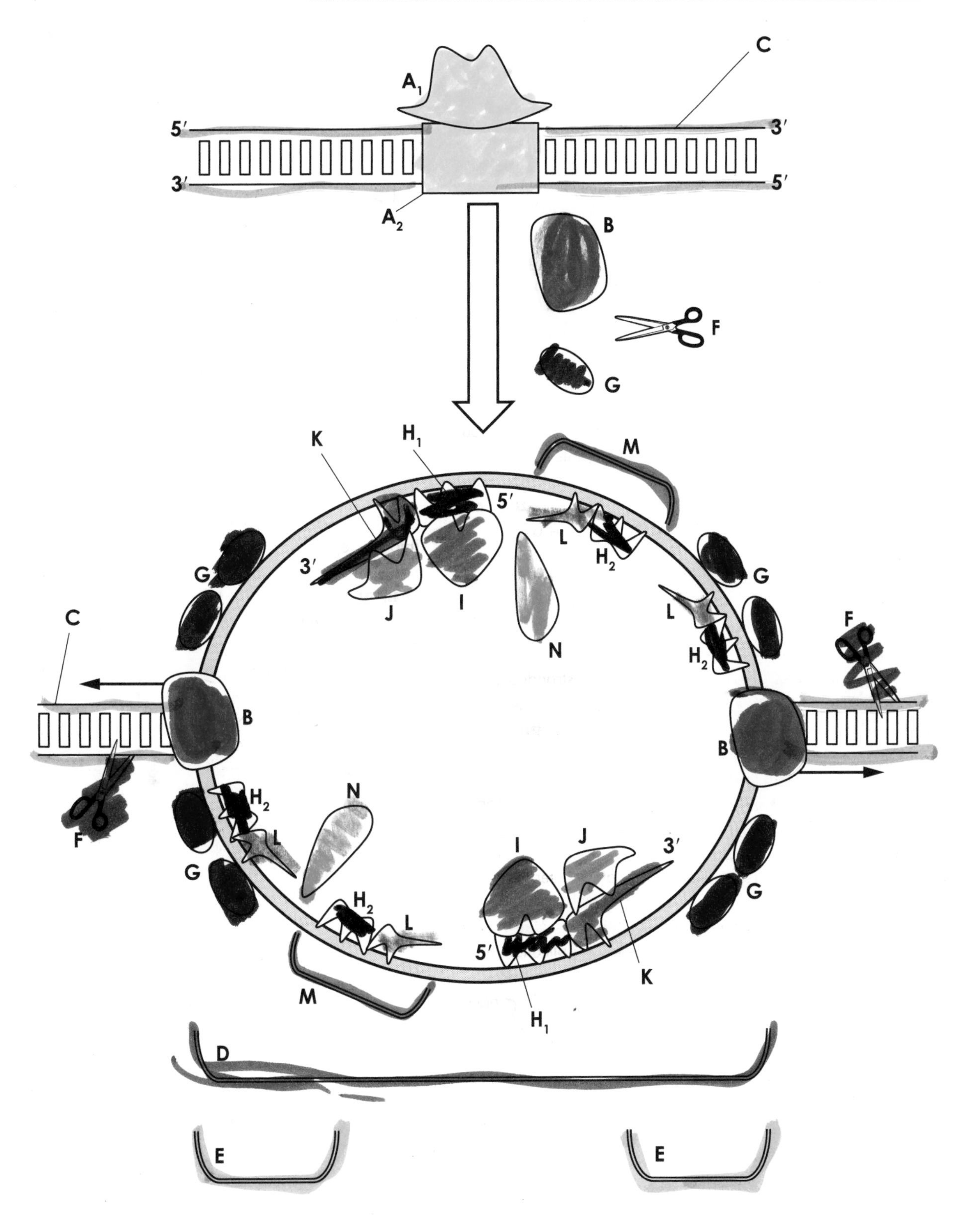
C
A1
5′
3′
3′
5′
A2
B
F
G
K
H1
M
5′
L
H2
3′
G
G
J
I
N
C
L
F
H2
B
B
H2
N
F
L
J
3′
G
I
G
H2
L
5′
K
M
H1
D
E
E

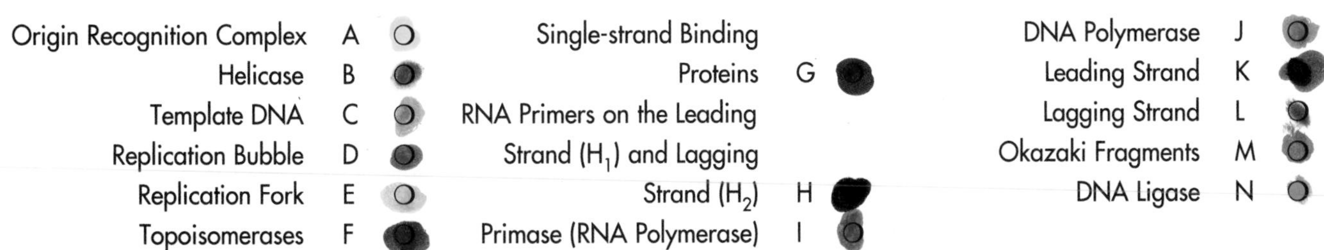
Origin Recognition Complex A
Helicase B
Template DNA C
Replication Bubble D
Replication Fork E
Topoisomerases F
Single-strand Binding Proteins G
RNA Primers on the Leading Strand (H1) and Lagging Strand (H2) H
Primase (RNA Polymerase) I
DNA Polymerase J
Leading Strand K
Lagging Strand L
Okazaki Fragments M
DNA Ligase N

DNA AND CHROMOSOMES

It is important to understand how DNA is packaged into genes and chromosomes since it will help you understand the cycles of condensation and unraveling that occur during mitosis and cell division. It is also important to understand how DNA is arranged into chromosomes because the spatial arrangement of DNA influences gene expression.

In addition, understanding the way in which DNA is coiled into chromosomes will help you understand how more than two meters (about six feet) of DNA fits into 46 chromosomes in a nucleus that's less than five micrometers in diameter. This plate explores the current model for chromosome organization in eukaryotic cells and shows how DNA is organized with protein in chromosomes.

Starting at the bottom of the plate, notice the molecule of DNA. The double-stranded molecule will become progressively more coiled and eventually forms the chromosome. We will describe that coiling as we proceed in the plate.

Electron microscopic studies and biochemical research have helped biologists understand how DNA associates with protein to form chromosomes. This packaging enables DNA to direct protein synthesis and replicate, but also prevents it from tangling in the process of mitosis.

Again note the double-stranded molecule of DNA. The **first sugar-phosphate backbone (D)** is at the top, and the **second sugar-phosphate backbone (B)** is at the bottom. These sugar-phosphate backbones should be colored different colors and traced in the DNA molecule until they are no longer visible in the diagram.

Associated with the sugar-phosphate backbones are the four nitrogenous bases of DNA. They are **adenine (A), thymine (T), cytosine (C),** and **guanine (G).** Four bold colors should be used to distinguish these **nitrogenous bases (E).** As you can see, they are initially distinct, but as the DNA molecule condenses, they are more difficult to pinpoint and color, so you might want to trace them.

We have reviewed the structure of the DNA molecule, and we now move on to show its association with histone proteins, the result of which is the nucleosome. Continue your reading below as you color.

In eukaryotic cells, chromosomes are associated with proteins known as histones. **Histone proteins (I)** occur in clusters of eight molecules, as the plate shows. These histones are small, basic proteins that facilitate DNA packaging. The eight histone proteins are shown distinctly at first; then for simplicity we show them collectively as **condensed histones (J).**

Note in the diagram that **two loops of DNA (H)** surround each histone. The product of this looping is a unit called a nucleosome. Several **nucleosomes (K)** are outlined by a bracket, which should be colored in a dark color. The nucleosome is the fundamental packing unit for DNA. You should try to color several of the condensed histones and their double loops of DNA.

Having studied the fundamental packing unit of DNA in the chromosome, we now examine how the nucleosomes combine with one another.

A particular type of protein locks the nucleosomes together so that DNA cannot unwind from its histone core; the nucleosomes remain strung together like beads on a necklace. The winding of DNA around the histones shortens its length considerably, but the strand must be further shortened if the chromosome is to fit in the cell's nucleus. This is accomplished when the nucleosomes are further packed into thick **coiled fibers (L).** A light color is recommended to avoid obscuring these coils.

The coiling of nucleosomes into coiled fibers produces thicker fibers that contain even more compact DNA. These fibers are collectively known as **chromatin (M).** During interphase and early prophase, the DNA of a cell exists as these ultramicroscopic fibers; during late prophase and metaphase, the chromatin condenses even further. This final compacting produces the traditional **chromosome (N).** The diagram shows two chromosomes joined at the centromere just before separation, during anaphase. You can see the **nucleus (O),** which indicates that it is a **eukaryotic cell (P).** (Few of the details of the cell are shown here since we are concentrating on the DNA and the chromosome.)

In bacterial cells, DNA exists without protein wrapping. Prokaryotes have a distinctive mechanism for making their single circular chromosome more compact and sturdy. An enzyme called DNA gyrase uses the energy of ATP to twist the gigantic circular molecule. Gyrase functions by breaking the DNA and twisting the two sides of the circle around each other. The resulting structure is a twisted circle that is composed of double-stranded DNA. As discussed earlier, the two strands are already coiled, forming a helix. The twists created by DNA gyrase are called supercoils, since they are coils of a structure that is already coiled.

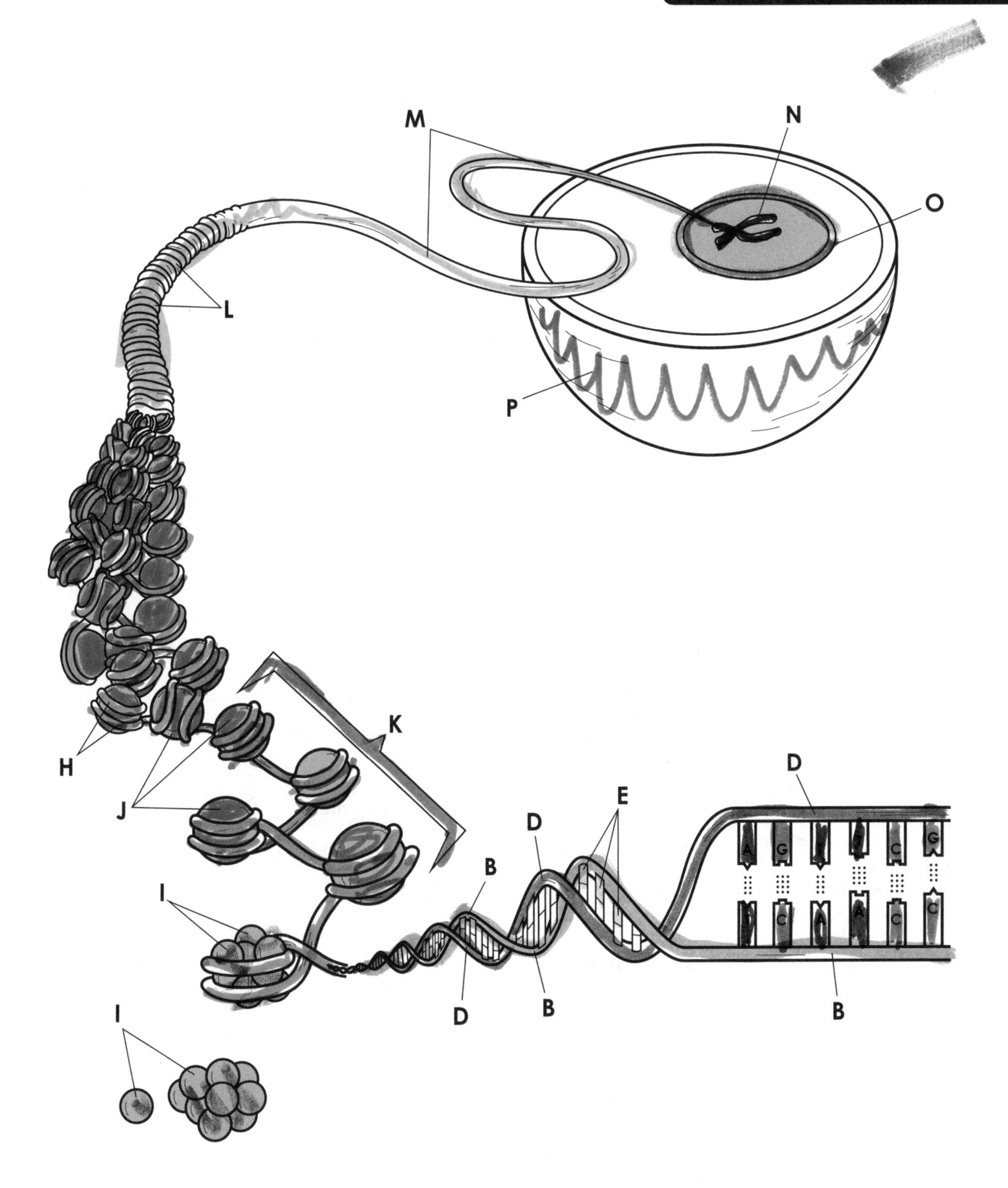

First Sugar-Phosphate Backbone	D	Cytosine	C
Second Sugar-Phosphate Backbone	B	Guanine	G
Adenine	A	Nitrogenous Bases	E
Thymine	T	Two Loops of DNA	H
		Histone Proteins	I
		Condensed Histones	J

Nucleosomes	K
Coiled Fibers	L
Chromatin	M
Chromosome	N
Nucleus	O
Eukaryotic Cell	P

DNA AND TRANSFORMATION

Modern biologists know that deoxyribonucleic acid (DNA) is inherited through generations, and that it is the carrier of genetic information. Over the decades, a wealth of evidence has accumulated attesting to this fact, but the relationship between DNA and gene expression was not recognized until the 1940s and 1950s. Prior to that time, scientists were uncertain where the processes of genetics took place, or what cellular components were involved. One of the first to draw conclusions about the role of DNA in heredity was Frederick Griffith, in 1928.

This plate contains diagrams detailing four experiments that were performed by Frederick Griffith in 1928. As a result of these experiments, Griffith managed to transform bacterial cells, converting them from nonpathogenic (non-disease causing) to pathogenic (disease causing) forms.

Frederick Griffith was a British bacteriologist who worked with the bacteria that cause pneumonia. These bacteria, which are called *Streptococcus pneumoniae,* are composed of a chain of cocci that are spherical. For now we will refer to this organism simply as "bacteria."

Diagram 1 shows one of the experiments performed by Griffith. (You may wish to use bold colors on this diagram.) The **bacteria (A)** are enclosed in **capsules (B).** These bacteria are pathogenic, meaning that they are disease causing. When the **pathogenic bacteria (C)** are placed in a syringe and injected into an animal, the animal gets sick. Note the **dead mouse (D).**

The second diagram shows another experiment performed by Griffith. Here we note that the bacteria (A) are not contained within capsules. These bacteria are harmless, or nonpathogenic. When a syringe is filled with these **nonpathogenic bacteria (E)** and the contents are injected into an animal, the animal lives. Note the **healthy mouse (F).** as you can see, the bacteria's capsule is the key to its pathogenicity; in its absence the bacteria are harmless.

We now know that the presence of a capsule is the difference between pathogenic bacteria and nonpathogenic bacteria. We now continue with Griffith's experiments as we concentrate on diagram 3.

The third experiment performed by Griffith is shown in diagram 3. Here the bacteria were killed by excessive heating. These **heat-killed bacteria (G)** are encapsulated, and a different color should be used for them. When the dead bacteria are injected into the animal, the animal lives. A **healthy mouse (F)** is a result of the third experiment.

We now focus on Griffith's most important experiment in which the phenomenon of transformation occurred. Focus on diagram 4 as you read below.

Griffith's key experiment in transformation was performed in the following way. He took live, unencapsulated bacteria (A) and mixed them with heat-killed dead bacteria (G) that were capsulated (B). This **bacterial mixture (H)** of live, unencapsulated and dead encapsulated bacteria should have been harmless to the mouse; this mixture was unlike the bacteria used in the previous three experiments. However, when a syringe full of the mixture was injected into a mouse, the animal died (D)!

Griffith questioned how a mixture of two kinds of nonpathogenic bacteria could kill the mouse. He obtained tissue from the mouse and examined it under his microscope. His observations were startling; he saw live bacteria (A) that were enclosed in capsules (B)! The mouse had died because it was attacked by live, pathogenic bacteria.

Griffith concluded that the live unencapsulated bacteria (A) were transformed into live, encapsulated bacteria through interaction with the dead encapsulated bacteria. Griffith hypothesized that the substance responsible for this transformation of genetic material was a protein, and called it the "transforming principle."

Griffith's work showed that bacteria could be transformed, but the method of the transformation was still unknown. Many years would pass before Oswald Avery and his group would identify the transforming substance as deoxyribonucleic acid (DNA). This occurred in 1944.

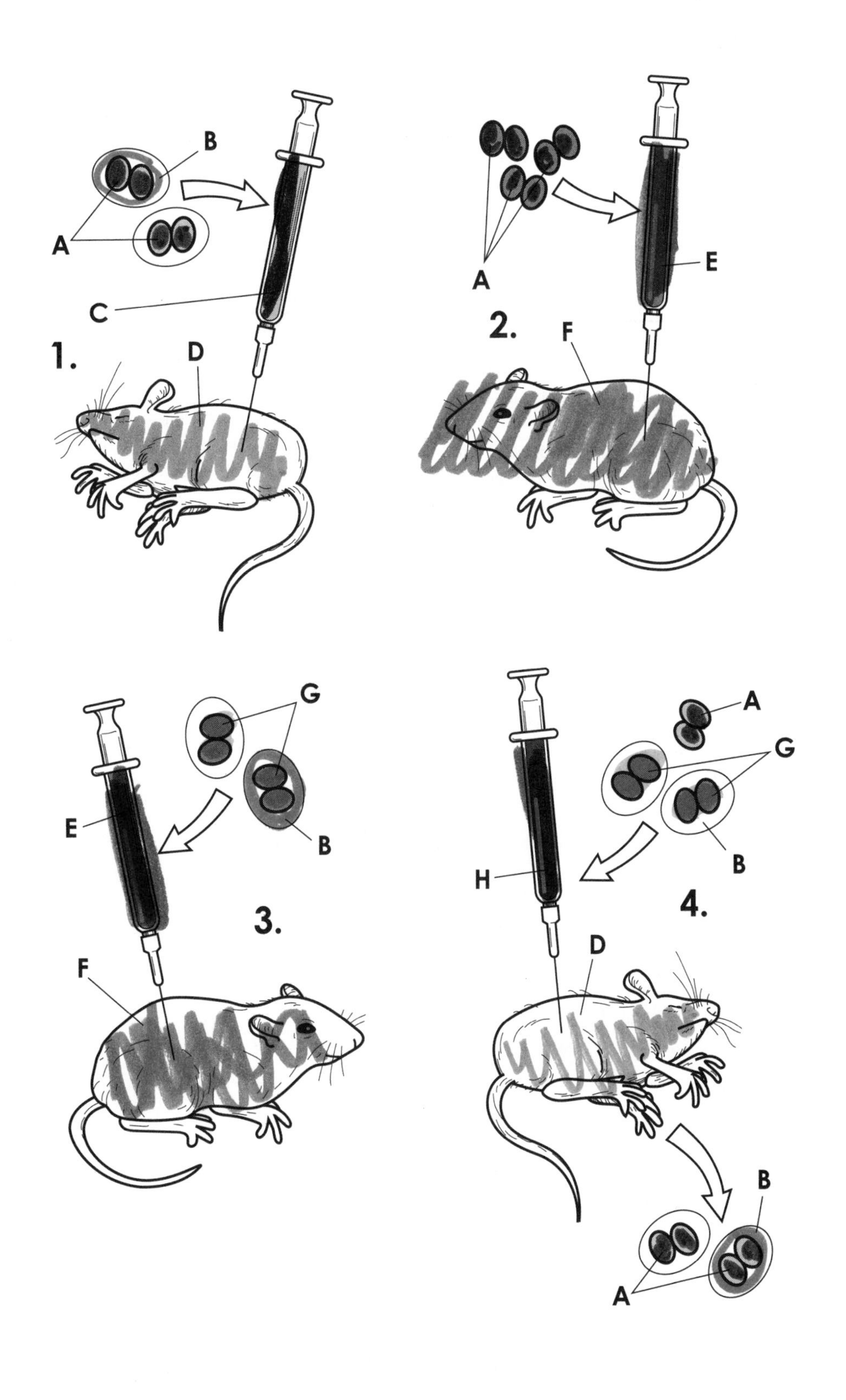

Bacteria A
Capsule B
Pathogenic Bacteria C
Dead Mouse D
Nonpathogenic Bacteria E
Healthy Mouse F
Heat-killed Bacteria G
Bacteria Mixture H

DNA AND PHENOTYPE

After the announcement of the structure of DNA by Watson and Crick, scientists set out to confirm that DNA is the basis for heredity, the source of genetic information in the cell. Molecules of deoxyribonucleic acid were identified as the material of which the gene is made, but no one knew how DNA transmitted genetic information.

After decades of careful research, scientists concluded that genes express themselves (through phenotype) by controlling the rate and amount of production of polypeptides. Because genes control polypeptide production, they also control the amounts and identities of proteins that are produced in a cell, which in turn decide its metabolic activity.

In this plate, we examine the relationship between DNA and phenotype, or expression of the gene, and see how proteins are intermediaries in the process of gene expression.

This plate displays the process by which the information contained in genes is expressed in an organism's phenotype.

The fact that genes direct the synthesis of proteins, and specifically enzymes, was first realized in the 1940s. As you may recall, enzymes are proteins that catalyze chemical reactions while they themselves remain unchanged, and in this plate we will see how the actions of a particular gene affect the production of an enzyme, which in turn determines the phenotype of a plant.

The process begins with a strand of **DNA (D),** which is enclosed by a bracket in our drawing that you should color in a bold color such as purple or blue. Color its **deoxyribose-phosphate backbone (B);** then color the four nitrogenous bases: **adenine (A), thymine (T), guanine (G),** and **cytosine (C).**

DNA expresses itself by serving as a template for the production of a molecule of **messenger RNA (H),** also known as mRNA. As you may recall, this process is called transcription. Messenger RNA is made up of a **ribose-phosphate backbone (E)** and linked nucleotide bases. The bases are similar to those in DNA, except that **uracil (U)** is present instead of thymine. Color the four types of bases, noting that in mRNA, uracil is paired with adenine.

Gene expression begins when a molecule of DNA is used as a template for the creation of a molecule of mRNA.

Each group of three bases in the strand of mRNA is called a **codon (I),** and each codon codes for a particular amino acid in a polypeptide chain. The box representing a codon should be colored lightly. The first specifies **amino acid 1 (J),** the second specifies **amino acid 2 (K),** the third specifies **amino acid 3 (L),** and the fourth specifies **amino acid 4 (M).** When the amino acids are joined by peptide bonds, a polypeptide results.

The result of translation is a polypeptide, and how this polypeptide functions will determine the phenotype of the organism.

In this case, the **polypeptide (N)** formed in response to the gene activity operates as an **enzyme (O)** involved in the production of pigment. The enzyme catalyzes a reaction that involves a preliminary **precursor pigment molecule (P).** Note that one of those precursor molecules is associated with the enzyme; the enzyme transforms these precursors into **pigment molecules (Q).** Remember that the enzyme in this process is the polypeptide that was formed, and that the DNA specified the amino acid sequence in the polypeptide.

Now gene expression occurs. The pigment molecules accumulate in the **flowers (R)** of the plant. The pigment molecules are red, so the flowers become red and this is one phenotypic trait that was decided by the plant's genetic makeup. Since the pigment molecules are produced by enzyme activity, and since the enzyme is a polypeptide specified by the DNA, we can see the relationship between DNA and phenotype.

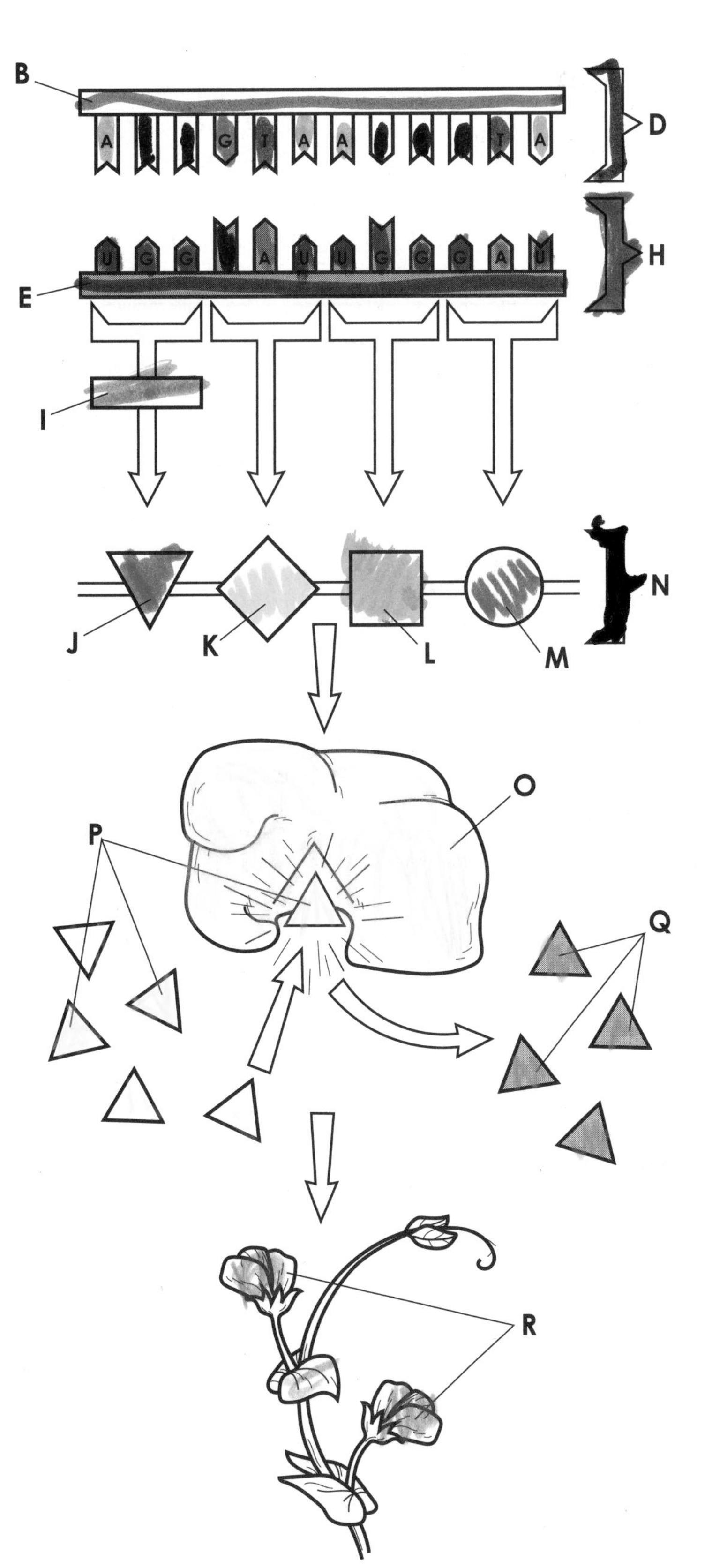
B
A
G
T
A
A
T
A
D
U
G
G
A
U
U
G
G
G
A
U
H
E
I
N
J
K
L
M
O
P
Q
R

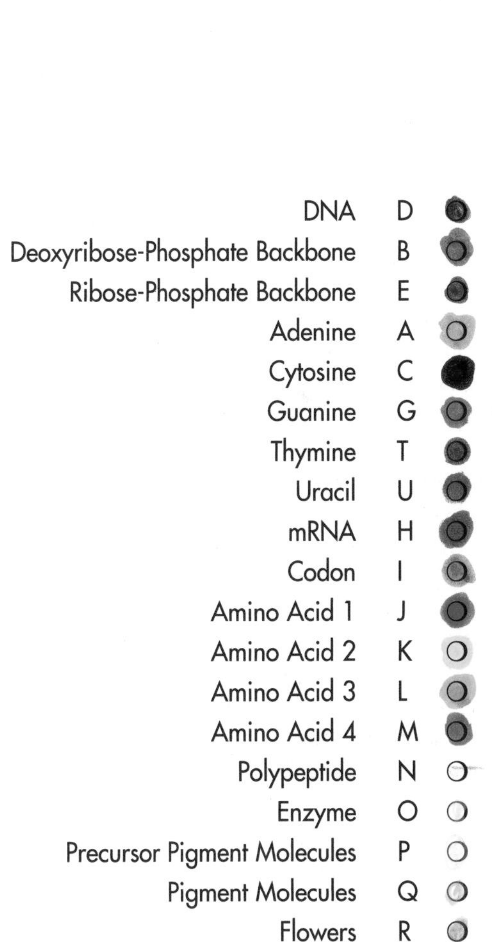
DNA D
Deoxyribose-Phosphate Backbone B
Ribose-Phosphate Backbone E
Adenine A
Cytosine C
Guanine G
Thymine T
Uracil U
mRNA H
Codon I
Amino Acid 1 J
Amino Acid 2 K
Amino Acid 3 L
Amino Acid 4 M
Polypeptide N
Enzyme O
Precursor Pigment Molecules P
Pigment Molecules Q
Flowers R

RNA STRUCTURE AND PHYSIOLOGY

To understand gene expression, biochemists needed to know whether the information in deoxyribonucleic acid (DNA) passed directly to an amino acid sequence in a protein, or whether its passage required an intermediary substance. One clue that indicated to them that there was an intermediary in the process was that the assembly of amino acids takes place in the cytoplasm, but DNA is confined to the cell's nucleus.

In the 1940s, biochemists reported that cells that are actively undergoing protein synthesis contain unusually high amounts of ribonucleic acid (RNA), which is a close relative of DNA. For this reason, it was theorized that RNA is involved in the production of protein from DNA. Now we know that RNA participates in gene expression in at least three important ways, as shown in this plate.

This plate examines the chemical composition of and the activities performed by ribonucleic acid (RNA). Direct your attention to the upper portion of the plate as you read the paragraphs below.

RNA is an important intermediary in gene expression. Although closely related to DNA, RNA differs in three important ways. In the diagram of RNA's structure, we see that the carbohydrate portion is made up of **ribose (R),** and not deoxyribose, as is the case in DNA. **Phosphate groups (P)** link the ribose molecules in the same way as they do in DNA.

The second important difference is the presence of the base **uracil (U).** Uracil does not exist in DNA; thymine is present in its place. Note that the other bases in RNA are the same as the ones in DNA; they are **adenine (A), guanine (G),** and **cytosine (C).**

A third important difference is that RNA is single-stranded, whereas DNA is usually double-stranded. In the right-hand portion of the top diagram, we see the single molecule of RNA with its **ribose-phosphate backbone (D)** and the four bases attached.

It was because of the many similarities between DNA and RNA that scientists believed that this molecule could receive a genetic message from DNA to use in protein synthesis.

Having studied the structure of RNA, we will now examine three different types of RNA that function in gene expression and protein synthesis. Continue reading below as your focus on the center and lower portions of the plate.

Biochemists soon discovered three types of RNA that participate in the process of protein synthesis. The plate shows a large molecule of DNA with its **deoxyribose-phosphate backbone (B)** and hydrogen bond-linked nitrogenous bases. Note that **thymine (T)** is present and that the DNA molecule is double-stranded. The first type of RNA we'll talk about is messenger RNA, also known as mRNA. This molecule initially copies DNA's genetic code, as we will explain in the next plate. It is a single-stranded **RNA molecule (K)** that contains a **ribose-phosphate backbone (D)** and the usual bases of RNA. Note that we have bracketed three bases (U-A-G). These three bases represent a **codon (H).** The sequence of codons in a strand of mRNA determines the identity of the protein that is created. Notice that the bases of the mRNA molecule complement the lower strand of bases of the DNA molecule above it.

The second type of RNA that functions in gene expression is ribosomal RNA (rRNA). A **small subunit (E)** of ribosomal RNA and peptides complexes with a **large subunit (F)** to form the active ribosome. Ribosomes are ultramicroscopic bodies found on the surface of the endoplasmic reticulum and floating freely in the cytoplasm. They act as the cell's "workbenches," at which proteins are synthesized.

We have discussed the structure of RNA and two important types of RNA that are used in gene expression. We now turn to a third type of RNA, transfer RNA. The information you receive in this plate will help you understand the synthesis of protein that is presented in the next few plates.

The third type of RNA we'll look at is transfer RNA, also called tRNA. There are dozens of different types of tRNA molecules floating in the cell's cytoplasm. They unite with specific amino acids and deliver them to the ribosome to be incorporated into proteins. The tRNA molecule is roughly the shape of a cloverleaf. A gray or pale shade should be used to shade it. At the upper end of the molecule, we see the site at which an **amino acid (J)** binds. This particular tRNA molecule binds to the amino acid serine.

At the bottom of the molecule, there is a three-base sequence (A-U-C) called an **anticodon (I).** This anticodon complements the codon on the mRNA molecule (U-A-G). During protein synthesis, this tRNA molecule transports the amino acid serine to the ribosome, where the mRNA molecule will match its codon to the tRNA's anticodon. This positions the amino acid at the correct location in the protein chain.

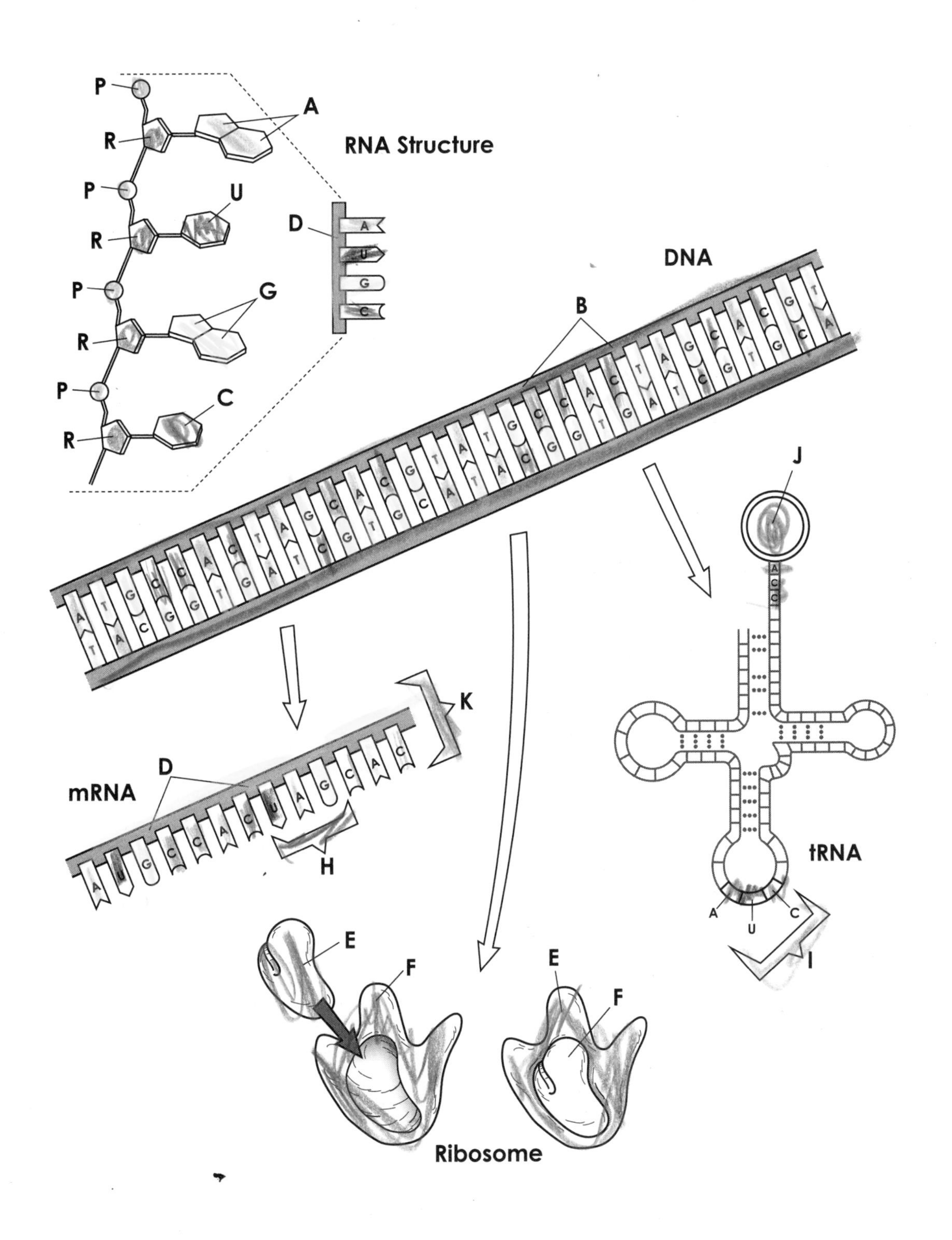

- Ribose R
- Phosphate Group P
- Uracil U
- Adenine A
- Guanine G
- Cytosine C

- Thymine T
- Deoxyribose-Phosphate Backbone B
- RNA molecule K
- Ribose-Phosphate Backbone D
- Codon H

- Small Subunit E
- Large Subunit F
- Anticodon I
- Amino Acid J

PROTEIN SYNTHESIS (TRANSCRIPTION)

The process of gene expression through protein synthesis is detailed and complex. Biochemists have uncovered many of the details of the process, but much more still needs to be learned. One of the first steps in protein synthesis is transcription. Transcription is the process in which the base sequence of DNA is transferred to a molecule of RNA; this RNA is then used for the construction of a protein molecule, in translation.

Two aspects of transcription are presented in this plate. In the upper part of the plate, we study the synthesis of a strand of messenger RNA. In eukaryotic cells, this molecule is then modified, and this is shown in the lower portion of the plate.

Biochemists now know that the process of gene expression begins when the DNA double helix starts to uncoil. In the upper part of the plate, we show the DNA molecule unwound at a central point (rather than at the end of the molecule). The **deoxyribose-phosphate backbone (B_1)** of both strands is shown, and the nitrogenous bases of DNA are labeled. They include **adenine (A), cytosine (C), guanine (G),** and **thymine (T).** Once the two strands of DNA are separated from one another, only one strand participates in the synthesis of a complementary mRNA strand. In this example, we will use the lower DNA strand, as the plate indicates.

As we mentioned in the last plate, a molecule of mRNA contains a **ribose-phosphate backbone (B_2),** and the bases of RNA include adenine (A), cytosine (C), guanine (G), and **uracil (U).** Remember that there is no thymine in RNA; it is replaced by uracil. The synthesis of RNA is mediated by an enzyme called **RNA polymerase (D).** A light color should be used to highlight this enzyme. Note that synthesis takes place at the right side of the molecule and, as synthesis continues, the RNA strand extends to the left.

Since there is no nucleus in prokaryotic cells, the mRNA exists in the cytoplasm along with the amino acids to be used for protein synthesis. In eukaryotes, the process of transcription takes place within the nucleus of the cell. Also in eukaryotic cells, such as human cells, the mRNA must be modified before it leaves the nucleus of the cell. How this occurs is discussed next.

After mRNA synthesis is complete, the two DNA strands recouple and the molecule recoils to assume its usual double helix.

We have examined the process of transcription, in which an RNA molecule must be processed further before leaving the cytoplasm. We will see how that is done in the following text. Bold colors should be used for this process because there are few details to be obscured.

In eukaryotic cells, the mRNA molecule produced in transcription is a preliminary mRNA molecule, or **pre-mRNA (H)**. Another name for this molecule is heterogeneous nuclear RNA (or hnRNA). This molecule contains a number of regions called **exons (E)** that will be expressed together as a single amino acid sequence in protein. Between these regions are a number of intervening regions of RNA that will not be expressed, called **introns (F).**

In the plate, we next see the introns (F) bulging out of the RNA molecule into loops. In the third diagram, the introns have been removed from the mRNA, which now consists solely of exons (E). This process takes place in the nucleus of the cell, before transport of mRNA to the cytoplasm. It is called splicing, and is mediated by the spliceosome, a complex that contains over 100 proteins and 5 small nuclear RNA (snRNA) molecules. (snRNA, rRNA and tRNA are all types non-coding RNA, so named because they do not code for proteins like mRNA does.)

The final modifying steps for the mRNA molecule include the addition, at one end of the molecule, of a **cap (I).** This cap consists of a molecule of 7-methylguanosine. At the right end of the molecule, a tail is added. The tail consists of a number of linked nucleotides, each of which contains the base adenine. This tail is therefore called a **polyadenine or poly-A tail (J).** After the addition of the cap and tail, the RNA molecule is complete, and is called a **processed mRNA (K).**

In eukaryotic cells, the processed mRNA leaves the nucleus and enters the cytoplasm. It approaches the ribosome and takes part in the process of protein translation, which is the second part of protein synthesis, and the subject of the next plate.

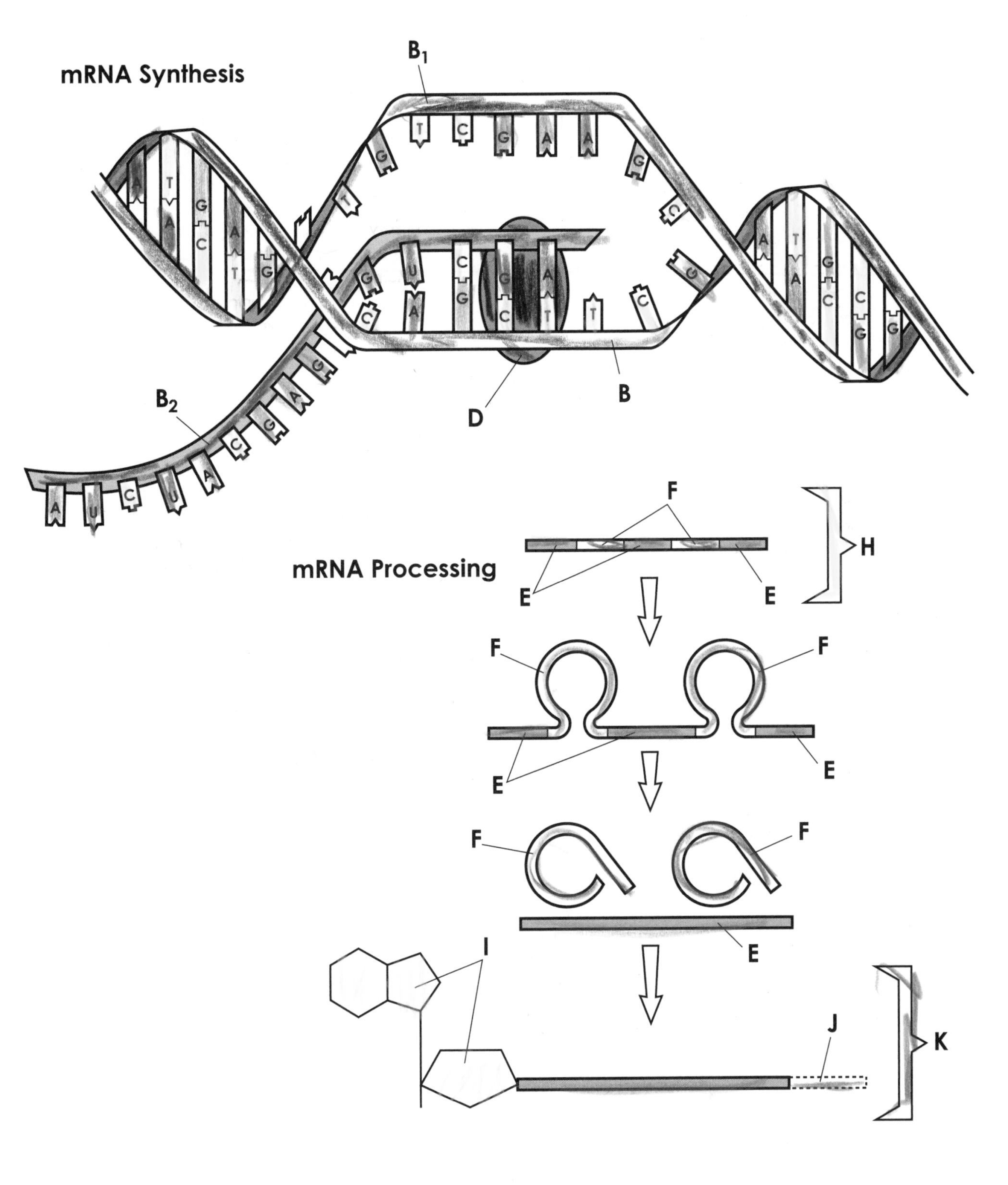

Deoxyribose-Phosphate Backbone B_1
Ribose-Phosphate Backbone B_2
Adenine A
Cytosine C
Guanine G
Thymine T
Uracil U
RNA Polymerase D
Exon E
Intron F
Pre-mRNA H
Cap I
Polyadenine Tail J
Processed mRNA K

PROTEIN SYNTHESIS (TRANSLATION)

Deoxyribonucleic acid (DNA) is the starting point for the important cellular process in which protein is constructed from single amino acids. The protein created can be an enzyme, structural material, or used for virtually any other cell need.

During transcription, the sequence of the nitrogenous bases of DNA directs the production of a strand of messenger RNA (mRNA). Then, in the process of translation, the strand of mRNA determines the sequence of amino acids, which determines the identity of the protein.

Translation is a complex process in which many things occur at the same time. To make the process clearer, we have depicted it as a sequential series of events taking place in the cytoplasm of the cell.

As you may recall, transcription begins with the production of a molecule of mRNA, which travels out into the cytoplasm to take place in translation.

Once it is in the cytoplasm, the mRNA strand complexes with a ribosomal complex. Take a look at the plate; first color the **ribose-phosphate backbone (B)** of the mRNA, and the nitrogenous bases: **adenine (A), cytosine (C), guanine (G),** and **uracil (U).** As you can see, the mRNA molecule is associated with the **ribosomal complex (D).** In the eukaryotic cell, ribosomes are located in the cytoplasm or associated with the endoplasmic reticulum; in prokaryotic cells, such as bacteria, ribosomes float freely in the cytoplasm.

The mRNA molecule is now anchored within the ribosomal complex of the cell. Each group of three bases in a strand of mRNA is called a codon, and each codon specifies a particular amino acid in a polypeptide chain. We will now see how these codons are involved in the production of polypeptides.

While the mRNA-ribosome complex is assembling, many other important activities are occurring in the cytoplasm. For instance, tRNA (transfer RNA) subunits are being connected to particular amino acids in energy-requiring processes. This is done by a family of enzymes called aminoacyl-tRNA synthetases.

We can see in the diagram, for example, that the amino acid **alanine (E)** has united with the **tRNA for alanine (E_1),** and one molecule of **lysine (F)** is floating free, while another has united with a **tRNA molecule for lysine (F_1).** Notice that different tRNA molecules have different three-base sequences. These are called anticodons. For instance, for alanine, the tRNA has an anticodon of A-U-G, while for lysine, the anticodon is C-C-C.

Also in the cytoplasm is a molecule of **tryptophan (H)** united with the **tRNA molecule for tryptophan (H_1),** and a **valine (J)** that is united with the **tRNA for valine (J_1).** Note that there is no tyrosine molecule shown for the **tRNA for tyrosine (K_1),** nor is there a leucine molecule available for the **tRNA for leucine (L_1).**

Having established the specificity of amino acids for their tRNA molecules, we now show how these complexes function in the translation process.

Once bound to their amino acids, the tRNA molecules travel to the ribosome where the mRNA molecule is anchored. The critical step is the matching of the codon on the mRNA molecule with the anticodon on the tRNA molecule. When this is accomplished, a certain amino acid is brought into position. For example, we see the amino acid lysine (F) attached to its tRNA molecule (F_1). The anticodon of the tRNA is complementary to the codon of the mRNA, and lysine is placed into position. Immediately to its left, an alanine (E) has been put into place by its tRNA (E_1). The alanine molecule is bonded to a **serine molecule (I),** which is in the process of breaking free from its **tRNA molecule (I_1),** and at the far left, an alanine molecule (E) has already detached from its tRNA. Note how the process unfolds as you color. Can you guess which amino acid will follow after the tryptophan?

As the arrow indicates, the ribosome moves along the mRNA, ensuring that each codon and anticodon matches. Enzymes create peptide bonds between the amino acids to form a polypeptide of increasing length. The translation process terminates when specific codons on the mRNA molecule signal to halt. At this point the polypeptide is released from the ribosome and is further processed to yield a functional protein.

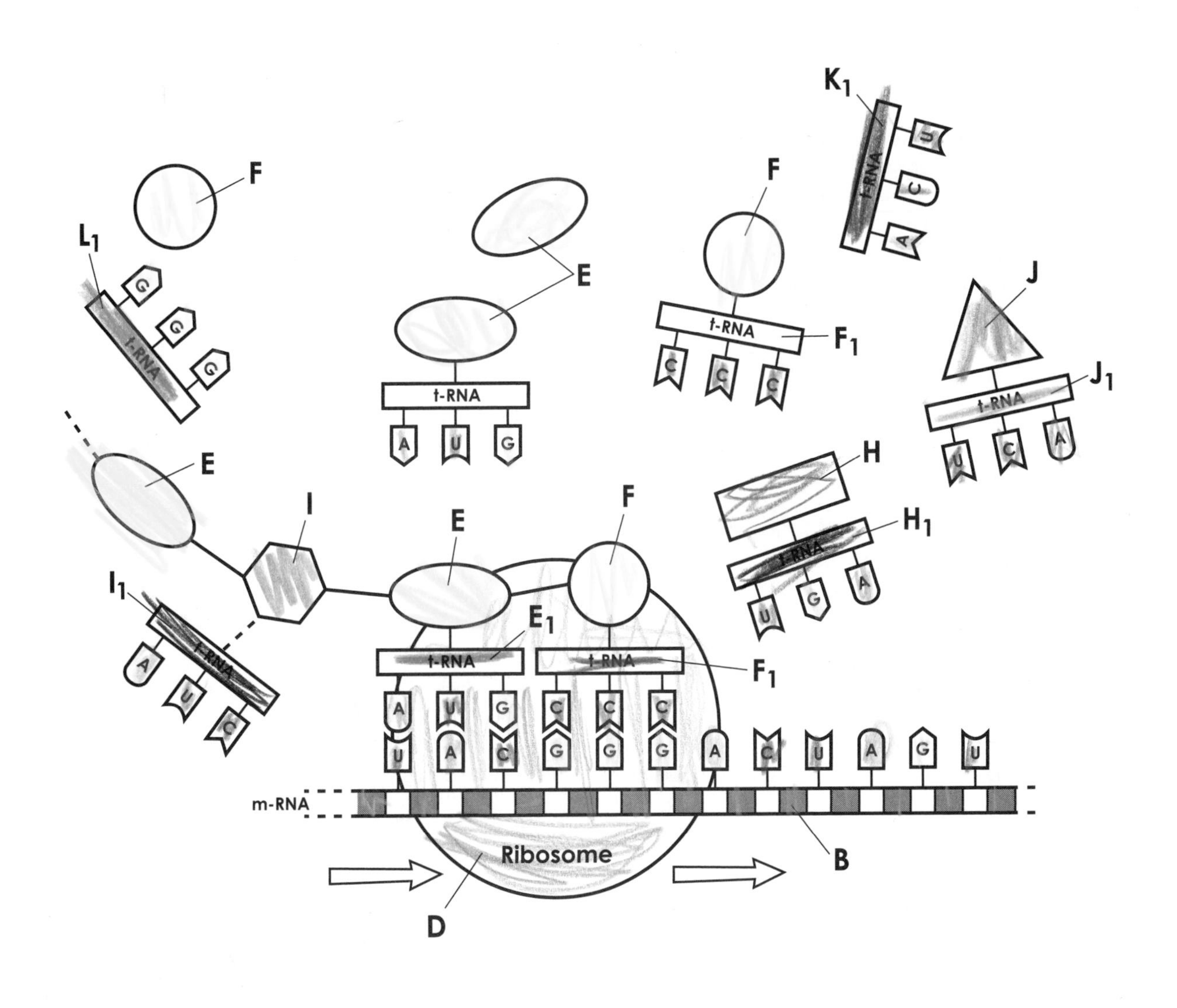

Ribose-Phosphate Backbone	B	○
Adenine	A	○
Cytosine	C	○
Guanine	G	○
Uracil	U	○
Ribosomal Complex	D	○
Alanine	E	○
tRNA for Alanine	E_1	○
Lysine	F	○
tRNA for Lysine	F_1	○
Tryptophan	H	○
tRNA for Tryptophan	H_1	○
Serine	I	○
tRNA for Serine	I_1	○
Valine	J	○
tRNA for Valine	J_1	○
tRNA for Tyrosine	K_1	○
tRNA for Leucine	L_1	○

GENE REGULATION (LACTOSE)

A human cell contains about 21,000 genes. Most of the time, the majority of these genes remain in a dormant state, while a small number are involved in the transcription process, which ultimately results in the production of proteins. A muscle cell, for example, will produce proteins that are needed for its metabolism that aren't the same as proteins required by other types of cells. Therefore, the other genes remain inactive.

The mechanisms for turning genes on and off are part of their regulation. Gene regulation has been studied at length in bacterial cells, which have relatively few genes, and where the biochemistry of regulation can be analyzed in depth. In this plate, we present a description of the mechanism for gene regulation in the colon bacterium *Escherichia coli*.

This plate contains two diagrams that illustrate the activity of genes first in the absence of the carbohydrate lactose and then in its presence. Lactose is a disaccharide made up of glucose and galactose, and when it is absent from the cellular environment, the enzymes needed to digest it are also absent. When it is present, the enzymes are produced to digest it. This phenomenon is possible because of gene regulation.

In the 1940s, the French investigators Francois Jacob and Jacques Monod researched gene regulation in bacteria and described the operon, a cluster of genes that work together to synthesize proteins. The operon they first described is called the lac operon, and it is frequently studied as a model of gene regulation.

In order to digest lactose, the cell must produce three enzymes. Look at the first diagram. These enzymes are encoded by **structural gene A (A), structural gene B (B),** and **structural gene C (C).** Next to the three structural genes on the chromosome is a region called an **operator region (D).** The operator region turns the transcription process of these structural genes on and off, and this determines whether their enzymes will be produced.

Next to the operator is a region of DNA called the **promoter region (E).** The promoter region is the section of the operon at which RNA polymerase binds. (You may recall that RNA polymerase is the enzyme that synthesizes RNA during transcription.) Further down the DNA molecule is a gene called the **regulator gene (F).** This part of the operon encodes a protein that represses transcription of the structural genes, as you will see next.

We have now described the components of the operon, which include the structural genes, operator, promoter, and regulator gene. We will now see how this operon works in the absence of lactose.

When lactose is absent from the environment, the regulator gene (F) encodes a strand of **mRNA (H).** We see this mRNA molecule associated with a **ribosome complex (I).** When translation of the mRNA takes place, a **repressor protein (J)** is produced. As the arrow indicates, the repressor protein then binds to the operator (D). Immediately to its left is a molecule of **RNA polymerase (G)** that's bound to the promoter. When the repressor protein is in place, it blocks RNA polymerase and prevents the enzyme from transcribing the structural genes. The structural genes remain dormant and fail to produce their enzymes, and that is why when lactose is absent in the environment, the cell produces no lactose-digesting enzymes.

But what happens when lactose is present in the cellular environment? Continue your coloring below, and you will find out.

We now turn our attention to the lower portion of the plate to see what happens when lactose enters the cellular environment. Let us suppose for example that a person drank a glass of milk, and the intestine is filled with lactose. The bacteria would now need the enzymes with which to digest the lactose.

In this situation, the **lactose molecule (K)** binds to the repressor and forms a lactose-repressor complex (J, K). Note that the structure of the repressor protein has changed because of this union, and its structure is such that it can no longer bind to the promoter. Thus, the RNA polymerase (G) is free to operate and the arrow shows it sweeping down the chromosome and transcribing the structural genes. As a result, the structural genes produce mRNA molecules that code for the three enzymes. There is an **mRNA strand for enzymes 1, 2, and 3 (L, M, and N).** These molecules are then translated and the results are the three enzymes for lactose digestion: **enzymes 1, 2, and 3 (L_1, L_2, and L_3).**

The lac operon gives an example of how gene regulation works. Its activity also shows the amazing biochemical efficiency of the cell.

a. Without Lactose

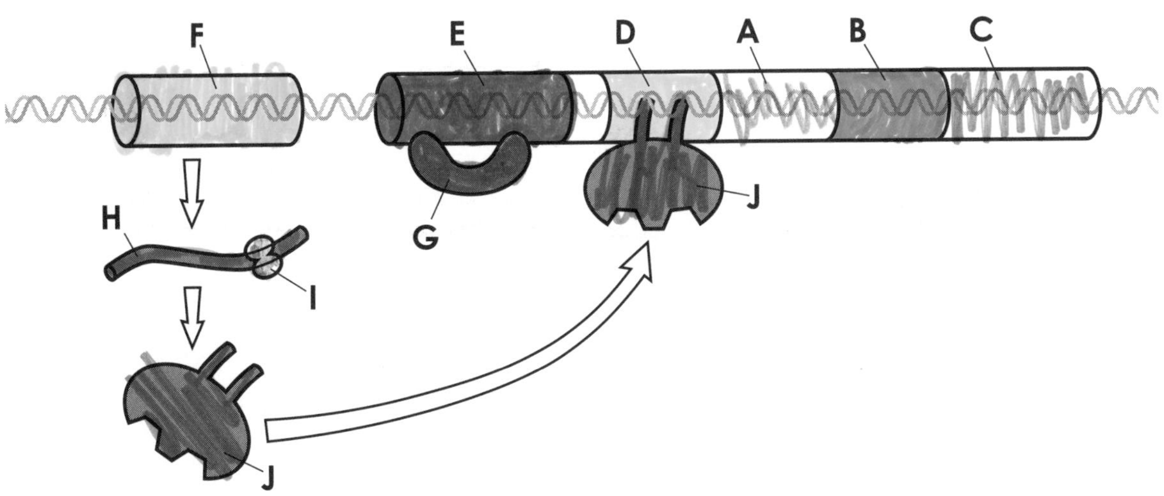

b. With Lactose

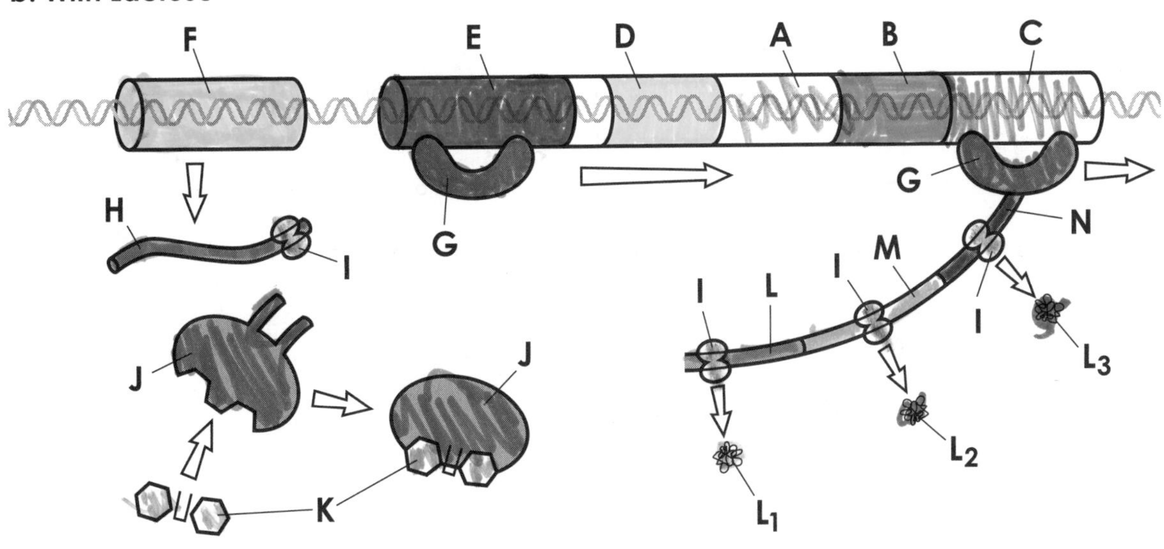

Structural Gene A	A	RNA Polymerase	G	mRNA for Enzyme 2	M		
Structural Gene B	B	mRNA for Repressor Protein	H	mRNA for Enzyme 3	N		
Structural Gene C	C	Ribosome Complex	I	Enzyme 1	L_1		
Operator Region	D	Repressor Protein	J	Enzyme 2	L_2		
Promoter Region	E	Lactose Molecule	K	Enzyme 3	L_3		
Regulator Gene	F	mRNA for Enzyme 1	L				

GENE REGULATION (TRYPTOPHAN)

The previous plate discussed the activity of the lac operon, which is a group of genes that work together to regulate a gene expression. Here we examine the operon that controls the synthesis of the amino acid tryptophan. This operon is known as the trp operon.

Looking over the plate, you will notice that we examine a series of genes and their operations in the presence and absence of tryptophan. We will show how the enzymes that participate in the synthesis of tryptophan are produced when tryptophan is needed but disappear when tryptophan is available.

Escherichia coli, a bacterium that's abundant in the colon, synthesizes tryptophan, an essential amino acid. *E. coli* uses five enzymes in its production of tryptophan. We will discuss the regulation of the synthesis of three of these.

The operon involved in the synthesis of the three tryptophan enzymes is shown in the plate. The three **structural genes A, B, and C (A, B, C)** provide the genetic codes for these enzymes. Adjacent to the structural genes is a region called the **operator region (D),** which turns the structural genes on and off. To the left of the operator is a region of DNA called the **promoter region (E).** The enzyme **RNA polymerase (G)** binds to the promoter and is involved in the synthesis of mRNA. Farther down the chromosome and somewhat removed from the other portions of the trp operon is the **regulator gene (F).**

Having described the genes involved in the trp operon, we will now see how they work together in the absence of tryptophan. Continue your coloring as you read below. The same colors used in the previous plate may be used in this one.

The first diagram shows the mechanism of gene regulation when **tryptophan (K)** is needed by the cell. Transcription takes place at the regulator gene (F) and a strand of **mRNA (H)** is produced. We see this strand complexed with its **ribosome (I).** Translation of this mRNA strand produces a **repressor protein (J).**

The repressor protein has no effect on the genes of the operon because it cannot bind to them. Therefore, RNA polymerase (G) is able to move down past the operator (D). These genes are transcribed and this results in transcribed strands of **mRNA from enzymes 1, 2, and 3 (L, M, N).** These mRNA molecules are translated, and the bacterium has the enzymes it needs to synthesize tryptophan **(L_1, M_1, N_1).**

As we mentioned, these three enzymes are among the five used in the synthesis of tryptophan (K). Note that in this mechanism, the trp operon is activated by the absence of tryptophan, while in the lac operon, activation occurs in the presence of lactose.

We now move to the second half of the plate and examine gene regulation in the presence of tryptophan. You will see that enzyme synthesis shuts down when tryptophan is available. Gene regulation occurs to control an unnecessary output of enzymes.

In the lower portion of the plate, we again see the structural genes A, B, and C, and the operator (D), the promoter (E), and the regulator gene (F). As before, the regulator gene is being transcribed to an mRNA molecule (H). At the ribosome (I), the repressor protein (J) is being produced.

But the situation is slightly different in this case; tryptophan is abundant in the cellular environment. A tryptophan molecule (K) unites with the repressor protein (J), which allows the tryptophan-repressor complex (J-K) to change its structure and lock itself into position at the operator (D). The complex blocks the path of the enzyme RNA polymerase (G) at the promoter (E), and RNA polymerase cannot transcribe the structural genes. Thus, no mRNA is made, and the enzymes necessary for the synthesis of tryptophan are not produced.

It makes biochemical sense to refrain from producing these enzymes when tryptophan is already abundant in the cell's environment, and is yet another example of biochemical efficiency. In this case, tryptophan is referred to as a corepressor since it enables the repressor protein to block the synthesis of enzymes. When the tryptophan from the environment is used up, the tryptophan-synthesizing enzymes are once again synthesized.

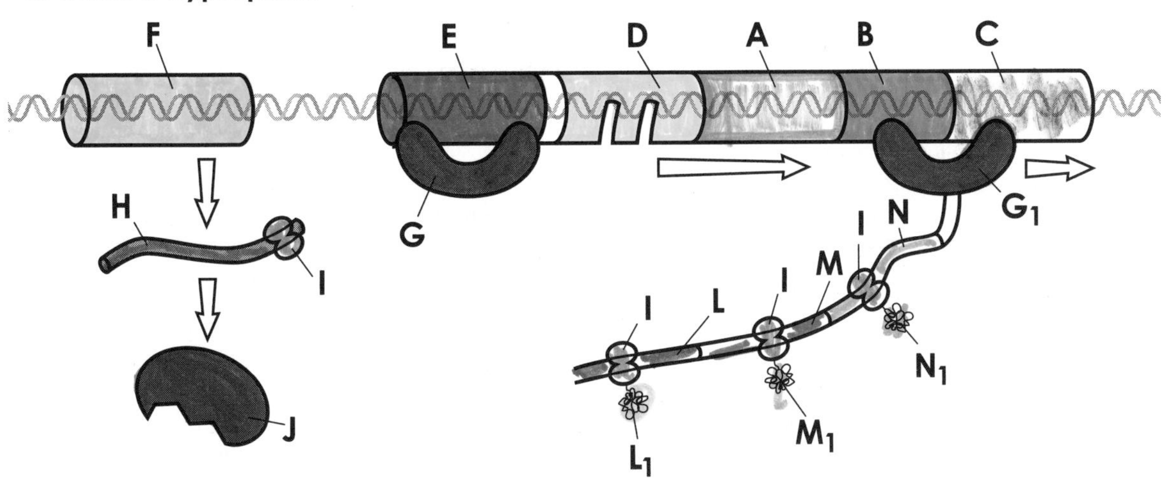

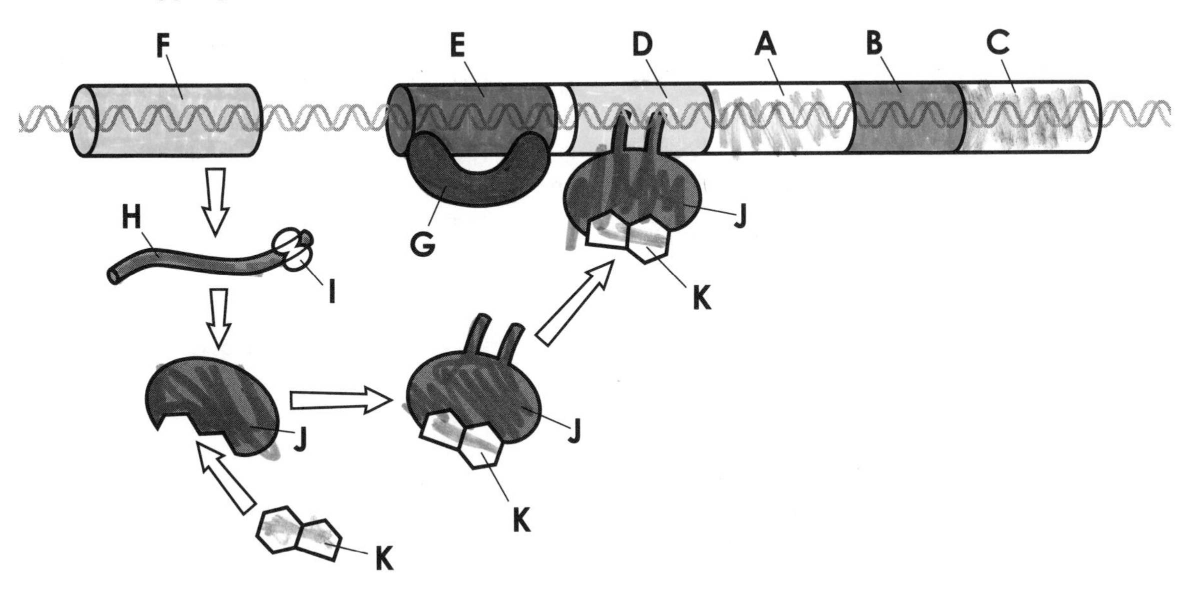

Structural Gene A	A	RNA Polymerase	G	mRNA for Enzyme 2	M
Structural Gene B	B	mRNA for Repressor Protein	H	mRNA for Enzyme 3	N
Structural Gene C	C	Ribosome	I	Enzyme 1	L_1
Operator Region	D	Repressor Protein	J	Enzyme 2	M_1
Promoter Region	E	Tryptophan	K	Enzyme 3	N_1
Regulator Gene	F	mRNA for Enzyme 1	L		

MUTATION AND GENE EXPRESSION

Mutations are unregulated alterations in DNA that may or may not affect the phenotype of individuals. Proteins encoded by these mutated genes are subsequently altered and they may fail to function. In this plate, we examine how a mutation in the DNA molecule affects the protein produced and ultimately, the phenotype of a plant. The plate contains some of the same information as an earlier plate entitled DNA and Phenotype; that plate shows this same gene without mutation.

> We are following the chain of events that begins with a molecule of DNA and ends in a plant. You will see that this particular mutation results in an altered phenotype. You may want to refresh your knowledge of genetic mutations by reviewing the plate titled Chromosomal Alterations in Chapter 3.

Mutations may be brought about by numerous agents in the environment. For instance, ultraviolet light and some chemicals are mutagens. Mutagens are agents that cause physical or chemical damage to the gene.

Let us assume that ultraviolet light has affected the DNA of a plant cell. It will alter the strand of **DNA (D).** You should color its **deoxyribose-phosphate backbone (B)** and nitrogenous bases: **adenine (A), cytosine (C),** and **thymine (T).**

The strand of DNA in this plate can be compared with the DNA molecule displayed in the plate entitled DNA and Gene Expression. You will note that all of the bases are identical, with one exception: Reading from the left, the fifth base in normal DNA is thymine (T). In this plate, the ultraviolet light has caused a mutation and thymine has been replaced by cytosine (C). This mutation affected only one base.

> We are now ready to display the effects of the DNA mutation. Continue your reading below as you color the plate.

When the strand of DNA is transcribed, a molecule of messenger RNA (mRNA) is formed. This **mRNA molecule (H)** is outlined by the bracket, which you should fill in with a dark color. Next color RNA's **ribose-phosphate backbone (E),** and note that the mRNA strand is nearly identical to the normal mRNA strand, but there is one important difference. Reading from the left, the fifth base should be adenine (A), but instead it is guanine (G) because of the initial mutation in the DNA.

As we mentioned earlier, each three-base group of mRNA constitutes a **codon (I)**, and the box representing a codon should be colored. The first codon, UGG, is normal. **Amino acid 1 (J)** is placed next in the developing polypeptide. The mutation has occurred in the next codon and the wrong amino acid is coded for. The effect, of course, is that the **wrong amino acid (K)** is placed in the polypeptide. The third codon is unaltered and **amino acid 3 (L)** is placed in correct position, as is **amino acid 4 (M).** Thus, the **polypeptide (N)** has only one incorrect amino acid in its sequence.

> We have now seen how an alteration in the base code of DNA is expressed in the resulting polypeptide. To see how the phenotypic change occurs, continue your reading.

The polypeptide formed in this plate is part of a protein that is involved in the production of the pigment molecules of the plant, but as we know, the polypeptide is flawed, and a **malformed enzyme (O)** results. The unmutated form of the enzyme would convert **precursor pigment molecules (P)** into pigment molecules, but because the enzyme is malformed, it cannot function. For this reason, precursor molecules accumulate, and there is no pigment formed.

The effect now appears in the **flower (R).** Without any pigment, the flowers are white. They should be left uncolored. Now you have seen how a mutation in DNA has altered the appearance of a flower.

In some cases, mutations can exert a lethal effect on the organism, but in other cases, defective information may lead to defective protein that cannot carry out its function. In still other cases, genetic mutation can have no affect on a protein. Finally, some mutations can have beneficial outcomes.

The mutations you learned about on this plate are called point mutations because they are single base pair substitutions (C in place of T, for example). There are three types of point mutations:

1. Missense mutation: This causes one amino acid to be replaced with a different amino acid.
2. Nonsense mutation: A stop codon replaces a regular codon and prematurely shortens the protein.
3. Silent mutation: A codon is changed into a new codon for the same amino acid, so there is no change in the protein's amino acid sequence.

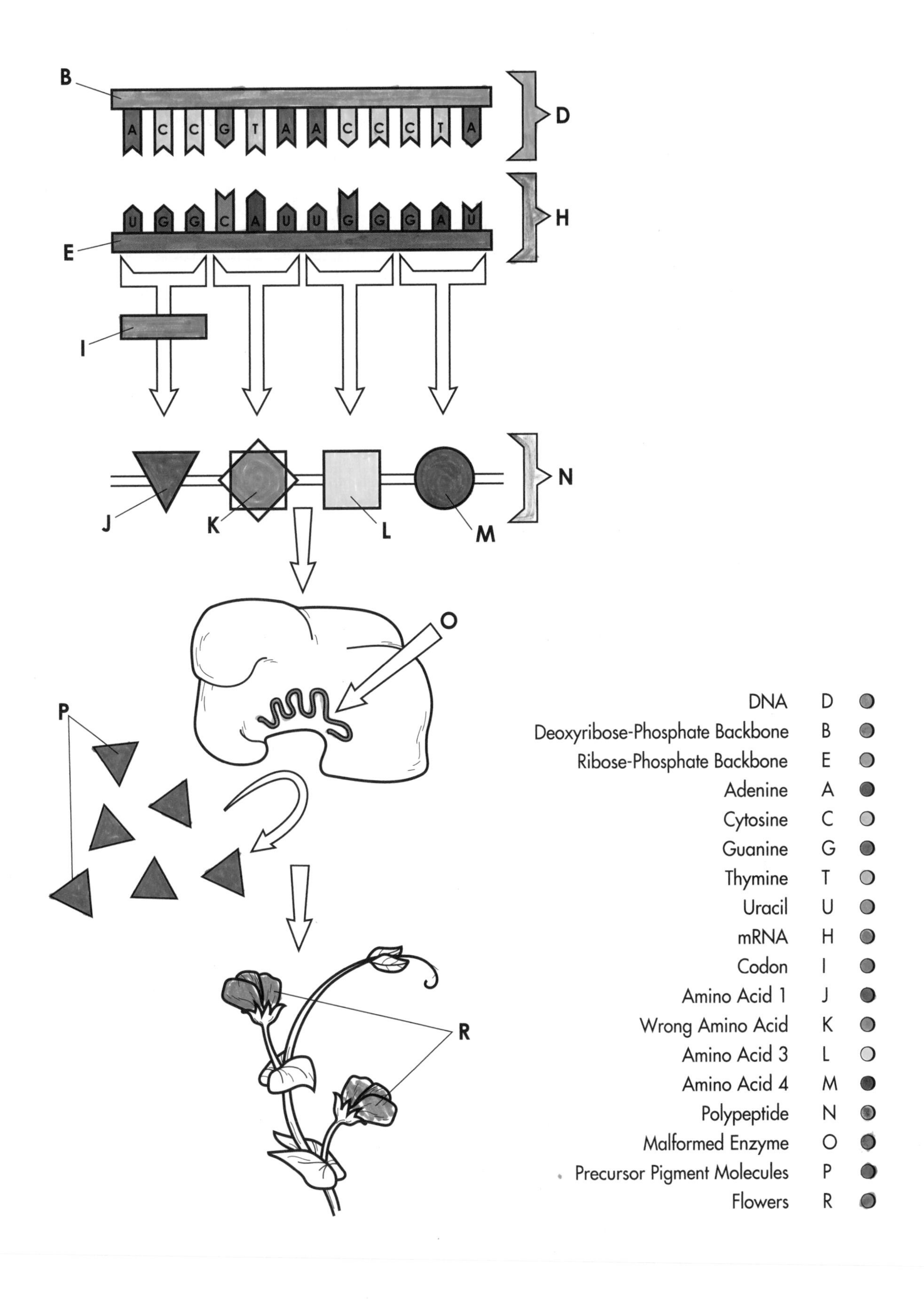
B
A C C G T A A C C C T A
D
U G G C A U U G G G A U
H
E
I
N
J
K
L
M
O
P
R
DNA D
Deoxyribose-Phosphate Backbone B
Ribose-Phosphate Backbone E
Adenine A
Cytosine C
Guanine G
Thymine T
Uracil U
mRNA H
Codon I
Amino Acid 1 J
Wrong Amino Acid K
Amino Acid 3 L
Amino Acid 4 M
Polypeptide N
Malformed Enzyme O
Precursor Pigment Molecules P
Flowers R

POLYMERASE CHAIN REACTION

Polymerase chain reaction (PCR) is a very quick and inexpensive method for detecting and amplifying specific DNA sequences. It is often the first step of a molecular biology experiment because the products of PCR can be used to screen for hereditary or infectious diseases, to clone genes, or for fingerprinting DNA. It is designed to generate many copies of a single template sequence. PCR thus allows the amplification and subsequent analysis of tiny samples of DNA. PCR is particularly important in forensic analysis because it allows crime scene investigators to replicate a very small amount of DNA into a large amount that can be tested in the lab.

In order to do PCR, you must have two primers. A **primer (A)** is a short piece (about 15 bases) of single-stranded DNA. It is designed so it will recognize and base pair with specific DNA sequences. Two PCR primers need to be designed—one at either end of the fragment **(B)** you are trying to amplify. One needs to be complementary and bind to one strand of DNA at one end of the fragment you're trying to amplify. The other primer needs to be complementary and bind to the opposite strand of DNA and at the other end. The DNA between the two primers is the fragment that will be amplified during the PCR reactions. One primer is called the forward primer, and the other is called the reverse primer.

On this plate, you're going to be coloring templates and primers many different times. Make sure all your template DNA is the same color, and make sure you use the same color for all the primers. See the legend for more information. Molecular biologists wear gloves when carrying out experiments, so be sure to color the hand in **(C)** as either blue or purple. PCR solutions are clear (like water), so you can use a pale blue or pale gray color for the actual solutions (which are about 50μL in volume—about the same volume as a drop of rain).

A few different things need to be added to a **PCR tube (C)** to make this solution. In addition to the **template DNA (D)** you're trying to amplify and both **primers (E),** you must also add lots of **deoxynucleotides (F),** such as deoxyadenosine triphosphate (dATP), deoxyguanosine triphosphate (dGTP), deoxycytidine triphosphate (dCTP), and deoxythymidine triphosphate (dTTP). These deoxynucleotides are the building blocks of DNA and will be used up as new DNA strands are built. You must add a heat-sensitive **DNA polymerase enzyme (G),** which will read the template DNA and build new complementary DNA off the primers. Many labs use Taq polymerase, which was first isolated from algae that thrive in hot springs. Finally, a buffer solution is added to keep the pH of the reaction steady and to provide ions required by DNA polymerase.

This mixture is placed into a PCR machine, which will carry out three basic steps:

Step 1: Initialization. The sample is heated to about 95°C. Heating the sample "melts" the hydrogen bonds that hold double-stranded DNA together. This turns the template DNA into two single-stranded pieces of DNA that will be read **(H).**

Step 2: Annealing. The sample is cooled to about 55°C. At this temperature, the **primers (I)** base-pair with the **template strands (J).**

Step 3: Elongation. The sample is heated to around 72°C. Using the **primers as starting points (K),** the **heat-sensitive DNA polymerase (G)** makes **strands of DNA (L)** that are complementary to each of the **template strands (M).** Each strand is polymerized in the 5′ to 3′ direction. Longer DNA targets take longer to synthesize, so the length of the elongation step depends on the length of the product DNA.

Each cycle of three steps takes between 30 seconds and 5 minutes, depending on the length of the target DNA product (and subsequent length of the elongation step). Most PCR reactions are run for 20–50 cycles. Because two new complementary strands are synthesized for each template strand in the sample, the PCR product grows at an exponential rate, yielding over a billion copies in just 30 cycles.

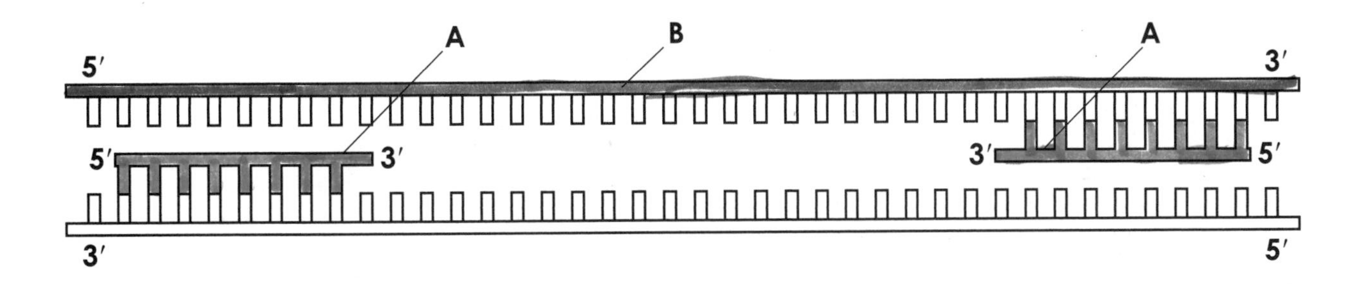

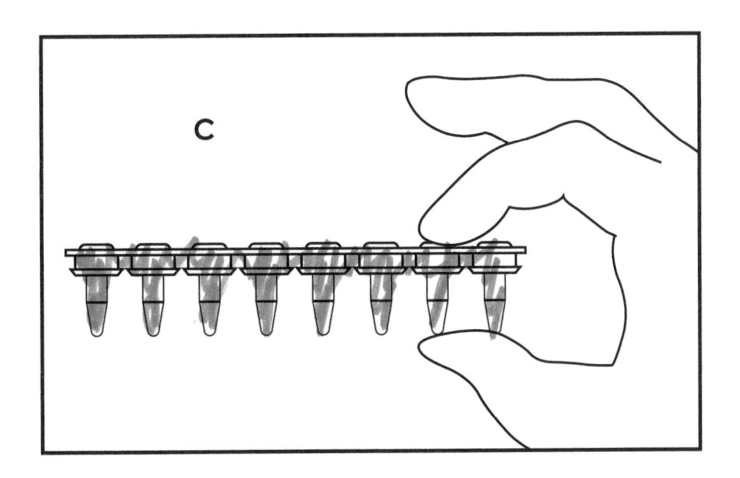

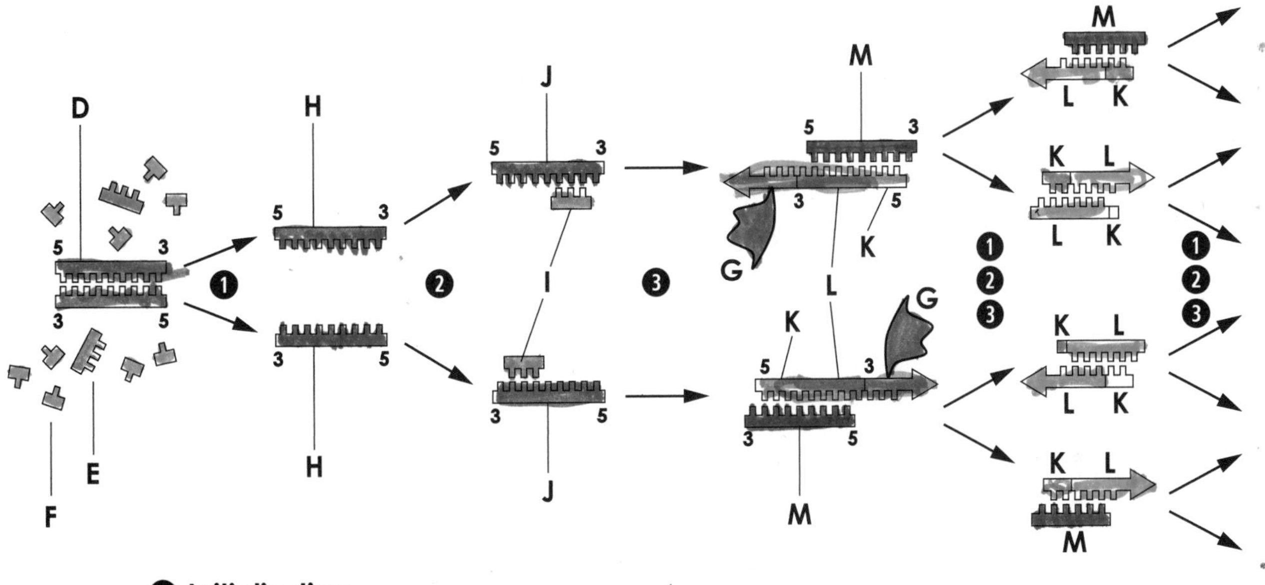

❶ **Initialization**
❷ **Annealing**
❸ **Elongation**

Primer	A, E, I, K	●
Template DNA	B, D, H, J, M	●
PCR Tubes	C	●
Deoxynucleotides	F	●
Heat-sensitive DNA Polymerase	G	●
Newly Synthesized DNA	L	●

RECOMBINANT DNA

A new technology was developed in the 1970s that allowed scientists to splice DNA fragments from one organism into another and then clone the new molecules. This technology came to be known as recombinant DNA technology. It is also known as genetic engineering.

Recombinant DNA technology has broad applications in the pharmaceutical, agricultural, and breeding industries. This plate discusses the series of steps by which recombinant DNA molecules are synthesized. The recombined DNA encodes for various proteins, as we will explain in the next plate.

Looking over this plate, you will notice that it consists of a progression of four diagrams that show the steps involved in the formation of a recombinant DNA molecule. The objective of the process is to synthesize a single molecule of DNA from two different DNA molecules.

Basically, recombinant DNA is formed by the introduction of foreign DNA into a DNA molecule. In diagram 1, we display **DNA molecule #1 (A).** You should focus on its middle portion, and a pale color should be used to color the brackets outlining the **recognition sites (B).**

At these recognition sites, enzymes called **restriction enzymes (C)** specifically cleave the strand of DNA. A medium color may be used to designate the enzyme in our diagram. Restriction enzymes are so-named because they operate on restricted points on the DNA molecule. In diagram 1 you can see the restriction enzyme cutting double-stranded DNA. The enzyme displayed is called EcoRI, it is a restriction enzyme that's derived from the colon bacterium *Escherichia coli*.

In diagram 2 we see that EcoRI has located two recognition sites and cut the DNA molecule. The result is a **DNA fragment (E)** that is left with single-stranded tails at each end. These tails can bond with DNA strands that have complementary tails, and for this reason the unpaired base sequences are called **sticky ends (F).** The bracket indicating the sticky ends should be colored with a dark color to highlight its presence.

We have now isolated a single DNA fragment. The remaining section of the DNA molecule will now be set aside as **discard DNA (D),** as the plate indicates.

We have now isolated a DNA fragment from the first DNA molecule. This DNA molecule has exposed bases at its sticky ends. Any DNA molecule with complementary bases can bind to these sticky ends. We continue this process by introducing new DNA, in the next diagram. Continue your coloring as you read below.

Now examine diagram 3, which shows a second DNA molecule, designated **DNA #2 (G).** This molecule has also been treated with EcoRI; it also has a sticky end (F).

We now bring the two DNA fragments together. The sticky end of DNA #2 overlaps with the sticky end of the original DNA fragment. The bases are complementary, that is, adenine (A) matches with thymine (T) and T matches with A. The cytosine and guanine bases are not involved in the matching.

We now move to diagram 4. A new enzyme called the **ligase enzyme (H)** is used at this point. Ligase is responsible for sealing any breaks in a DNA molecule in living cells. Biochemists use this enzyme to seal DNA #2 to the DNA fragment; the ligase connects the DNA molecule to the DNA fragment.

The result of the ligase activity is shown in diagram 4. We see DNA #2 (G), which has combined with the DNA fragment (E); the result is called **recombinant DNA (I).** The bracket should be colored in a dark color.

The technology of recombinant DNA permits scientists to attach one or more genes to a carrier molecule of DNA. This recombined molecule can be placed in an environment in which the DNA fragment will be expressed. For example, the recombinant DNA molecule can be placed in a bacterium where the foreign DNA fragment will encode a protein that is not normally produced by that bacteria. If the DNA fragment included the gene responsible for the production of the hormone insulin, then the bacteria would produce insulin.

Products of recombinant DNA technology include vaccines, research chemicals, medical proteins, and genetically altered organisms used for such purposes as resolving pollution problems, increasing soil fertility, and killing insect pests.

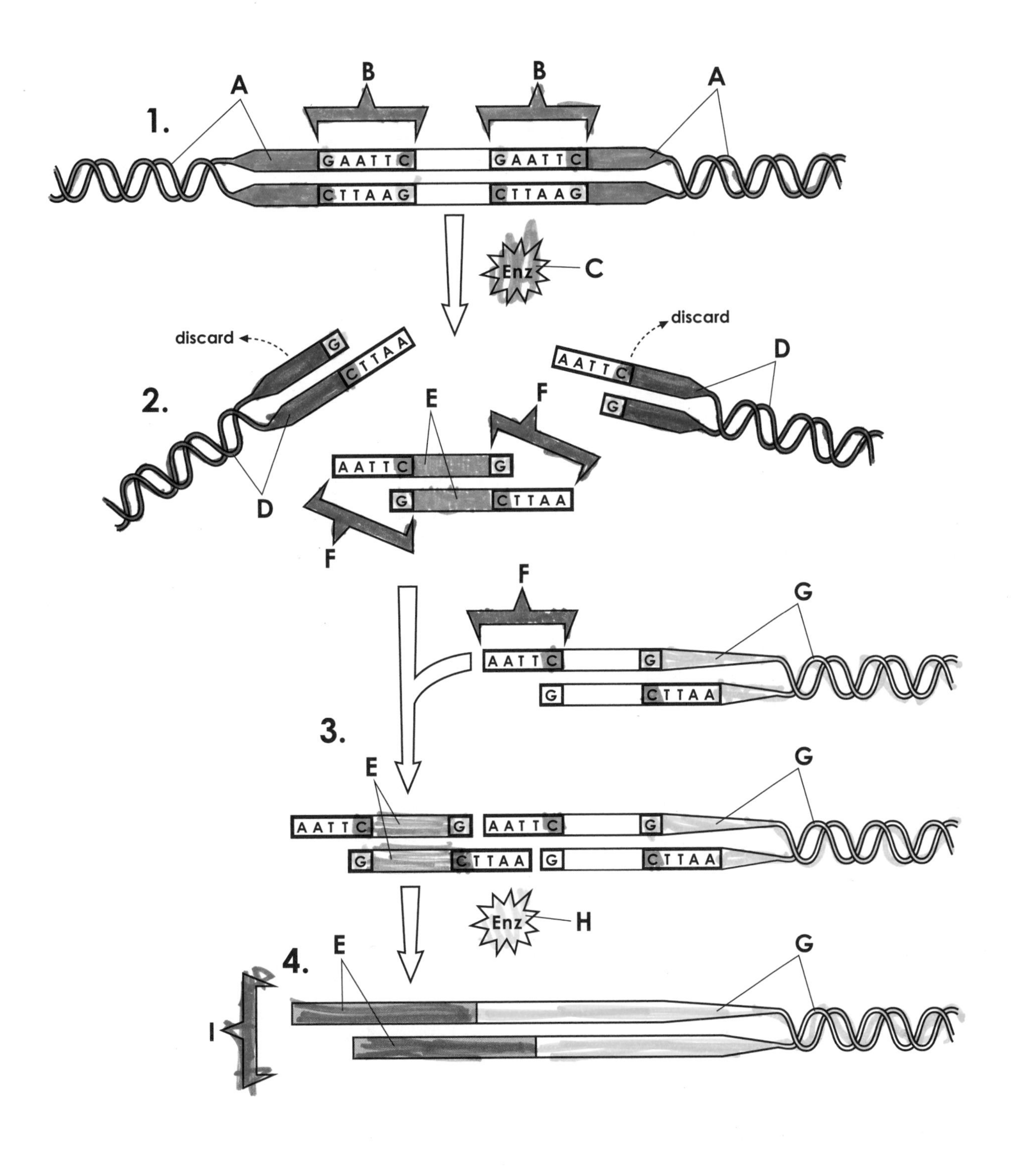

DNA #1	A	●
Recognition Sites	B	●
Restriction Enzyme	C	●
Discard DNA	D	●
DNA Fragment	E	●
Sticky Ends	F	●
DNA #2	G	○
Ligase Enzyme	H	○
Recombinant DNA	I	○

GENETIC ENGINEERING

The calculated manipulation of DNA is a relatively recent accomplishment of science; it was only in the 1970s that scientists began experimenting with recombinant DNA. They found that when the recombined DNA molecule was placed in new cells, these cells would synthesize the proteins encoded for by the newly introduced genes. In this plate we will study how this occurs.

This plate continues to describe the process of genetic engineering, or recombinant DNA technology. A segment of DNA from a foreign cell is first inserted into a fragment of DNA, and the resulting recombinant DNA is then placed into fresh cells, where the genes are expressed.

The process of genetic engineering caused a revolution in biotechnology. Genetic engineering techniques are currently being used to study processes such as gene regulation, to develop and produce products such as human hormones, to create disease-resistant plants, and to diagnose and treat genetic disorders. In this plate we study how genetic engineering can be used to synthesize large quantities of human insulin from bacterial cells. Insulin is a pancreatic hormone in which diabetics are deficient; it facilitates the uptake of glucose by cells.

We begin the genetic engineering process with bacterium and a human cell. The **bacterium (A)** is shown at the top left. A light color should be used to shade it. The **bacterial chromosome (B)** appears as a highly coiled loop of DNA. Also necessary for the process of genetic engineering is a tiny loop of DNA called a **plasmid (C).** A plasmid is a small, circular, double-stranded DNA molecule that is different from a cell's chromosomal DNA. Plasmids are replicated independently from the genome and occur naturally in bacterial cells, and some eukaryotes. Since plasmids are small, they often encode only a few genes, and it is common for these genes to provide the host cell with some sort of genetic advantage (such as antibiotic resistance). Plasmids are often used in genetic engineering.

On the right you can see the **human cell (D),** which should be colored lightly. The **nucleus (E)** in this diagram contains only one **human chromosome (F)** for the simplicity. This particular chromosome is the site of the insulin-encoding genes.

We have established the two sources of DNA for genetic engineering. The bacterium supplies plasmid DNA, while the human cell supplies chromosomal DNA.

The next step in this process is to produce recombinant DNA. First, a **restriction enzyme (G_1)** is used to open the plasmid at the left. The enzyme acts at a restricted point (explained in the previous plate) and creates an **opened plasmid (J).** At the points at which the enzyme has spliced the plasmid, the DNA has **sticky ends (I).** You should color the arrow that points to the sticky ends.

Next, the same restriction enzyme (G_1) is used to cut the human chromosome, and the result of this digestion is that the **insulin gene (H)** is isolated. This DNA fragment also has sticky ends (I), and these are indicated by the arrow.

Now the recombinant DNA molecule is created. Both the opened plasmid (J) and the insulin fragment bear DNA that has sticky ends, and these sticky ends are complementary. This means that when they are brought together, they overlap, and then **ligase (G_2)** is used to seal the ends together. This ligase creates two hydrogen bonds between adenine and thymine bases and three between guanine and cytosine bases. The result is a **recombined plasmid,** also known as a **chimera (K).**

The genetic engineering process has now progressed to the point where the insulin gene has been inserted into a bacterial plasmid. The recombined plasmid (chimera) will now be inserted into new cells.

Recombined plasmids can be inserted into new bacterial cells by alternately heating and cooling the cells. Electric current can also be used to introduce plasmids into cells. Using either method, this process is called transformation, and when completed, the recombined plasmid (K) is inside a **new bacterium (L).**

Now the bacterial cell is allowed to grow and multiply by placing it in an enriched environment. As these **metabolizing bacteria (M)** grow and increase in number, they carry on their normal metabolic activities. All bacteria have the insulin genes in their new, recombined plasmids, and these genes now operate and encode the protein insulin. Before long, the bacterium will begin producing **insulin (N).** We therefore have the unusual situation of a bacterial cell that produces human insulin!

For many decades insulin was isolated from the pancreases of animals, and diabetics depended on this insulin to allay the symptoms of their disease. It is now possible to obtain large amounts of insulin from metabolizing bacteria. This is one example in which genetic engineering has made a substantial impact on human welfare.

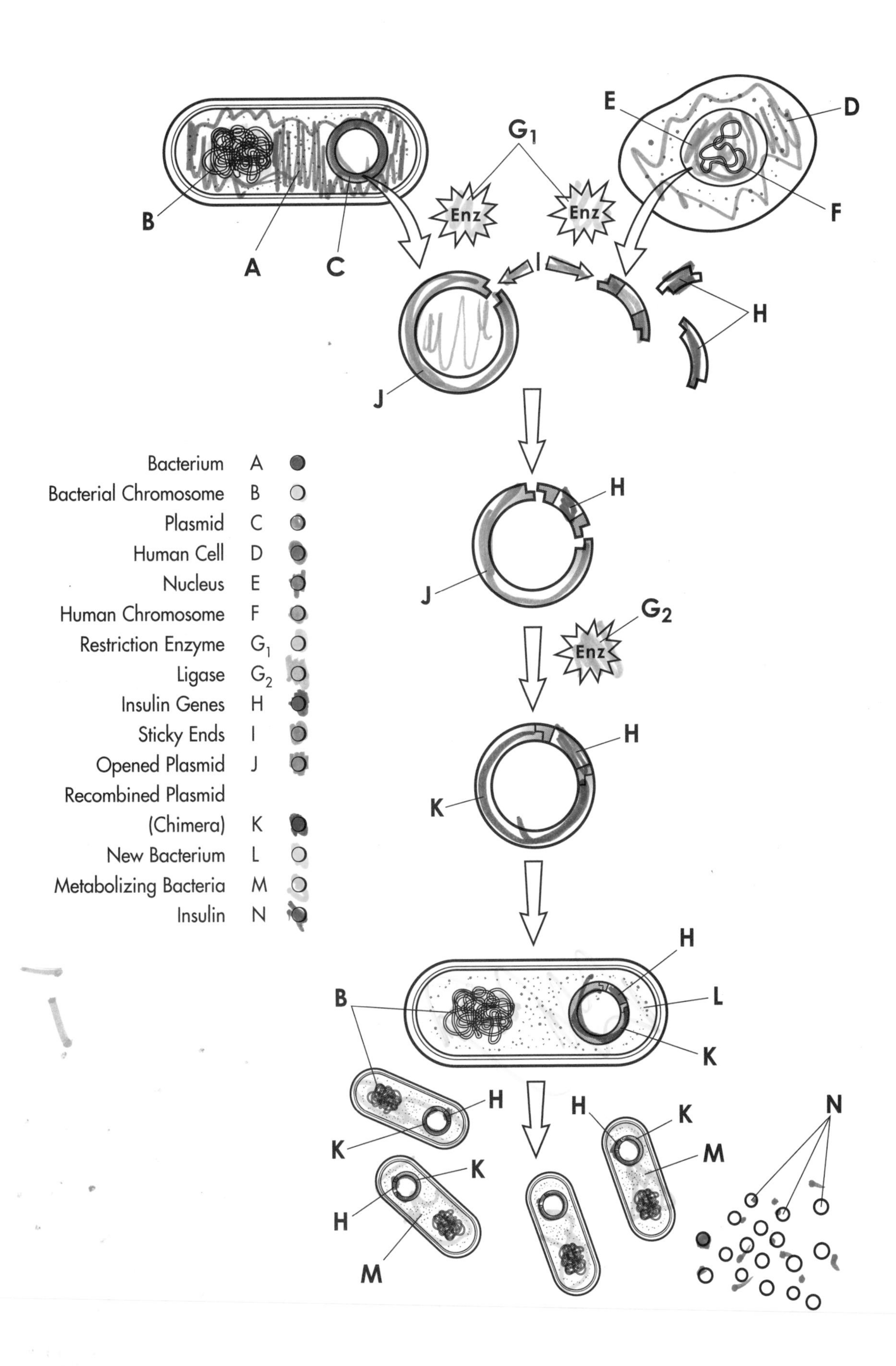

G1
Enz
Enz
A
B
C
D
E
F
I
H
J
J
H
G2
Enz
K
H
B
L
H
K
H
K
M
N
K
H
M
H
K
Bacterium A
Bacterial Chromosome B
Plasmid C
Human Cell D
Nucleus E
Human Chromosome F
Restriction Enzyme G1
Ligase G2
Insulin Genes H
Sticky Ends I
Opened Plasmid J
Recombined Plasmid (Chimera) K
New Bacterium L
Metabolizing Bacteria M
Insulin N

GENE PROBES

Many of the important procedures in genetic engineering research involve what is known as the gene probe. A gene probe is a relatively small, single-stranded molecule of DNA that can recognize and bind to a complementary strand on a larger DNA molecule.

Gene probes have become an important research tool in biotechnology. In this plate, we will first see how a gene probe is used to determine the presence of the insulin gene.

This plate contains two diagrams: In the upper half of the plate, a gene probe is used to determine whether or not an insulin gene is present in a strand of DNA. In the lower portion of the plate, three different gene probes are used to determine the identity of a bacterium. These are two examples of how gene probes are useful in research.

A gene probe is a single-stranded DNA molecule that can specifically bind to (hybridize) a target molecule of DNA. The process of the binding of gene probe and target molecule sends out a recognizable signal.

First the gene probe needs to be created. This is done using the sequence of the target gene. Since many genomes have been sequenced, this type of information is readily available online.

In the first diagram, we see one or more copies of seven different DNA molecules. These DNA molecules have been obtained from the nucleus of pancreas cells and we wish to know whether the gene for the hormone insulin is among them.

We will call the seven possible insulin-containing gene sequences **unknown DNA molecules one to seven (A to G).** you should use seven different colors to distinguish them. Each of the seven unknown DNA molecules has different base sequences, and all of these DNA fragments are single-stranded.

Now we take a look at the **gene probe (H).** This fragment of DNA also has an identified base sequence, and it has been prepared specifically for the insulin gene. The gene probe is attached to a **radioactive signal (H_1).**

The question is this: Which unknown DNA molecule has a base sequence that complements the base sequence of the gene probe? To answer this question, you should examine the base sequence of the gene probe and the base sequences of the unknown strands and determine which unknown DNA is complementary to the gene probe. When the two strands combine, a radioactive signal is given off, which identifies it as containing the insulin gene.

Having examined how gene probes work, we now focus on a case in which three different gene probes are used to determine the identity of an unknown bacterium. Continue your reading below and color the gene fragments as you proceed.

Another possible use for a gene probe is in the identification of an unknown bacterium. For instance, if a bacterium contaminates a body of water, it is useful to identify it so that a procedure for its elimination can be decided upon. In this case, the gene probe technique is used.

In the top half of the diagram, you can see the **DNA from the unknown bacterium (I).** This bacterium may be the colon organism *Escherichia coli*, the food poisoning organism *Staphylococcus aureus*, or the intestinal organism *Salmonella*. To determine its identity, we employ three gene probes: The first is the **gene probe for *E. coli* (J).** It is attached to a radioactive isotope that gives a **red signal (J_1)** upon binding. For *Staphylococcus aureus*, we use the **gene probe for *S. aureus* (K),** which gives a **blue signal (K_1)** on binding. The third is the **gene probe for *Salmonella* (L),** which emits a **green signal (L_1)** on binding with DNA. Notice that all three gene probes have different base sequences.

To identify the unknown DNA, we must identify which gene probe has a base sequence that is complementary to the DNA from the unknown bacterium. The base sequence may be complementary to any sequence along the DNA strand. You should examine the three gene probes and determine which one complements the unknown DNA. Which bacterium is present, and what color will the signal be when the binding has taken place between the probe and DNA molecule?

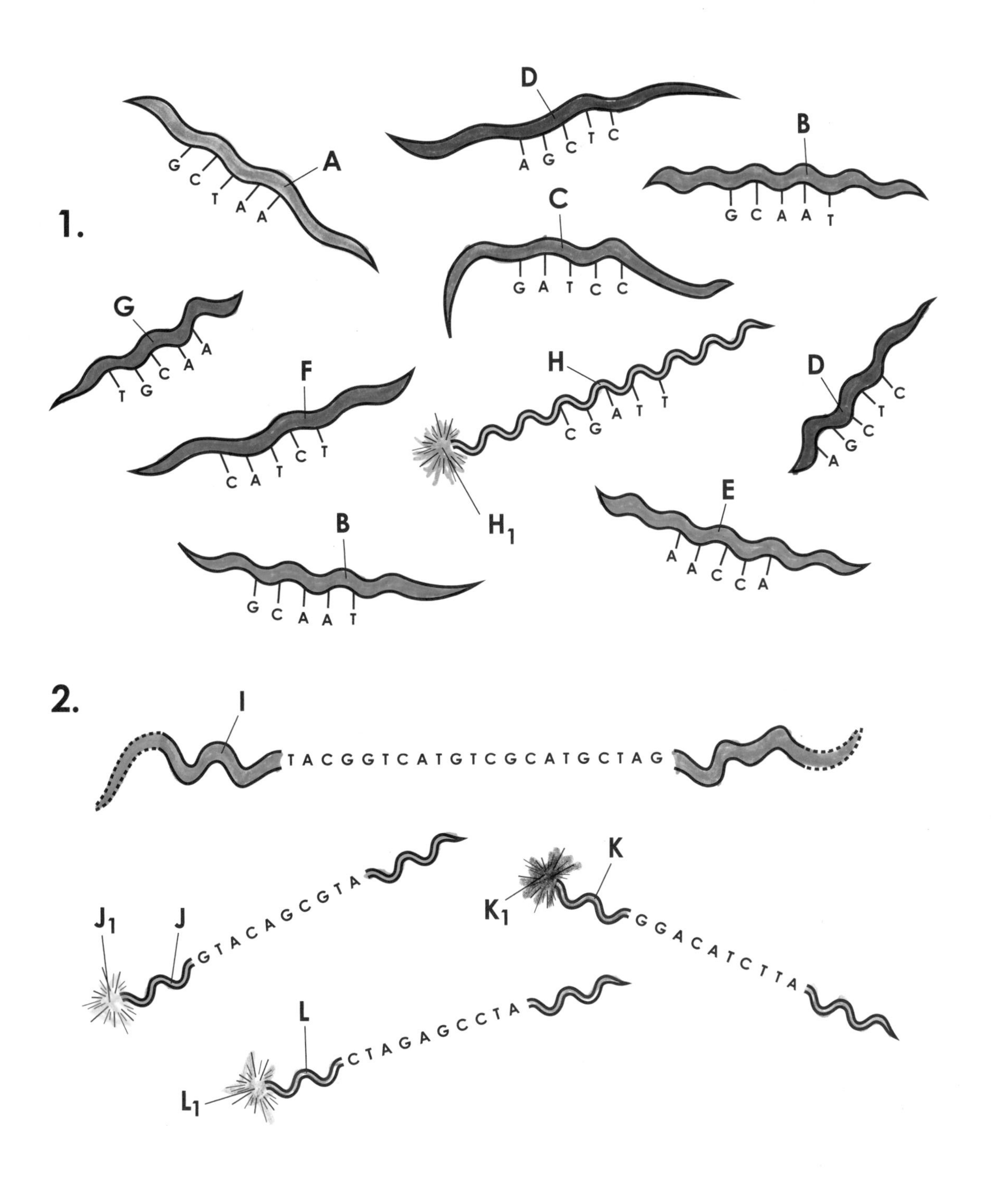

Unknown DNA #1 A
Unknown DNA #2 B
Unknown DNA #3 C
Unknown DNA #4 D
Unknown DNA #5 E
Unknown DNA #6 F
Unknown DNA #7 G
Gene Probe H
Radioactive Signal H_1
DNA from Unknown
Bacterium I
Gene Probe for *E. coli* J
Red Signal J_1
Gene Probe for *S. aureus* K
Blue Signal K_1
Gene Probe for *Salmonella* L
Green Signal L_1

DNA FINGERPRINTING

DNA fingerprinting is an extraordinarily useful investigative technique. It is a DNA identification procedure that requires only a single drop of blood, a few cells, or a hair fiber from a crime scene. From samples of this size, biochemists can remove a full set of genes. Should this DNA match a suspect's DNA, the prosecution has important evidence of the suspect's involvement. This plate will display the biochemical basis for DNA fingerprinting.

This plate consists of four diagrams that outline the procedure for DNA fingerprinting. We will proceed around the plate and examine the various materials and procedures that are used to match one of two suspects with the DNA from a crime scene. The DNA used has been obtained from a hair fiber collected at the scene.

In DNA fingerprinting, a minute sample of body tissue is subjected to DNA analysis and its owner can be identified with an exceptionally high degree of certainty. The procedure begins as shown in diagram 1. Several hair fibers have been found at a crime scene, and DNA has been extracted from them. This is the **DNA from the crime scene (A).** The lab has also obtained **DNA from suspect #1 (B)** and **DNA from suspect #2 (C).** White blood cells are the source of the suspects' DNA.

The first procedure in the technique is to cut the three samples of DNA into fragments with **restriction enzymes (D).** These enzymes were described in an earlier plate entitled Recombinant DNA Technology. Restriction enzymes cut the DNA samples into collections of different-sized fragments. The arrow (D) shows the places at which the restriction enzymes make their incisions.

Continue your reading to learn how the lab analyzed the DNA fragments of the two suspects to see if either one matches the DNA at the crime scene.

The DNA fragments that result from the restriction enzyme activity are known as restriction fragment length polymorphisms, or RFLPs. The RFLPs are analyzed by gel electrophoresis. We have the **RFLPs from the crime scene (E),** the **RFLPs from suspect #1 (F),** and the **RFLPs from suspect #2 (G).** They are all placed in different lanes on a gelatinous sheet of material called **agarose (H).** At opposite ends of the agarose sheet are **electrical poles that produce a current (I).** When the current is on, the RFLPs move from the negative pole to the positive pole at a rate that depends upon their size. The smaller RFLPs move more quickly and thus farther than the larger RFLPs in a given period of time. Thus, the gel acts as a molecular sieve and RFLPs spread out on the agarose.

Having separated the RFLPs, we are now ready to analyze the agarose and find out how far they have moved. Continue reading below, and color the diagrams as you proceed.

At this point, the RFLPs cannot be observed directly. The agarose sheet (H) is overlaid with a **nylon membrane (J),** and the RFLPs move out of the agarose gel and into the nylon membrane. The RFLPs stick to the membrane and preserve their separation pattern.

Next comes the identification step. A sample of **mixed DNA probes (K)** is added to the nylon membrane. DNA probes react specifically with the DNA fragments that they complement, as the previous plate indicates. The probes carry radioactive signals that cause dark bonds to occur where matches are made.

Now the location of the DNA fragments from the three sources can be identified. In the first lane, radioactive bands appear in two places. These bands represent fragments from the DNA that was recovered at the crime scene (A). Now look at the bands in lanes two and three. Which DNA matches the DNA from the crime scene? It is clear that DNA from suspect #1 (B) has fragments similar to the ones at the crime scene. By contrast, the DNA from suspect #2 (C) has a different pattern. From this observation, it may be concluded that suspect #1 was at the crime scene.

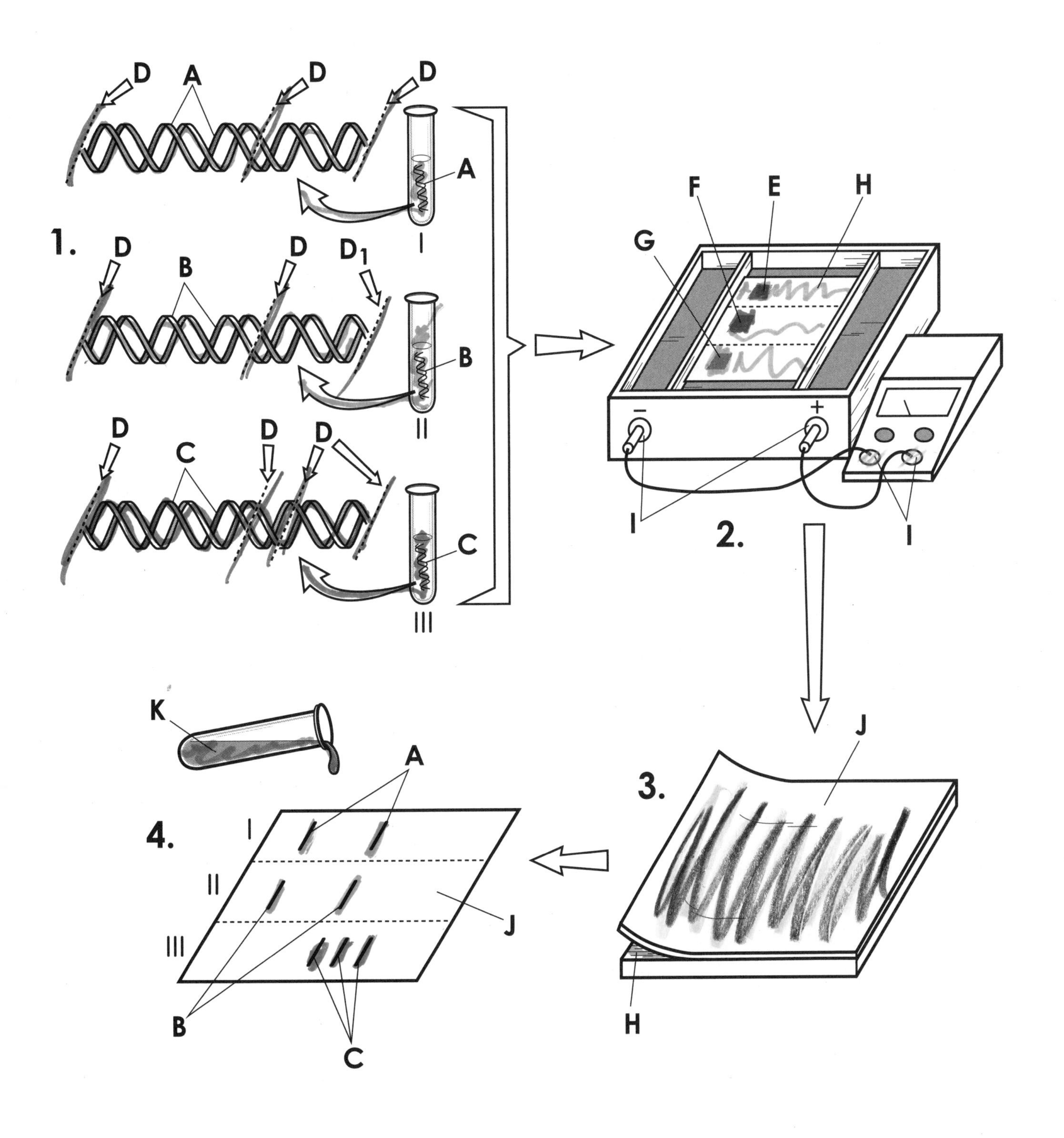

DNA from Crime Scene A
DNA from Suspect #1 B
DNA from Suspect #2 C
Restriction Enzyme D
RFLPs from Crime Scene E
RFLPs from Suspect #1 F
RFLPs from Suspect #2 G
Agarose H
Electrical Poles and Current I
Nylon Membrane J
Mixed DNA Probes K

GENE THERAPY

Gene therapy is a medical procedure in which cells are taken from a patient, altered by the addition of genes, and then replaced. These new genes code for the protein or proteins that that the patient is lacking. There are two types of gene therapy. In germline gene therapy, sperm or eggs (also called germ cells) are modified by the introduction of functional genes into their genomes. Modifying a germ cell causes all the organism's cells to contain the modified gene, and the change is therefore heritable and passed on to later generations. Many countries have regulations controlling this type of research, especially on humans.

In somatic cell gene therapy, therapeutic genes are transferred into any cell other than a germ cell. This means the modifications will affect only the individual patient, and are not inherited by offspring. This technique can be used to treat genetic disorders, especially those caused by one gene. Hemophilia, some immunodeficiencies, thalassemia, and cystic fibrosis are examples. Many clinical trials have been done using somatic cell gene therapy.

Substituting working copies of genes to replace defective ones is a practical outgrowth of recombinant DNA technology and genetic engineering, and a number of methods for transferring genes into human cells are under investigation.

This plate shows the process of gene therapy that involves a number of steps. As you color the diagrams, read the paragraphs below.

There are several ways of replacing a patient's defective genes with healthy ones. Some methods involve laboratory-cultivated cells and some involve synthetic carriers. In this plate, we demonstrate a method that uses cells that are derived from the **patient's body (A).** The **patient's cells (B)** are represented by one large cell in our diagram. Within the **cell nucleus (C)** are the 46 human chromosomes; here we show only three of the 46 **patient's chromosomes (D).**

Preparing genes for gene therapy is a lengthy and difficult procedure. First, the functional gene must be located, isolated, and amplified. PCR and a form of DNA fingerprinting (both discussed in previous plates) are used to do this. The **gene for insertion (E)** is shown on the diagram.

In order to transport the healthy gene into the cell, a carrier, or vector, must be used. Certain gene therapy experiments use **retroviruses (F)** as carrier particles. Retroviruses have an icosahedral coating and a genome made up of RNA. (Viruses are discussed in a separate plate in this book.

As you can see, the process of gene therapy begins when cells are isolated from the patient who is to receive the therapy. At this point, the functional gene has been isolated and amplified, and a suitable vector must be chosen to transport it into the cell. Continue your coloring as you read below. Light colors are probably best, since many of the details are quite small.

Retroviruses have the ability to cross the membrane of human body cells. In the plate, we show a retrovirus carrying a gene into one of the isolated patient's cells (B).

Once the virus has entered the cell, its **retroviral RNA (G)** is released. In the lower cell, we see the remaining capsid of the retrovirus. In the cytosol, the retroviral RNA (G) is used as a model for the synthesis of a molecule of **complementary DNA (H).** Normally DNA is used as a model to make RNA. Here, however, the RNA is used as a model for the synthesis of a DNA molecule. For this reason, the virus is called a retrovirus. The enzyme that brings about the synthesis of DNA is called **reverse transcriptase.** DNA made from an RNA template is called complementary DNA (cDNA). If cDNA is made from mRNA in a cell, it will contain only the exons of a gene and will thus be shorter than the version of the gene in the nuclear genome. The genomic version of genes also contains introns, as you learned about in a previous plate. Our cells don't normally express reverse transcriptase, but it is commonly used in molecular biology labs and is expressed by many viruses.

Let's return to the experiment on the plate. The new strand of complementary DNA inserts into a human chromosome, and when it does so, it carries along the inserted gene. Thus, in the second cell, we see the patient's chromosomes plus the **inserted gene (I).**

The final step in the process occurs when the **gene-altered cells (J)** are loaded into a syringe and then injected back into the patient's body (A). Here they will multiply, grow, and produce the proteins encoded for by the new genes.

One area in which gene therapy has been very helpful is in the treatment of cystic fibrosis. A major problem of patients with cystic fibrosis is the buildup of sticky mucus in the lungs, which is a result of the accumulation of ions within cells of the lungs. Normally, these ions pass out of the cells through a protein channel, but this protein is lacking in patients with cystic fibrosis because the genes that code for them are defective. In gene therapy, normal genes are placed in the patient and the protein is produced. This allows the ions to pass out of the cells and the sticky mucus begins to dissipate.

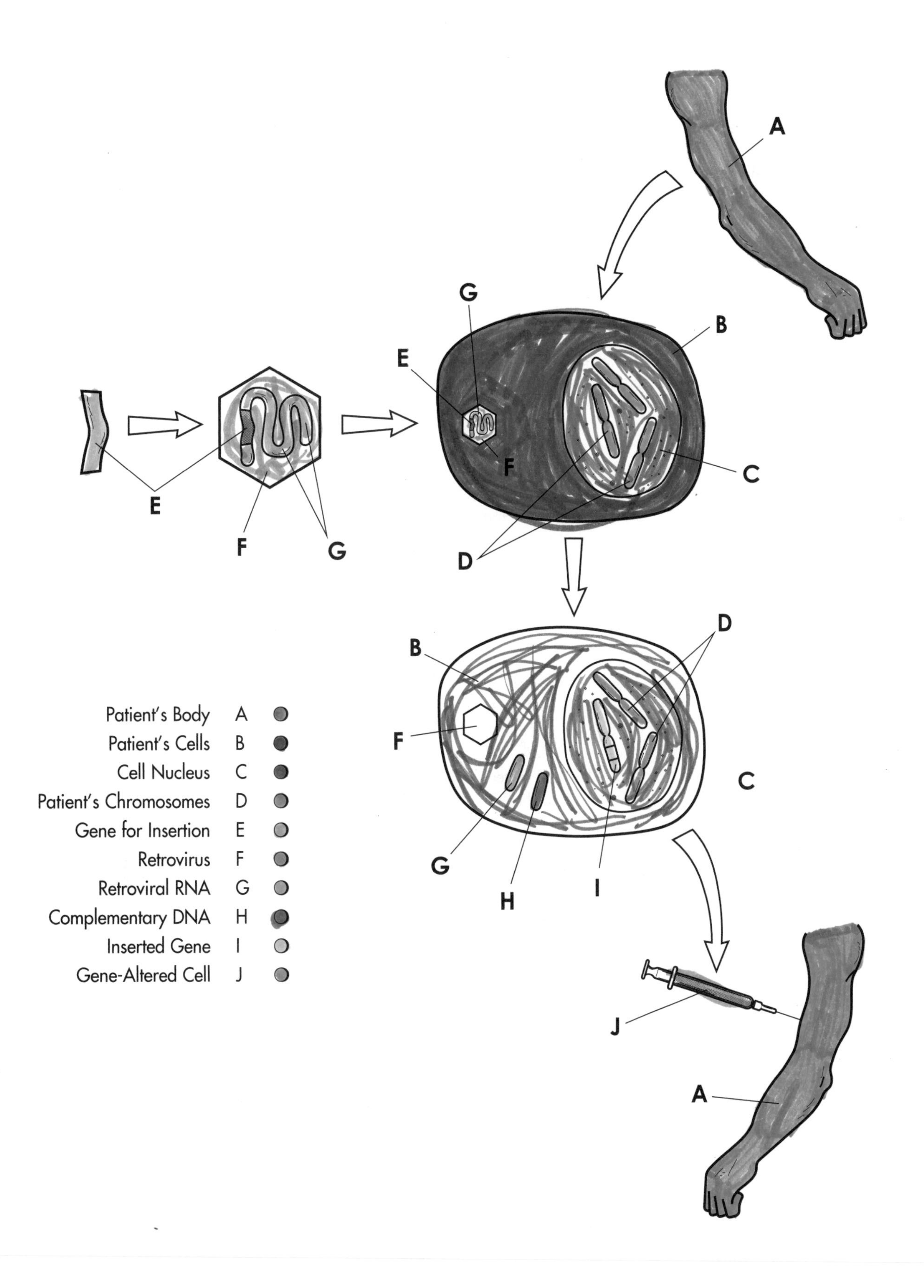
A
G
B
E
E
F
C
E
F
G
D
D
B
Patient's Body A
Patient's Cells B
Cell Nucleus C
Patient's Chromosomes D
Gene for Insertion E
Retrovirus F
Retroviral RNA G
Complementary DNA H
Inserted Gene I
Gene-Altered Cell J
F
C
G
H
I
J
A

ANTISENSE TECHNOLOGY

Through the decades, scientists have developed a large number of different antibiotics that unite with and then eliminate various microorganisms, which in turn cures disease.

Contemporary biotechnologists are trying out a new type of antibiotic. This type works within the cell to interrupt the biochemical mechanism of protein synthesis; it is called the antisense molecule. Antisense strands can be used to alleviate diseases and to produce certain agricultural products.

This plate contains two views of a cell; the first cell operated in a normal fashion, and the second cell has been treated with an antisense molecule. We will show how an antisense molecule is produced and how it can be used for practical benefit.

An antisense molecule is a synthetic RNA molecule that combines with the messenger RNA molecule found in a cell and renders it inactive. The synthetic mRNA has a base sequence that is complementary to the cell's regular mRNA.

To understand how an antisense molecule works, we quickly review the process of protein synthesis. This process takes place within the cell **cytoplasm (A).** A pale shade or gray color should be used to color the cytosol, in order to avoid obscuring other structures in the cytoplasm.

Within the nucleus of the cell is a double-stranded **DNA molecule (B).** This molecule provides the genetic code for the synthesis of a molecule of messenger RNA (mRNA). This **mRNA (C)** is produced in the nucleus of the eukaryotic cell, and then enters the cytoplasm. Here it provides the codons needed for the linking of **amino acids (D)** in translation, and the result of translation is the **protein (E).** We show the protein composed of different amino acids aligned in a chain.

Normally a protein benefits the cell that produces it in some way, but what is a benefit to the tomato plant may not necessarily be a benefit from the point of view of an eager farmer. For example, a certain protein causes tomatoes to ripen quickly; the enzyme polygalacturonase digests pectin in the cell walls of tomato plants and hastens their decay. This enzyme, not harmful to the plant, may not be beneficial to the tomato grower who wishes to slow ripening.

Thus far, we have reviewed the process of protein synthesis and mentioned that all proteins produced may not be beneficial in some practical situations. We will now use antisense technology to show how certain proteins can be eliminated from the cell. Continue your reading below as you color the diagram.

In antisense technology, the mRNA of a cell is first obtained and analyzed. Its base sequence is determined, and a synthetic mRNA is prepared that has complementary bases. Sequencing data, now readily available online for many different organisms, can help speed up the process of figuring out the sequence of an RNA molecule.

In the plate, we see a molecule of mRNA. We call it **sense RNA (F)** because it encodes for a protein. Biochemists synthetically produce a molecule of **antisense RNA (G).** Note that its nitrogenous bases are complementary to the bases in the sense RNA.

We have seen that an antisense molecule is prepared that complements the sense molecule. We will now see how the molecule has practical value in agriculture. Continue your coloring as you read the paragraphs below.

Let us now return to the tomato cell, and look at diagram 2. Once again we see the cytoplasm (A) and the DNA (B). A molecule of mRNA is encoded, and this RNA (F) leaves the nucleus and enters the cytoplasm.

But the scientist has added molecules of antisense RNA (G) to the cell, and their bases complement the bases in sense RNA, so that the two strands combine with one another. This eliminates the mRNA molecule.

The effect of this elimination is quickly seen. The amino acids (D) do not form the particular protein; the enzyme that encourages rotting is no longer produced. Without cell wall digestion, rotting slows considerably, and the tomato can be left on the vine to ripen longer. Furthermore, it will not undergo as much rotting during shipping and will remain ripe for about twice as long as the conventional tomato. Antisense technology has also been used in medicine, with practical benefits. For example, the human immunodeficiency virus (HIV) multiplies in human cells. With the use of antisense molecules, the mRNA used to synthesize some viral particles can be eliminated. As the amount of mRNA decreases, the symptoms of HIV infection are eased.

Many organisms regulate gene expression using a similar process, which is called RNA interference (RNAi). Some genes encode RNA molecules called microRNA (miRNA) and small interfering RNA (siRNA). These RNAs are not made into proteins. Instead, they bind messenger RNA (mRNA) molecules in the cytoplasm. These double-stranded RNA molecules (mRNA with miRNA, or mRNA with siRNA) are degraded by the cell. This is a way to inhibit gene expression; because the mRNA is broken down, the protein it codes for will not be made.

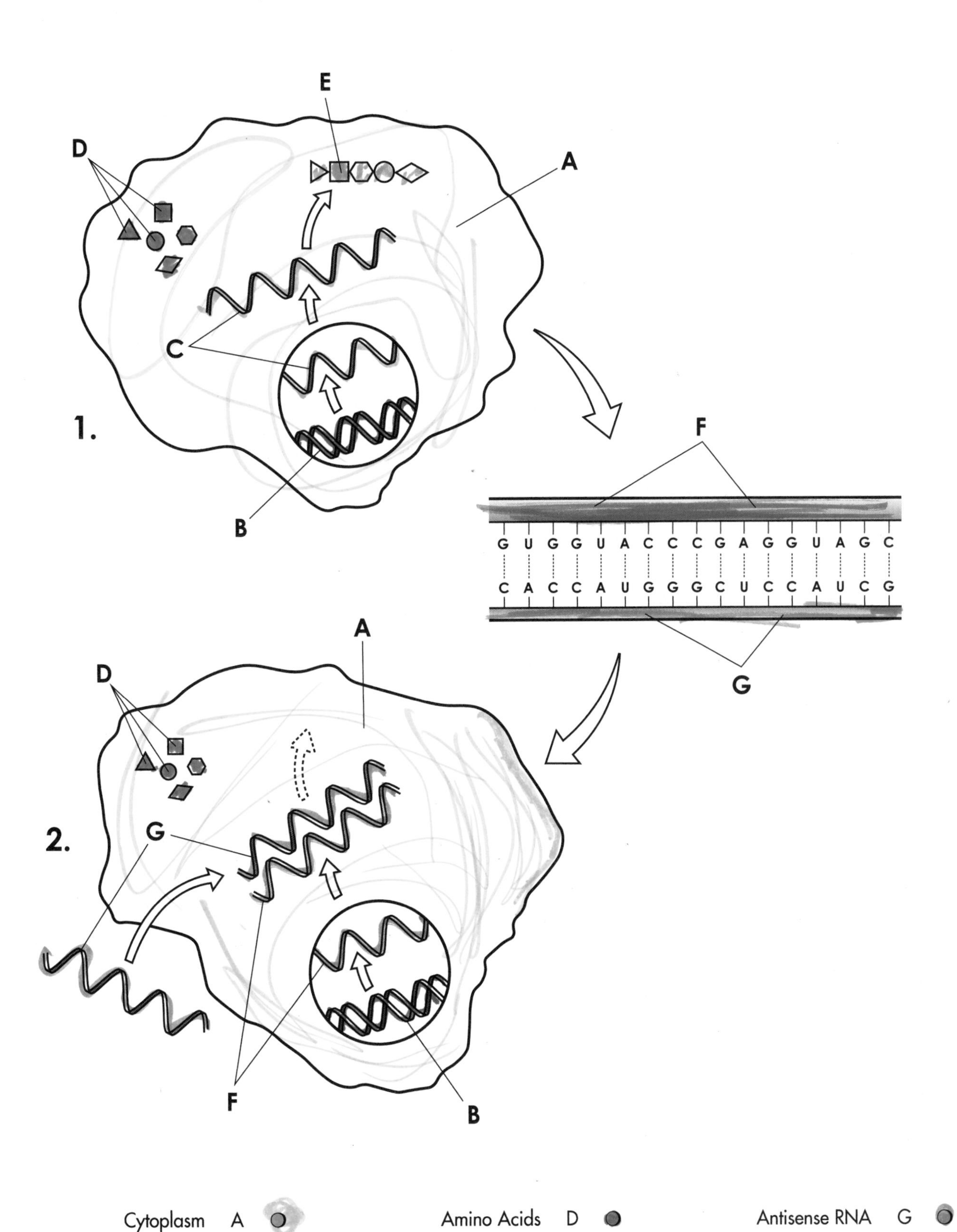

Cytoplasm A
DNA B
mRNA C
Amino Acids D
Protein E
Sense RNA F
Antisense RNA G

BLOTS: ANALYTICAL TECHNIQUES IN MOLECULAR BIOLOGY

Scientists use blotting experiments to study biological macromolecules. Simply put, blotting is the transfer of nucleic acids or proteins from an electrophoresis gel to a membrane. Once transferred, further experiments can be run to isolate or detect a particular nucleic acid fragment or protein (called "probing"). Blotting is classified by the type of molecule being probed.

Southern Blotting

Southern blotting allows you detect the presence of specific sequences within a heterogeneous sample of DNA. This process also allows you to isolate and purify target sequences of DNA for further study.

Step 1: Separate the **DNA fragments (A, B)** on an electrophoresis gel.

Step 2: Transfer the fragments to a **paper membrane (F)** using a **salt solution (C).** The salt solution passes through the **gel (E)** and membrane (F) on its way to the **paper towels (G).** It moves the DNA from the gel to the membrane as it does this **(H).**

Step 3: The membrane is "probed" for the target DNA sequence. **Hybridization probes (I)** are short single-stranded sequences of nucleic acid (usually DNA) that have two important features:

1. They are complementary to (and thus will base-pair with) a portion of the **target DNA sequence (J).**
2. They are constructed with radiolabeled nucleotides, which allows for the visualization of the target sequence with special film.

Step 4: Wash off unbound probe.

Step 5: Expose the X-ray film to the membrane, thus producing an **autoradiogram (K)** showing where the DNA of interest (bound to the radioactive probe) is located.

Northern Blotting

Northern blotting is almost identical to Southern blotting, except that RNA is separated via gel electrophoresis instead of DNA. The rest of the process is the same.

Western Blotting

Western blotting allows you to detect the presence of certain proteins within a sample and also serves as a diagnostic tool. You could determine, for example, whether cancer cells express certain tumor-promoting growth receptors on their surface. Here are the steps:

Step 1: Separate a **heterogeneous protein sample (L)** on an electrophoresis gel.

Step 2: Separated proteins from the gel are transferred to a membrane, usually using electric current.

Step 3: The **membrane (O)** is probed for the **target protein (P).** Probing for proteins in western blotting differs from probing in Southern or Northern blotting in that antibodies are used as the probes rather than nucleic acids. A **primary antibody (Q)** is used first, which will recognize only the target protein. Then, an **enzyme-linked secondary antibody (R, S)** is used to recognize the primary antibody. The enzyme on the secondary antibody will fluoresce when a **detection substrate (T)** is added, and this **detection signal (U)** can be photographed with special film. The target protein will show up as a band with an intensity that is proportional to the abundance of the protein in the sample.

Southern Blotting

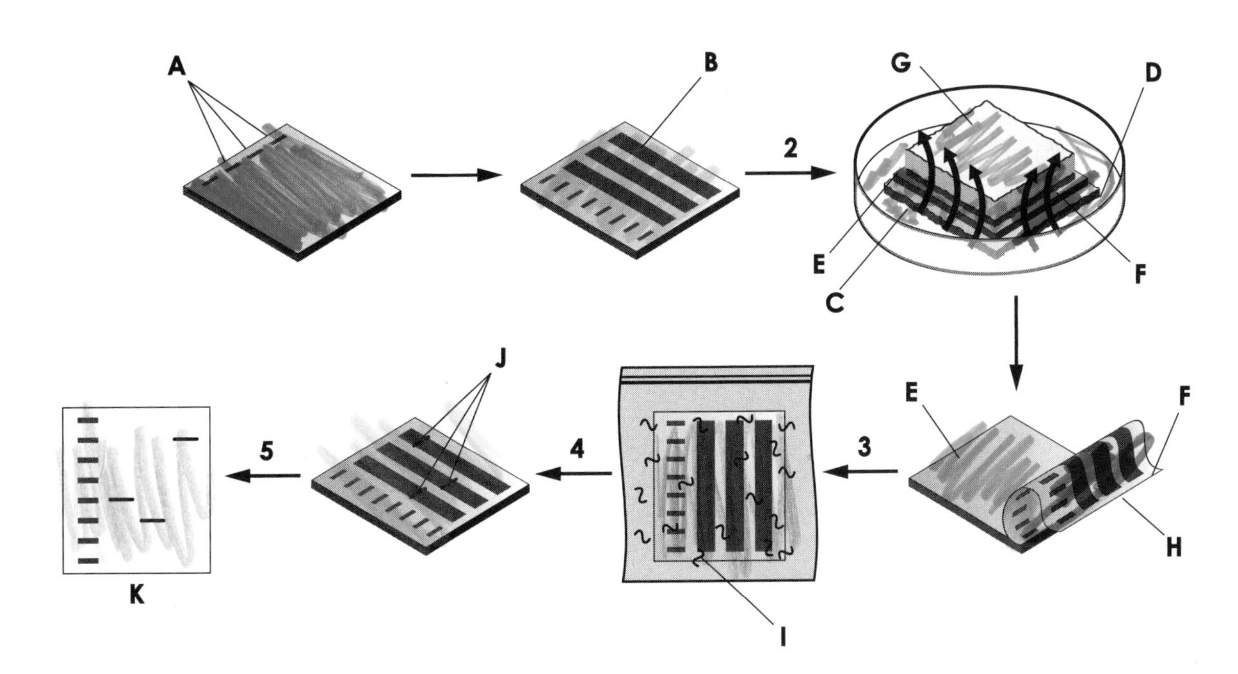

Western Blotting

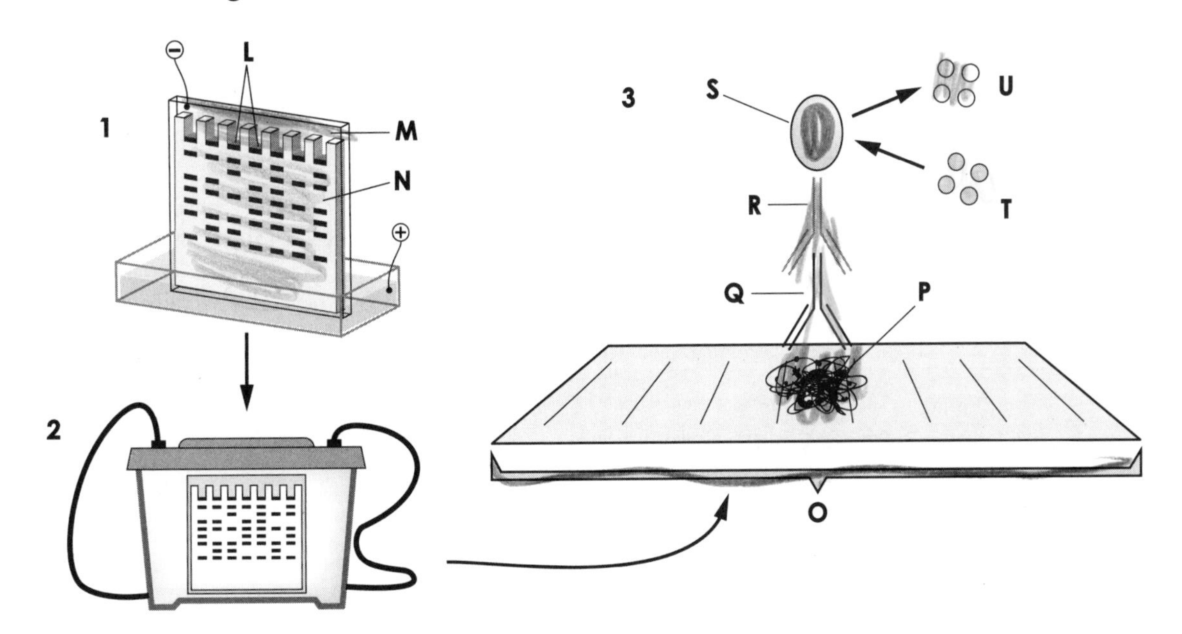

DNA Sample A
Separated DNA B
Salt Solution C
Sponge D
Gel E
Membrane F
Paper Towels G

DNA Transferred to Membrane H
Probe I
Hybridized Probe J
Autoradiogram K
Protein Sample L
Buffer M
Gel N

Membrane O
Target Protein P
Primary Antibody Q
Secondary Antibody R
Enzyme S
Detection Substrate T
Detection Signal U

TRANSGENIC PLANTS

DNA technology has allowed us to dramatically increase plant resistance to disease and grow plants under conditions that were once believed impossible. With DNA technology, the size of edible plant parts such as seeds, fruit, and storage roots may be significantly increased. In addition, foods may be rendered more nutritious by the alteration of their amino acid content, and it may also be possible to convert plants, which are notable for their production of carbohydrates, into producers of protein.

A genetically altered plant is called a transgenic plant. All the cells of a transgenic plant are derived from one cell so that all of the plant's cells express the same genetic information as that cell. For many reasons, plants are difficult to work with, but the results of genetic engineering have been impressive.

This plate displays the series of events involved in the production of a transgenic plant. Focus on the first diagram as you begin your reading below.

In order to develop transgenic plants, DNA technologists had to devise an insertion method by which DNA could be brought into plant cells. They focused their research on the **bacterium *Agrobacterium tumefaciens* (A).** The **chromosome (B)** of this bacterium is shown in the plate. Because the chromosome is large and difficult to work with, biochemists focus on a plasmid in the cytoplasm called the **Ti plasmid (C).** A light color should be used to color the cell, and dark colors should be used to trace the chromosome and plasmid. Together, *A. tumefacien* and the Ti plasmid cause plant tumors called crown galls; crown galls develop when a portion of the Ti plasmid inserts into a plant cell chromosome.

The Ti plasmid of the bacterium *A. tumefaciens* is the main vector for genes in transgenic plants. We shall now see how the Ti plasmid is used in DNA technology. Continue your reading below as you color. Light colors are recommended, since the structures are small.

The Ti plasmid (C) is isolated from the bacterium, and you should color it. You should use a different color to color the region of **T-DNA (D).** Shown in diagram 2, the T-DNA region leaves the Ti plasmid and inserts in the chromosome. A dark color such as blue or red is recommended for this region.

To prepare a recombined chromosome, technologists next obtain a **gene for insertion (E)** from a suitable donor organism. For example, they might want to enhance the nitrogen trapping capabilities of a plant, and use a gene that encodes an enzyme that traps nitrogen from the atmosphere. They would then biochemically combine this gene with the T-DNA region on the Ti plasmid. Diagram 3 shows the gene inserted within the T-DNA region. The **recombined Ti plasmids (C_1)** are then inserted into **fresh bacteria (F)** of the species *A. tumefaciens.* We see the plasmids in these bacteria in diagram 3.

Having recombined the plasmids and inserted them into the bacterium *A. tumefaciens,* we are now ready to transform a regular plant into a transgenic plant. Continue your reading as you focus on diagram 4 of the plate.

To produce a transgenic plant, biochemists must be able to cultivate an entire plant from a single, nonreproductive cell. Scientists have been able to cultivate plants such as cabbage, carrots, and potatoes in this way, but technology for the cultivation of other plants is less advanced.

In DNA technology, a series of plant cells are combined with the recombined bacteria, in a sterile medium of carefully balanced plant nutrients and hormones (diagram 4). Soon the cells develop into a random mass of cells called a leaf disk, also known as a **callus (G).** A few days or weeks later, the forerunners of roots, stems, and leaves appear on the callus, and soon **growing plants (H)** can be seen on the callus. Diagram 5 shows these plants.

These miniature plants are now transferred to individual containers until they achieve sufficient size and hardiness for planting outdoors. When the biochemist takes a sample of the **plant cells (I),** a **nucleus (J)** can be isolated, and a **plant chromosome (K)** can be examined. The chromosome contains the T-DNA region (D) of the Ti plasmid that was derived from the bacterium, and within this region is the inserted gene (E). A successful gene alteration has taken place in the plant, and the plant is now considered a transgenic plant.

Enhancing the nitrogen content of plants, developing plants' resistance to cold temperatures and insects, and producing new biotech foods are but a few of the possible uses of this type of biotechnology.

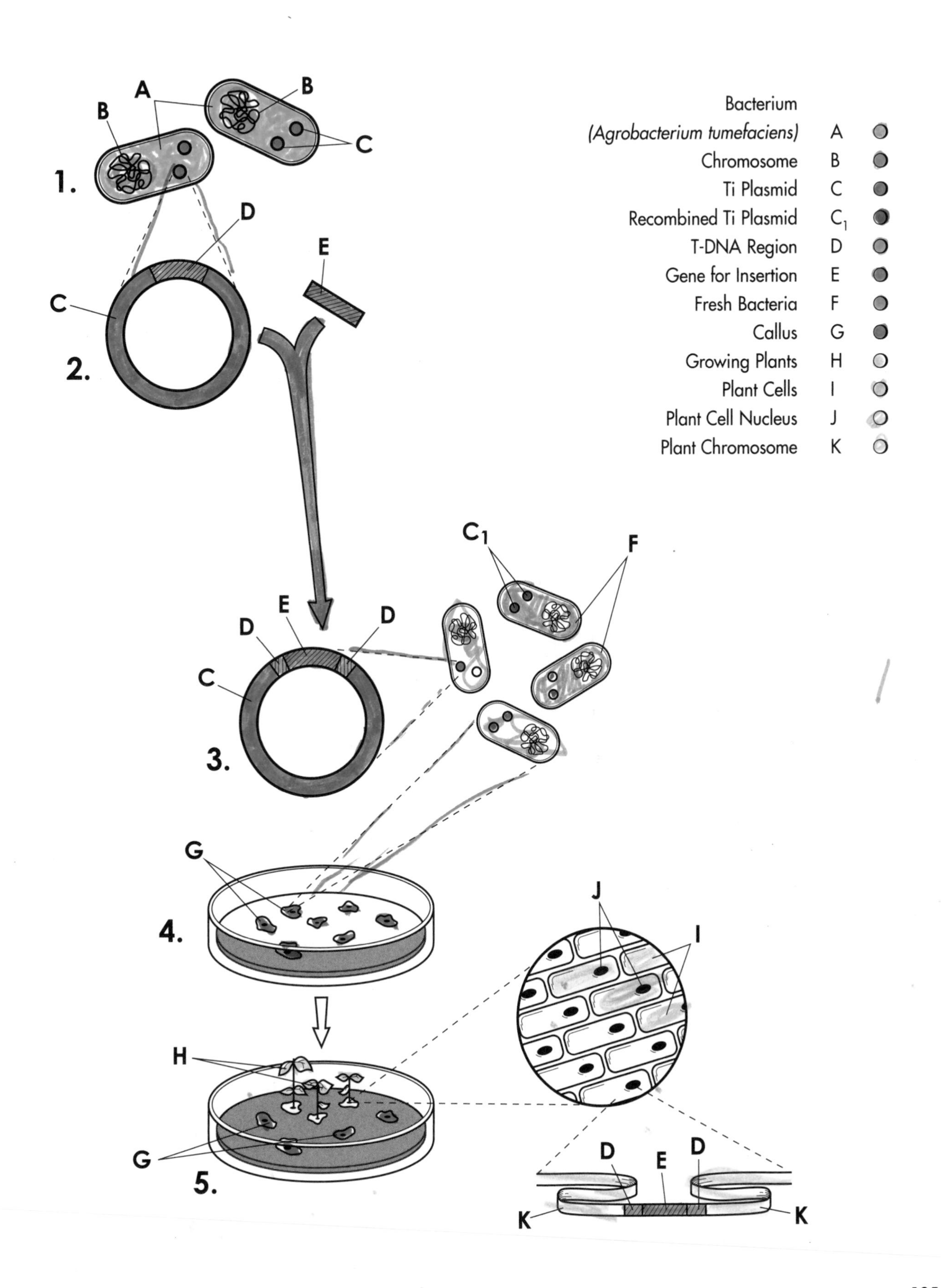

Bacterium
(Agrobacterium tumefaciens) A
Chromosome B
Ti Plasmid C
Recombined Ti Plasmid C1
T-DNA Region D
Gene for Insertion E
Fresh Bacteria F
Callus G
Growing Plants H
Plant Cells I
Plant Cell Nucleus J
Plant Chromosome K
1.
2.
3.
4.
5.
A
B
C
D
E
C1
F
G
H
I
J
K

TRANSGENIC ANIMALS

Stem cells have many important uses in biology. For example, therapy using **embryonic stem cells** (or ESCs) could revolutionize regenerative medicine and alleviate human suffering. Many diseases could be treated using these cells, such as blood and immune system genetic disorders, many cancers, spinal cord injuries, Parkinson's disease, juvenile diabetes, and blindness. The basic idea behind these stem cell therapies is to manipulate ESCs to become other cells for use in treatment. For example, ESCs induced to become oligodendrocytes have been used to treat patients with spinal cord injuries.

ESCs can also be used to make transgenic animals, which are often used to model human diseases. These animals allow scientists to understand how diseases develop and also to test potential cures. Embryonic stem cells from model organisms (such as mice and rats) can be isolated, grown, and manipulated in the lab. First, blastocysts are collected. A **blastocyst (A)** is a structure formed in the early development of mammals. It has an **inner cell mass (B)** and an outer layer called the **trophoblast (C).** The cells of the inner cell mass are isolated and are the source of **ESCs (D),** because they are pluripotent. This means they have the ability to become any cell in an adult organism. Color the inner cell mass (B) cells and embryonic stem cells (D) the same light gray color. This represents the fur color of the mouse they were collected from.

Next, ESCs are manipulated with transgenic or **recombinant DNA (E).** The procedure of transferring foreign DNA into animal cells is not perfect, so this results in a **mixed population of cells growing in the culture dish (F).** Color some of these cells the same light gray you used for ESCs (these represent normal cells) and color others a darker gray or black (these represent targeted or altered cells). Scientists must select the correctly targeted cells to continue with the experiment. The **targeted cells (G)** are then grown in the lab to make sure there are lots of cells to use for the next step. All the cells in this dish should be dark gray or black.

These **targeted ESCs (H)** can then be aggregated with a normal **morula (I),** a cluster of cells found earlier in development than the blastocyst. Color the targeted ESCs (H) dark gray or black, and the normal morula (I) brown. Again, this represents the fur color of the mouse the morula came from. The **chimeric (or mixed) morula (J)** is then injected into the uterus of a female animal that is given hormone injections to simulate pregnancy. She acts as a **recipient (L)** or surrogate mother for the transgenic animals. Once in the uterus, the chimeric morula develops into a **chimeric blastocyst (K)** and eventually a chimeric embryo. Color the cells in the chimeric morula and the chimeric blastocyst either dark gray/black or brown. It doesn't matter which cells you make which color; both structures contain a random mixture of both types of cells. Color the recipient female mouse brown.

A few weeks later, **chimeric pups (M)** are born, which are a mix of targeted stem cells (black) and normal stem cells (brown). This means they will have both fur colors; color the pups in (M) with a mix of black and brown fur. Finally, scientists look for "founder" animals, where the germ line was derived from the targeted ESCs. In this way, new transgenic lines can be generated and used for study. For example, a knock-in mouse could be made which over-expresses a gene of interest. Models are often made using tissue-specific promoters, so studies can be done on certain tissues without affecting all cells in the animal.

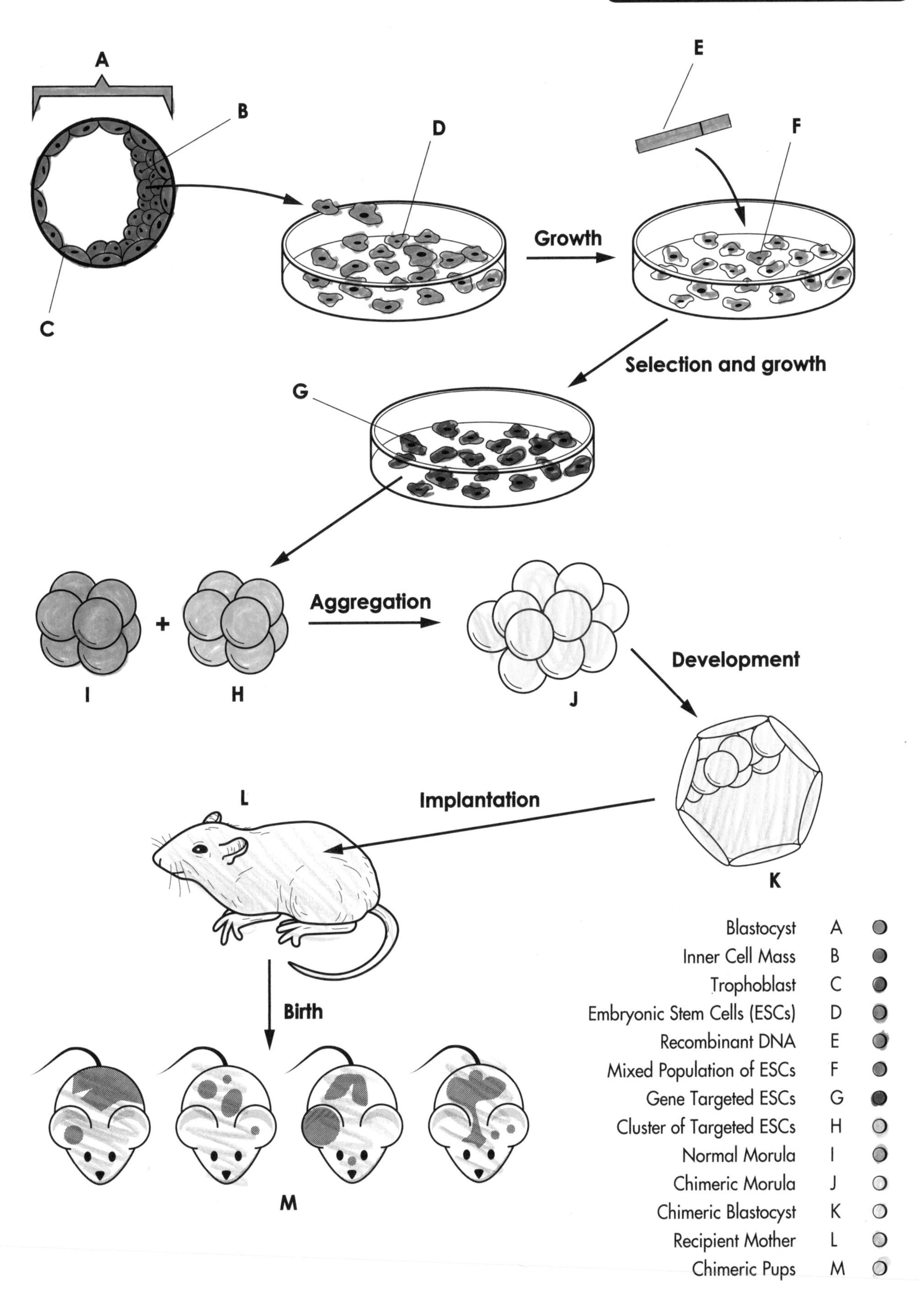
A
B
C
D
E
F
Growth
Selection and growth
G
I
H
Aggregation
J
Development
K
L
Implantation
Birth
M
Blastocyst A
Inner Cell Mass B
Trophoblast C
Embryonic Stem Cells (ESCs) D
Recombinant DNA E
Mixed Population of ESCs F
Gene Targeted ESCs G
Cluster of Targeted ESCs H
Normal Morula I
Chimeric Morula J
Chimeric Blastocyst K
Recipient Mother L
Chimeric Pups M

CHAPTER 5:

PRINCIPLES of EVOLUTION

INTRODUCTION TO EVOLUTION

The work of Alfred Russell Wallace and Charles Darwin resulted in the concept of evolution. This concept is one of the foundations of modern biology. Evolution refers to the change that occurs over long periods of time in the characteristics and diversity of biological organisms. One of the driving forces behind evolution is the process of natural selection.

This plate contains two diagrams that represent two theories for evolution: the Theory of Acquired Characteristics and the Theory of Natural Selection. In both cases, changes occurring in a giraffe population are presented. You may use either dark or light colors, depending on the detail of the giraffes that you wish to preserve.

The concept of evolution was considered in ancient times by Greek philosophers and since then, the idea that different kinds of organisms might have arisen from ancestral forms has appeared in historical and biological writings. For example, during the mid-1700s, the Swedish botanist Carolus Linnaeus cataloged thousands of different plants and animals, and speculated on their origins.

In 1809, the French scientist Jean-Baptiste Lamarck concluded from his observations that complex organisms evolve from less complex organisms. He proposed the idea that individuals acquire traits during their lifetimes that help them adapt to their environments. His theory, now discredited, came to be called the theory of acquired characteristics. These new characteristics, said Lamarck, are passed on to the offspring and result in generational changes.

Lamarck suggested that giraffes evolved their long neck during the course of their lives, in order to compete successfully for food. The diagram at the top describes his theory: Originally, **giraffes had short necks (A)** and could reach vegetation only in the lower branches of **trees (B).** However, as the vegetation in lower branches became sparse, the **giraffes stretched their necks (C_1)** to reach the higher branches. As this vegetation became more sparse, the **giraffe stretched even further (C_2).** Their longer necks were genetically passed on to their offspring.

Scientists do not believe that the theory of acquired characteristics is a suitable explanation of evolution. For example, if one were to cut the tail off a mouse and breed it, the newborn mice would still have full-length tails; the acquired characteristic would not be passed along. Furthermore, the theory suggested that organisms developed certain traits as they needed them, which is not plausible to modern scientists.

We now concentrate on the second theory for evolution, the Theory of Natural Selection. This theory, proposed by Darwin and Wallace, is now accepted. As you focus on the diagrams of this theory, read the paragraphs below and continue your coloring.

Charles Darwin and Alfred Russell Wallace traveled separately throughout the world, observing animals and plants in their natural settings. Independently, they developed ideas of evolution based on their observations and set the stage for the concept of natural selection. Wallace explored South America and Indonesia. Darwin's 1859 work, *On the Origin of Species*, remains the definitive book describing evolution.

The Darwin-Wallace concept of evolution concerns changes that occur over long periods of time. It proposes that all present-day species have evolved in a slow, gradual process from a few early, primitive organisms. The key to Darwin's argument as proposed in *On the Origin of Species* is the theory of natural selection. This theory proposes that variability exists in a population, and that the environment acts on this variation so that only the most fit individuals survive and reproduce.

The theory of natural selection is shown in the second diagram. In the first view, we see a population of giraffes. Some are **short giraffes (E)** and others are **tall giraffes (F).** The tall ones reach the leaves and vegetation above and obtain enough food to survive. In the second view, we see how the short giraffes have starved to death because their short necks did not permit them to reach as much food. The tall giraffes have been naturally selected to reproduce.

In the third view, we see that the tall giraffes survive and will continue to breed and reproduce to yield offspring that have long necks. This is what is meant by the phrase "survival of the fittest." Along with other processes, natural selection is a major factor in evolution. In succeeding plates, we will examine more of the evidence for evolution and review other processes by which evolution occurs.

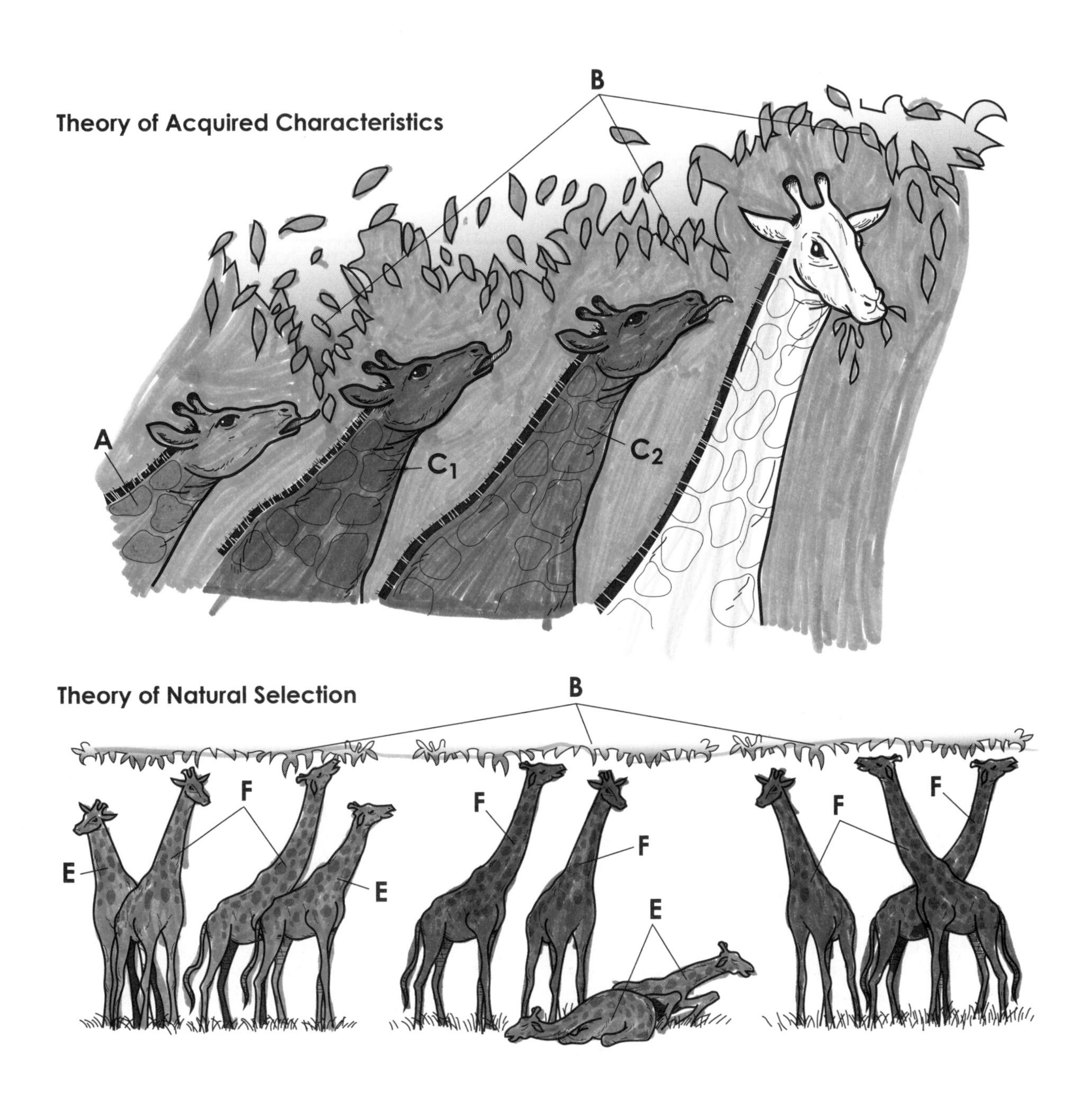

Theory of Acquired Characteristics

Short-necked Giraffe A 

Vegetation B

Intermediate-necked Giraffe C_1 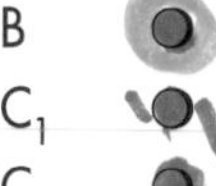

Long-necked Giraffe C_2

Theory of Natural Selection

Short Giraffe E

Tall Giraffe F 

DARWIN'S FINCHES

Perhaps the most significant event in the life of Charles Darwin was his appointment as naturalist in 1831 on the British survey ship, the HMS *Beagle*. During his five-year trip on the *Beagle*, Darwin visited Australia, several Pacific Islands, and South America. The ship also stopped at the Galapagos Islands, a remote chain of volcanic islands about 600 miles off the coast of South America. Darwin recorded much information on plants, animals, rocks, climate, and the native peoples he encountered in his now-famous notebooks. He took particular note of the island's population of birds, especially the finches. The observation of these finches would inspire him to form his theory of descent by modification, which we call evolution. In this plate, we examine the finch population as observed by Darwin, and note how they illustrate the concept of evolution.

> The finches in this plate serve as an example of an evolutionary pattern known as adaptive radiation. In this pattern, there is an evolutionary explosion in the number of closely related species from a common ancestor.

Finches are notoriously poor distance fliers, so Darwin was surprised to see 13 different species of finches on the **Galapagos Islands (A),** which lie approximately 600 miles west of **Ecuador (B)** on the western border of South America. Darwin categorized the different species of finches based on the bird's eating habits and the shapes and sizes of their beaks.

It is postulated that a small number of finches found their way to the Galapagos Islands many thousands of years ago, perhaps rafting over on floating debris. They may have been the first birds on the island, and with little competition, they multiplied into a large population. As they adapted to changing conditions on the island, the finches underwent natural selection and developed into different species.

One species of Galapagos finch is the **warblerlike finch (C).** This bird resembles a warbler very closely, but its eggs, nest, and courtship behavior are more similar to those of finches. The warblerlike finch evolved to resemble a warbler even though it is not of the same species; the absence of competition allowed its evolution. If true warblers had been present on the island, they would have occupied the niche normally held by warblers and this finch probably would not have appeared.

> With the warblerlike finch, we introduce the 13 species of finches seen on the Galapagos Island and nowhere else. Continue your reading below and notice how different finches have characteristics that vary according to their particular needs.

A second finch observed by Darwin was the **insect-eating finch (D).** This finch lives in trees and has a heavy beak that allows it to grasp insects. Its body shape is adapted for life in the trees. Notice how its beak and general size compare to that of the warblerlike finch.

Another interesting species observed by Darwin was the **woodpeckerlike finch (E).** This species has a beak like a woodpecker's (long and narrow) and it uses its beak for drilling holes in wood. However, it has no tongue for removing insects from the wood, and instead uses a cactus spine to dig insects out of the wood. If woodpeckers were present on the island, they would have occupied the niche for this type of animal and this finch would not have appeared.

The next finch is the **plant-eating finch (F).** Again, notice the different shape of the beak. This animal lives in trees and eats only plants. Its beak is designed for grabbing and tearing plant stems and roots.

> We have now seen four different types of finches found only in the Galapagos Islands. Differences are apparent in their sizes and beak shapes. We conclude the plate by observing two other finches that live on the islands. Continue your reading below as you color the appropriate pictures.

One of the more interesting finches is called the **ground finch (G).** This bird has a massive head and beak, as the diagram shows. It feeds on large seeds and nuts and its beak is adapted for crushing their hard shells.

Another finch that feeds on the ground is the **cactus-eating finch (H).** This bird subsists on cactus seeds. It uses its bill for crushing, but since cactus seeds are relatively soft, its bill is smaller.

The evolution of the 13 species of finches from a common ancestor is an example of adaptive radiation. In adaptive radiation, different types of species radiate out as they adapt to different environmental conditions. As you'll see in the next plate, another type of adaptive radiation occurred when animals first invaded the land masses of the Earth.

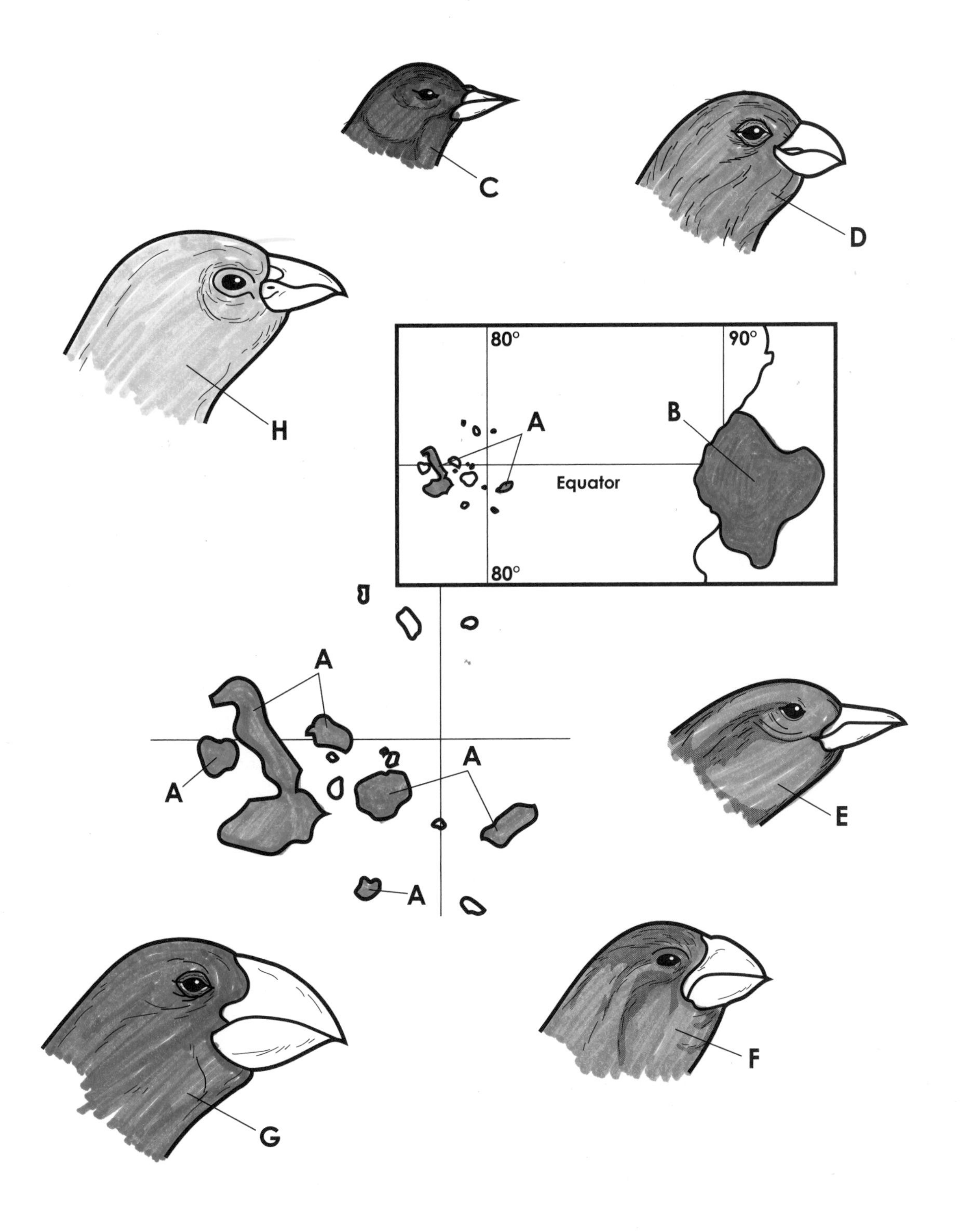

Galapagos Islands A
Ecuador B
Warblerlike Finch C
Insect-eating Finch D
Woodpeckerlike Finch E
Plant-eating Finch F
Ground Finch G
Cactus-eating Finch H

EVIDENCE FOR EVOLUTION

According to the Theory of Evolution, all existing biological species are derived from previous ones; that is, all organisms, past and present, share a common ancestry. Data derived from various fields including paleontology, genetics, biochemistry, anatomy, and embryology provide an overwhelming mass of evidence consistent with evolutionary theory. Two important pieces of evidence for evolution are found in comparative anatomy and comparative embryology, as we see in this plate.

As you look over the plate, you will notice that it has two sections: one devoted to comparative anatomy and one to comparative embryology. Read about comparative anatomy below, and focus your attention on the upper portion of the plate.

Comparative anatomy is basically the science of comparing the physical features of present-day organisms. The plate presents the forelimb of a human, dog, bird, and whale. The upper forelimb in the human is the **humerus (A),** which is also found in the forelimb of the dog, bird, and the flipper of the whale. The same dark color should be used for each. Below the humerus is the **radius (B),** and similar bones are found in the dog, bird, and whale. Parallel to the radius is the **ulna (C)**, and again, we find the same bone in all four animals.

Next come a set of small bones in the human wrist called **carpals (D);** similar bones are found in the three other species. The carpals are followed by the **metacarpals (E),** and all four creatures have the same bones. In the human, the metacarpals are found in the palm of the hand. The forelimb is completed with a set of **phalanges (F),** which are also found in all four species.

The science of comparative anatomy shows us that homologous structures composed of the same bones are found in four different animals. At first glance, the forelimbs of humans, dogs, birds, and whales appear to have little in common. But from an evolutionary standpoint, the similarity of these homologous structures shows that the basic structure of the forelimb has been modified through natural selection into arms in humans, front legs in dogs, wings in birds, and flippers in whales. These anatomical similarities provide evidence for the evolutionary descent of the four animals from a common ancestor. Modifications for different purposes have occurred through time, but the supporting bones remain very similar.

We now turn to the second piece of evidence for evolution: comparative embryology. Your focus should be on the lower half of the plate, where you will see four columns of embryonic, fetal, and newborn illustrations. As you read about comparative embryology, color the appropriate structures in the plate.

Evidence for evolution can also be seen by comparing the embryos of different animals. Looking across the first row from left to right, the plate shows the **embryos of a fish, tortoise, chick, and human (G_1–J_1).** The same color can be used for each. Note the great similarity between the four embryos. (The embryo is the name of the structure from the early hours after fertilization until the point at which the organs are fully formed.)

After the embryonic stage comes the fetal stage. The **fetuses (G_2–J_2)** of the four animals are shown in the second row. The same color may be used for the four to indicate the fetal stage. Notice that the fetuses appear different from one another. The fetal stage extends from the time that the organs are fully formed to the time of birth.

In the bottom row, we see the **newborn stage of the fish, tortoise, chick, and human (G_3–J_3).** The same light color may be used for all four. At this point we can see how extremely different the newborns are from one another.

Early embryos of fish, tortoises, chicks, and humans all display fishlike structures, including arched blood vessels and gill slits. In fish, these gill slits develop into gills, while in animals such as humans, the gill slits never become functional.

The similarities of these embryos demonstrate that certain developmental processes remain constant during the evolution of animals. The similarities show that in the process of evolution, pre-existing structures were adapted to serve new functions. The parallels in embryonic structures would be difficult to account for in any way except through evolution.

Comparative Anatomy

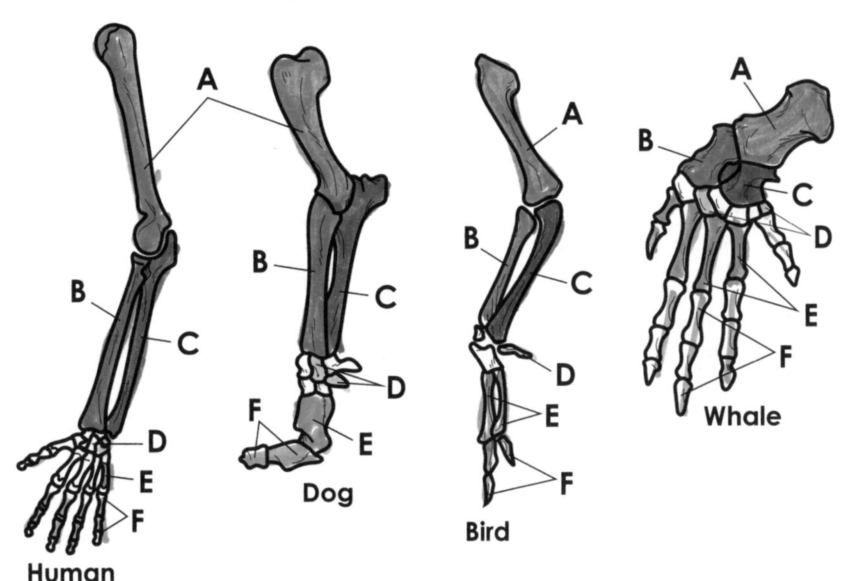

Comparative Embryology

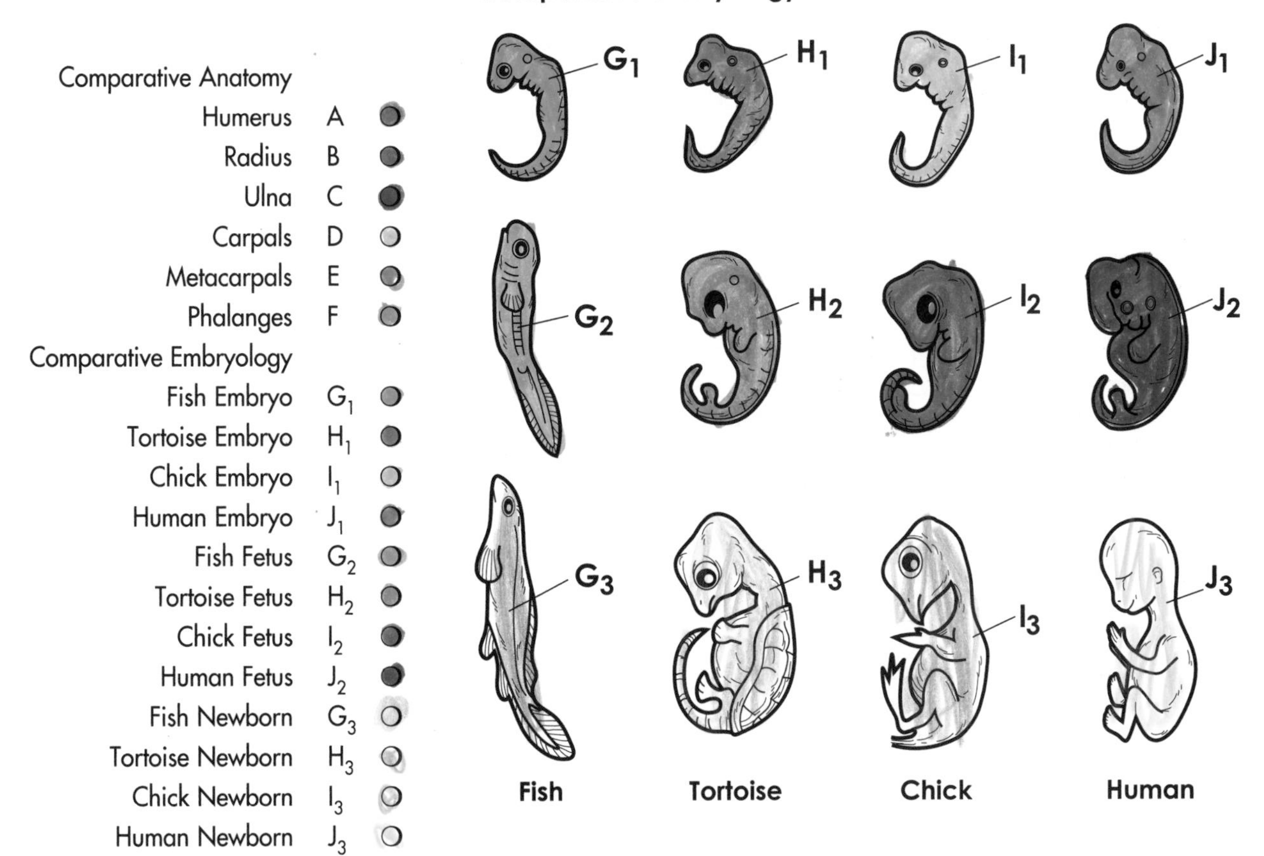

AN EXAMPLE OF EVOLUTION

Evolution suggests that existing biological species are derived from previous ones by descent; that is, all organisms now on Earth share a common ancestry. For this reason, evolution is a unifying theme in biology. It provides a framework for the organization of diverse species into a linked pattern.

The evolution of the horse is a well-documented example of evolution. By comparing the bone structures and size of ancestral horses from several different periods, scientists have been able to trace the evolution of the modern horse from its ancient predecessor. In this plate, we will trace its evolution.

This plate shows four ancestors of the modern horse. You can see an overall increase in its size, which was determined through a study of the animals' fossilized bones. Begin your study at the bottom of the plate and work your way up.

The oldest known predecessor of the modern horse is ***Eohippus* (A).** This animal should be colored in a light color, and the same color should be used for the animal's forelimb and skull. This horse-like creature was a small mammal that was approximately two feet tall and lived in forests where it hid from its predators. Its forelimbs ended in four toes, and its feet lay flat against the ground. This animal lived during the **Eocene Epoch (A_1),** which occurred approximately 60 million years ago, and during this period it apparently diverged into two other **species that resembled horses (B).**

We will now turn our attention to a relative of *Eohippus* and point out the evolutionary changes that took place in *Eohippus* to give rise to a new species of ancestral horse.

After the Eocene Epoch, a new ancestral horse appeared, called ***Miohippus* (C).** This animal arose during the **Oligocene Epoch (C_1),** approximately 40 million years ago, and was only slightly larger than Eohippus. Miohippus also inhabited the forest, and it had three functional toes, all of which touched the ground. The fourth toe originally present in *Eohippus* had regressed, and the horse's skull was larger.

At this point, environmental pressures caused *Miohippus* to diverge and a number of horse-like relatives arose. One of these horse-like relatives evolved into the third ancestral horse that we'll look at in this plate.

The third species was ***Merychippus* (D).** This horse arose during the **Miocene Epoch (D_1).** Its skull was larger than the skull of its predecessor, *Miohippus*. This animal was pony-sized and had three toes, with only one functional center toe. *Merychippus* lived in grasslands, which meant that it could not hide from its predators and had to outrun them. Evolution therefore favored a species that had longer and stronger legs, and could run quickly.

We have seen that environmental changes caused the development of two successive types of horse from *Eohippus*. The remaining two types of horse are seen in the final sections of this plate. Continue your reading and color the animals in the plate as you proceed.

The **Pliocene Epoch (E_1)** occurred seven million years ago. By this time, *Merychippus* had evolved into a larger animal with a larger skull, called ***Pliohippus* (E).** The foreleg remained a single hoof with two lateral toes, and the horse was a running animal that grazed in the extensive grasslands of its environment.

The modern horse, ***Equus* (F)** arose during the **Pleistocene Epoch (F_1),** which occurred approximately two and a half million years ago. In the plate, you can see the modern horse with one large central toe, which is what we call a hoof. The front limb, however, retains vestiges of the lateral toes in the form of pairs of bones that are not prominent externally.

Through their study of a series of fossils, scientists have traced the evolution of the modern horse from its ancient ancestors. Natural selection caused changes in this animal that allowed it to adapt to its changing environment.

AN EXAMPLE OF EVOLUTION

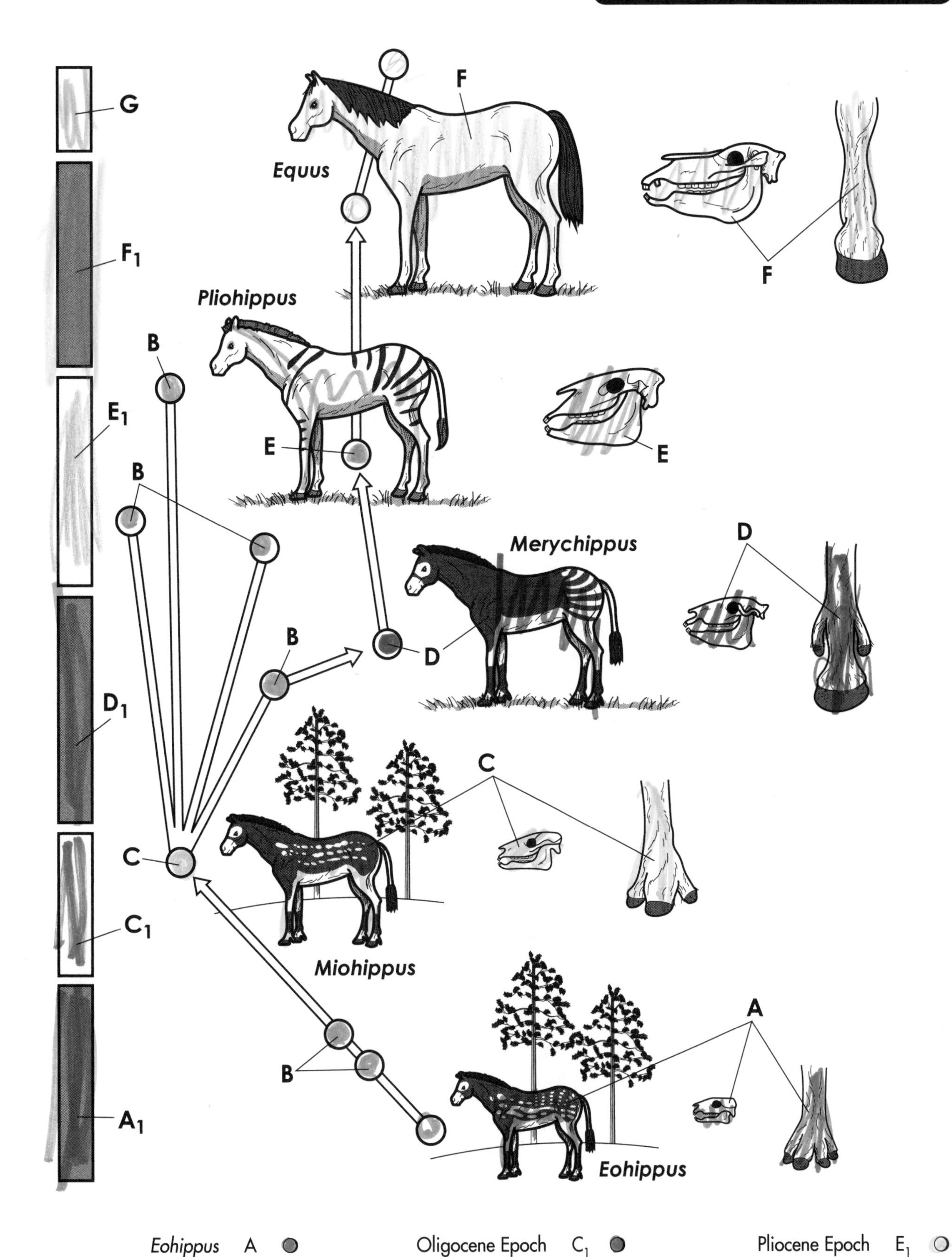

Eohippus A
Eocene Epoch A_1
Horse-like Species B
Miohippus C
Oligocene Epoch C_1
Merychippus D
Miocene Epoch D_1
Pliohippus E
Pliocene Epoch E_1
Equus F
Pleistocene Epoch F_1
Recent Epoch G

THE GENE POOL

One key to understanding evolutionary mechanisms is to determine the factors that create genetic variation among members of a species. It is these variations that are acted on in the process of natural selection that bring about evolutionary change. Without variation, natural selection is not possible.

Genetic variation is brought about by the processes of gene flow and genetic drift, which are discussed in other plates. In this plate, we will explore how variation can be produced by variations in the gene pool.

In this plate, we will follow the history of a herd of animals that feed on a grassy plain. We will show how the animals are affected by a catastrophe and how this results in genetic variation. Once variation has taken place, natural selection may or may not follow.

The phenomenon of evolution is measured in the study of individuals, but individuals do not evolve; species do. Herds and populations, for example, are subject to the forces of variation and natural selection, both of which are agents of evolution. Populations undergo changes in genotype and phenotype, and are the smallest units of living organisms that undergo evolution.

Within populations, individuals possess different combinations of genes, and all members of the population contribute their genes to what is called a gene pool. In other words, the gene pool is the sum of all the individual genes in a population. The forces of evolution change the composition of the gene pool and in doing so, change the characteristics of the members of the population.

Variations of a gene are called alleles, and different combinations of alleles produce different phenotypes in individuals. In diagram 1, we see a large group of animals with different phenotypes. **Dark-skinned animals (A)** have two dominant B genes, **medium-skinned animals (B)** have a dominant B allele and a recessive b allele, and **light-skinned animals (C)** have two recessive b alleles. The different animals should be colored with three different bold colors. A light color should be used to outline the box and shade the gene pool. Notice that the gene pool consists of both B and b genes.

Having defined the gene pool, we will now see what happens when a catastrophe occurs that causes a change in the pool, which in turn affects the population. Your attention should now be directed to the second diagram.

Say that, at some point in the history of the population, a catastrophe strikes. We will use the example of a fierce rainstorm that eliminates all but two members of the population. We see, in diagram 3, that two light-skinned animals (C) have survived. We also note that the gene pool now consists solely of b genes.

An event such as this, in which many individuals are killed, leaving only a small number of survivors, is known as **the bottleneck effect (D).** As you can see, the gene pool of the survivors is very different from the gene pool of the larger population. Both animals are light-skinned, reflecting a single phenotype. The small number of individuals surviving are called founders, and the phenomenon is called the founder effect. It need not result from catastrophe. Instead, two individuals might wander off from their herd, to found a new population.

Focus your attention on the fourth diagram to see the effect of the change on the gene pool.

Now a large amount of **time passes (E).** In the art, this passage of time is represented by an arrow. One of the two surviving animals is male and one is female, and they breed to eventually produce a new population of animals. The gene pool of this new population is different from that of the original population, in that it consists solely of b genes. Moreover, the animals are all light-skinned. This may affect the population; for instance, predators might now be able to see the animals more easily, which makes them easier prey. Adaptations may be useful, detrimental, or they may have no effect on a population at all.

Dark-skinned Animals A
Medium-skinned Animals B
Light-skinned Animals C
Bottleneck D
Passage of Time E

GENE FLOW

The forces of evolution take place within a population, which is defined as a group of interbreeding members of a species living in a geographical area. The collection of genes within this population is referred to as the gene pool. Gene flow is the addition of an immigrant's genes to the gene pool of a population that leads to altered allele frequencies in the population. In this plate, we examine how gene flow takes place in a population of fish as we explore its history.

The history of the fish population is seen in four views, represented by the four rows of diagrams in the plate. As the history proceeds, we will see how new genes arise in a population, changing the character of it. Plan to use medium colors, and watch for the arrows, which are an integral part of the plate.

The process of gene flow begins with an original **fish population (A)** living in a lake. The fish are all of the same species, so the same medium color should be used for all of them. The gene pool in this population is stable and the phenotypes of all fish appear the same.

At some point in the history of the population, a climatic change occurs, and the lake begins to dry at a high area across its center. When this occurs, a **separation of the population (B)** occurs, indicated by the arrows. The result of this is seen in the second diagram, where the original fish population is split into two populations. You should use the same color that you used in the first diagram because the gene pool remains the same, despite the population division. One group has now been geographically isolated from another group of the same species.

Now we have seen how members of a population can become geographically isolated by an environmental change. Geographic isolation results when two populations are unable to interbreed. Note, however, the gene pool is still the same, despite the separation. We now continue with the history of the population and investigate the introduction of new genes.

The two fish populations continue to live separately. At some time in the future, it is possible that a mutation may occur in the population on the left. This is the **first mutation (C_1).** While this is happening, a **second mutation (C_2)** may take place in the second fish population and a new gene will be introduced to both groups. In the lake on the left, there is **a new population (D),** while in the lake on the right the **second new population (E)** arises. Different colors should be used for the fish, because they now have different genetic compositions. The gene pools of the two populations are different. It should be noted, however, that new species of fish have not arisen. Instead, some new genes have entered the population. The species remains the same because the fish in the two lakes can still breed with one another.

Having established two different gene pools in two different populations of fish, we are now ready to examine the process of gene flow. Your attention should be focused on the fourth diagram as you continue your reading below. The same colors used in the third diagram should be used in the fourth, because the gene pool has not changed.

Let us suppose that at some time in the future, a climatic change causes a channel to form between the two lakes, so that the fish are no longer isolated geographically. As a consequence of the union, a fish of the second population swims into the first population in a **first exchange (F_1).** Similarly, a fish of the first population swims to the second population, in a **second exchange (F_2).** The addition of each immigrant fish brings its genes into the populations. Genes are exchanged between the two groups during interbreeding. If the stranger brings variations into the population, then the effect of gene flow is similar to that of a mutation. In this way, variability is introduced to the population and new combinations of genes are possible in the gene pool.

Contribution of gene flow to evolutionary change depends on the rate of immigration and the amount of genetic difference between the immigrants and the host population. For example, when the immigration rate is low (one fish has entered the population in our example), the effect on the host population's gene pool is small. By contrast, if many immigrants enter the population, the effect will be great and the contribution of gene flow to evolutionary change will be substantial.

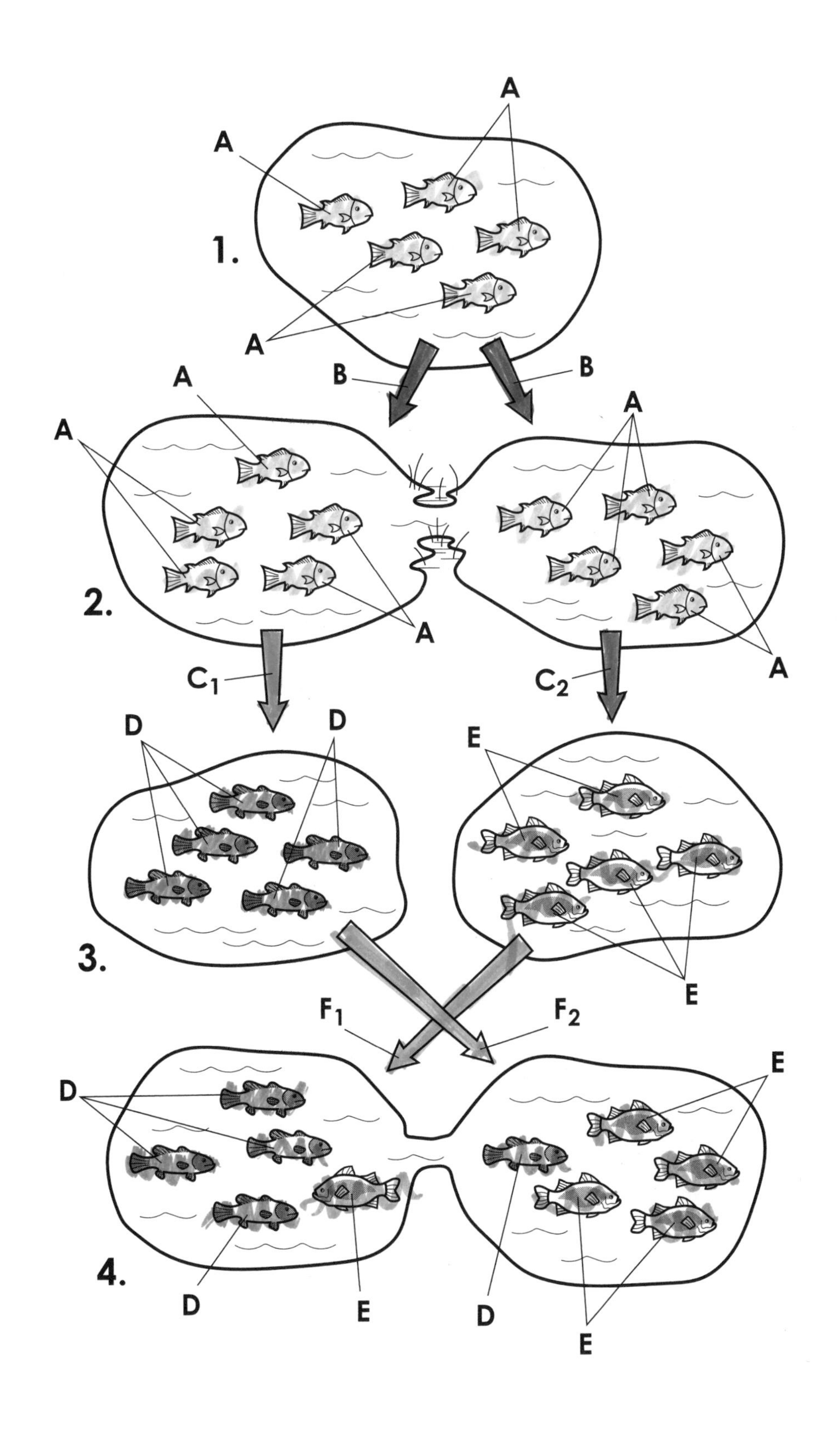

Original Fish Population A
Separation of Population B
First Mutation C_1
Second Mutation C_2
First New Population D
Second New Population E
First Exchange F_1
Second Exchange F_2

GENETIC DRIFT

The total number of genes in a population, also called the gene pool, can undergo changes that lead to variations in the physical characteristics of a population. One of the forces that changes the gene pool and leads to evolutionary change is genetic drift.

When the gene frequencies of a population change, genetic drift has taken place. The evolution that accompanies genetic drift occurs most often in small, isolated populations, as we will show in this plate. Larger populations are less susceptible to the force of genetic drift. The changes brought about by genetic drift are not necessarily adaptive, because pure chance brings them about. Genetic drift may work to the detriment of a population, rendering it more poorly adapted to its environment.

In this plate, we examine how genetic drift can change the gene pool of a population, and note how the rate of genetic drift is much greater for a smaller population than a larger one.

As you look over the plate, you will note that we are presenting two populations of fish in separate lakes. One population is relatively large, and the other is relatively small. We see three events in the history of the populations: the original populations in the first diagram, a removal of genes from the populations in the second diagram, and the result of the genetic drift in the third diagram. Only two colors should be used for this plate. We recommend red for the darker fish, and yellow for the lighter fish.

We begin our study of genetic drift by noting the two populations of fish. In the first lake, we see a small population of fish with two **red fish (A)** and four **yellow fish (B).** In this case, the ratio of dark fish to light fish is 2:4. This ratio may be expressed as 0.5.

Looking at the second lake, we see a much larger population of fish. In this case, there are eight red fish (A) and sixteen yellow fish (B). The ratio of red to yellow fish is 8:16, or 0.5, the same as it was in the first lake. Thus, the ratio of red genes to yellow genes is the same in the two lakes.

Having established the gene pools of the two populations of fish, we will now see how a chance event affects each population. The random event removes a certain number of genes from each population. Focus your attention on the second diagram of the plate.

In the second diagram, we see that a fisherman has come to each lake. In each case, the fisherman has caught a red fish (A). This random event changes the characteristics of the gene pools in each lake and may change the evolutionary direction of the population.

In the third diagram, we see the effects of the random event. Because red genes have been removed from the populations, changes in the gene pools occur.

The loss of one fish in diagram 2 has resulted in changes in both populations. In the small population, there is now one red fish and four yellow fish, in a ratio of 1:4, or 0.25. Thus, the ratio of red to yellow genes has dropped from 0.5 to 0.25.

In the larger fish population, there are now seven red fish (A) and sixteen yellow fish (B). The ratio of red to yellow genes is now 7:16 or 0.4375. The ratio of genes has therefore dropped from 0.5 in the original population, to 0.4375. You can easily see that the ratio of the genes in the large population was affected much less by this chance occurrence than was the small population.

Notice that genetic drift has produced an alteration in the gene pool of the population but has not affected the population's chance for survival. Indeed, if a predator comes along that is attracted to red fish, the predator may snap up the remaining red fish from the smaller population, and the red genes will disappear entirely. The remaining population will then contain only yellow genes and will have undergone an evolutionary change. By contrast, the larger population is less likely to lose its red genes because the predator is less likely to consume all the red fish in the population. Thus, variability is less likely to be completely lost in the larger population.

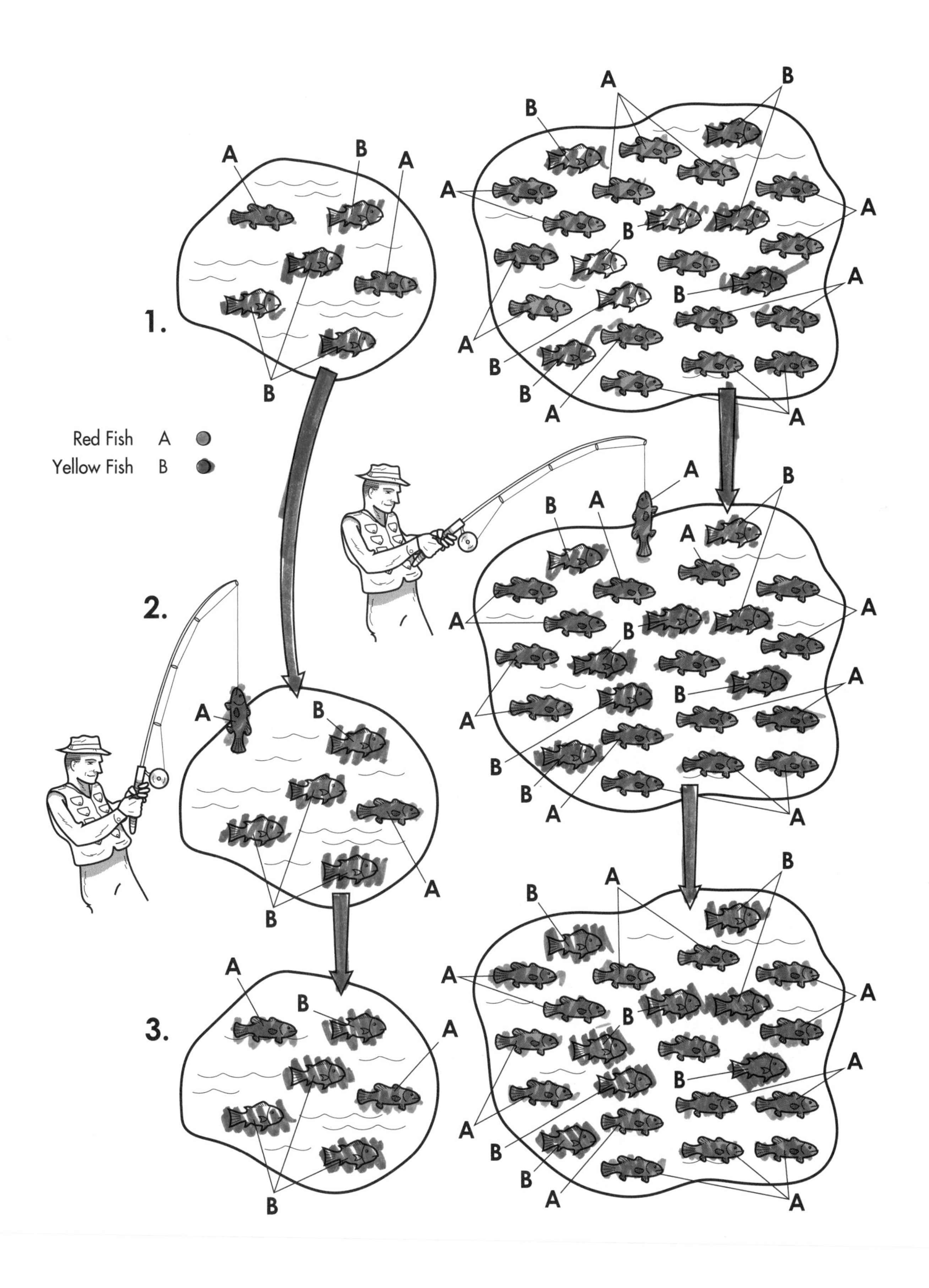
1.
2.
3.
Red Fish A
Yellow Fish B
A
B

MUTATION AND NATURAL SELECTION

Variation in the gene pool of a population occurs due to genetic drift, gene flow, and mutation. The effects of mutation will be examined in this plate.

Once variations are present in a population, natural selection occurs and shapes the evolution of organisms. Selective pressures brought about by the environment on a population of individuals effectively remove unfit individuals. Over long periods of time, the fit individuals multiply, as we will see in this plate.

This plate presents the history of a population of fish. We will see how a mutation occurs in the population and how the mutation results in fish that are better adapted to the environment. These better adapted individuals survive conditions in which others perish. This is the "survival of the fittest" phenomenon.

We begin with a large population of **original fish (A).** These fish have lived for a long time in a body of water with a **cool temperature (G).** The population is relatively stable, and all fish are of essentially the same type in the population. A dark color may be used to color them, but a light color should be used for the thermometer.

As diagram 1 indicates, a mutation occurs in the population. A mutation is an alteration in the DNA in an organism, which may or may not affect its characteristics. In this particular case, the **mutation occurs in one of the fish (B)** in the population. The **mutation (C)** can be caused by many factors, including an episode of ultraviolet light from the Sun. Mutations occur within **strands of DNA (D),** bringing about changes in the base sequence of a particular gene. They result in specific locations in DNA that are then considered **mutated (E).** The fish that results is a **mutated fish (F)** with a new phenotype. In this case, the mutation is not lethal, so the fish will survive and remain in the population.

We have seen how a mutation brings about an alteration in the genes of one member of a population. Because the fish is alive and healthy, there is no apparent overall effect of the mutation on the population. However, that will change as the environment exerts a selective pressure, as we will see in the next few paragraphs. Continue your reading below as you continue to color.

Over long expanses of time, climatic shifts take place on the surface of Earth. Suppose that a long **passage of time (H)** has occurred. During this time, the temperature of the water has increased substantially and the water is now quite **hot (I).** At this higher temperature, the original fish (A) produce their **eggs (J),** but they produce fewer than the normal number because of the temperature change. Meanwhile, the mutated fish is well adapted to warmer temperatures. Part of its adaptation is that it can produce many more eggs than the original fish at the higher temperature. Thus it is fitter than the original fish and the environment has selected it out from the original population by favoring its reproduction.

At this point, we have seen how mutation and natural selection take place. The mutation changed the characteristics of the fish, and natural selection resulted in its increased reproductive capacity. We now see the effect of natural selection in the third diagram at the bottom of the plate. Here we see how the history of the population has changed. Continue your coloring as before and note the effects of natural selection.

Now the population experiences a brief passage of time (H). During this time, the **adapted fish eggs (K)** become new individuals, and the environment is now filled with adapted fish. Only one of the original fish (A) remains. The warmer water is the environment's selective pressure and the emergence of the adapted fish shows the result of the natural selection. In effect, this mutation has made the fish better prepared for the changing environment.

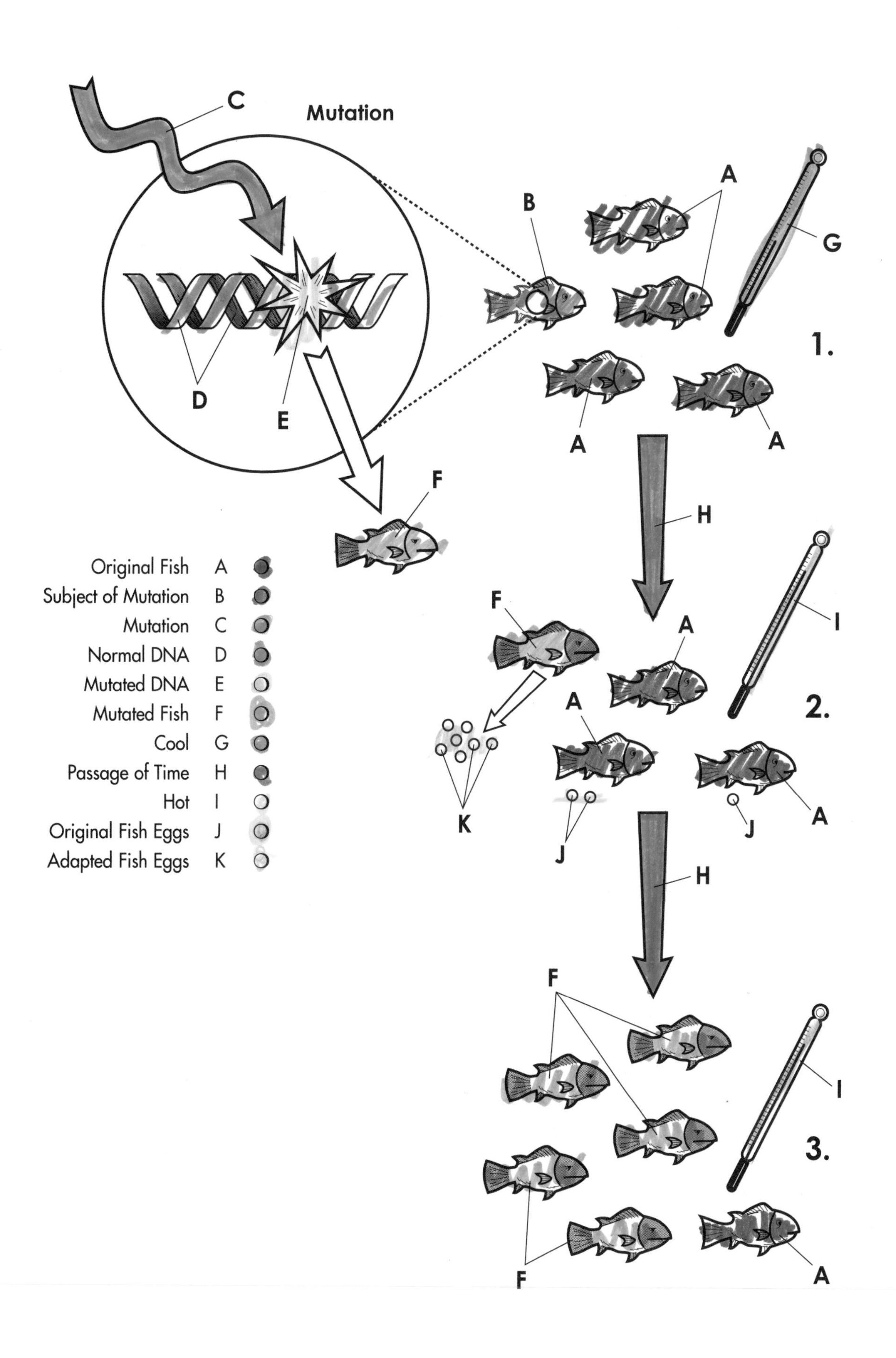
C
Mutation
A
B
G
D
E
1.
A
A
F
H
F
A
I
A
2.
K
J
J
A
H
F
I
3.
F
A
Original Fish A
Subject of Mutation B
Mutation C
Normal DNA D
Mutated DNA E
Mutated Fish F
Cool G
Passage of Time H
Hot I
Original Fish Eggs J
Adapted Fish Eggs K

NATURAL SELECTION

Darwin and Wallace recognized that members of a population of organisms have unequal chances for survival and reproduction. They must compete for food and mates, safety from predators, safety from disease, and numerous other factors. In these struggles for existence, some individuals in a population will be more successful than others because of variations in their physiology, body structure, or behavior. The reproduction of fitter organisms is the basis for natural selection.

Both Darwin and Wallace believed that natural selection is the primary mechanism in the evolution of new species. Contemporary biologists attribute evolution to other phenomena such as gene flow and genetic drift, but natural selection remains an important factor.

In this plate, we examine the process of natural selection, and we show how a population changes as a result of it. We will see how fitness is a measure of an individual's ability to survive and reproduce and how some members of a population are more fit than others.

Look over the plate and color as you read the paragraphs below. There are four parts to be colored in this plate: an isolated island on which natural selection occurs, two types of animals from the population of the island, and a predator. You will need only four colors. We will follow the story of the island through five diagrams, beginning at the top of the page.

Historically, biologists wondered how animals and plants adapted, uniquely fitting to their environments, but they now know that variation and natural selection are responsible. Variations give certain individuals advantages over others, and they survive and reproduce as noted above. Natural selection encourages survival of the better-adapted individuals and acts on variations present in individuals within the population, rather than on the population as a whole.

Take a look at the first diagram. Two different animals exist on the **isolated island (I).** The **first animal (A)** lives solely on the ground, has a pointed snout, and digs holes in the ground. The **second animal (B)** has a blunt snout and cannot dig holes, but can climb trees to obtain food. The first animal is unable to climb trees. Thus, we see a variation in both the physical characteristics and lifestyles of the two animals.

Contrasting colors should be used for the first animal (A) and the second animal (B). We have established a population of mixed animals in a geographically isolated environment. We will now see how natural selection causes a change in the gene pool of the population. Focus on the second diagram and use the same colors that you've been using.

As we see in the second diagram, a **predatory bird (C)** now enters the environment. The first animals, which have pointed snouts (A), quickly dive into their holes in order to escape from the bird. The second animals (B), however, have blunt snouts and cannot dig holes into which they can escape.

In diagram 3, we see that the birds (C) are carrying off the less fit animals (B), while the burrowing animals (A) are in their holes. Natural selection is now taking place; animals that are less fit for that environment are removed from the population and the more fit ones survive. It is a chance event based on the anatomical advantage of one of the animals.

In diagram 4, we see the close of the episode of predation. The bird (C) cannot get at the animals that have pointed snouts (A). They have survived because they are suited to the environment and can use their qualities to their advantage. Soon, the birds will no longer hunt on this island since they will have exhausted the supply of tree-climbing animals that can be easily reached.

We close our story by looking at the last diagram in the plate. Some years have passed and we see the effect of the predatory event. Continue to use the colors you used before.

After time passes, the only animals present on the island will be those that have pointed snouts. These animals have reproduced and passed their characteristics to their offspring.

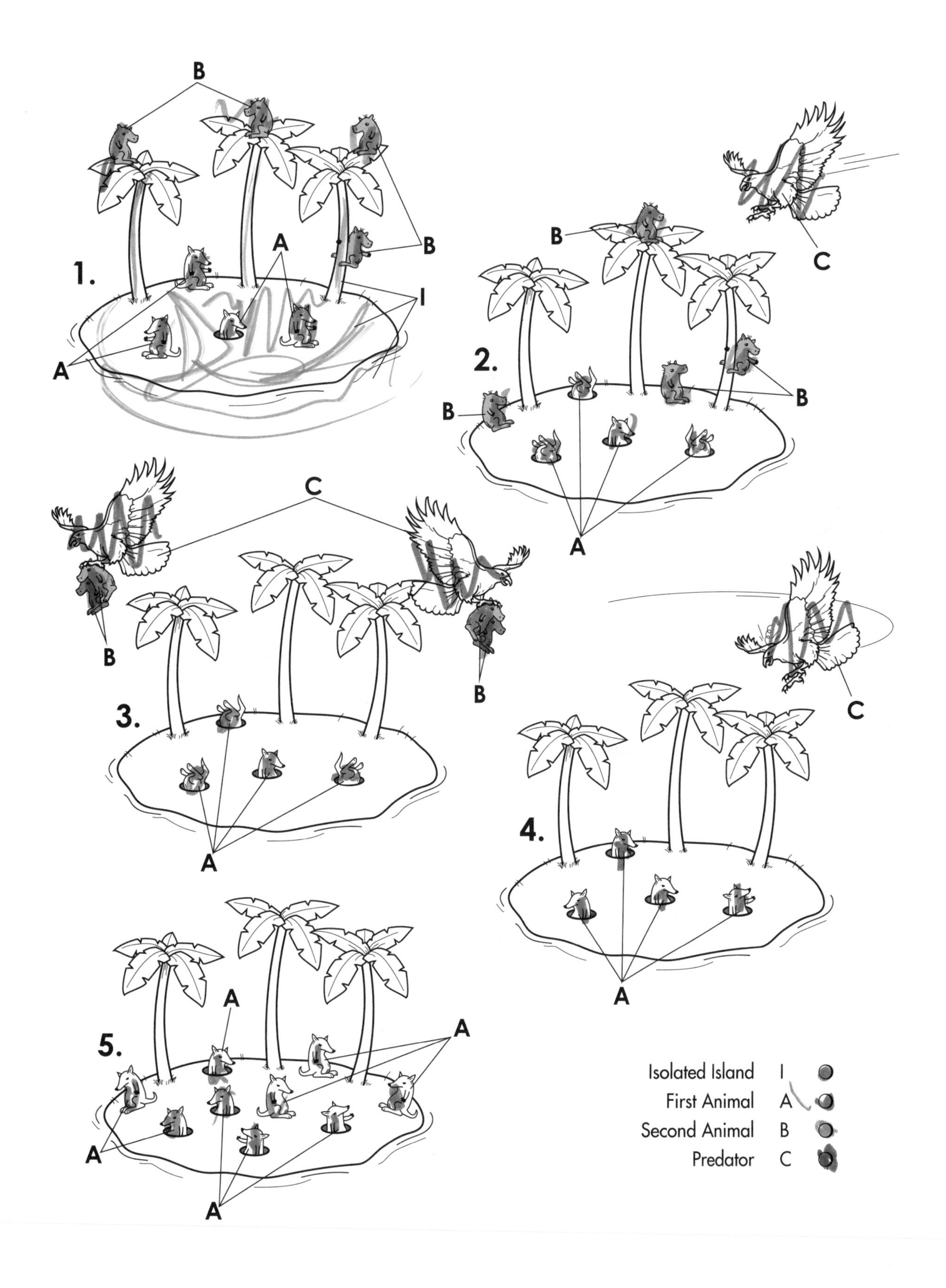
B
B
A
B
1.
I
A
B
C
2.
B
B
A
C
B
B
3.
A
C
4.
A
5.
A
A
A
A
Isolated Island I
First Animal A
Second Animal B
Predator C

ALLOPATRIC SPECIATION

The study of population genetics has produced the following definition of a species: an interbreeding group of individuals that share a gene pool. The gene pool is the sum total of all the genes in a population. Each species is reproductively isolated from every other species in nature, meaning that species do not interbreed. Gene flow occurs between members of a species, but not between members of different species.

Speciation is the mechanism by which a new species is formed. To produce a new species, substantial genetic changes between two populations arise that make their mating impossible. These genetic changes can result from mutations or genetic drift. Speciation usually depends on the isolation of two populations as well as their genetic divergence. The two types of speciation are allopatric and sympatric speciation.

As you look over the plate, you will notice that we are presenting three views of the history of a population. A substantial amount of time passes between each of the views and each one shows the development of a species that ultimately results in speciation.

Allopatric speciation is speciation that occurs when two populations are geographically isolated from one another. The populations must be separated either by distance or an impassable barrier. In diagram 1, this barrier has not yet developed, and we see an **original animal species (A)** that lives near a **river (B).** The river is traveling along its original course. The **environment (C)** contains the same type of plants throughout the diagram, and you should color them all the same color. Under these conditions, gene flow occurs between all the members of the population, and there is not much variance in the population.

Now focus on the second diagram in the plate, and notice that a substantial alteration has taken place. A geographic barrier has developed, and this will cause speciation. Continue reading below and focus on the second diagram.

As a result of a climatic variation, the river changes from its original course to a **new one (D),** splitting the population into two isolated populations between which genes cannot flow. The original animals (A) are present, but a **diverging animal species (E)** begins to develop on the far side of the river. It may be a mutation that gives the animals some new characteristics. A new color should be used for these animals.

The word *allopatric* means "having a different fatherland." In diagram 2, we see one way in which two populations can be isolated from one another. As the animal species diverge, natural selection exerts its pressure, and the original population is split to the extent that the isolated animals cannot interbreed with the original animals.

In the third diagram, you will notice that the animals have been reunited, but that a new species has emerged.

With the **passage of time (G),** climatic conditions may cause the river to return to its original course (B) as shown in diagram 3, and the animals will be reunited. However, a **new animal species (F)** now exists. These animals have undergone sufficient genetic divergence so that they cannot breed with the original animals (A). Perhaps the new animals become nocturnal, while the original animals remained active during the day. Because they are active at different times, they are reproductively isolated even though their geographic barrier has been removed. Speciation is complete and the environment now has two different species of animals.

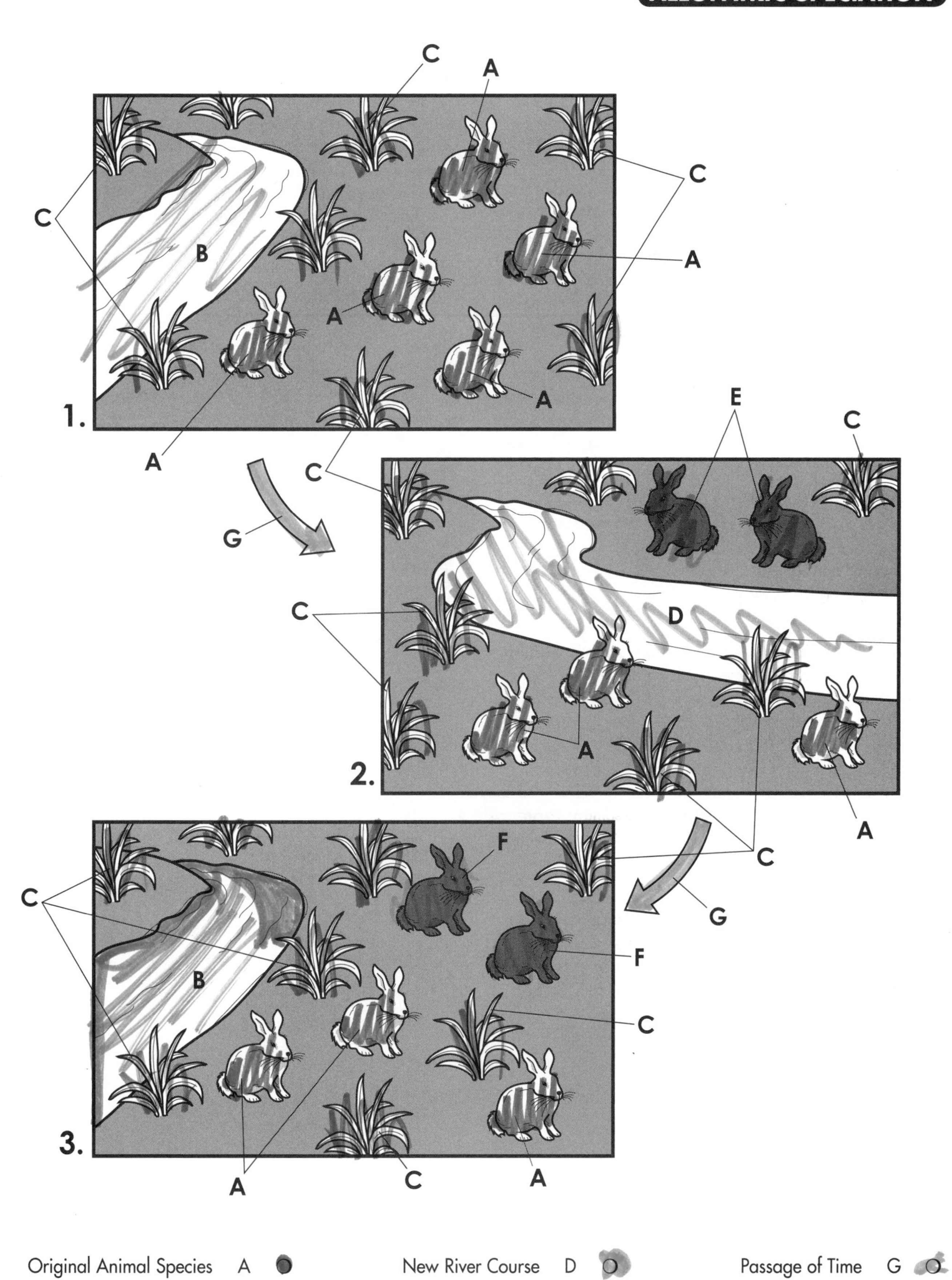

Original Animal Species A
River B
Original Environment C
New River Course D
Diverging Animal Species E
New Animal Species F
Passage of Time G

SYMPATRIC SPECIATION

A species is generally defined as a group of interbreeding individuals in a population, and the process of species formation is called speciation. Species are distinguished from one another by their inability to interbreed. It is believed that most species arose as a result of populations becoming isolated from one another. The split populations slowly diverged as mutation, genetic drift, and natural selection caused different sets of characteristics to accumulate. Eventually barriers to reproduction emerged and mating was impossible even when the divergent populations came into contact.

Speciation is the result of two events: a splitting of a population and the isolation of the two groups and their genetic divergence. The two types of speciation are allopatric and sympatric speciation, and this plate discusses the mechanism of sympatric speciation.

In this plate, you will notice that we present three periods in the history of a population. An original population diverges into two species through sympatric speciation.

Sympatric speciation is the process of speciation in which a new species develops when members of a population develop a genetic difference that prevents them from reproducing with members of the original species. The word sympatric comes from the Greek *syn,* which means "together," and the Latin *patria,* which means "homeland." A single population experiences speciation within a single geographic area. In the diagram we see the **original animal species (A)** in its **original environment (B).** The environment is stable and gene flow occurs between members of the population.

We now focus on the second diagram in the plate. Here we see that a genetic mutation has occurred in three individuals, which has erected a reproductive barrier between them and the parent population. To see how this occurs, focus your attention on the second diagram and continue your reading. Bold colors such as reds, greens, and blues should be used.

As we have mentioned, sympatric speciation occurs when reproduction is prevented between the three original members of the population because of a chromosomal mutation. In diagram 2, we see the appearance of three new **divergent animal species (C).** Notice that the original species (A) still exists.

The population is now heterogeneous; it contains two distinctly different species. There is no barrier between these different species as there is in allopatric speciation, but even so, under these new conditions one group of animals has diverged.

In the third diagram, we see a mixing of the two new species. To see the effects of this, continue your reading below, and color the plate accordingly. You may continue to use bold colors.

In the third diagram, we see the original animal species (A) mixing with members of the new species. The two species are reproductively isolated, and factors such as genetic drift can cause additional divergence between the two gene pools even though they are experiencing the same pressures.

One prominent example of sympatric speciation is found in plants, when a phenomenon called polyploidy takes place. In polyploidy, the multiplication of a chromosome occurs, and plants that have multiple copies of these become new species. For example, the cells of a particular plant may have four copies of each chromosome instead of two. These tetraploid plants are usually healthy, and many can undergo reproduction among themselves. Many common vegetables and grains such as wheat are the result of polyploidy.

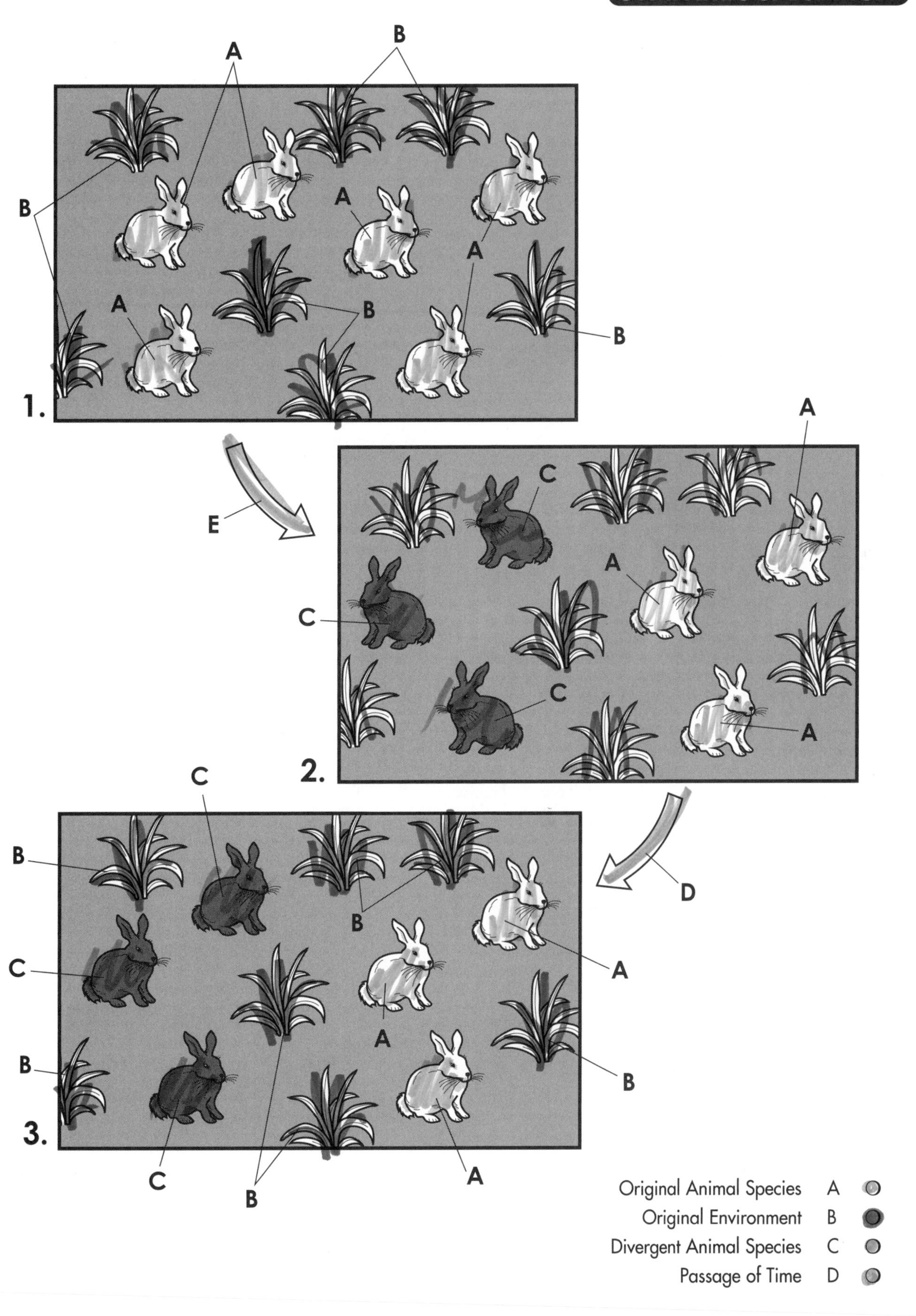
1.
2.
3.
A
B
C
D
E
Original Animal Species A
Original Environment B
Divergent Animal Species C
Passage of Time D

ADAPTIVE RADIATION

The forces of mutation, natural selection, gene flow, and other phenomena on our planet have been at work for over three billion years. They have shaped and molded the many life-forms on Earth and have been responsible for the development of billions of species. A great number of these species have become extinct as conditions on Earth changed, and many left behind evidence of their existence in the form of fossils. By studying the fossil record, scientists have pieced together the histories of many of these creatures.

Repeatedly, scientists have found that organisms move into and colonize new habitats, utilize new resources, and occupy new geographic areas. These movements give rise to an evolutionary pattern known as adaptive radiation. In this plate, we will describe this process.

Notice that we are presenting a series of now-extinct species as well as the species into which they evolved. You should use bold colors such as reds, greens, blues, and purples for this plate, since these structures are relatively large.

Some adaptations involve the movement of organisms into new geographic areas where they may find more resources and fewer predators. This is referred to as adaptive radiation. As the species spread out they discover niches that suit them and exploit new areas for growth and reproduction. The processes of speciation and isolation lead to the introduction of species into areas.

This plate describes adaptive radiation as it occurred in prehistoric times to create creatures that are now extinct. The pattern begins in one corner of the art with an **ancestral reptile (A),** and in the opposite corner, with an **ancestral mammal (B).** As we see in the plate, the ancestral reptile and the ancestral mammal gave rise to similar animals that existed in similar environmental niches.

The **adaptive radiation (C)** of both the ancestral reptile and the ancestral mammal are indicated by the curved arrows. Note that the ancestral reptile gave rise to a creature that successfully invaded the air. This creature is the **pterosaur (D_1).** When the pterosaur and its relatives became extinct, this geographic niche was left open and the ancestral mammal, the **bat (D_2),** took it over.

We have begun our study of adaptive radiation by showing how both an ancestral mammal and reptile gave rise to animals that filled a geographic niche available in the air. We now continue with other examples of adaptive radiation.

Both the ancestral reptile and the ancestral mammal gave rise to animals that inhabited the sea. The ancestral reptile gave rise to an animal called the **ichthyosaur (E_1).** When the ichthyosaur became extinct and its niche was available, the ancestral mammal gave rise to the **porpoise (E_2).** Modern animals that share this niche include the dolphin, shark, seal, penguin, and fish. All of the animals that evolved to inhabit the ocean share some characteristics—the evolution of similar characteristics that enable animals to inhabit the same environment is called convergent evolution. Because they are subjected to the same environmental forces, the porpoise (a mammal), shark (a fish), penguin (a bird), and seal (a mammal) independently evolved similar shapes. For instance, they all possess streamlined bodies that enable them to move quickly through water.

Adaptive radiation also occurred from the ancestral reptile to yield a **triceratops (F_1),** which was a plant-eating animal. After it became extinct, its environmental niche was left open, and the ancestral mammal gave rise to the **rhinoceros (F_2).**

We have now shown that adaptive radiation leads to the evolution of new species, and that convergent evolution produces organisms that, though they are different species, share certain characteristics. We will now look at a final example of adaptive radiation.

One ancestral reptile gave rise to the familiar dinosaur ***Tyrannosaurus rex* (G_1).** *T. rex* was a carnivore (a meat-eating animal) and after it became extinct, about 65 million years ago, its niche was left open. Adaptive radiation eventually gave rise to an animal that filled the niche of *T. rex*, the **lion (G_2).** Note that much convergent evolution did not occur; the lion is not nearly the size of *T. rex*; however, it performs the same function in the community because it is a carnivore.

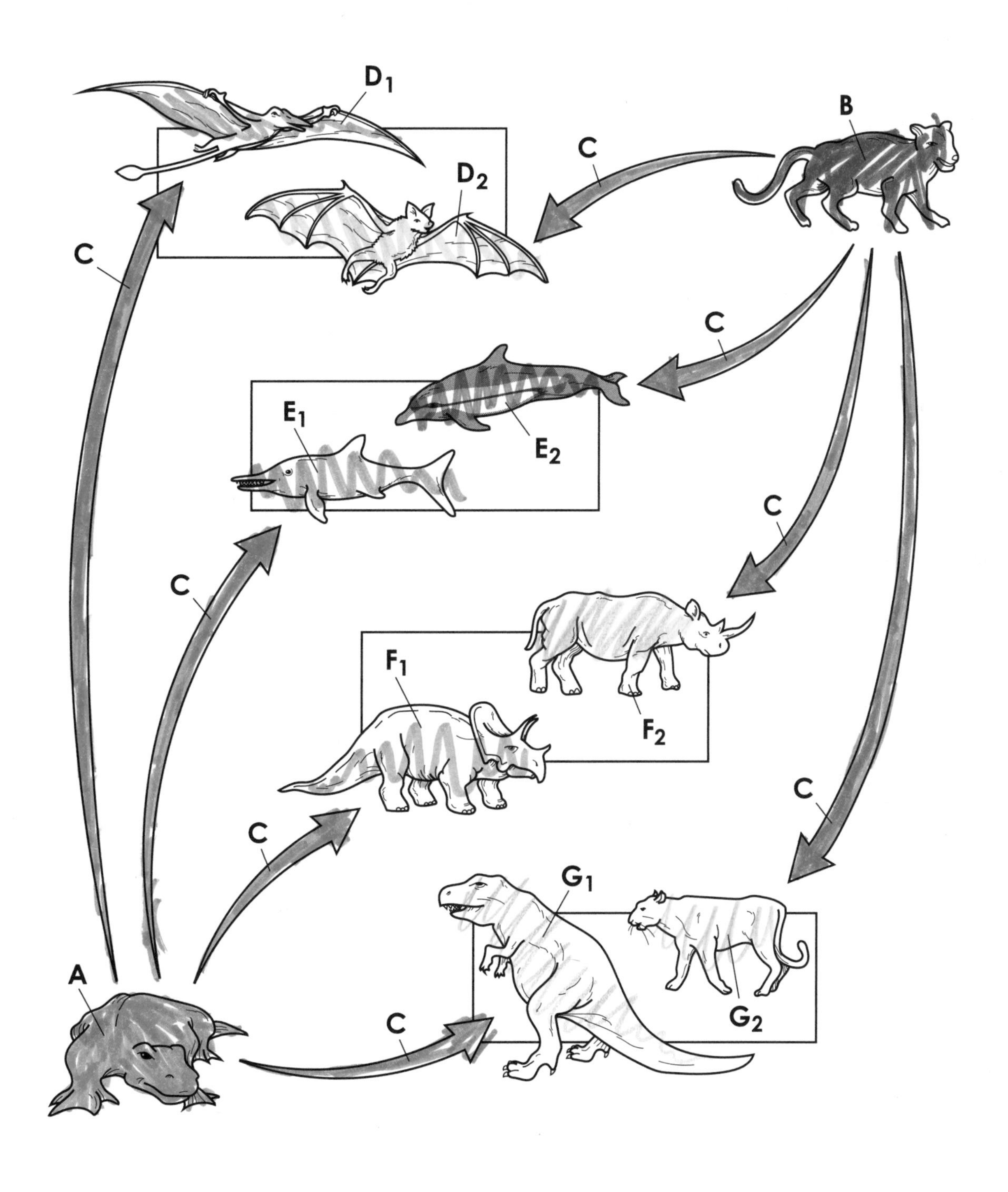

Ancestral Reptile	A	Bat	D_2	Rhinoceros	F_2
Ancestral Mammal	B	Ichthyosaur	E_1	*Tyrannosaurus rex*	G_1
Adaptive Radiation	C	Porpoise	E_2	Lion	G_2
Pterosaur	D_1	Triceratops	F_1		

THE EVOLUTION OF PLACENTALS AND MARSUPIALS

Biologists have hypothesized that there are more than five million different species on the Earth today, but only approximately two million of them have been named. The process by which new species are formed is called speciation. It occurs when a population diverges to the extent that its members can no longer interbreed.

The most common evolutionary pattern is divergent evolution, in which two or more species evolve from a common ancestor and continue to become increasingly different over time. Monkeys and apes, for example, diverged from a common ancestor. When members of a species move into an area that contains many diverse environments, new species can rapidly evolve. The opposite of divergent evolution is a process called adaptive radiation; we began our study of adaptive radiation in the last plate, and will elaborate on it here.

This plate consists of diagrams of ten animals. The animals in the left-hand column are all placental, while those in the right-hand column are marsupial. We will explain this difference as we discuss adaptive radiation.

Approximately 225 million years ago, the super-continent Pangea was in existence—all of the major continents of the Earth were joined in one large land mass. As millions of years passed, Pangea broke apart and eventually the continents that we know today took shape. While the continents were still locked together as Pangea, the first mammals evolved. Many of these mammals were concentrated in what is now Australia.

Adaptive radiation is a diversification process; it occurs when a group of organisms branches out to form many new species that occupy a diverse set of environments. For instance, when the Australian landmass separated from Pangea, it carried along ancestors of the present-day marsupials. Marsupials, like kangaroos, are animals that bear their young in an external pouch. The animals left on the remaining landmasses (Africa, South America, North America, Asia, and Europe) developed into placental mammals, which complete their embryonic development in the mother's uterus.

As new species move into previously unexploited environments, evolution adapts them to particular niches. The marsupials of Australia and the placentals in the rest of the world have evolved to fill many of the same niches, but on different continents. In the plate, we see the **placental ocelot (A_1),** a cat found in Africa. The **marsupial cat (A_2)** looks similar to the ocelot. Regions of Africa are home to a **placental anteater (B_1),** and Australia is home to a **marsupial anteater (B_2).** Note that in both of these cases, the different species have adapted to similar environments so that they are strikingly similar in appearance.

We have now seen how adaptive radiation has given rise to similar animals in isolated parts of the Earth. The animals are members of different species, but they have adapted to similar ecological niches and therefore resemble one another.

Adaptive radiation has given rise to numerous other animals in various parts of the world. One species you are probably familiar with in our environment is the **placental mouse (C_1).** In Australia, there is a very similar **marsupial mouse (C_2)** that occupies the same niche; the resemblance between the two is the result of convergent evolution. As you can imagine, convergent evolution often leads people to misinterpret the evolutionary history of species.

Now let's look at two flying squirrels. The **placental flying squirrel (D_1)** is on the left, and the **marsupial flying phalanger (D_2)** is seen on the right. The similarity between these two animals is obvious; both use flaps of tissue that are attached to their appendages for flying, but detailed studies of their other structures show that they are very different animals.

A final example of adaptive radiation and convergent evolution is the wolf. The **placental wolf (E_1)** is found in North America, and the **marsupial Tasmanian wolf (E_2)** is a native of Australia. They occupy similar niches in the forest, but close study shows that they, too, are different species.

Placental Ocelot A_1
Marsupial Cat A_2
Placental Anteater B_1
Marsupial Anteater B_2
Placental Mouse C_1
Marsupial Mouse C_2
Placental Flying Squirrel D_1
Marsupial Flying Phalanger D_2
Placental Wolf E_1
Marsupial Tasmanian Wolf E_2

EVOLUTION AND THE SHIFTING EARTH

Charles Darwin proposed that evolution was a gradual process of change that occurred over a vast expanse of time. He believed that major changes, such as the development of wings in birds, resulted from the accumulation of thousands of minute ones.

It is now apparent that the history of the Earth has been characterized by periods of rapid change interrupted by long stretches of little change. Biologists refer to this method of change over time as punctuated equilibrium. As we will demonstrate, evolution is closely entwined with the changing Earth.

In the art, you can see four periods during the geologic history of the Earth. The purpose of this plate is to show how the shifting landmasses have influenced evolution. Focus your attention on the first view of the Earth.

The Earth's crust is composed of huge landmasses called plates. The continents of the world ride on these plates so that as the plates move, the continents are carried along with them. The study of these movements is called plate tectonics.

During the early Triassic Period, approximately 225 million years ago, all the continents were united in one enormous landmass known as **Pangea (A).** A light color should be used to color it. The formation of this landmass by the fusion of the continents destroyed the habitats of many marine species. During this time, biologists estimate that over half of the existing marine species became extinct. It is also theorized that the interior region of this super-continent must have experienced extreme temperatures and rainfall because it was isolated from the ameliorating effects of large bodies of water.

We now move to a second view of Earth, much later in its history. Your attention should now be focused on diagram B, which shows Earth during the Cretaceous Period. As you color this diagram, continue your reading below.

The landmass of Pangea remained whole until the Cretaceous Period, about 135 million years ago. At this point, Pangea split into separate masses, which began to slowly move apart, geographically isolating species from one another. The two initial landmasses that resulted from the splitting up of Pangea were **Laurasia (B)** and **Gondwana (C).** Different colors should be used to color them.

You should now focus on the third view of the Earth, during the Paleocene Epoch. The configuration of continents as we know it has begun to appear.

The continual, slow movements of the continents on their plates brought about dramatic shifts in the environment and chances in Earth's climate. The third diagram shows the continents during the Paleocene Epoch, which occurred approximately 65 million years ago. You can see that **Eurasia (D)** and **Africa (E)** are becoming distinct, and that **India (F)** is a completely separate island. Also seen as a separate landmass is **Madagascar (G),** off the coast of Africa, and you can see that **Australia (H)** and **Antarctica (I)** have moved apart considerably. **North America (J)** and **South America (K)** are also widely separated from one another. This breakup of continents isolated untold numbers of plant and animal populations and caused them to develop into new species, so that, over millions of years, each continent has evolved its own plant and animal life.

The final focus is on the Earth as it appears today. Look at the diagram showing the present day Earth. Use the same colors as you used in diagram C and continue your reading below.

Today the continents look much as they did during the Paleocene Epoch. One important difference is that India (F) has crashed into Asia to create the Himalayan Mountains. On the now isolated continents, ancestral groups have branched into many new life forms that occupy diverse environments. This process of diversification is called adaptive radiation, and you have encountered it in earlier plates. It is easy to see that the shifting Earth has substantially influenced the evolution of species.

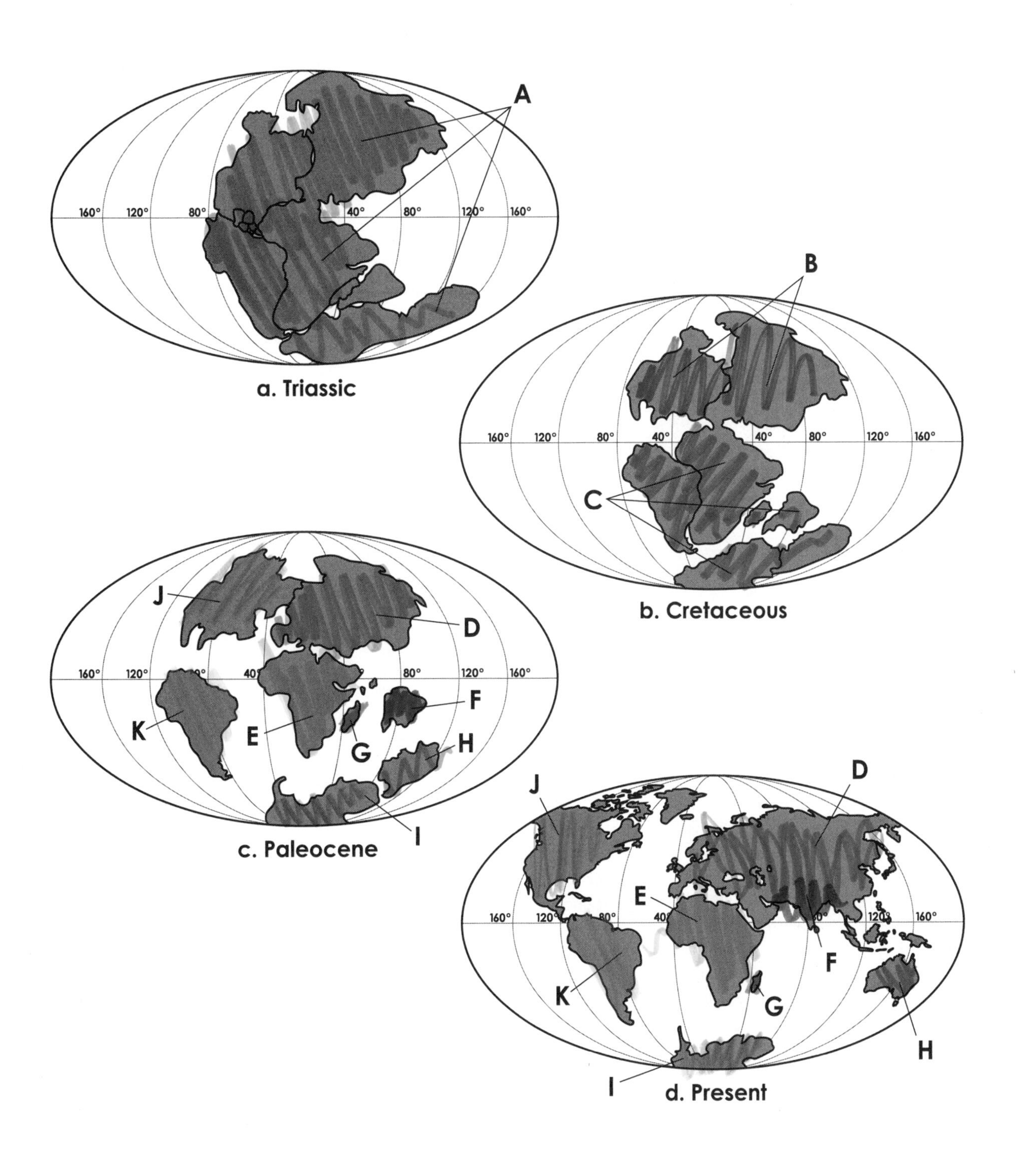

Pangea A
Laurasia B
Gondwana C
Eurasia D
Africa E
India F
Madagascar G
Australia H
Antarctica I
North America J
South America K

GRADUALISM VS. PUNCTUATED EQUILIBRIUM

Charles Darwin and his followers believed that the rate of evolution was very slow and that it took millions and millions of years of gradual changes to bring about the evolution of a new species. The concept that evolution takes place continuously, gradually, and over a long period of time is called gradualism. A second model for evolution, proposed in the 1970s, is called punctuated equilibrium. According to this theory, adaptations in species arise very rapidly, quickly giving rise to new species. These periods of rapid adaptation are interspersed with periods in which little or no change takes place.

As you can see, two concepts have been proposed to describe the rate of evolution. We will examine both of those concepts in this plate.

In this plate, we present the two main models for the process of evolution. Gradualism is presented in the first diagram in the plate, and punctuated equilibrium is presented in the second. Light colors are probably best for the species and gene pools, and darker colors should be used for the mutation and time periods.

Scientists have detected subtle differences in the physical features of different specimens of particular organisms that have been fossilized over vast expanses of time. The accumulation of changes can result in the development of a whole new species over time, and this evidence substantiates the Theory of Gradualism. **Species A (A)** has a gene pool that's made up of dominant and recessive genes. You should use the same color for the box and the gene pool. Over a vast expanse of **time (D),** species A gives rise to **species B (B).** Notice that the gene pool has changed. The agent of change was **mutation (E).** This slow divergence of the population through gradual modifications is the hallmark of gradualism.

We will now focus on punctuated equilibrium and show how it differs from gradualism. Focus your attention on the second diagram of the plate, and continue your reading below as you color.

As early as the 1940s, some biologists began to challenge the theory of gradualism. They pointed out that the fossils of some species remained virtually unchanged over millions of years, which conflicts with the concept of gradualism. Moreover, some organisms seem to have appeared out of nowhere in the fossil record, appearing suddenly and changing rapidly. In 1972, biologists Niles Eldredge and Stephen J. Gould proposed the idea of punctuated equilibrium to explain these phenomenon. Punctuated equilibrium describes evolution as a process that occurs in spurts of rapid change that are followed by long periods with little or no evolutionary change. This is shown in the second diagram. As in the first diagram, species A undergoes change into species B over a long period of time; the same colors can be used as were used in diagram 1. Notice, however, that at one point in the organism's history one large mutation (E) occurs. This sudden dramatic mutation brings about a burst of change that quickly creates the new species. As the new species is formed, the population grows and a new gene pool emerges. **Species C (C)** now exists, and you should notice that its gene pool consists of both dominant and recessive genes.

The history of life on Earth has now been proven to contain several events that add credence to the Theory of Punctuated Equilibrium. For instance, about 600 million years ago, all invertebrates (animals without backbones) seemed to have come into existence within a relatively short period of time. At the present time, however, neither gradualism nor punctuated equilibrium is universally accepted, and each side has its supporters.

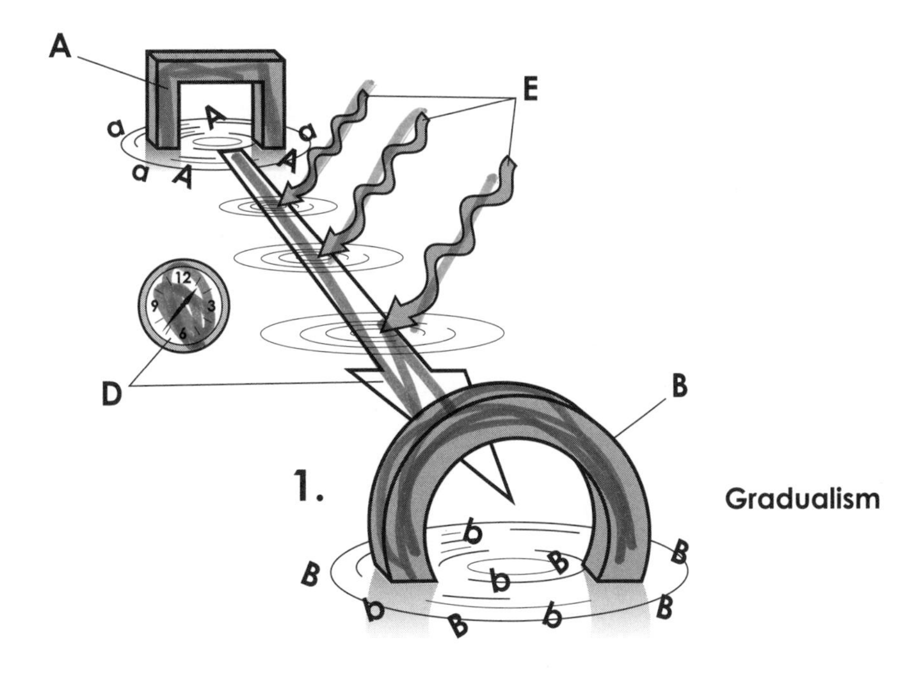

Gradualism

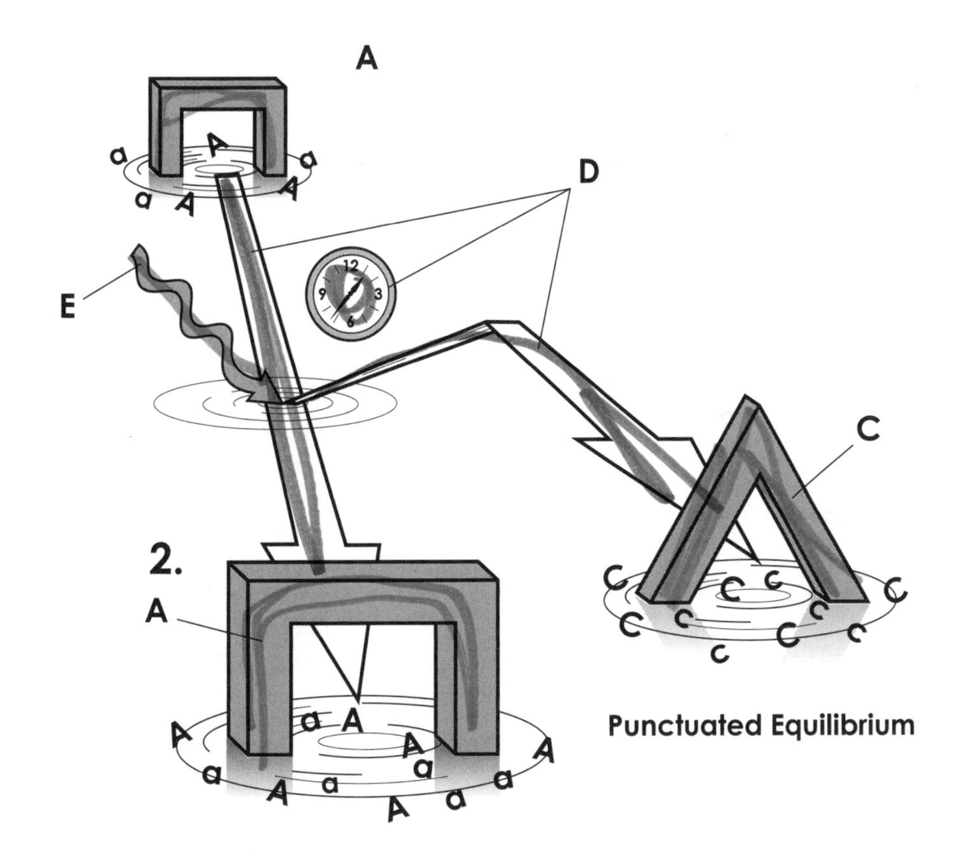

Punctuated Equilibrium

Species A	A	○
Species B	B	○
Species C	C	○
Time	D	○
Mutation	E	○

HUMAN EVOLUTION

The first fossil records of prokaryotic cells were found in rocks that date to about 3.5 billion years ago. Eukaryotic cells date back about 1.5 billion years and multicellular organisms first appeared about a billion years ago. The first multicellular organisms appeared in the sea; they were adapted to move quickly and had improved senses. Approximately 500 million years ago, the first land organisms, plants, appeared. The first land animals appeared 350 million years ago. These first land animals were amphibians, which later evolved into reptiles, and subsequently into birds and mammals. One group of mammals evolved into tree-dwelling primates, and about 25 million years ago, some descended from the trees and evolved into apes and humans. About 4 million years ago, organisms that were beginning to show human characteristics arose in Africa.

In this plate, we will compare modern humans with two ancestors to show how the human skull evolved.

The fossils of primates are relatively rare compared with those of other animals for several reasons. Most early primates did not live in habitats that readily preserved fossils, were relatively small so predators and scavengers were likely to break their bones into unrecognizable fragments, and lived in relatively small populations. The first primates were shrew-like creatures that ate insects and existed about 80 million years ago. Forward-facing eyes and large brains facilitated their hand-eye coordination, depth perception, and balance, and ancestral hominids evolved from this line. As the hominids diversified on the Earth, one line gave rise to humans.

The first true hominids (humans and their bipedal ancestors) arose about 4.5 million years ago. These individuals were called australopithecines. In Africa, the footprints of bipeds dating close to 4 million years ago were discovered by Mary Leakey. This discovery, as well as the shape of their pelvic bones, indicated that australopithecines walked upright.

One of the most ancient australopithecines was *Australopithecus afarensis*, fossils of which were discovered in the Afar region of Ethiopia in 1974. The skeleton, named Lucy, was a female who was approximately 3.5 feet tall. The skull of this human ancestor is shown in diagrams 1a and 1b. The skull of ***Australopithecus afarensis* (1a, 1b)** is similar to that of a small ape in that the brain case is small, but it has humanlike jaws and teeth.

Australopithecus afarensis shared characteristics with both apes and humans. This hominid probably lived in a woodland environment, walked erect, and used its forelimbs for gathering food and its hind limbs for locomotion. Although she walked upright as we do, Lucy's arms were proportionately very long for her size, suggesting that this hominid was still an able tree climber. Lucy was believed to weigh about 30 kilograms, and had a very heavy jaw, which you can see in the diagram.

We will now concentrate on the second set of diagrams and point out the difference between the australopithecines and the ancestors of modern humans.

The oldest known fossils of the human genus come from *Homo habilis*. This hominid was present on the Earth about 2 million years ago and was believed to have arisen from *A. afarensis*. *Homo habilis* was probably the first hominid to use stone tools.

About 1.8 million years ago, ***Homo erectus* (2a, 2b)** appeared. *Homo erectus* probably descended from *Homo habilis*. Its brain was as large as the smallest adult human brain—about 1000 cubic centimeters. Compare the skull in diagram 2a to that in diagram 1a. The face of *Homo erectus* was also notably different from that of the australopithecine. It had large bony ridges above the eyes, its face protruded less, and its chin was much reduced. Also, the nose of *Homo erectus* projected more than that of *Australopithecus*, an adaptation for hot, dry climates.

Homo erectus was much taller than *Australopithecus*; females were about 5 feet tall. Their robust and highly muscular skeletons still retained some of the *Australopithecus* features, and *H. erectus* is believed to be the first hominid to venture out of Africa and into Europe. The well-known Java man and Peking man are now considered to be the extinct *Homo erectus*. *Homo erectus* survived until 300,000 years ago, and were the first to use fire and build advanced tools, such as heavy teardrop-shaped axes and cleavers.

We will now focus on the third skull in the collection, that of modern humans, *Homo sapiens*.

Approximately 300,000 years ago, modern *Homo sapiens* evolved from *Homo erectus*. A popular theory suggests that ***Homo sapiens* (3a, 3b)** evolved in Africa and then migrated to Europe and Asia at about 100,000 years ago. The skull of modern *Homo sapiens* is domed, its eyebrows are smooth, and its chin is prominent. Notice these three features in the third diagram and compare them to the second and first sets of drawings. The forehead is high, the brain case is large, and the face is flat, compared to *Homo erectus* and *Australopithecus*.

Early European *Homo sapiens*, the Cro-Magnons, hunted in groups and may have been the first to use language. Their culture included art, typified by the paintings of animals found on cave walls in France and Spain. They also sculpted figurines out of animal bones. Their remains were first discovered in France in the Cro-Magnon region.

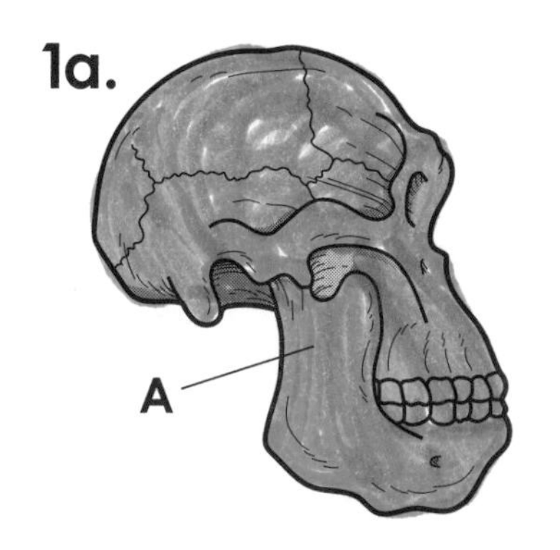

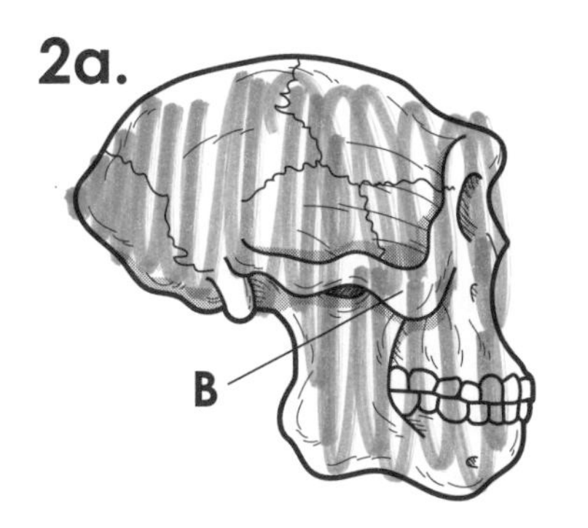

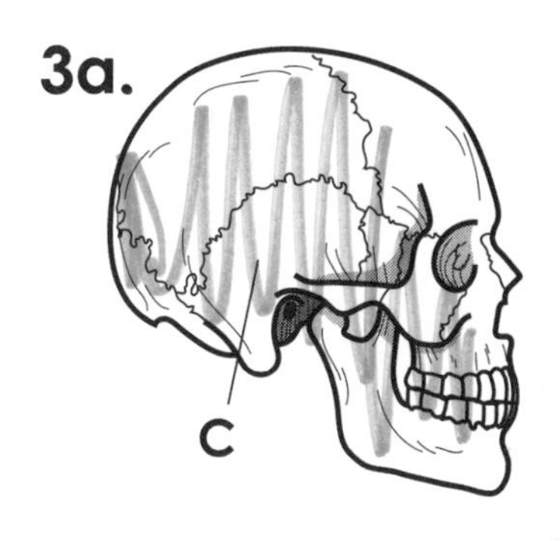

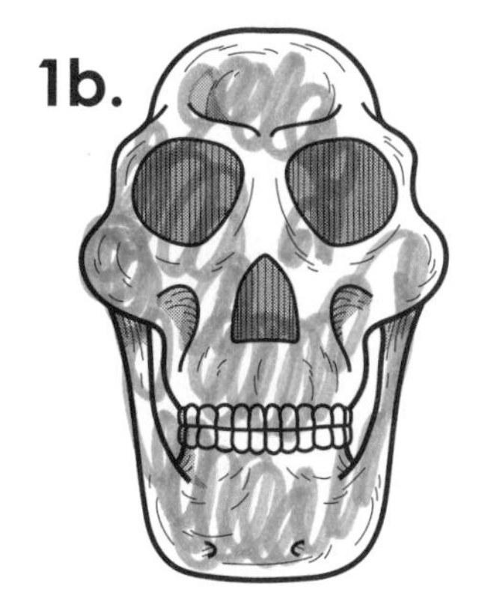

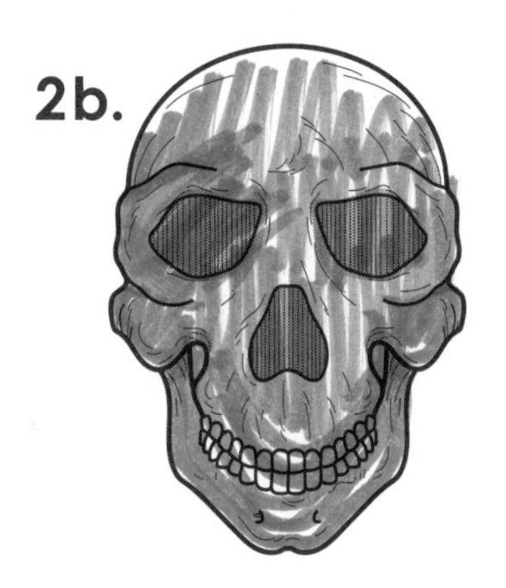

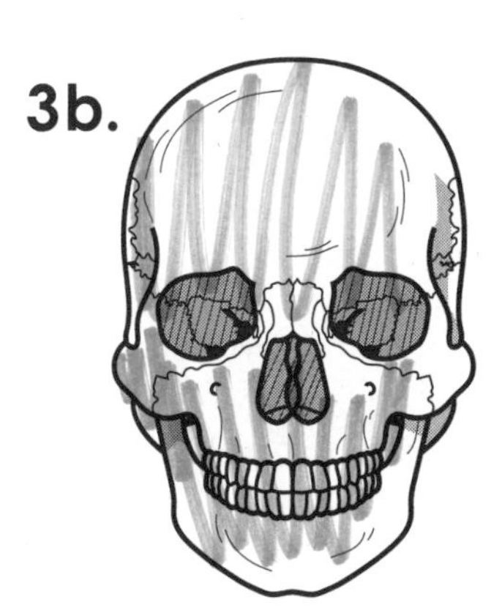

Australopithecus afarensis 1a, 1b

Homo erectus 2a, 2b

Homo sapiens 3a, 3b

CHAPTER 6:

The ORIGIN of LIFE and SIMPLE LIFE FORMS

THE ORIGIN OF ORGANIC MOLECULES

The modern theory that explains the origin of life on Earth is centered on the idea of the primordial cloud. According to this theory, the primordial cloud was actually many clouds of cosmic dust and gases that condensed into planets to form the solar system. A series of thermonuclear reactions caused the largest mass in the center to become the Sun, and other areas of concentrated matter to become the planets. Over eons of time, organic molecules developed on primordial Earth, after which came the first living things. This plate considers one of the theories for the origin of organic molecules.

This plate illustrates a theory first proposed in 1953 that accounts for the origin of organic molecules on Earth, and we use a laboratory setup to simulate primitive Earth conditions to demonstrate it.

Billions of years ago, as the primordial Earth traveled through space, its gases contracted to form a hot, dense core that had a temperature of several thousand degrees. When the planet began to cool four billion years ago, a primitive atmosphere formed from gases that escaped from the Earth's core through volcanic action. We see a **volcano (A)** in the plate; the lava should be colored dark red, and a lighter color should be used for its remainder.

In 1953, Stanley Miller and Howard Urey at the University of Chicago tackled the question of how these volcanic gases could transform into our present atmosphere. They poured water into a flask and placed a **flame (B)** under it. The heat caused the water to **boil (C).** Soon **water vapor (D)** filled the area above the flask much as hot vapors probably filled the early atmosphere.

Earth's early atmosphere contained a number of different **atmospheric gases (E).** A pale color should be used to shade the sky. Miller and Urey filled a chamber with the gases that were believed to exist at that time, which included **methane (F), ammonia (G),** and **hydrogen (H).**

Notice that there was no free oxygen on the Earth at this time, so that organic molecules could not have been created from diatomic oxygen. We will now continue describing the experiment that Miller and Urey performed. Continue your reading below as you color.

Miller and Urey asked themselves how methane, ammonia, and hydrogen could be transformed into organic molecules. They theorized that the energy source that drove chemical reactions might have been ultraviolet light that was bombarding the primitive Earth. The source of this ultraviolet light would have been **lightning (I).** To simulate these conditions, Miller and Urey used electrical energy from **electrodes (J).** A dark color should be used for the electrodes, which provided an **electrical spark (K).**

Miller and Urey questioned whether electrifying the primordial gases would indeed lead to the synthesis of organic molecules. They found their answer when they analyzed the products of the reaction.

Miller and Urey then passed the gaseous mixture through a **condenser tube (L).** A **condenser coil (M)** carried cold water, which caused the vapors to condense. The scientists reasoned that early in the history of the Earth, **rains (N)** were unrelenting for years at a time, and the effects of this rain were simulated by the condenser.

Droplets of water formed in the condenser, as Miller and Urey anticipated, and after analysis, they found that these drops of water contained a number of **organic molecules (O).** For example, they found amino acids, which are the building blocks of protein. They also found acetic acid, which is a simple acid widely encountered in organic chemistry, and urea. These analyses indicated that organic molecules could indeed form from primitive Earth gases when charged with electricity. These early oceans were probably full of organic molecules, and it was in them that living things first appeared.

Volcano A
Flame B
Boiling Water C
Water Vapor D
Atmospheric Gases E
Methane F

Ammonia G
Hydrogen H
Lightning I
Electrodes J
Electrical Spark K
Condenser Tube L

Condenser Coil M
Rain N
Organic Molecules O
Seawater P

THE ORIGIN OF LIFE

In the course of the evolution of life on Earth, gases in the atmosphere combined to form simple organic molecules. It is conceivable that these simple organic molecules joined to form more complex molecules in a process called polymerization, but in order for these molecules to polymerize, they first needed to be isolated from the atmosphere. The process of the polymerization of molecules eventually led to the creation of cells, and this plate demonstrates the most popular theory about how cells first came into existence.

This plate contains three sections, each of which diagrams a stage in the origin of life on Earth. Focus your attention on the first diagram, which is entitled Coacervate Activity, and begin your reading below.

One of the first steps in the formation of the cell was the isolation of molecules from the environment. This would have isolated and concentrated the molecules, which would have made it possible for them to combine more frequently in chemical reactions.

There are several theories that describe the first precells that might have occurred. These precells probably had water-repellent, membrane-like shells, that were semipermeable and served to concentrate their organic molecules even further.

One of the first models of a precell was created by A.I. Oparin in 1938. Oparin proposed the early existence of microscopic droplets called **coacervates (A).** He created these by shaking a solution of nucleic acids, polypeptides, and polysaccharides, and found that the droplets formed were similar to fat droplets. Since then it has been discovered that interesting things take place when enzymes are added to coacervates. For instance, the arrow represents the enzyme **phosphorylase (B).** When **glucose-phosphate (C)** is added, the enzyme acts on its substrate, transforming it into **starch (E)** and, in the process, releasing a **phosphate group (D).** Another enzyme, amylase, then breaks down the starch into **maltose (F).**

Other types of droplets have been proposed; in 1959 Sidney Fox worked with small protein spheres called proteinoid microspheres. In recent years, some scientists have also proposed that microscopic lipid droplets called liposomes could have been precells.

We will next examine what are thought to be some of the first complex molecules on Earth. Your attention should be directed to the second diagram entitled RNA Formation.

When primitive cells first came into existence, they most likely used energy that was derived from adenosine triphosphate (ATP) to metabolize organic compounds. The breakdown of carbohydrates may have provided this energy, and fermentation was probably another way in which cells obtained ATP.

The hereditary information of the first cells was contained in nucleic acids. Recent experiments that are believed to prove that RNA was the first genetic material show that **RNA nucleotides (G)** spontaneously bond with one another to form a single-stranded **molecule of RNA (H).** It is also known that RNA is self-replicating; a strand of RNA called a ribozyme acts as a template that dictates the order of addition of RNA nucleotides to produce a new, complementary RNA chain. This is shown in the second diagram. RNA is essential to the synthesis of protein since it provides the information necessary for the assembly of amino acids into proteins. Thus, RNA has some enzyme capabilities and serves as a carrier of genetic information.

We will now show how RNA could have been used to form DNA, which is the genetic information in the modern cell.

Scientists have theorized that it was not until about four billion years ago that DNA evolved from RNA to become the primary hereditary material. It was discovered that certain microorganisms contain the enzyme reverse transcriptase, which has the ability to use RNA as a template for the synthesis of DNA. Look at diagram 3. Here, a molecule of RNA (H) provides a template and, with the help of the enzyme reverse transcriptase (not shown), a number of **DNA nucleotides (I)** are combining to form **single-stranded DNA (J).** The single-stranded DNA molecule then serves as a model for the synthesis of another strand of DNA. The two complementary strands then intertwine to form **double-stranded DNA (K).**

Scientists know that the first true cell must have contained DNA (K) as well as RNA (H). **Ribosomes (L)** are other structures associated with, and necessary for, the synthesis of protein, so they must also have been present. These materials were presumably suspended in some sort of **cytoplasm (M)** and enclosed by a **cell membrane (N).** The cells we have just described are similar to the most primitive bacteria now found on Earth. These organisms are prokaryotes, and they contain little more than the parts just mentioned, but they are considered functionally complete cells.

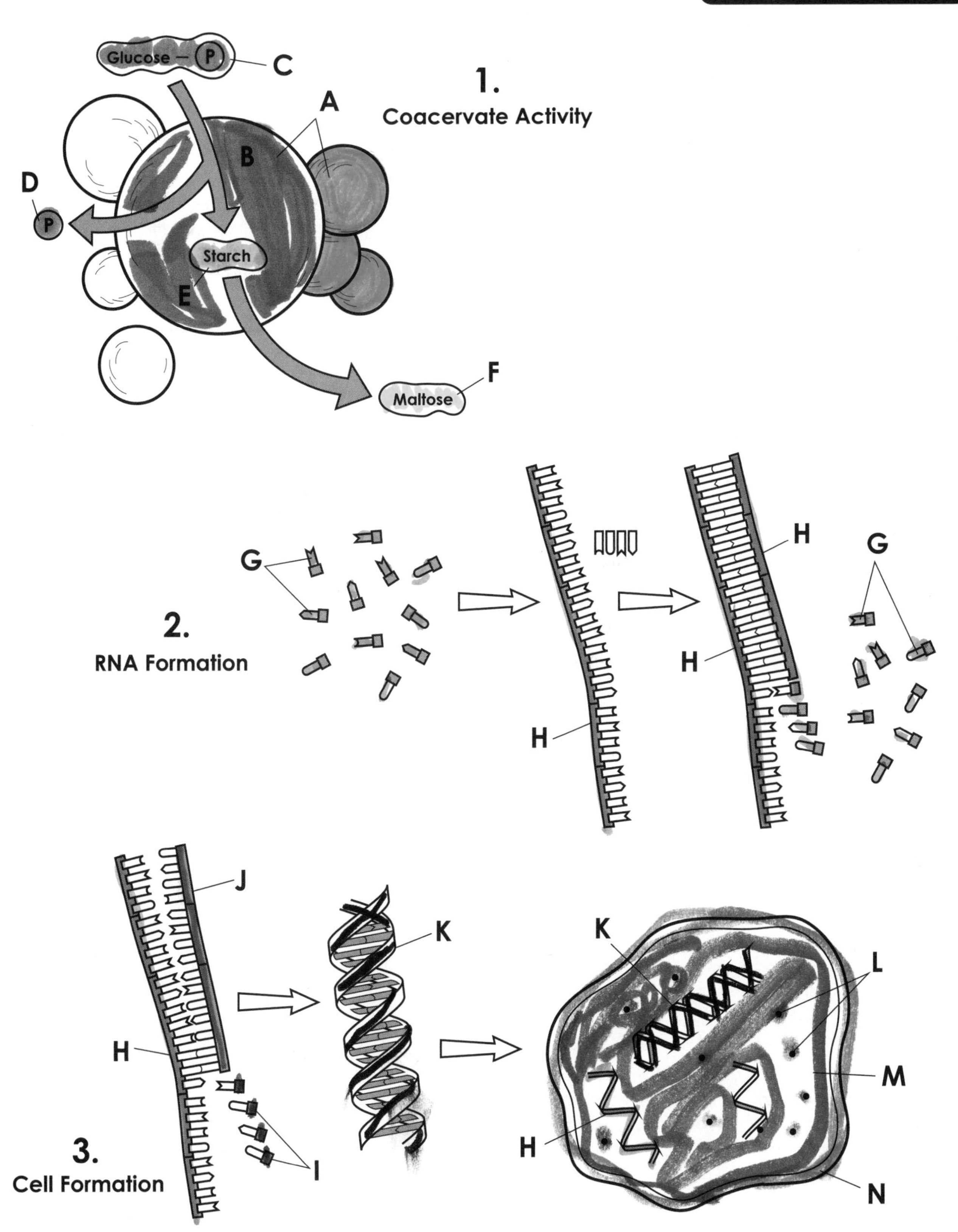

Coacervate	A	●
Phosphorylase	B	●
Glucose-Phosphate	C	●
Phosphate	D	●
Starch	E	●
Maltose	F	●
RNA Nucleotides	G	●
RNA	H	●
DNA Nucleotides	I	●
Single-Stranded DNA	J	●
Double-Stranded DNA	K	●
Ribosomes	L	●
Cytoplasm	M	●
Cell Membrane	N	●

THE FIRST EUKARYOTIC CELLS

The first cells to exist on Earth were very simple prokaryotic cells that were similar to today's bacteria. They are thought to have existed for approximately three and a half billion years, and for about two billion years they were the only cells on Earth. These prokaryotic cells lived in an anaerobic environment and assimilated organic molecules from their surroundings. They had no nuclei or organelles, did not reproduce by mitosis, and each possessed only a single strand of DNA.

Eukaryotic cells, by contrast, are much more complex. They possess nuclei, nuclear membranes, organelles, multiple chromosomes (often occurring in pairs), and they reproduce by mitosis. Eukaryotic cells appeared in the fossil record about one and a half billion years ago; scientists believe that they arose from prokaryotic cells in a process described by the Endosymbiont Theory.

This plate displays a multistep process in which a prokaryotic cell evolved into a eukaryotic cell. Focus on the first diagram.

Let's first take a look at a prokaryotic cell. A prokaryotic cell has **cytoplasm (A),** which is enclosed in a **cell membrane (B).** As the first diagram shows, the **DNA (C)** in the cell consists of a single, long molecule arranged in a ring. A light color should be used to trace this molecule. Scattered throughout the cytoplasm are molecules of **RNA (D).**

The endosymbiont theory is one of a few theories that describe how eukaryotic cells may have arisen from prokaryotic cells. This theory proposes that **aerobic bacteria (E)** are taken into a prokaryotic cell by phagocytosis. In the second diagram, we see that the cell membrane **invaginates (F),** and several aerobic bacteria are taken into the cell to become **symbiotic bacteria (G),** which take up permanent residence within the cell. The DNA (C) moves to the center of the cell, and the cytoplasm (A) and cell membrane (B) are still clearly seen.

Now take a look at diagram 3. Here we encounter the first organelles in the evolving cell. Continue your coloring as you read the text below.

In diagram 3, you can see that the symbiotic bacteria have evolved into **mitochondria (H),** several of which are shown. Mitochondria are the sites of cellular respiration in the modern cell. Support for the endosymbiont theory comes from the fact that the genetic material found in the mitochondria is very similar to that found in prokaryotes. Furthermore, there is no histone protein associated with either prokaryotic DNA or mitochondrial DNA.

The emerging eukaryotic cell also shows two membranous organelles. The DNA (C) is grouped at the center of the cell and a **developing nuclear membrane (I)** encloses it. Also, at this point, the invaginating cell membrane begins to develop into **endoplasmic reticulum (ER) (J).** The ER is a complex of interconnected membranes that's continuous with the nuclear envelope; its associated ribosomes are the sites of protein synthesis and the ER is responsible for the distribution of these new proteins.

On the bottom right, we see the nonphotosynthetic, contemporary eukaryotic cell. The **nucleus (O)** contains DNA, and a **nuclear membrane (N)** encloses it. Mitochondria (H) and **endoplasmic reticulum (M)** are seen in the cell.

Focusing on the cell on the bottom left, you can see that chloroplasts, which do not exist in animal cells, are present in plant cells. Now we will see how the endosymbiont theory accounts for their development.

Cyanobacteria (K) are bacteria that have photosynthetic pigments in their cytoplasm. Scientists propose that these bacteria were taken into the cytoplasm of eukaryotic cells, and that they remained in them to become what we now call **chloroplasts (L).** This theory is supported by the fact that there are several similarities between chloroplasts and prokaryotic DNA. Notice the other major features of the eukaryotic cell, including the nucleus (O), the nuclear membrane (N), mitochondria (H), and the endoplasmic reticulum (M). One notable structure of the photosynthetic plant cell is the **cell wall (P),** which is found outside the cell membrane.

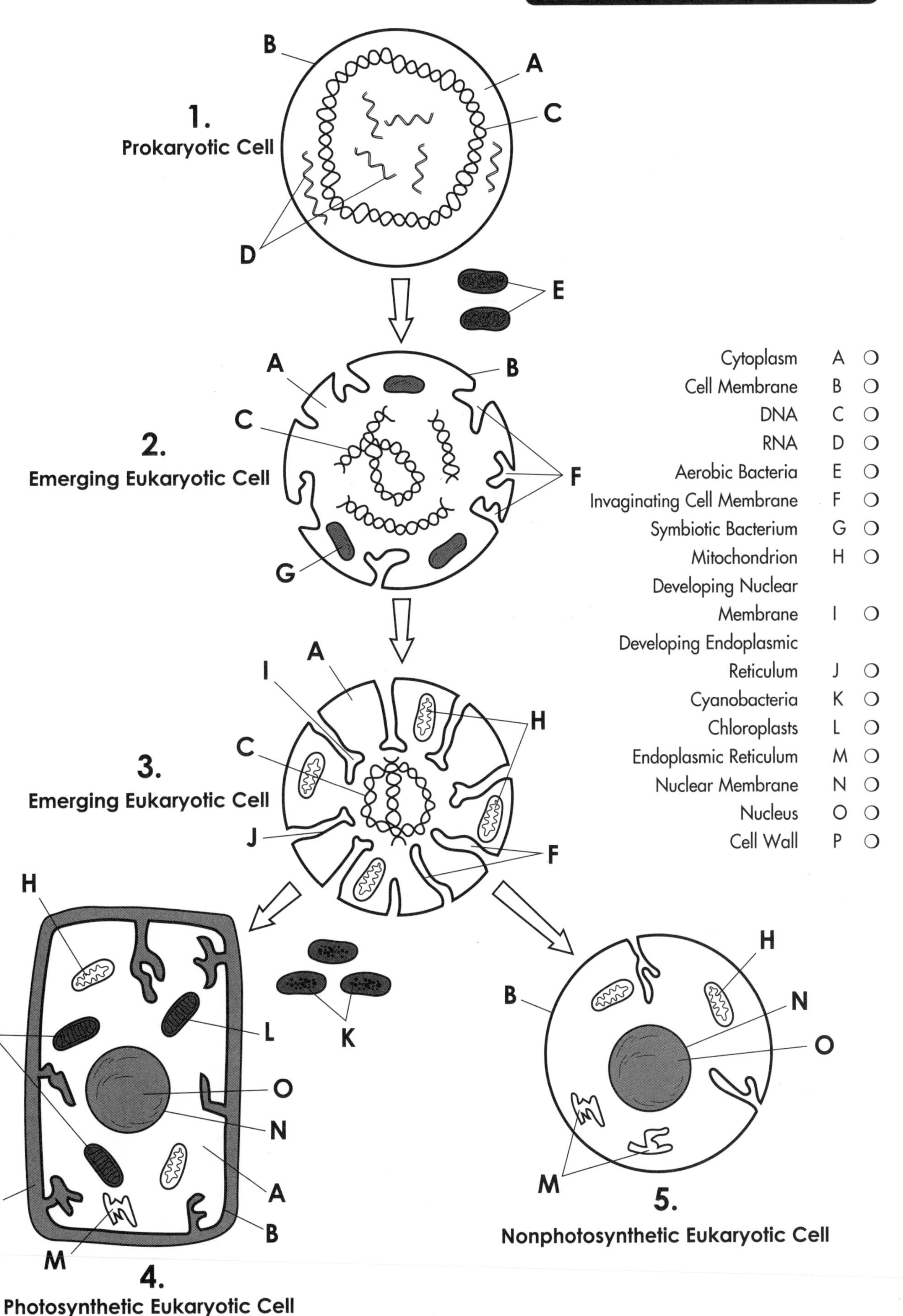

Cytoplasm	A	❍
Cell Membrane	B	❍
DNA	C	❍
RNA	D	❍
Aerobic Bacteria	E	❍
Invaginating Cell Membrane	F	❍
Symbiotic Bacterium	G	❍
Mitochondrion	H	❍
Developing Nuclear Membrane	I	❍
Developing Endoplasmic Reticulum	J	❍
Cyanobacteria	K	❍
Chloroplasts	L	❍
Endoplasmic Reticulum	M	❍
Nuclear Membrane	N	❍
Nucleus	O	❍
Cell Wall	P	❍

HISTORY OF LIFE ON EARTH

Life on Earth began with the origin of organic molecules and continued with the formation of cells. As time passed, various life forms appeared and disappeared, and great geographical changes occurred on the planet.

Geologists divide the Earth's 4.6 billion years of existence into four eras. The eras are subdivided into periods, whose beginnings and endings are marked by major events, such as episodes of mass extinction. The most recent era is also divided into epochs that last a few million years. In this plate, we will discuss the main events in the history of life on Earth from the earliest known period to the present day.

This plate consists of an overview of the various periods on Earth and the major events that occurred during them.

The time span before the eras shown in the plate is known as the Precambrian Era. It began with the origin of the Earth, about 4.6 billion years ago, and ended approximately 570 million years ago (MYA). During this time, the first prokaryotic cells appeared, the first eukaryotic cells appeared, and multicellular organisms came into existence. Bacteria were prevalent, photosynthesis began, and oxygen appeared in the atmosphere. Algae also appeared, as did soft bodied marine invertebrates (invertebrates are organisms that have no backbones). The Precambrian Era is also called the Proterozoic Era.

We will begin our study of the history of life on Earth in the **Paleozoic Era (A),** which extended from 570–245 MYA. A dark color should be used to shade the block labeled (A). At the beginning of this era, **marine invertebrates (B)** developed and flourished. Then, during the Ordovician Period, the first vertebrates (animals with backbones) evolved, the **fish (C).** Along with the mollusks and arthropods, they dominated the sea. During the next period, the Silurian, primitive vascular plants appeared, as did the first jawed fish. In the Devonian Period, seed ferns evolved and the first **amphibians (D)** appeared in the seas.

The next period of the Paleozoic Era was the Carboniferous Period, in which club mosses and ferns were plentiful. Amphibians flourished and **insects (E)** evolved and spread over the Earth. During the Permian Period, conifers appeared, amphibians declined in number, and **reptiles (F)** diversified. At the end of the Permian Period, approximately 245 MYA, a mass extinction took place.

We will continue our survey of the history of life on Earth with the next era, the Mesozoic. Note that a mass extinction ended the first era, and marks the beginning of the second.

The **Mesozoic Era (G)** is divided into three periods. During the Triassic, the first dinosaurs and mammals appeared, mollusks dominated the seas, and forests of ferns and gymnosperms were prevalent on land. The next period, the Jurassic, is also known as the age of the **dinosaurs (H).** The first birds appeared at the end of this period (they evolved from reptiles), a few species of mammals were present, and gymnosperms continued to flourish on the land. During the Cretaceous Period, placental **mammals (I)** appeared and insect groups continued to evolve. Flowering plants covered the Earth, and the dinosaurs continued to flourish.

At the end of the Cretaceous Period, about 65 MYA, another mass extinction episode took place, and most of the dinosaurs disappeared. This mass extinction has been theoretically linked to the crash of a huge meteorite in the Yucatan Peninsula of Mexico. This meteorite may have raised a huge dust cloud that blocked the Sun's radiation, leaving the Earth in cold and darkness. The mammals (I) present at this time survived the extinction episode, and would soon occupy the niche once held by the dinosaurs.

We will complete our perusal of the history of life on Earth by examining the third and final period. This period includes the present era. Continue your coloring and read the paragraphs that follow.

The **Cenozoic Era (J)** began approximately 65 MYA. At the start of the Cenozoic Era, the dinosaurs had disappeared and mammals began to dominate the land. The age of mammals (I) began early in the Cenozoic Era and continues today, and insects increased dramatically in diversity and number during this time. By the Tertiary Period, birds were huge in number and flowering plants had spread over many parts of the Earth. Drift brought the continents to their current positions, and the climate became similar to today's. Toward the end of the Tertiary Period, forests declined and grasslands spread, bony fish became abundant in the seas, and all of the modern genera of mammals were present. The end of the Tertiary Period also saw the evolution of hominids on Earth.

The most recent period is the Quaternary. This is the age of **humans (K),** who appeared in their modern state at this time. Flowering plants continued to spread and diversify. Many of the earlier, giant mammals became extinct, and climatic conditions became similar to present ones.

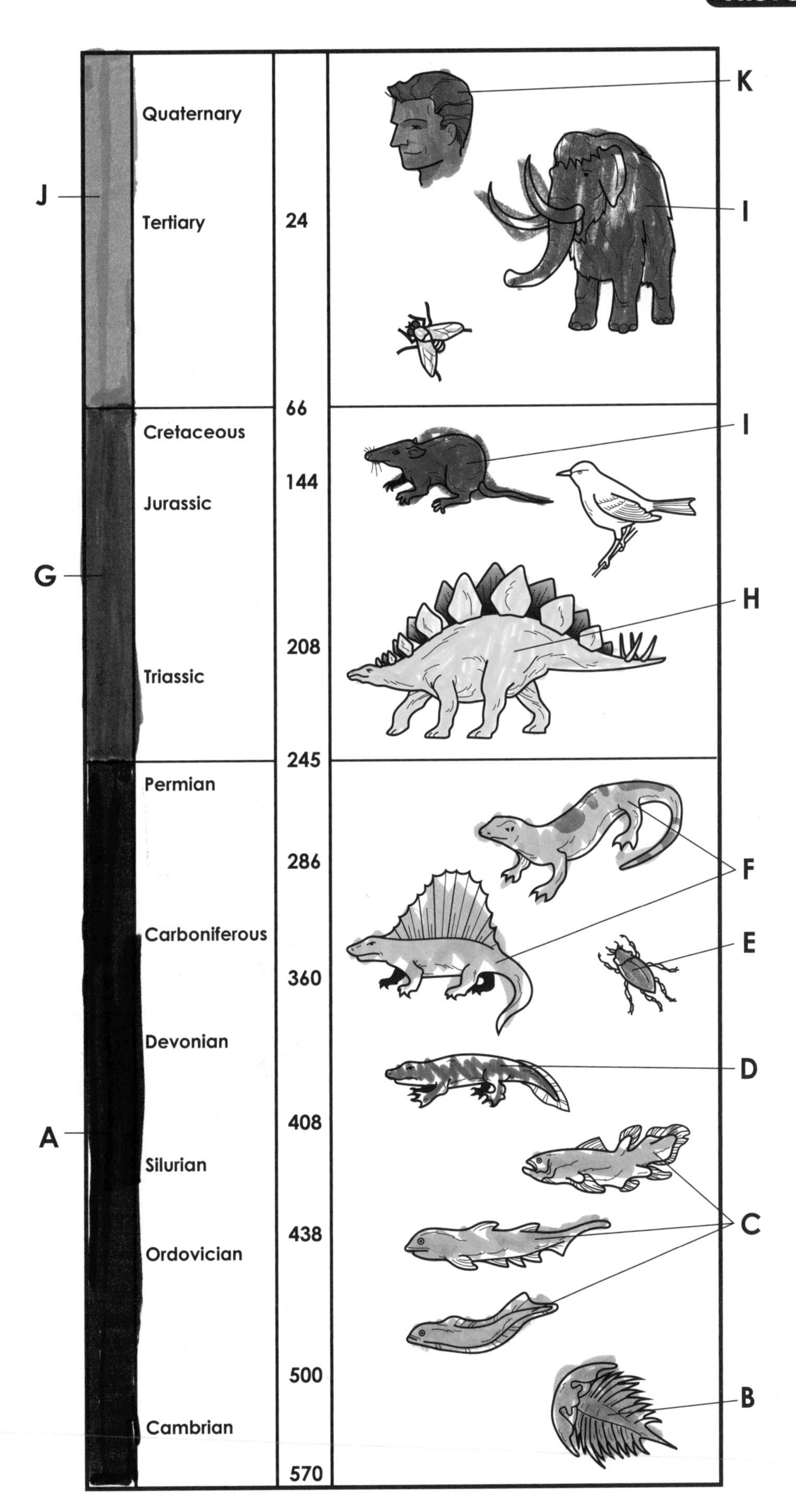

Paleozoic Era A
Marine Invertebrates B
Fish C
Amphibians D
Insects E
Reptiles F
Mesozoic Era G
Dinosaurs H
Mammals I
Cenozoic Era J
Humans K

THE CLASSIFICATION SCHEME

Scientists are aware of the existence of millions of different kinds of organisms. Because of this huge number, scientists need an elaborate classification scheme to organize them. The science of classifying and naming organisms is known as taxonomy. Linnaeus, who was perhaps the first taxonomist, was a Swedish botanist who lived in the mid-1700s. Over a period of many years, he developed the scheme of classification that is still used today. Binomial names are assigned to species; the names consist of the genus (one specific category of living things) to which the organism belongs (e.g., *Homo*) and an adjective describing the organism, which is called a specific epithet (e.g., *sapiens*). Thus, our binomial name is *Homo sapiens*. We will study the classification system in this plate.

This plate shows the categories used to classify organisms. The categories are arranged in a hierarchy from the broadest group, the kingdom, to the most specific group, the species.

Beginning in the 1950s, biologists recognized that all living things fall into five broad categories called **kingdoms (A).** These kingdoms are discussed in succeeding plates. They include the **Monera (A_1), Protista (A_2), Fungi (A_3), Plantae (A_4),** and **Animalia (A_5).** Human beings fall into the kingdom Animalia, and we will begin our study with animals.

We will begin our study of classification by looking at human beings, which are categorized in the kingdom Animalia. Note that this kingdom extends outward in the plate.

The kingdom Animalia is subdivided into a number of categories called phyla. Within these phyla are animals that lack backbones, including sponges, hydras, flatworms, segmented worms, arthropods, and starfish. Humans are classified in the **phylum Chordata (B),** and this phylum projects outward in the plate. Some of the animals in the phylum Chordata are vertebrates, and all of the members of this phylum have a dorsal nerve chord, a notochord, and a backbone. The **other phyla (B_1)** lack these properties.

The phylum Chordata contains mammals (mammalia), birds (aves), reptiles (reptilia), amphibians (amphibia), and some classes of fishes (agnatha, placodermi, chondrichthyes, and osteichthyes). Humans belong to the **class Mammalia (C),** and this class extends outward. Mammals nurse their young with mammary glands and generally have hair on their body. The **other classes (C_1)** should be also colored.

Within the class Mammalia are several orders. For example, the order we belong to is the **order of primates (D),** which includes monkeys, apes, and humans. They have large brains, short snouts, and well-developed binocular vision, together with complex social behavior. The **other orders (D_1)** do not share these characteristics.

We have now progressed from the animal kingdom to the primate order. Each group is successively more specific than the previous one. We now come to the human species. Continue your reading below and color the plate as you proceed.

Within the primate order, there are several families. One of these is the family **Hominidia (E).** This family contains the members of the human family. These animals have (and had) larger brains than monkeys and other **families (E_1)** of the order primata.

The family Hominidia contains at least two genera of human beings and their ancestors. One is the genus ***Homo*** **(F).** This is the group to which modern species belong. The other genus is *Australopithecus*, an early hominid that shares many characteristics with modern humans. The plate on human evolution discusses this genus in more detail. **Other genera (F_1)** may exist in addition to *Australopithecus*, but they have not yet been identified.

Within the genus *Homo* is the species ***Homo sapiens*** **(G).**This is the modern human being. *Homo habilis* and *Homo erectus* are **other species (G_1)** in this genus. The plate on human evolution discusses these species in more depth. Now we have completed the classification of the human from the broadest category, the kingdom, to the most detailed category, the species.

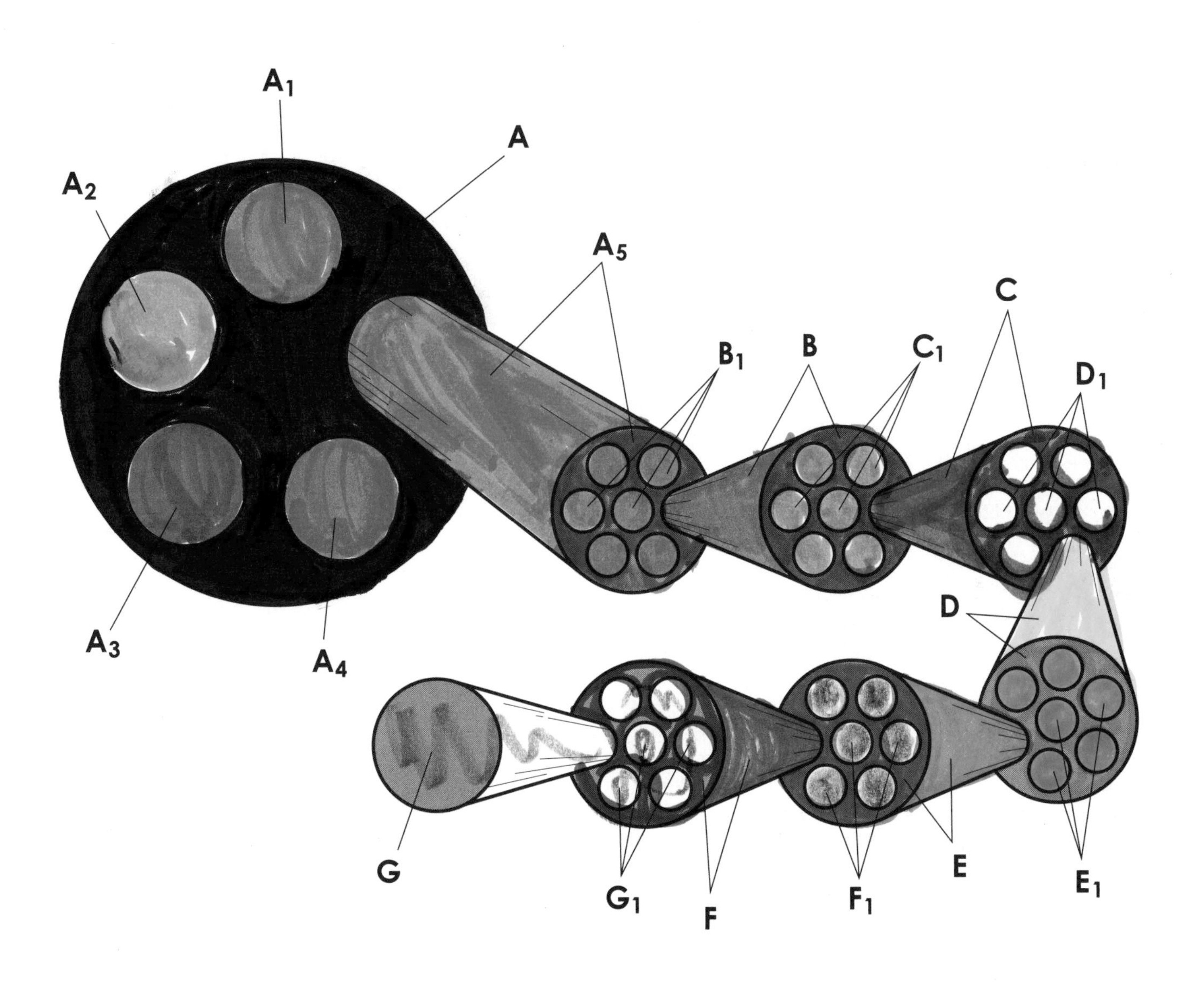

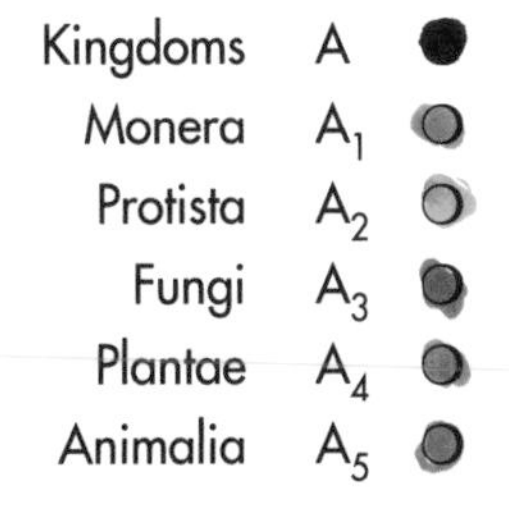

Kingdoms	A	
Monera	A_1	
Protista	A_2	
Fungi	A_3	
Plantae	A_4	
Animalia	A_5	
Phylum Chordata	B	
Other Phyla	B_1	
Class Mammalia	C	
Other Classes	C_1	
Order Primata	D	
Other Orders	D_1	
Family Hominidia	E	
Other Families	E_1	
Genus *Homo*	F	
Other Genera	F_1	
Species *Home Sapiens*	G	
Other Species	G_1	

THE FIVE KINGDOMS

In 1969, Robert H. Whittaker of Cornell University proposed a classification system that was made up of five kingdoms. The previous classification scheme used to identify plants and animals did not include certain organisms such as bacteria and mushrooms.

Whittaker's five-kingdom classification is based on two principal criteria: whether an organism is unicellular (single-celled) or multicellular (many-celled); and the type of nutrition that is practiced by the organism (photosynthesis, ingestion, or absorption). Whittaker's system has been widely accepted by the scientific community, and we will describe it in this plate.

As you look over the plate, you will notice that it contains a single large diagram that's divided into five sections that represent the five kingdoms. Organisms typifying each kingdom are shown.

At the base of Whittaker's five-kingdom system is the **kingdom Monera (A).** This kingdom contains a number of modern bacteria, cyanobacteria, and ancient bacteria called archaebacteria. Collectively, they are known as **monerans (A_1).** All of these organisms are prokaryotes, which means that they are single-celled and lack distinct nuclei and membrane-bound organelles. Bacteria are the most diverse and abundant organisms on Earth. They occupy a wide array of habitats and some are photosynthetic, while others are nonphotosynthetic. Cyanobacteria, once called blue-green algae because their chlorophyll is located in their membranes and not chloroplasts, are in this group. Most of the organisms in this kingdom are decomposers.

According the Whittaker system, the monerans evolved to give rise to members of the **kingdom Protista (B),** which contains **protistans (B_1).** Protistans include the protozoa, single-celled algae, and slime molds—all of which are eukaryotes. The cells of protistans have distinct nuclei and membrane-bound organelles, and most are unicellular. Heterotrophic and photosynthetic organisms exist in this kingdom and certain of them, such as slime molds, share characteristics with plants, protozoa, and fungi. Many of these organisms are producers in marine and freshwater environments, and there are some parasitic protistans.

We have seen two of the kingdoms of the Whittaker system. Note that the kingdoms seem to be becoming increasingly complex. We will now look at the remaining three kingdoms. Continue your coloring as you read the paragraphs below.

In the Whittaker system, three kingdoms arose from the protista. The first is the **kingdom Fungi (C),** which contains **molds and yeasts (C_1).** These organisms are eukaryotic, heterotrophic, and usually multicellular. Unicellular yeasts are one exception. Most of these organisms are decomposers, but some are parasites that coexist with animals and plants.

To the left, you can see the fourth kingdom, the **kingdom Plantae (D),** which includes the **plants (D_1).** According to the system, plants evolved from protistans. All plants are eukaryotic and multicellular, and all are adapted for photosynthesis. During their development, these organisms pass through distinct developmental stages, and display alternation of generations (discussed in Chapters 6 and 7). Almost the entire ecosphere depends on plants as the primary producers of oxygen. Also included in the kingdom Plantae are mosses, ferns, seed-bearing plants, and flowering plants.

In the center of the plate is the final kingdom, the **kingdom Animalia (E),** which contains the wide variety of **animals (E_1),** all of which are eukaryotic. They are multicellular and heterotrophic, and most move by muscular contraction and respond to stimuli with specialized nervous tissue. One notable member of this kingdom is the human being.

As we complete this plate, you should be aware that there are exceptions to many of the general characteristics of each kingdom. For example most, but not all, animals move, and most, but not all, have complex organ systems.

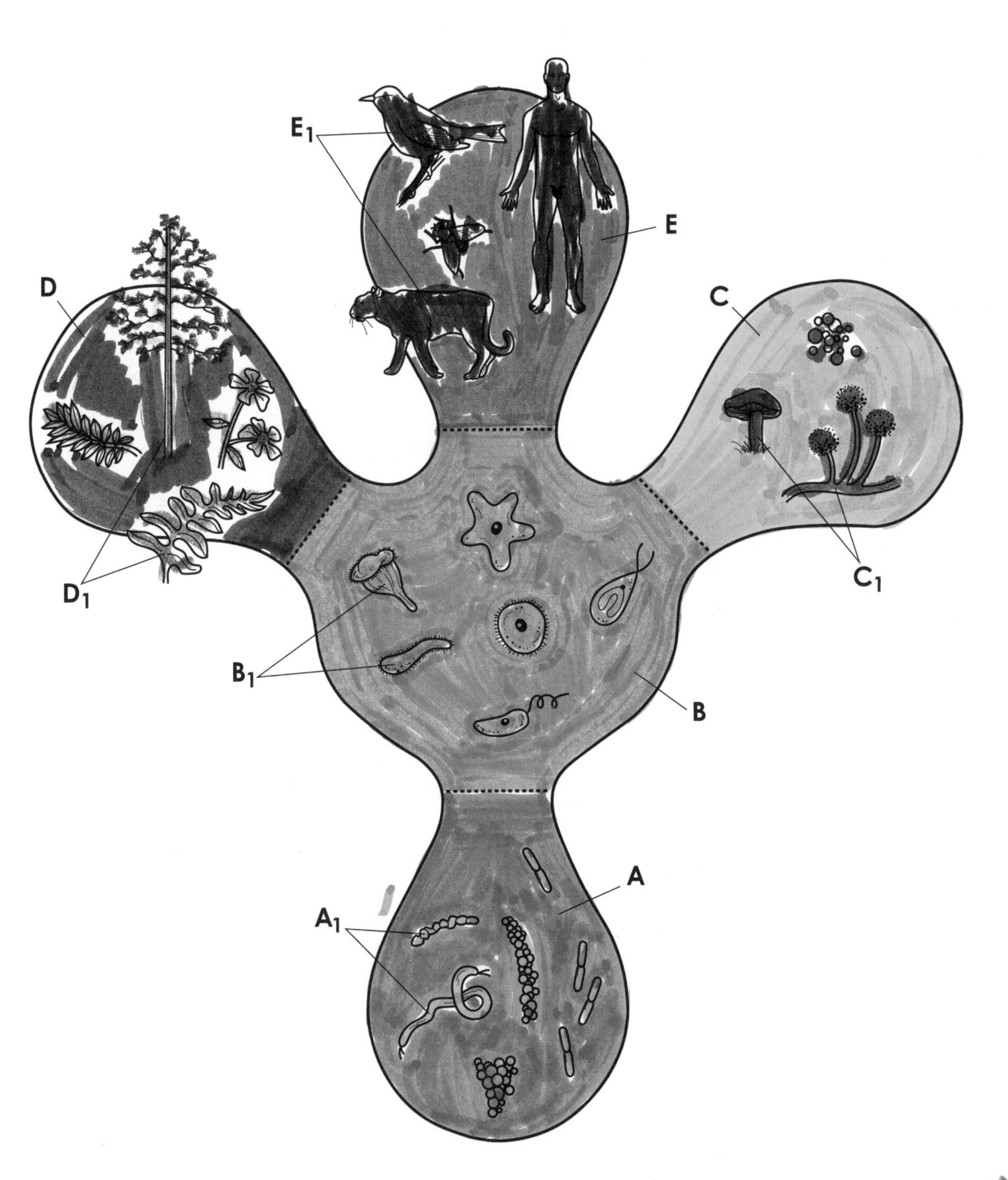

Kingdom Monera A
Monerans A_1
Kingdom Protista B
Protistans B_1
Kingdom Fungi C
Molds and Yeasts C_1
Kingdom Plantae D
Plants D_1
Kingdom Animalia E
Animals E_1

DOMAINS OF LIFE

For hundreds of years scientists have been classifying the diversity of life into categories. For the most part, this was done by comparing features we can see and observe: structure and function. Over the last few decades, DNA sequencing has allowed scientists to rethink how living organisms are categorized. A huge number of species have had their genomes fully sequenced, and because of this data, many people are rethinking the 5-kingdom classification system you just learned about. The most recent method of classification uses domains instead of Kingdoms as the top of the classification hierarchy. The details of this system are still being modified and this discussion will likely continue for many years, as molecular biologists obtain more genome sequencing data. This is a great example of how biology is an ever-evolving field.

The prokaryotic kingdom has been split into two domains: **Archaea (A)** and **Bacteria (B).** Both domains include microbes that are unicellular prokaryotes and these organisms are similar in size and shape. However, Archaea have genes and several metabolic pathways that are closely related to eukaryotes. Some use sunlight as an energy source and others can fix carbon. Bacteria and Archaea also have different cell membrane and cell wall structures. Color **(A)** with purple and **(B)** with blue. Where the two meet, you can blend these two colors together.

Archaea were initially viewed as extremophiles living in harsh environments (such as salt lakes or hot springs), but scientists now know Archaea can live in many different environments. They are common in oceans, and play important roles in the nitrogen and carbon cycles.

All eukaryotic organisms are now grouped in domain **Eukarya (C).** This domain includes three kingdoms of multicellular eukaryotes: plants, fungi, and animals. The easiest way to distinguish between these kingdoms is how they obtain nutrients:

- **Plants (D)** perform photosynthesis and thus produce their own sugars and food. Color this region green and where plants meet with bacteria, blend the two colors together. This is likely where photosynthetic bacteria such as cyanobacteria evolved.
- **Fungi (E)** absorb dissolved nutrients from their surroundings; this is called absorptive feeding. Color this region yellow.
- **Animals (F)** obtain food by eating and digesting other organisms; this is called ingestive feeding. Color this region pink.

Protists (G) are also included in domain Eukarya; however the classification of these organisms is still being debated. Most protists are unicellular eukaryotes and most live in damp environments. However, recent evidence suggests that some are more closely related to plants, animals, or fungi than they are to other protists. Color this region orange and where it meets the other domains and kingdoms, blend the colors together. This blending shows that different groups of organisms are similar and likely related evolutionarily. Some scientists have proposed that kingdom Protista should be separated into several kingdoms, or possibly split into a new domain. Right now, most people agree that protists belong in domain Eukarya, since protist-like organisms were likely the evolutionary link between prokaryotes and other eukaryotic organisms.

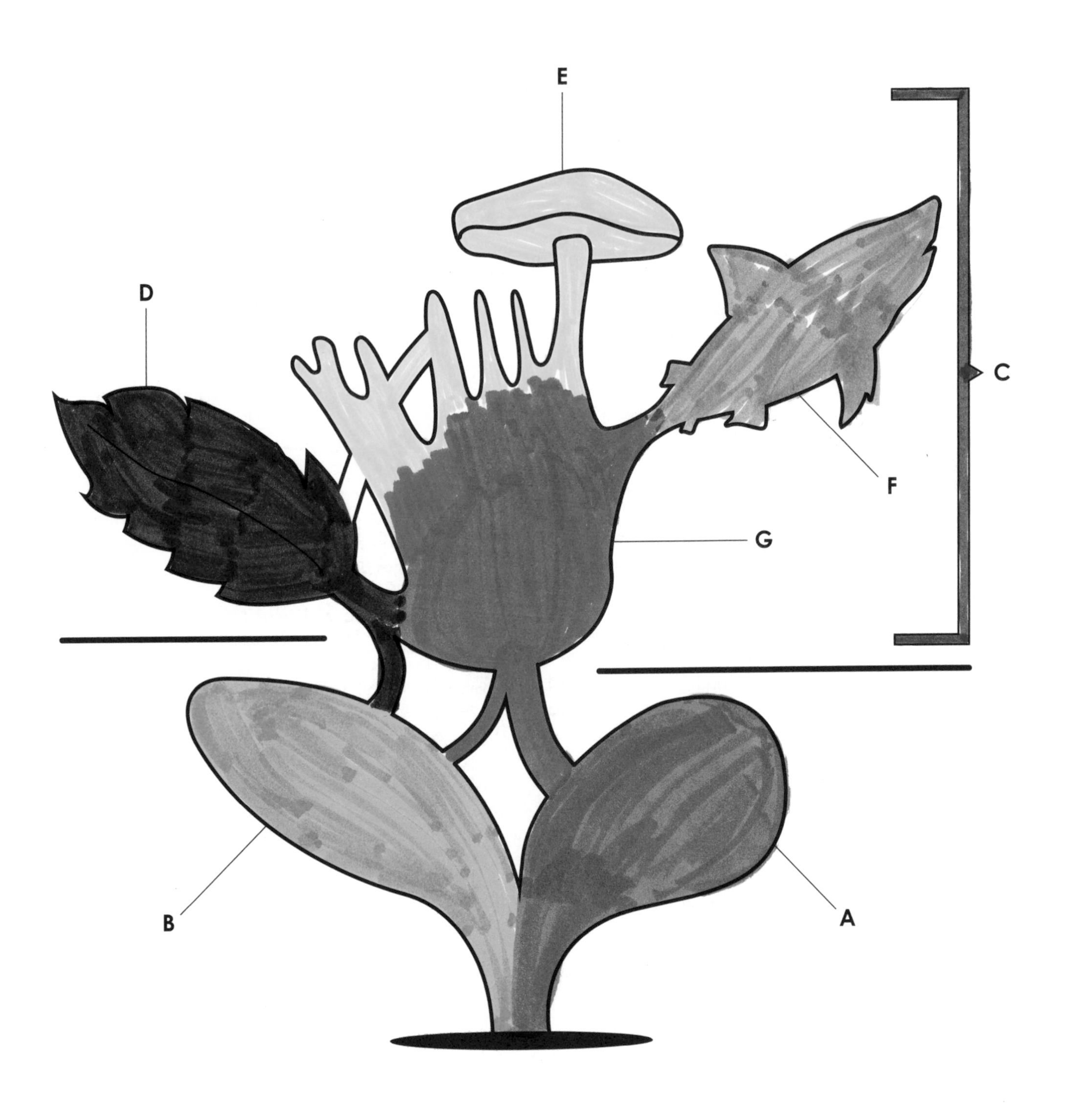

Domain Archaea A
Domain Bacteria B
Domain Eukarya C

Plants (Kingdom Plantae) D
Fungi (Kingdom Fungi) E

Animals (Kingdom Animalia) F
Protists (Kingdom Protista) G

VIRUSES

Viruses are tiny agents that can infect humans, animals, and plants. Scientists have not yet decided if they should be considered alive, for they do not grow, undergo metabolism, or evolve, and they cannot reproduce themselves.

Viruses are very simple organisms. They consist of little more than nucleic acid enclosed in a coating of protein. The nucleic acid genome can be made of DNA or RNA, can be double-stranded or single-stranded, and can be linear or circular. In this plate, we will describe how viruses replicate.

This plate discusses some of the different types of viruses, as well as the viral replication process. Begin reading the text below.

Viruses fall below the range of vision of the light microscope, but they are visible under the electron microscope.

There are a few types of viruses; one is the icosahedral virus. It consists of a fragment of DNA or RNA (but never both), which is known as the **genome (A).** A dark color should be used to trace over the genome in the diagram. The genome may consist of a closed loop of nucleic acid or a linear fragment, and is enclosed in a layer of protein called a **capsid (B).** The capsid is made up of smaller identical units called **capsomeres (C).** One face of the capsid in our diagram is shown as capsomeres, but all of the faces of the capsid are made up of these tiny units. In the icosahedral virus, the capsid is organized into a 20-sided figure called an icosahedron, in which each side is an equilateral triangle, so that there are 12 points and 12 edges.

All viruses have a genome and capsid. Many, but not all types of viruses are enclosed in a **membranous envelope (D).** This envelope is similar to the cell membrane of a eukaryotic cell, but it contains components specific to each virus. For instance, this envelope may have projections called spikes. The viruses that cause herpes simplex, infectious mononucleosis, chickenpox, and acquired immune deficiency syndrome (AIDS) are all icosahedral viruses.

The second major viral type is the helical virus. There are two types of helical viruses. In the first, the genome (A) is wound in a helix, and the shape of the capsid conforms to the shape of the genome. You can also see the capsomeres (C) and the envelope (D) of the helical virus. The rabies virus is an example of a helical virus.

In a second type of helical virus, the genome (A) is found in fragments that are mixed together; this virus has a more circular shape. The influenza virus is an example of the second helical virus.

A virus that attacks bacteria is the bacteriophage. Bacteriophages have icosahedral heads that contain the genome (A) and icosahedral capsids. They also have extended **tails (E)** that show series of rings, and sets of **fibers (F)** at the end of their tails. Bacteriophages are very complex in comparison to other types of viruses.

Notice that the only components of a virus are the fragment of nucleic acid and the coat of protein. An envelope is found in some, but not all viruses. We will now examine the method of viral replication. Look at the second part of the plate and continue reading.

Viruses need the machinery of living cells in order to replicate. In diagram 1, we show a simplified cell; you can see the **cytoplasm (G)** and **nucleus (H).** At its membrane is a virus displaying the **viral genome (I)** and the **viral capsid (J).**

In diagram 2, a hole has opened in the cell membrane and the viral genome (I) has entered the cytoplasm. In this case, the viral capsid (J) has remained outside the cell, but in other cases, the entire virus enters the cytoplasm by phagocytosis, and the viral envelope blends with the cell membrane. The key process here is that the viral genome is released into the cytoplasm of the cell.

In the third diagram, the viral genome has directed the synthesis of new viral parts, and you can see a group of **new genomes (K)** and a group of **new capsids (L).** Note that the cell's nucleus has disappeared; during the process of viral replication, some of the cell structures are consumed and destroyed.

In diagram 4, **new viruses (M)** are constructed within the cell cytoplasm (G). About one hour has passed since the viral genome was released in the cell cytoplasm, and now the cytoplasm is filled with hundreds of thousands of new viruses.

In diagram 5, viral replication is complete, and new viruses (M) are leaving the cell cytoplasm (G). In some cases, the viruses force their way through the cell membrane to the exterior. This is called the viral productive cycle and is how viruses with an envelope acquire this structure. In other cases, the cell bursts in a process called lysis, releasing the new viruses. This is called the viral lytic cycle.

Viral Types

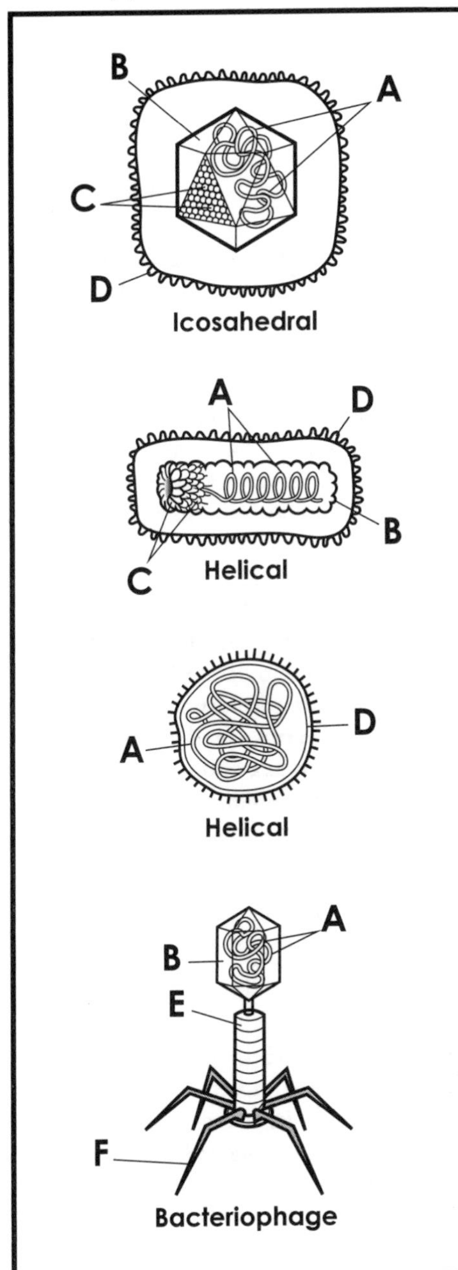

Viral Replication

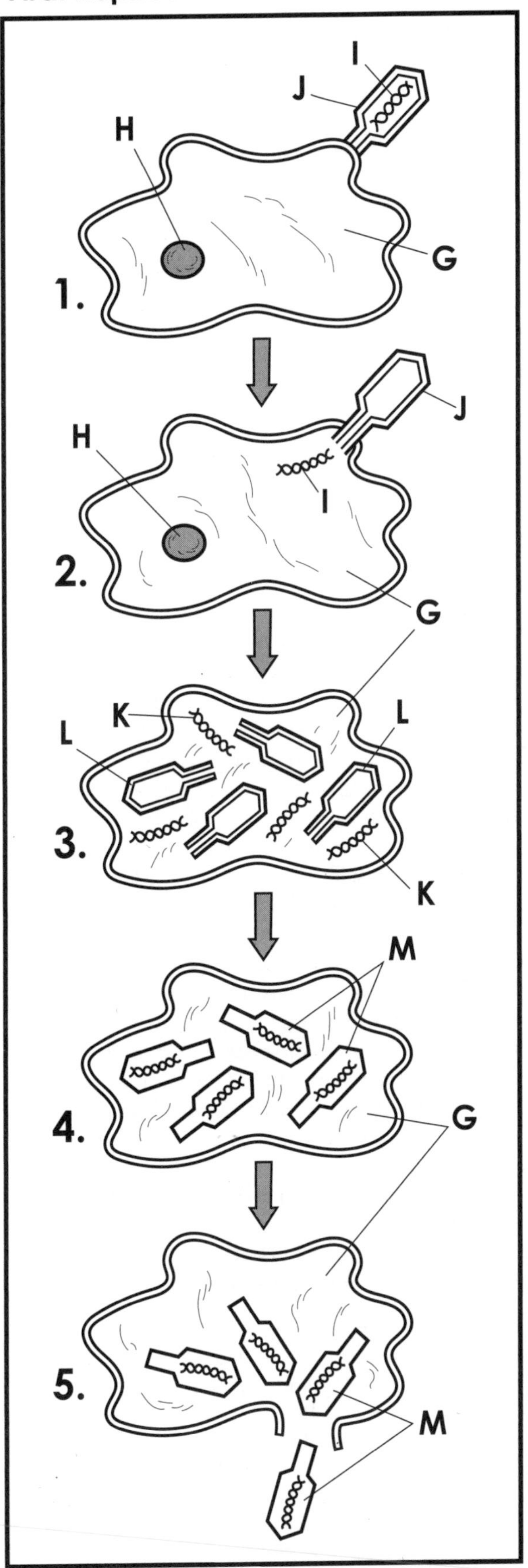

Genome A
Capsid B
Capsomeres C
Membranous Envelope D
Tail E
Fibers F
Cytoplasm G
Nucleus H
Viral Genome I
Viral Capsid J
New Genomes K
New Capsids L
New Viruses M

BACTERIA

In the Whittaker system of classification, the first kingdom is the kingdom Monera, which contains the bacteria. This kingdom is also known as Prokaryotae, because all of its members are prokaryotes. Traditionally, the kingdom has included all bacteria, but a contemporary hypothesis separates the ancient bacteria into a domain of their own. These ancient bacteria are called archaebacteria and were introduced in the plate called "Domains of Life."

In this plate, we will examine bacteria. These are organisms that have neither organized nuclei, nor membrane-bound cytoplasmic organelles. They reproduce by the simple process of binary fission and do not have any of the structures necessary for mitosis. They also contain only a single chromosome.

Look over the plate and notice that it contains three sections: one that shows bacterial types, one that shows bacterial structure, and a third that describes bacterial reproduction. Focus your attention on the first section and begin your coloring.

Bacteria are among the oldest living organism—scientists have found fossils of bacteria that are over 3.5 billion years old. Bacteria are the most common organisms on the Earth, and it has been calculated that the mass of bacteria on our planet outweighs the mass of all other living organisms combined. A single pinch of rich soil, for example, can contain over a billion bacteria.

Despite their immense numbers, bacteria fall into only three general shapes. The first shape is the rod, known as the ***bacillus* (A).** Bacilli are used in many ways by humans, including scientific research and the pharmaceutical production of amino acids and vitamins. Bacilli are also directly involved in the cycles of nitrogen, carbon, sulfur, and other minerals on Earth, and some are known to cause human, animal, and plant diseases.

The second major shape is the bacterial sphere known as the coccus. Variations of cocci exist; for example, the ***diplococcus* (B)** consists of pairs of cocci. Gonorrhea, pneumonia, and a form of meningitis are caused by diplococci. Cocci in a chain form ***streptococcus* (C).** Streptococci are used for making yogurt, and certain species cause tooth decay and strep throat. An irregular cluster of cocci is the ***staphylococcus* (D).** The "staph" infection is caused by a type of *staphylococcus*.

The final bacteria type is the ***spirochete* (E).** The spirochete is a spiral-shaped organism, one agent of which causes syphilis.

We will now turn our attention to a "typical" bacterial cell and highlight some of its structures. These structures are intimately associated with bacterial functions. As you read about the structures below, color them in the diagram.

Many bacterial species have the ability to move independently, using a long rotating structure called a **flagellum (F).** The bacterium in the diagram has two flagella, but some bacterial species have a dozen or more.

Many species of bacteria also possess hairlike structures called pili, also known as **fimbriae (G).** Many disease-causing bacteria infect animal tissue by attaching to it with their pili. Many bacterial species are surrounded by polysaccharide structures called **capsules (H).** Capsules provide protection to the bacteria by shielding them against sunlight, chemicals, and other harsh environmental factors.

Almost all bacteria have an intricate **cell wall (I)** that contains the substance peptidoglycan. The cell wall lends rigidity to the bacterium, helping it to retain its shape. Inside the cell wall is the **cell membrane (J)**, which is similar to the cell membrane of eukaryotic cells.

The **cytoplasm (K)** of the bacterium contains proteins, fats, enzymes, carbohydrates, and other materials normally found in cytoplasm. As is the case in eukaryotic cells, bacteria possess **ribosomes (L).** These ultramicroscopic bodies are the sites of protein synthesis.

As the diagram illustrates, the bacterium contains a single, closed-loop **chromosome (M).** The chromosome is in the cytoplasm; there is no nuclear membrane. You can also see a small, closed loop of DNA called a **plasmid (N).** These plasmids are key in the field of genetic engineering.

Some bacterial species produce a structure called an **endospore (O).** These bacteria replicate their DNA and one copy is stored inside this resistant cell, which is able to survive trauma that the bacteria itself is not.

We now turn to a study of bacterial reproduction. The process of bacterial reproduction is called binary fission.

In the diagram we see a single coccus representing the **first generation (P).** The coccus divides, yielding two cocci in a **second generation (Q).** Cocci of the second generation metabolize for a period of time and then undergo binary fission to yield a **third generation (R).** At this point, four cocci have resulted from the initial cocci. These third-generation cocci (R) metabolize, and then undergo binary fission to yield the **fourth generation (S).**

Before binary fission can occur, the DNA genome must be replicated. During fission, each daughter cell gets one copy of the genome and this is a tightly regulated process. In contrast, plasmids sort randomly into daughter cells. This means that if a parent cell has only a few copies of a plasmid, one daughter cell can receive this plasmid while the other may not. Finally, remember that binary fission and mitosis are very different processes. Both allow a cell to replicate asexually, but they happen differently and involve different steps.

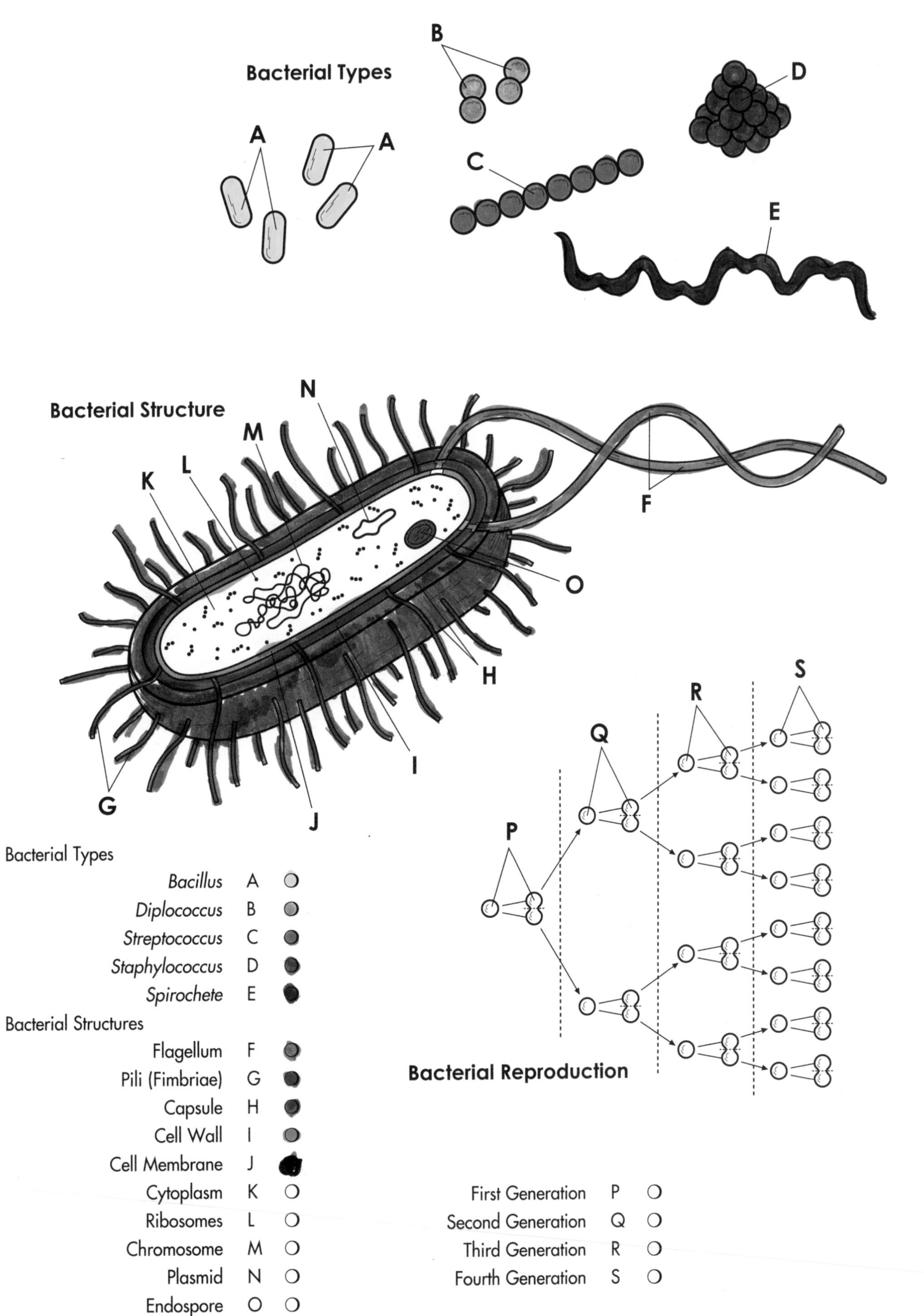
Bacterial Types
A
B
C
D
E
Bacterial Structure
F
G
H
I
J
K
L
M
N
O
Bacterial Reproduction
P
Q
R
S
Bacterial Types
Bacillus A
Diplococcus B
Streptococcus C
Staphylococcus D
Spirochete E
Bacterial Structures
Flagellum F
Pili (Fimbriae) G
Capsule H
Cell Wall I
Cell Membrane J
Cytoplasm K
Ribosomes L
Chromosome M
Plasmid N
Endospore O
First Generation P
Second Generation Q
Third Generation R
Fourth Generation S

PROTOZOA

The kingdom Protista includes three major groups: protozoa, slime molds, and single-celled (unicellular) algae. In this plate, we will examine the four types of protozoa. Protozoa are all eukaryotic, which means that they have nuclei and organelles, that they reproduce by mitosis, and that they have multiple chromosomes. As this plate will show, they are divided into groups according, in large part, to their method of locomotion.

This plate shows a collection of different types of protozoa, organized into four groups based on their method of locomotion. Focus on the first group, the sarcodines.

Sarcodines are a group of amoeboid protozoa. They are blob-like and asymmetrical, and can assume an infinite variety of shapes. Among the sarcodines are two **shelled amoebae (A)** called *Foraminifera* and *Radiolaria*, which have hard shells. The foraminiferan shell resembles a snail, while the radiolarian shell is rounder, with spikes.

One typical sarcodine is the amoeba, which has a sizeable amount of **cytoplasm (B),** a clear **nucleus (C),** and several **food vacuoles (D).** The distinctive characteristic of the amoeba is its false foot, or **pseudopodia (E).** This extension of the cytoplasm permits the amoeba to move independently. Notice the **cell membrane (F)** enclosing the cytoplasm.

We now move to the second and third groups of protozoa and note the two types of motility in these groups. Continue your coloring as you read the paragraphs below.

A second group of protozoa are the flagellates; these organisms all possess a whip-like flagellum. One typical flagellate is *Trypanosoma*, which is an elongated, worm-like protozoan that has the characteristic cytoplasm (B) and nucleus (C) enclosed by the cell membrane (F). A very thin sheet of tissue called the **undulating membrane (G)** extends from its cytoplasm, and its **flagellum (H)** whips back and forth, propelling it. This protozoan causes sleeping sickness in humans.

Another flagellate is *Giardia*. In *Giardia* we see the standard parts of the protozoan as well as multiple flagella (H). This organism is distinct because of its two nuclei (C).

A third flagellate is *Trichomonas*. Again you see multiple flagella (H), and you can also see the long, spine-like rod that runs the length of the cell, called the **costa (I).** *Trichomonas* causes a sexually transmitted disease called trichomoniasis.

The last flagellate we'll look at is *Euglena*, which is distinguished by its **chloroplasts (J).** *Euglena* is a photosynthesizing organism, and its chloroplasts contain green chlorophyll pigment. *Euglena* has a **nucleolus (K)** within its nucleus (C), and an **eyespot (L),** which is a concentration of photosensitive cells. A **contractile vacuole (M)** allows this organism to eliminate excess water from its cytoplasm (B).

A third type of protozoa is the ciliate, and a characteristic ciliate is *Ballantidium*. This organism has a number of hair-like **cilia (N)** on its surface, which wave in a synchronous motion and enable the organism to move about. A contractile vacuole (M) and a very large nucleus (C) are present in this organism.

Perhaps the most familiar ciliate is *Paramecium*. *Paramecium* has numerous cilia (N), and food enters this organism through an oral groove, forming food vacuoles (D) that move about the ciliate as the food is digested. A contractile vacuole (M) is responsible for removing excess fluid from the cytoplasm (B). *Paramecium* has a **macronucleus (O)** as well as a **micronucleus (P).**

We will complete our survey of protozoa by looking at a fourth group. The members of this group have no organs for locomotion. Continue your coloring as you read the paragraphs below.

The members of the final group of protozoa, the sporozoans, have no organs for locomotion in their adult form. They are very complex organisms, and many are parasites of animals, plants, and humans. One example of this type is *Babesia*. This organism exists within the cytoplasm of **red blood cells (Q).** It is a simple organism that has the usual cytoplasm (B) and nucleus, but when it destroys red blood cells, a malaria-like illness called babesiosis occurs.

The organism that causes malaria is *Plasmodium*. This complex organism is shaped like a ring, and can be seen within red blood cells (Q). The nucleus (C) of this organism gathers at one side of the cell, while the cytoplasm fills most of the rest of the space. When *Plasmodium* multiplies, it causes the blood cell to burst, or lyse, and a form of anemia results.

The final organism we will consider is *Toxoplasma*. These organisms appear as crescents and they infect liver cells, white blood cells, and brain cells. They cause liver destruction, blood cell insufficiency, and brain disease, and are transmitted from domestic house cats to humans through contact.

Sarcodines

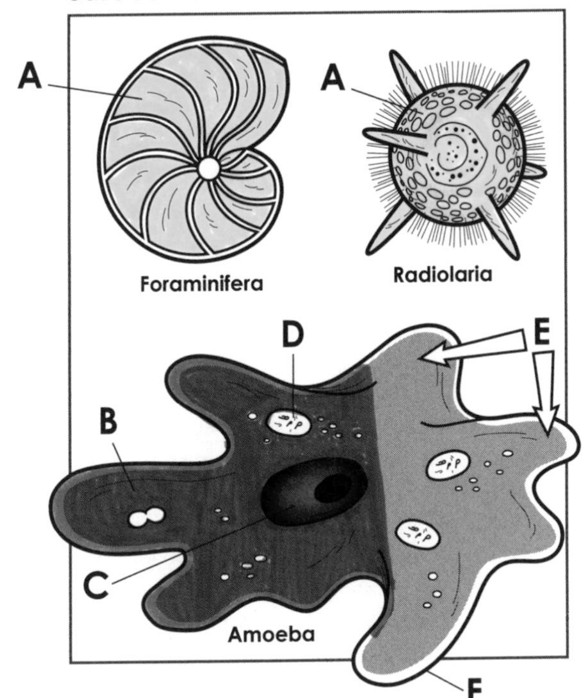

Flagellates

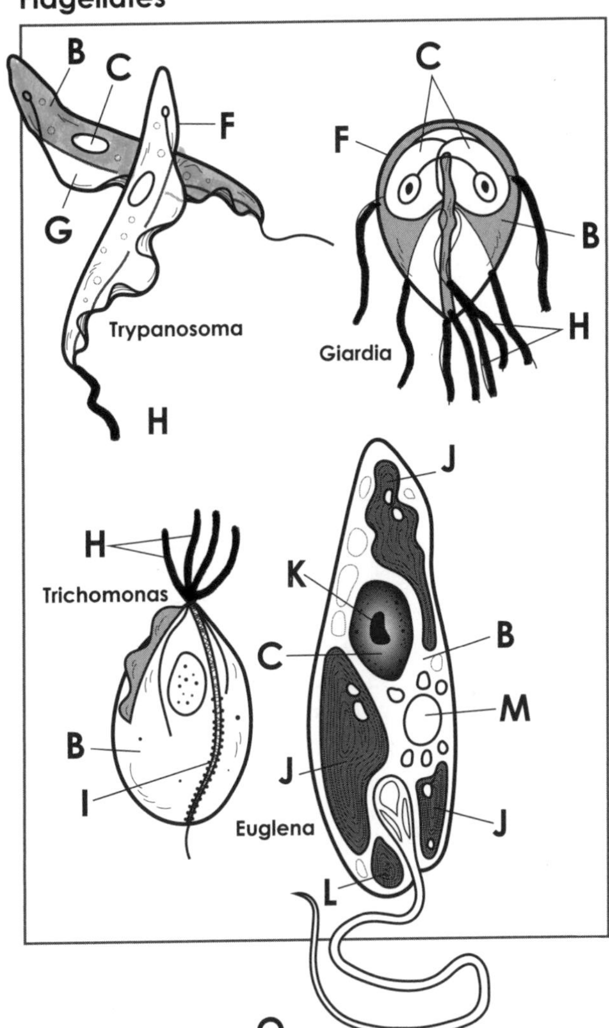

Ciliates

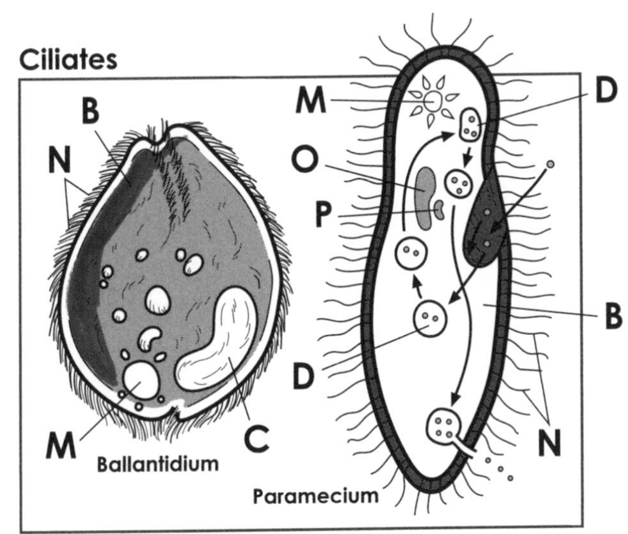

Sporozoans

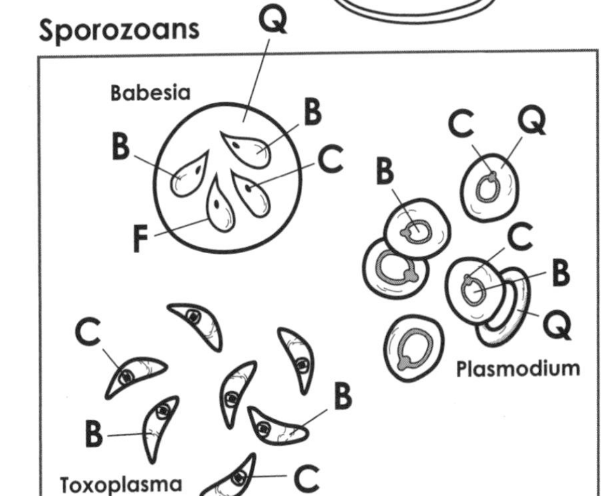

Shelled Amoebae	A		Undulating Membrane	G		Contractile Vacuole	M	
Cytoplasm	B		Flagellum	H		Cilia	N	
Nucleus	C		Costa	I		Macronucleus	O	
Food Vacuole	D		Chloroplasts	J		Micronucleus	P	
Pseudopodia	E		Nucleolus	K		Red Blood Cell	Q	
Cell Membrane	F		Eyespot	L				

SLIME MOLDS

The kingdom Protista includes a number of fungus-like organisms called slime molds. Slime molds are nonphotosynthesizing organisms that are usually found growing in damp areas or on decaying vegetation.

There are two recognized types of slime molds: acellular slime molds and cellular slime molds. Acellular slime molds consist of masses of cytoplasm, called plasmodium, that may spread thinly over several square meters. Their nuclei are not confined to discrete cells (hence the term acellular). Cellular slime molds live in nature as independent, haploid cells that form a mass called a pseudoplasmodium, as this plate will describe. (A haploid cell contains a single set of chromosomes.) Each cell in the pseudoplasmodium is a separate, individual cell. We will examine the life cycle of a cellular slime mold in this plate.

In this art, we show the life cycle of a cellular slime mold. The forms of its existence are quite diverse, as the plate indicates. Focus on the fruiting body at the top end of the life cycle and begin your reading.

We will begin our study of the life cycle of cellular slime mold with the fruiting body. This structure consists of a **base (A),** which is attached to a suitable anchoring device, such as a rotting log. Rising from the base is the **stalk (B),** and at the tip of the stalk is a structure called the **spore cap (C).** A number of spore caps are shown in the plate. Numerous haploid spores are produced in the spore cap through the process of mitosis.

As the plate shows, the spore cap bursts to release a number of haploid **spores (D).** The spores now undergo a developmental process called **germination (E).** In germination, each spore releases a tiny, haploid **amoeba (F),** which feeds by phagocytosis and undergoes repeated mitotic division to produce a large number of haploid offspring. These **feeding amoebae (G)** are shown in the figure.

We now focus on the second process in the life cycle. When the local food supply becomes scarce, the amoebae begin to come together, or aggregate. In the plate, we present an **aggregate of amoebae (H),** which still consists of individual haploid cells.

A new form in the life cycle develops as a result of the aggregation of amoebae. Continue your reading below as you color the plate. Light colors should be used in order to prevent obscuring the details.

The amoebae aggregate in response to a chemical signal that they release when food is scarce, and the result of the aggregation is a mass called a **pseudoplasmodium (I).** The pseudoplasmodium consists of individual cells, and it behaves like a multicellular organism. As it moves, the migrating **pseudoplasmodium resembles a slug (J).** The slug-like slime mold migrates toward light, but eventually becomes stationary, and changes into a **differentiating slime mold (K).** A base forms at one end, and a stalk soon begins to emerge. At the tip of the stalk, a spore cap develops, and haploid spores develop within it. The life cycle is now complete; the organism has undergone a series of changes that led it back to its original form.

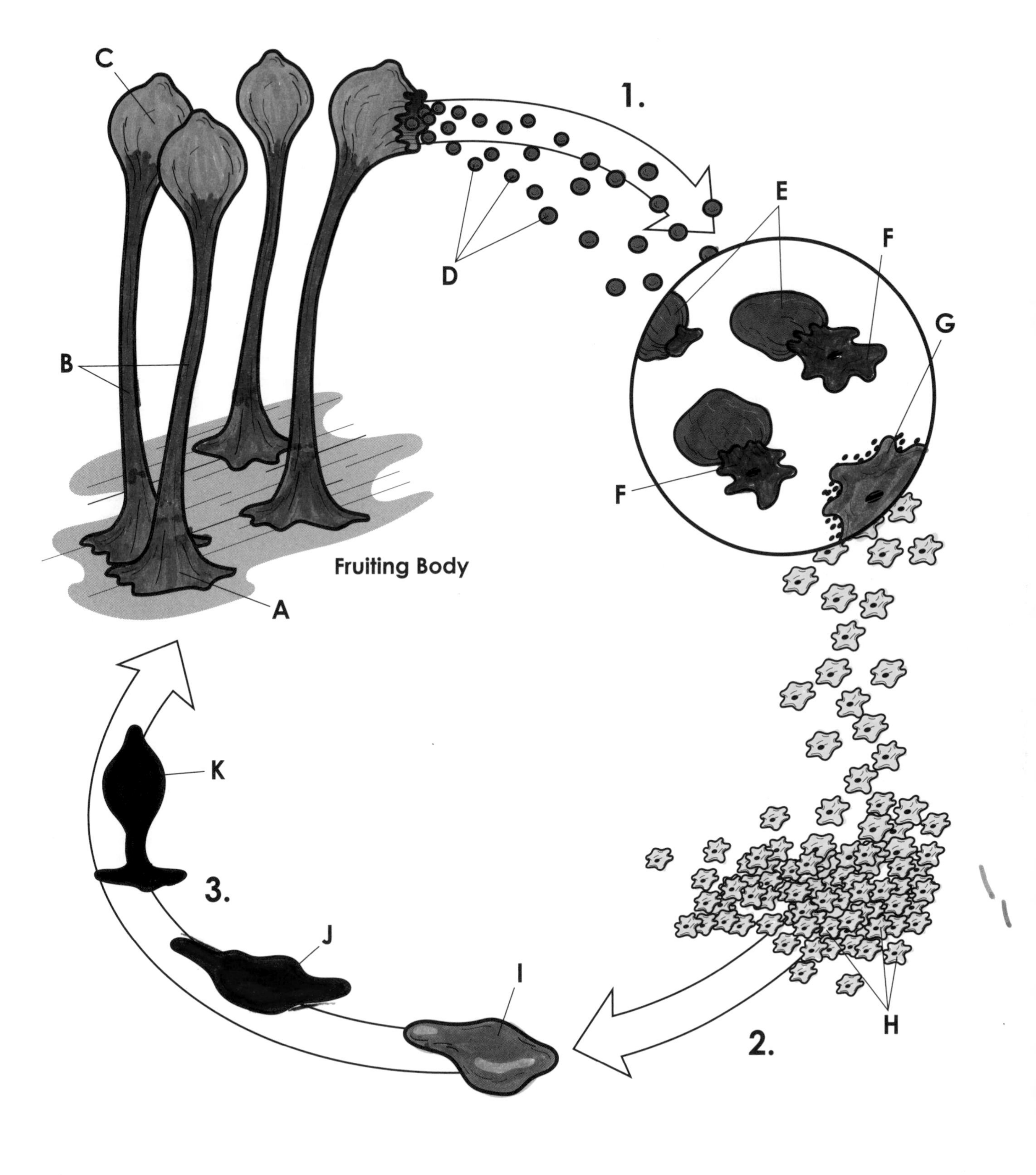

Base	A	●	Germination	E	●	Pseudoplasmodium	I	●
Stalk	B	●	Amoeba	F	●	Pseudoplasmodium Resembling a Slug	J	●
Spore Cap	C	●	Feeding Amoebae	G	●	Differentiating Slime Mold	K	●
Spores	D	●	Aggregate Amoebas	H	○			

SIMPLE ALGAE

The kingdom Protista is made up of protozoa, slime molds, and simple algae. Many scientists believe that complex plants evolved from simple algae. All simple algae organisms are eukaryotic, but can be unicellular (single-celled) or multicellular. In this plate, we will describe various types of simple algae.

Looking over the plate, you will notice that we show examples of simple algae, along with the life cycle of one alga.

Alga is the common name for a large number of chlorophyll-containing simple organisms; they are photosynthetic, but have no roots, stems, or leaves. Most algae live in the ocean, but there are many freshwater forms. Some species are unicellular, others appear as colonies of cells, and some are truly multicellular, but with little differentiation of cells.

One group of simple algae is the pyrrhophytes. They are unicellular, have photosynthetic pigments, and are propelled by movement of their flagella. The example we show here is the **dinoflagellate (A).** Through photosynthesis, dinoflagellates and other pyrrhophytes synthesize energy-rich carbohydrates. Dinoflagellates move by means of **flagella (B)** and are covered by armor-like plates.

Another simple alga is the **diatom (C).** A light color should be used to color the diatom, in order to preserve its details. Diatoms are chrysophytes, and have walls made up of silicon dioxide, so they appear glassy. As photosynthetic organisms, diatoms are among the most important producers of carbohydrates in the oceans, and are at the base of many food chains.

We have introduced the simple algae, and will now move to larger forms that can be seen with the naked eye. Continue your reading below, and color the organisms.

The alga ***Chondras*** **(D)** is a type of red alga, also called rhodophytes. Red algae contain red pigments as well as the usual blue-green and green ones, and all but a few are aquatic; most are marine forms, and they comprise a large portion of the sea weeds. They are true multicellular organisms, and some red algae (kelps) grow to be 50 meters in length. The thickening agent, agar, is produced from red algae.

The brown algae are known as phaeophytes. They have characteristic brown pigments in addition to the usual chlorophyll pigments. One example of a brown alga is ***Laminaria*** **(E),** which is an ocean kelp, and another is ***Fucus*** **(F),** the common rockweed found near the beach. *Fucus* has **air bladders (G)** that enable it to float.

A final group of simple algae is the green algae, also called chlorophytes. They exist mostly in freshwater forms; for example, the colonial organism *Volvox* is a chlorophyte. Another green alga is ***Spirogyra*** **(H),** in which the **chloroplasts (I)** exist in a spiral formation through the cells of the filament.

In the final portion of the plate, we will look at a unicellular green alga and show its life cycle. Continue your coloring as you read the paragraphs below.

The life cycle of simple algae is illustrated here by the unicellular green alga *Chlamydomonas*. In our diagram, this organism has a prominent chloroplast (I) and **nucleus (K).** The cell also contains a **starch granule (L)** and an **eyespot (M)** where photosensitive cells conjugate.

Chlamydomonas can reproduce asexually; through mitosis, this organism produces a number of haploid zoospores. They are represented by the **haploid zoospore (N)** in the diagram. Soon a number of **zoospores (O)** form and each develops into a mature *Chlamydomonas* cell.

In the sexual process of *Chlamydomonas* reproduction, the cell first forms a number of **developing gametes (P).** (Remember that gametes are haploid cells.) The **gametes fuse (Q)** at the beginning of the diploid stage so that there are two sets of chromosomes per cell: At this point the cell is called a zygote. The zygote forms a **zygospore (R),** in which meiosis takes place. Emerging from the **meiotic zygospore (S)** are a number of zoospores (O), each of which is haploid. The zoospores develop into new *Chlamydomonas* cells. In the life cycle of *Chlamydomonas,* both haploid and diploid generations exist. This process is called alternation of generations, and is common in complex plants. Alternation of generations will be discussed in a few future plates about plants (see "Life Cycle of a Moss" and "Life Cycle of a Fern" for more information).

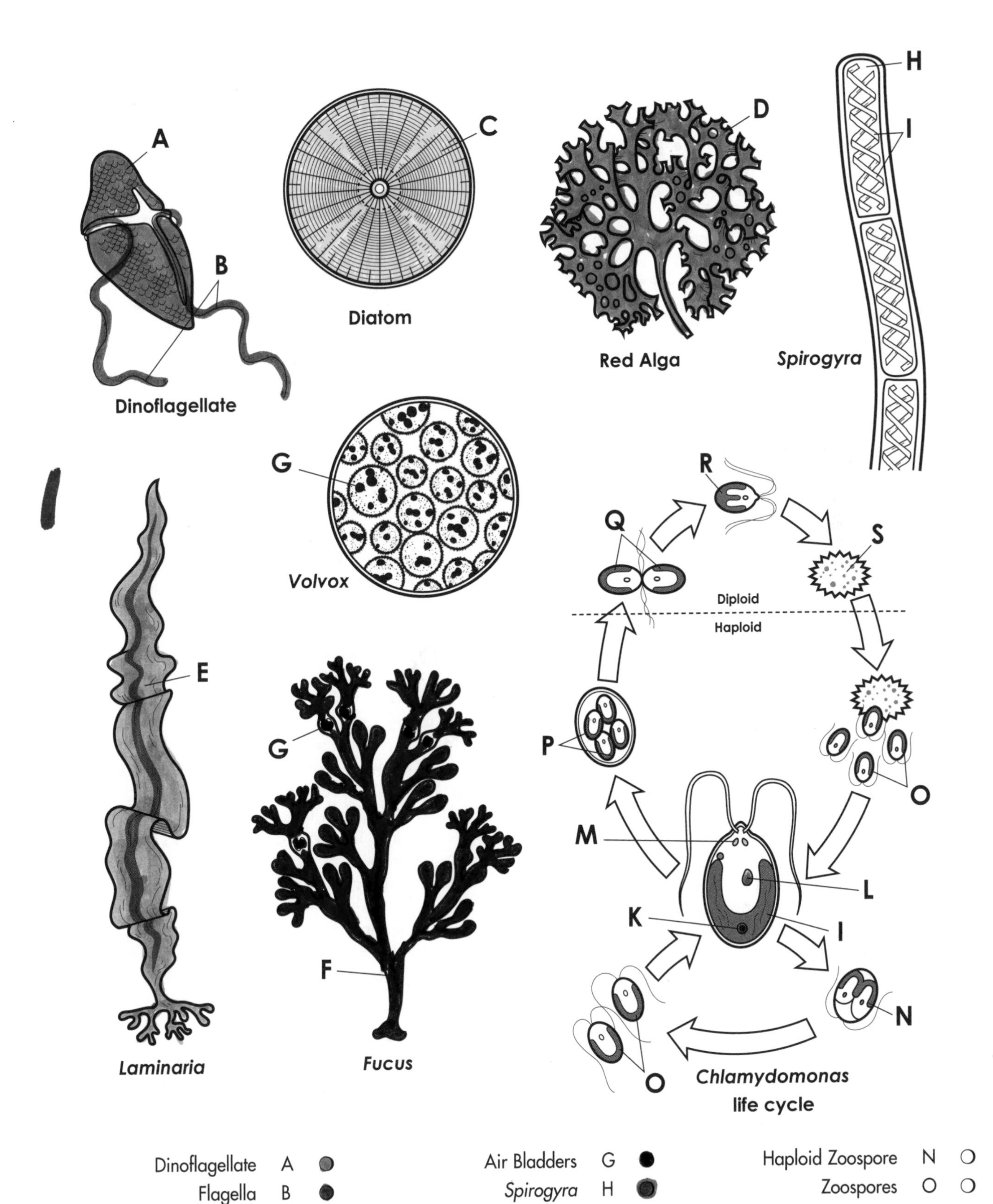

Dinoflagellate A	Air Bladders G	Haploid Zoospore N
Flagella B	*Spirogyra* H	Zoospores O
Diatom C	Chloroplast I	Developing Gametes P
Chondras D	Nucleus K	Gametes Fusing Q
Laminaria E	Starch Granule L	Zygospore R
Fucus F	Eyespot M	Meiotic Zygospore S

KINGDOM FUNGI

Although they were once classified as plants, members of the kingdom Fungi are not at all plant-like. For example, plants engage in photosynthesis, but fungi are nonphotosynthetic and consume other organisms for food. Also, most fungi have filamentous bodies that penetrate animal waste, soil, or other forms of organic matter. In many cases, the only visible portions of the fungi are its reproductive structures. Fungi are eukaryotic organisms that have cell walls made of chitin, and digestive patterns suited to the absorption of nutrients.

Scientists recognize five divisions of fungi. The fungi we will study here produce sexual, diploid zygospores, and are called Zygomycota. This phylum includes bread and other common molds.

This plate illustrates the sexual and asexual reproductive forms of a typical fungus. Fungi are relatively simple and are commonly found on foodstuffs. All of the structures typical of a fungus are displayed on this plate.

With rare exceptions (e.g., yeasts), fungi have filamentous bodies that are composed of eukaryotic cells. They live primarily in acidic environments that are near to room temperature and reproduce by both asexual and sexual methods, as we will show in this plate.

We will begin our study with a slice of **bread (A).** A gray color should be used to shade the slice of bread, including the surface upon which the fungus is growing. The basic unit of this fungus is a long, filamentous, branching set of eukaryotic cells called a **hypha (B);** these are shown throughout the plate. Specialized branches of the hyphae called **rhizoids (C)** penetrate the surface of the bread and absorb nutrients that sustain the fungus.

Arising from the surface of the bread, certain hyphae form specialized stalks called **sporangiophores (D).** At the tip of each sporangiophore there is a spore-bearing structure called a **sporangium (E).** Many sporangia are shown in the plate, and one is shown in detail. You should notice that the open, cross-sectioned **sporangium is releasing its spores (F).** These spores are **asexual (G);** they are produced by the cells at the tip of the sporangiophore, which undergo mitosis. These spores are genetically identical.

Having pointed out the major structures of the fungus, we will now focus on the process of asexual reproduction. Continue your coloring as you read the text below.

The first type of reproduction displayed by the fungus is asexual reproduction. In this process, the asexual spore produced by mitosis lands on the surface of the bread (A), to become a **germinating spore (H).** The spore opens, and a new hypha develops. Eventually the hypha will form sporangiophores and sporangium, and new, asexual spores will form within the sporangium, completing the cycle.

We will now look at the second type of reproduction displayed by most fungi, sexual reproduction.

Sexual reproduction takes place when hyphae from opposite mating types meet and form mating structures. In the plate, we see **sexually opposite hyphae (I)** that are distinguished by a plus and minus. In the cutout, the sexually opposite hyphae (I) grow toward one another, forming gamete-bearing sacs called **gametangia (J).** Eventually the **gametangia fuse (K)** and the nuclei of the cells join to form a diploid zygote. This mature, thickened cell is the **zygospore (L),** and later, the zygospore will **germinate (M).** The zygotes have undergone meiosis and reverted to the haploid state, and the zygospore forms a sporangium (E) and releases **sexual spores (N).** Each sexual spore is capable of landing on the bread or another suitable environment and reproducing new hyphae.

Sexual reproduction results in fewer spores than asexual reproduction, but the sexually produced spores are genetically different from one another, which introduces variation into the fungus and makes evolution possible.

We concentrated on one phylum of fungi in this plate but a few others are worth mentioning. Ascomycota are also called sac fungi. This phylum includes a range of organisms from unicellular yeast to multicellular truffles and cap fungi. These fungi get their name from the production of sexual spores (called ascospores) in sac-like structures called asci. Finally, Basidiomycota are also called club fungi, and this phylum includes puffballs, mushrooms, and shelf fungi. They are important decomposers of wood and other plant material. Many are edible. Most species in this phylum produce sexual spores (called basidiospores) on basidiocarp (or basidia) structures located on the fruiting body.

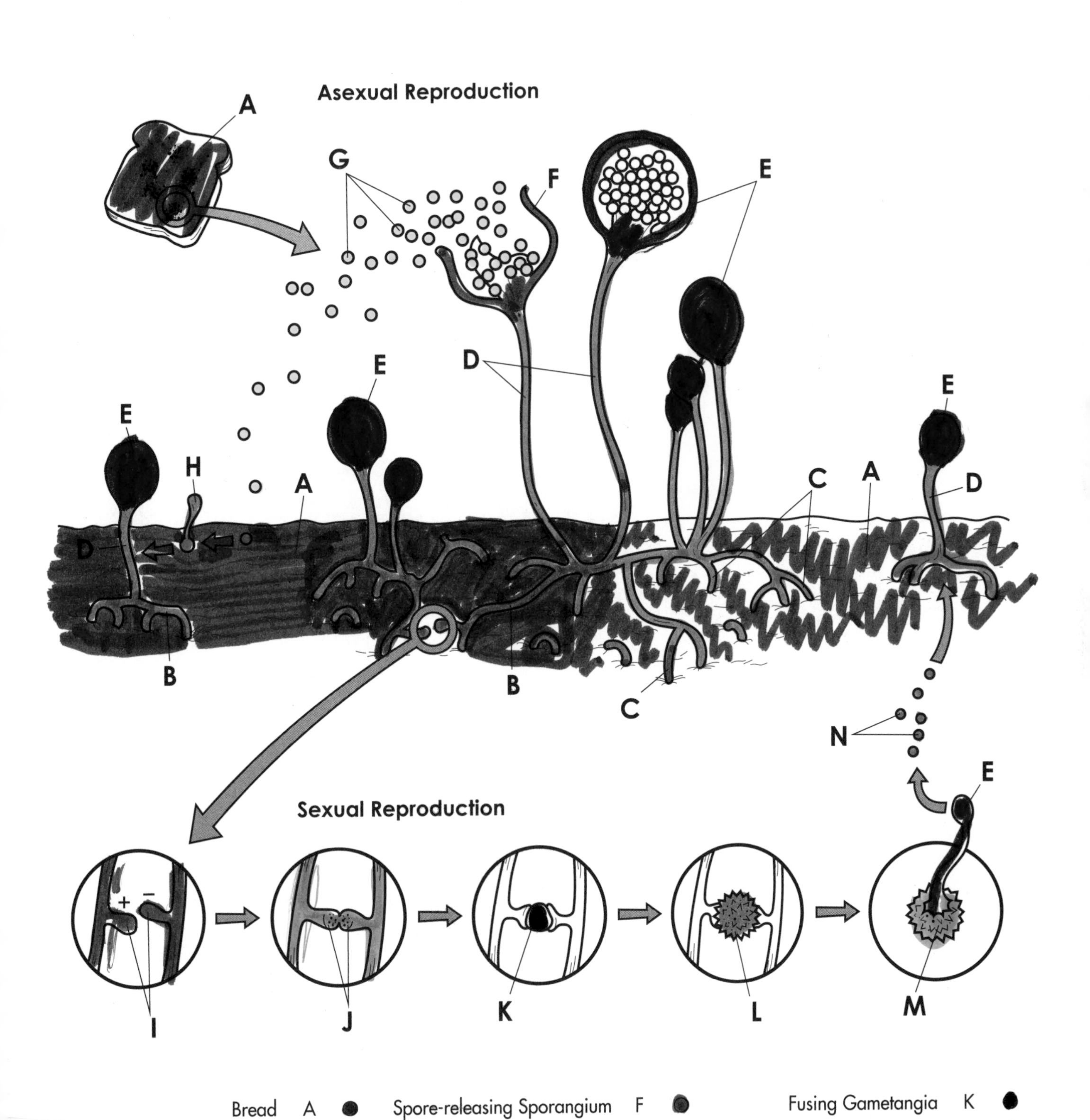

Bread	A		Spore-releasing Sporangium	F		Fusing Gametangia	K	
Hypha	B		Asexual Spores	G		Zygospore	L	
Rhizoids	C		Germinating Spore	H		Germinating Zygospore	M	
Sporangiophores	D		Sexually Opposite Hyphae	I		Sexual Spores	N	
Sporangium	E		Gametangia	J				

CHAPTER 7:

BIOLOGY of the PLANTS

STRUCTURE OF A FLOWERING PLANT

Many species of flowering plants differ in how they obtain resources from the environment, and all are well adapted to their particular niche on Earth. But all flowering plants have several structural features in common that make them self-sufficient and able to thrive in their environment. In this plate, we will examine the general structure of a flowering plant.

Looking over the plate, you will notice that it contains a single large diagram of a typical plant. This composite plant shows structures found in virtually all flowering plants. We will survey the structures and their functions in some detail. Bold colors such as reds, greens, purples, and oranges should be used for the plant since there are many brackets to be colored and the structures are fairly large.

The body plan of the common flowering plant such as the rose, tulip, vegetable plant, or tree includes two main systems: the shoot system and the root system. The **shoot system (A)** is surrounded by a large bracket, which should be colored in a bold color. This part of the flowering plant is above the ground, and is composed of several organs.

One important organ of the shoot system is the **stem (B),** some of which is enclosed by a bracket. Stems are the trunks and branches of trees and shrubs, the vines of grapes and ivy, and the leaf-bearing stalks of herbs and grasses. Stems support the plant and distribute its leaves so that it can take maximum advantage of available sunlight. The step is divided into areas called **nodes (C),** and the regions between the nodes are called **internodes (D).**

We now turn to another organ of the shoot system, the **leaf (E),** which you should color green. A typical leaf consists of a thin, flattened main portion called the **blade (F),** which is joined to the stem by an extension called the **petiole (G).** Vascular tissue (xylem and phloem) runs through the petiole into the blade, where it forms the parallel veins of monocots and the netlike veins of dicots. These two terms are explored in the next plate.

Another important organ of the plant is the **flower (H),** which is actually a leaf that has been modified for reproductive purposes. Monocot and dicot plants vary in the number of petals in their flowers, as the next plate will show.

One of the characteristic features of the flowering plant is that the seed is encased in a **fruit (I).** The fruit is a swollen reproductive structure that acts as an ovary in which seeds are nourished until they can survive in the environment. The fruit develops after fertilization of the flower has taken place.

Returning to the shoot system, we see a point just above where the leaf attaches to the stem called the **lateral bud (J).** This bud is a young, inactive shoot that will eventually grow into a new leaf-bearing branch. The upper portion of the shoot, or the main stem in this case, is the **shoot apex (K).** This is the terminal portion of the stem at which growth occurs.

Although stems usually rise straight from the soil, the stems of some plants, like strawberries and Bermuda grass, grow horizontally. Stems that grow horizontally and underground are called rhizomes; ferns and potatoes are examples of plants that have rhizomes.

We have surveyed some of the main parts of the shoot system of the flowering plant, and will now turn our attention to the root system.

We complete the survey of the plant structures by focusing on the **root system (L);** the large bracket that encloses it should be colored in a dark color. The primary roles of the root are to provide anchorage for the shoot system and to absorb water and minerals from the soil. In some crop plants such as beets, carrots, and turnips, roots are modified to store water and nutrients.

After germination, the first portion of the root to emerge is the **primary root (M).** This root may be a fibrous root or a taproot, as you will see in the next plate. As the primary root continues to grow, smaller **lateral roots (N)** emerge. Lateral roots branch out to the sides and provide anchorage and absorption until a tangled mass of roots results.

STRUCTURE OF A FLOWERING PLANT

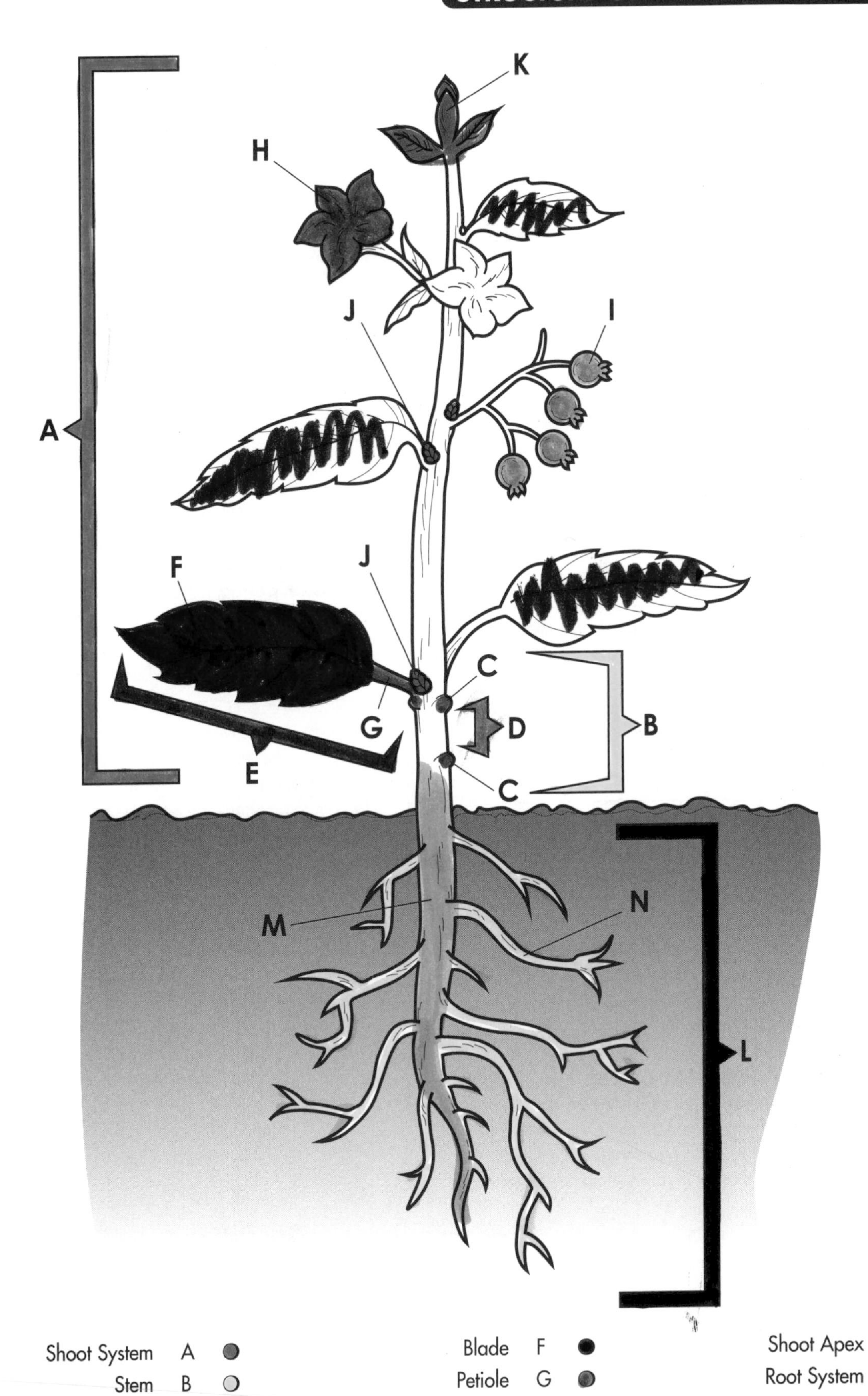

Shoot System	A	Blade	F	Shoot Apex	K
Stem	B	Petiole	G	Root System	L
Node	C	Flower	H	Primary Root	M
Internode	D	Fruit	I	Lateral Root	N
Leaf	E	Lateral Bud	J		

MONOCOTS AND DICOTS

The evolution of plants in the terrestrial environment required adaptations for avoiding water loss and providing supportive structures for the plant. Among these adaptations was a vascular system of tubules capable of absorbing moisture and nutrients and transporting them to all parts of the plant. Most plants found on the Earth today are vascular plants, and most vascular plants have seeds found either in cones (the cone-bearing gymnosperms) or within fruits (the flower-bearing angiosperms). Angiosperms are the most familiar and common plants on Earth and are the focus of this and subsequent plates.

Botanists are scientists who study the physiology and anatomy of plants. They recognize two major classes of flowering plants: the dicotyledons, also known as dicots; and the monocotyledons, also called monocots. Botanists have recognized about 65,000 species of monocots and about 170,000 species of monocots. The major differences between these two classes of flowering plants are the focus of this plate.

This plate compares five characteristics of the monocots and dicots: seed, leaf, stem, flower, and root. The first row of diagrams displays the monocot characteristics and the second row displays dicot characteristics. Start the plate by focusing on each of the five categories as we compare the two groups of plants.

The monocots include such plants as grasses, lilies, and palm trees. Dicots include most of the common trees (with the exception of cone-bearing trees), as well as flowers and crop plants. One major difference between the two groups is in their seeds. Monocots have only one **cotyledon (A),** while dicots have two cotyledons. A cotyledon is an embryonic leaf that expands during seed development and accumulates a storehouse of nutrients that supports the growing seedling after germination. In monocots, the **endosperm (B)** remains as the cotyledon expands, while in the dicot, the endosperm withers.

A second major difference in monocots and dicots can be seen in their leaves, and in particular the **veins (C)** of the leaves. They should be traced in a dark color. Notice that in the monocots, the veins run parallel to one another in the leaf, while in the dicot, the veins are netlike and branched. These veins are the vascular tissue of the leaf.

We have begun our comparison of the monocots and dicots by focusing on their seeds and leaves. We will continue our comparison by showing the distribution of vascular tissue in the stems of both types of plants. Continue your reading as you focus on the monocot and dicot stems.

The stem contains the supporting and conducting vascular tissues of the plant, including xylem and phloem. Xylem transports water and minerals up from the soil, while phloem transports sugars throughout plant tissues. The stem of the monocot contains random **vascular bundles (D),** which are collections of xylem and phloem tubes. They are scattered through the stem in a complex arrangement. By comparison, the vascular bundles of the dicot stem are arranged in a ring, and in some cases, there are several concentric rings of vascular tissue beneath the outer epidermal layer of the stem.

Another major difference between monocot flowering plants and dicot flowering plants occurs in the flower itself. The outer parts of the flower are specialized leaves called petals, and **monocot petals (E)** usually occur in threes or multiples of threes. For example, an iris or lily has three petals and, in the diagram, we show a monocot plant that has six petals. By comparison, **dicot petals (F)** occur in fours or fives or multiples of fours or fives, and we show a dicot flower that has five petals. Beneath the five petals are five leaf-like **sepals (G).** One example of a dicot is the common rose.

Now that we have compared four characteristics of monocots and dicots, we will focus on the root of the plant. Continue your coloring as you read below. Darker colors should be used since there are few details to be obscured.

Our final point of comparison is the root. As the diagram shows, monocots have a **fibrous root (H)** with many side branches that reach out into the soil and absorb water and minerals.

By comparison, the dicot plant has a single, long **taproot (I).** This root extends deep into the ground, providing strong anchorage and support for the plant. Crop plants such as corn, wheat, and rye depend upon the anchorage provided by their deep taproots.

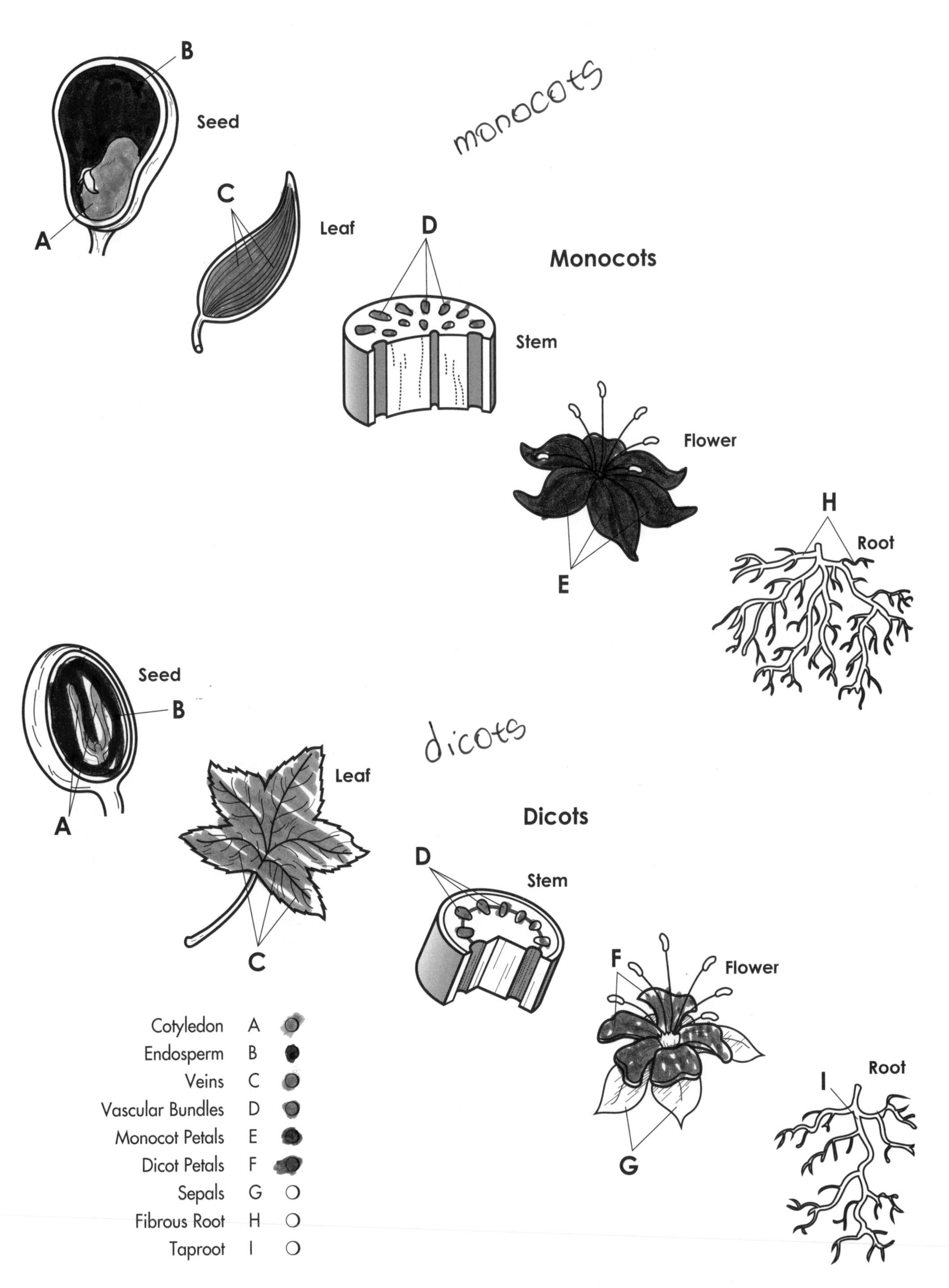
monocots
Monocots
B
Seed
A
C
Leaf
D
Stem
Flower
E
H
Root
dicots
Dicots
Seed
B
A
Leaf
C
D
Stem
F
Flower
G
I
Root
Cotyledon A
Endosperm B
Veins C
Vascular Bundles D
Monocot Petals E
Dicot Petals F
Sepals G
Fibrous Root H
Taproot I

LIFE CYCLE OF A FLOWERING PLANT

Scientists believe that terrestrial plants evolved from multicellular green algae, retaining the pattern of alternation of generations. Alternation of generations refers to the cycle of repeating sporophyte and gametophyte generations. As plants adapted to dryer terrestrial life, the gametophyte stage became shorter and the sporophyte stage became dominant. Another evolutionary trend was the development of a root-stem-leaf system with specialized tissues for the transport of water and nutrients.

A third trend in plant evolution was the development of the seed and the fruit that encases it. This is a distinguishing feature of the angiosperms, or flowering plants. Flowers and seeds play an important role in the life cycle of angiosperms, as we will see in this plate.

In this plate, we will examine the life cycle of a flowering plant. Beginning in the center and upper middle section of the art, we will follow the development of egg and sperm cells.

During its reproductive phase, the flowering plant produces reproductive structures within its flowers. The first floral organs to develop are leaf-like **sepals (A),** which envelop the flower bud. When the sepals fold back, the **petals (B)** are exposed. The petals are the most attractive parts of the flowers and are often colorful and display patterns that attract pollinating birds and insects.

In the upper right portion of the plate, we show details of the flower. The male reproductive organs are the **stamens (C),** enclosed by a bracket. Stamens consist of a four-lobed sac called an **anther (C_1),** seen in the close-up on the top left, and it is held on a slender **filament (C_2).**

The female reproductive organs in the plant consist of the **pistil (D),** which is enclosed by a bracket. At the top of the pistil is the sticky, hairy surface called the **stigma (D_1)** where pollen grains are trapped, and beneath the stigma is a slender stalk called the **style (D_2).** At the bottom is a flask-shaped, enlarged base, the **ovary (D_3),** which is opened in the enlarged view to show the **ovules (D_4)** inside.

We have examined the major reproductive parts of the flower of angiosperms, and we will now move down the left side of the plate and examine the development of the reproductive cell, a pollen grain. Continue your reading below as you color.

Take a look at the upper left-hand part of the plate. Meiosis takes place within the anther (C_1) and yields a number of **microspores (E).** Once these microspores have formed, they multiply by mitosis to become **pollen grains (F),** which are the male gametophytes. Pollen grains make contact with the stigma and are held to it by a sticky substance (shown in the bottom left). The pollen grain forms a **pollen tube (G)** that extends down through the style. The **sperm nuclei (L)** of the pollen grains enter the tube and move toward the ovule.

Now take a look at the right side of the plate where we show the development of the female gametophyte, the egg cell. We will also show the results of a union between sperm and egg in the flowering plant. Continue your reading below as you color the plate.

Events that take place in the ovary (D_3) lead to the development of the female gametophyte. An ovule differentiates, and a single large cell undergoes meiosis to produce four haploid **megaspores (H).** Three of these degenerate and one develops into the female gametophyte. The nucleus of the surviving megaspore undergoes mitosis and produces a number of nuclei within the **female gametophyte (I).** One nucleus becomes the **egg nucleus (J),** two others become **polar nuclei (K),** and the rest degenerate. The female gametophyte is ready for fertilization.

The pollen tube grows down toward the ovule, and two sperm nuclei enter the female gametophyte and participate in a double fertilization. One sperm nucleus combines with the egg nucleus (J) to form a diploid **zygote (M),** and the second sperm nucleus combines with two polar nuclei (K) to form an **endosperm cell (N).** The endosperm cell contains three sets of chromosomes; it is triploid (3N). Both the zygote and endosperm cell are within the female gametophyte.

At this point, regular development continues. The zygote becomes a multicellular **embryo (O),** which is the immature plant and a young sporophyte. The endosperm cell becomes **endosperm tissue (P),** which provides nutritional support for the embryo. Soon a layer of cells forms a **seed coat (Q)** that encloses the endosperm and embryo. The combination of endosperm tissue, embryo, and seed coat is the **seed (R).** The embryo continues to digest the endosperm to obtain nutrients as it develops. Eventually, the seed germinates and a new **seedling (S)** emerges.

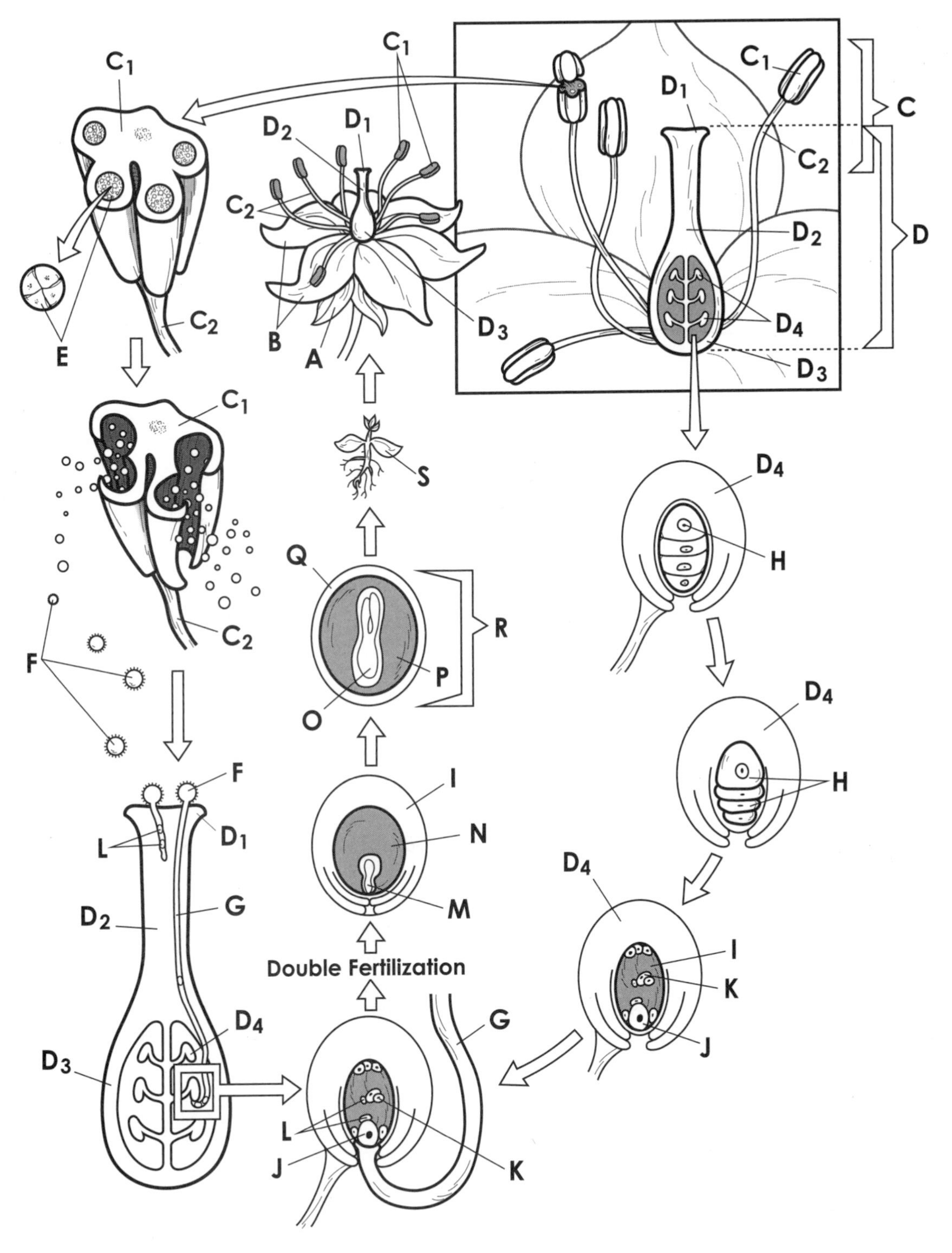

Sepals	A	❍
Petals	B	❍
Stamen	C	❍
Anther	C_1	❍
Filament	C_2	❍
Pistil	D	❍
Stigma	D_1	❍
Style	D_2	❍
Ovary	D_3	❍
Ovule	D_4	❍
Microspores	E	❍
Pollen Grain	F	❍
Pollen Tube	G	❍
Megaspores	H	❍
Female Gametophyte	I	❍
Egg Nucleus	J	❍
Polar Nuclei	K	❍
Sperm Nuclei	L	❍
Zygote	M	❍
Endosperm	N	❍
Embryo	O	❍
Endosperm Tissue	P	❍
Seed Coat	Q	❍
Seed	R	❍
Seedling	S	❍

LIFE CYCLE OF A MOSS

A wide variety of plants are classified in the kingdom Plantae. Members of this kingdom are eukaryotic, multicellular, photosynthetic organisms that have cell walls composed of cellulose. The life cycle of plants consists of two stages: a haploid stage in which there is a single set of chromosomes per cell and a diploid stage in which there are two sets of chromosomes per cell. This life cycle is referred to as alternation of generations, and we will see how it operates as we examine the life cycle of a simple plant, the moss.

As you look over the plate, notice that it contains diagrams of a life cycle that takes us through the various stages of the moss plant. Read below as you start to color. Use light colors since the structures tend to be small and detailed.

Plants exhibit two types of multicellular bodies in two generations that alternate in the life cycle. During the first generation, the sporophyte generation, the plant cells are diploid (there are two sets of chromosomes per cell), but then the plant produces haploid spores, which have one set of chromosomes per cell. These spores germinate to form the next generation, called the gametophyte generation, and cells in the gametophyte generation are also haploid. Male and female gametes (sex cells) are formed at the end of this generation, and when male and female gametes unite, the diploid sporophyte is formed again.

Among the simple members of the plant kingdom are the bryophytes. They are nonvascular plants that are believed to have been the first land plants. Most bryophytes live near water or in moist land areas and they include the liverworts, hornworts, and mosses.

We will begin our study of the life cycle of the moss with a collection of **spores (A).** Dots of color should be used to color these haploid cells, and we will demarcate the haploid condition with the letter N.

The first structure of the gametophyte generation is a filament of cells called the **protonema (B).** At the right side of the diagram, the protonema is seen emerging from the spore and growing until a leafy gametophyte begins to develop with a **gametophyte bud (C).** Eventually, the gametophyte buds form enlarged leafy gametophyte plants. Some of the plants are **male gametophytes (D),** while others are **female gametophytes (E),** but all of the cells of the gametophyte are haploid. A series of root-like structures called **rhizoids (F)** anchor the plant and absorb water and minerals from the soil.

We have seen that tiny spores yield haploid plants called gametophytes, and that rhizoids anchor the plants to the soil. We will now follow the reproductive pattern of the gametophyte plant. Continue your coloring below.

The gametophyte generation predominates in the life cycle of the moss; it is the leafy structure found near moist aquatic environments. Mosses lack true roots, stems, and leaves, and they require water for fertilization. For this reason, they are found on the dark, moist sides of trees and the sheltered forest floor.

The gametophyte generation of the moss plant produces oval, sperm-producing structures called **antheridia (G),** and hundreds of motile, flagellated **sperm cells (H)** are enclosed in these structures.

The female gametophyte produces flask-like, egg-producing structures called **archegonia (I),** and each archegonia contains an **egg cell (J).** Both the sperm and egg cells are haploid (N).

During reproduction, the sperm cells swim through water to the archegonia, as the arrow indicates. The sperm cells fertilize egg cells and a diploid zygote is formed. This marks the beginning of the sporophyte generation.

We have seen that sexual fertilization produces a diploid zygote within the archegonium of the moss plant. We will now follow the development of the sporophyte generation. Continue your coloring as you read the paragraphs below.

Within the archegonium, the zygote undergoes mitosis and begins producing a multicellular structure called the **sporophyte (K).** In the diagram, you can see the sporophyte growing out of the archegonium in the female gametophyte (E). As the sporophyte grows, it develops an elongated **stalk (M),** and at the end of the stalk, a swollen structure called a **capsule (L)** forms. Some cells within the capsule undergo meiosis to produce a great number of haploid spores (A), and when the capsule opens, the spores are released into the air where they are dispersed by the wind. The gametophyte generation is ready to begin again.

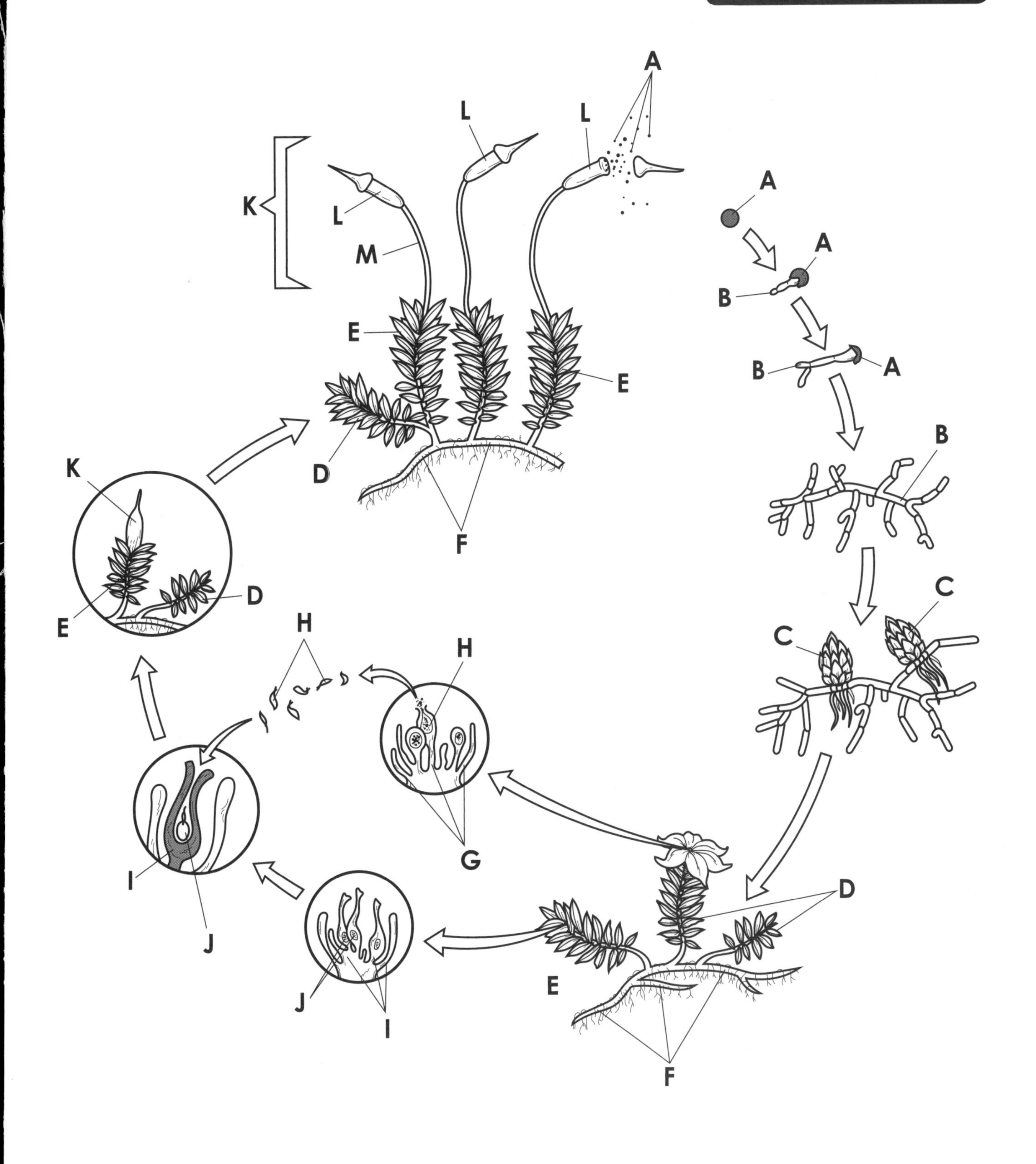

Spore A ❍
Protonema B ❍
Gametophyte Bud C ❍
Male Gametophyte D ❍
Female Gametophyte E ❍

Rhizoids F ❍
Antheridia G ❍
Sperm Cells H ❍
Archegonia I ❍
Egg Cell J ❍

Sporophyte K ❍
Capsule L ❍
Stalk M ❍

LIFE CYCLE OF A FERN

The photosynthetic, eukaryotic members of the kingdom Plantae use sunlight, water, carbon dioxide, and inorganic minerals to synthesize complex carbohydrates and release oxygen into the atmosphere. Their life cycles have evolved so that a haploid stage alternates with a diploid stage in a process called alternation of generations. It occurs in simple plants such as mosses, as you saw in the previous plate, as well as in vascular plants, as this plate will show.

In order for plants to thrive on land, they first had to find a way to prevent desiccation. Vascular tissue evolved in order to transport moisture and nutrients throughout the plant, and with it came a system of roots that could absorb water and minerals from the soil. Plants also developed leaves that could reach into the air for sunlight and gas exchange, and rigid stem systems that provided the plant with support. All of these adaptations are seen in vascular plants, including primitive ones such as the ferns.

Looking over the plate, you will see that we present the life cycle of a vascular plant, a fern. Ferns, like mosses, undergo alternation of generations. If you compare the plants, you will notice that, in ferns, the sporophyte generation is much more complex than the gametophyte generation. Concentrate on the paragraphs below as you begin your study of this plant.

Ferns are found primarily in the tropics, but they also grow in other moist, temperate regions near streambeds and roadsides. Some ferns are the size of a thumbnail, while others grow to be many feet in height.

The conspicuous generation of the fern is the **sporophyte (A),** which is the recognizable fern plant that's enclosed by the bracket. A dark color should be used for the bracket. The cells of the sporophyte are diploid (2N), and the plant has a system of **roots (B)** that provide anchorage and absorb water and minerals. The plant also contains a horizontal root-like structure called the **rhizome (C).** Though it is not a true stem, the rhizome contains vascular tissue that links the roots and leaves. The most obvious portion of the fern is the **leaves (D),** which are upright and contain vascular tissue. Large fern leaves are known as fronds.

Having introduced the anatomical features of the fern plant, we will now turn to its reproductive pattern. Continue reading below as you color the structures noted in the text.

Our focus is now on the underside of a fern leaf, or frond, where there are a number of small brown dots called **sori (E).** Sori are made up of a group of spore-producing structures called **sporangia (F).** (A single sporangium is displayed.) Within the sporangium, spore mother cells undergo meiosis and produce large numbers of haploid spores (G). These spores are carried away by the wind and fall to the ground. The **spore (G)** is shown germinating and a small structure called a **prothallus (H)** emerges. This marks the beginning of the gametophyte generation.

With the production of the prothallus, the sporophyte generation comes to an end and the gametophyte generation begins. In this generation, the cells are haploid (N).

The prothallus develops into a small, heart-shaped structure and eventually becomes the **gametophyte (H_1),** which is about the size of a fingernail and is made up only of haploid cells. The mature gametophyte gives rise to two reproductive structures; one is the **archegonium (I),** shown in an expanded view, and within the archegonium, a haploid **egg cell (J)** forms. The second structure is the **antheridium (K).** In the expanded view, the antheridium is shown producing a number of haploid **sperm cells (L),** which then swim to the egg cell and fertilize it, producing a **zygote (M).** The zygote undergoes mitotic division within the archegonium and begins to form the sporophyte generation.

As the sporophyte develops, all of its cells undergo mitosis to form a mass of diploid (2N) cells. Soon a young sporophyte grows on the underside of the gametophyte from the archegonium, and we see the beginning of leaves and roots. The gametophyte (D) begins to shrivel and die as the sporophyte continues its independent existence, and soon becomes an adult diploid sporophyte. This is the mature fern plant.

LIFE CYCLE OF A FERN

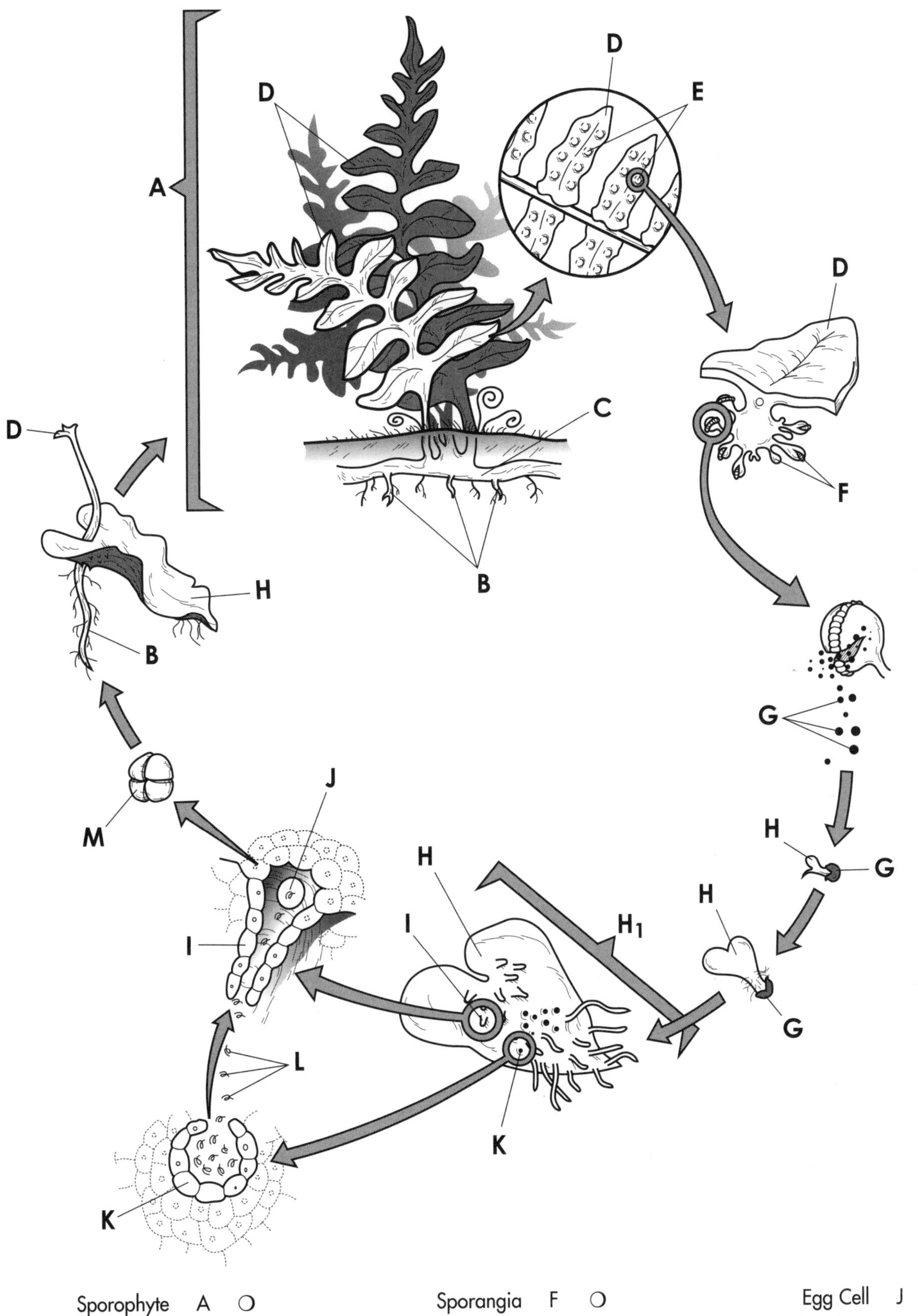

Sporophyte A ❍
Roots B ❍
Rhizome C ❍
Leaves D ❍
Sori E ❍

Sporangia F ❍
Spore G ❍
Prothallus H ❍
Gametophyte H_1 ❍
Archegonium I ❍

Egg Cell J ❍
Antheridium K ❍
Sperm Cells L ❍
Zygote M ❍

LIFE CYCLE OF A PINE

Among the more adapted and complex members of the kingdom Plantae are the seed-producing vascular plants. These plants have true roots, stems, and leaves, as well as an enclosed embryonic sporophyte known as a seed. Seeds are an advantage in terrestrial environments because they can withstand droughts and very cold temperatures in their dry, inactive state. When favorable conditions prevail, the embryo uses the stored food in the seed and becomes a seedling and eventually an adult.

Plants that produce their seeds on surfaces with no covering are called gymnosperms. These plants have cones rather than flowers; their seeds are not enclosed in fruits as they are in flowering plants. Among the three groups of gymnosperms are the conifers, which include the pines, firs, cedars, spruces, hemlocks, redwoods, and yews. These plants provide the great majority of the lumber used in the United States today.

In this plate we will study the life cycle of a pine as a representative gymnosperm and seed-bearing plant. The plate contains several views of the life cycle, which may be compared to that of the fern and moss in the previous plates. You should notice that the sporophyte generation is more complex than the gametophyte generation.

We begin the life cycle of the pine with a **sporophyte (A),** which is outlined by a bracket. The sporophyte is the familiar pine tree; it consists of true **roots (B),** the tree trunk itself which is the **stem (C),** and a number of needle-shaped **leaves (D).** These leaves are an adaptation to the harsh, windy environments in which pine trees often live.

Conifer trees such as the pine produce two kinds of cones. The male cone is the smaller **pollen cone (E),** seen here among a number of leaves (D). Also known as a staminate cone, it is often found dropped around the base of the pine tree and produces pollen grains that develop within a structure called the **sporangium (F)** at the lower surface of the cone scale.

Within the sporangia, a microspore mother cell undergoes meiosis to produce a number of haploid **microspores (G),** each of which develops into a **pollen grain (H),** outlined by the bracket. The pollen grain consists of a **male gametophyte (I)** bound by two **air sacs (J)** that add buoyancy when the grain is released into the air. The pollen grain is equivalent to the spore of the sporophyte generation, and is haploid (N). Pollen grains are released by the millions, and may land on female cones.

We have examined the production of male gametophytes in the form of pollen grains, and we will now see how female gametophytes are produced in a second type of cone. Continue your reading below as you color the appropriate structures. For purposes of simplicity, we are omitting some of the details in female gametophyte production.

The second type of cone associated with the pine tree is the **ovulate (seed) cone (K);** in our drawing, a single cone is shown among a number of pine leaves (D). Three years are required for the cone to reach seed-bearing maturity, and pollination takes place late in the first year of development. The ovulate cone contains an **ovule (L),** in which nutritive tissues and the spore mother cell are stored. The spore mother cell undergoes meiosis to produce a number of megaspores, each of which has a single set of chromosomes and is haploid (N). A megaspore develops into the **female gametophyte (M)** shown within the ovule, and after a complex series of transformations, each female gametophyte develops two or more **egg cells (N).**

Pollination is the process in which the male gametophyte comes in contact with the female structures of the pine tree, but it is not the same as fertilization, which is the joining of a sperm and egg cell. The pollen grain is trapped by the sticky fluid in the ovulate (seed) cone, and a long period of time passes. Pollen grains drawn into the seed cone grow **pollen tubes (O).** On reaching the egg cell, the pollen tube releases a sperm nucleus, fertilization takes place, and a zygote develops. The zygote has two sets of chromosomes and is diploid (2N).

We have now seen how fertilization occurs, and will begin our study of the sporophyte generation of the pine tree.

After another year or so, the zygote develops into an **embryo (P).** Embryonic development continues, and a hard **seed (Q)** eventually takes shape. Extending from the end of the seed is a **wing (R),** which aids in dispersal of the seed. When the seed falls onto a suitable site and germinates, it gives rise to a new sporophyte. This small seedling eventually becomes the pine tree, which is the mature sporophyte.

LIFE CYCLE OF A PINE

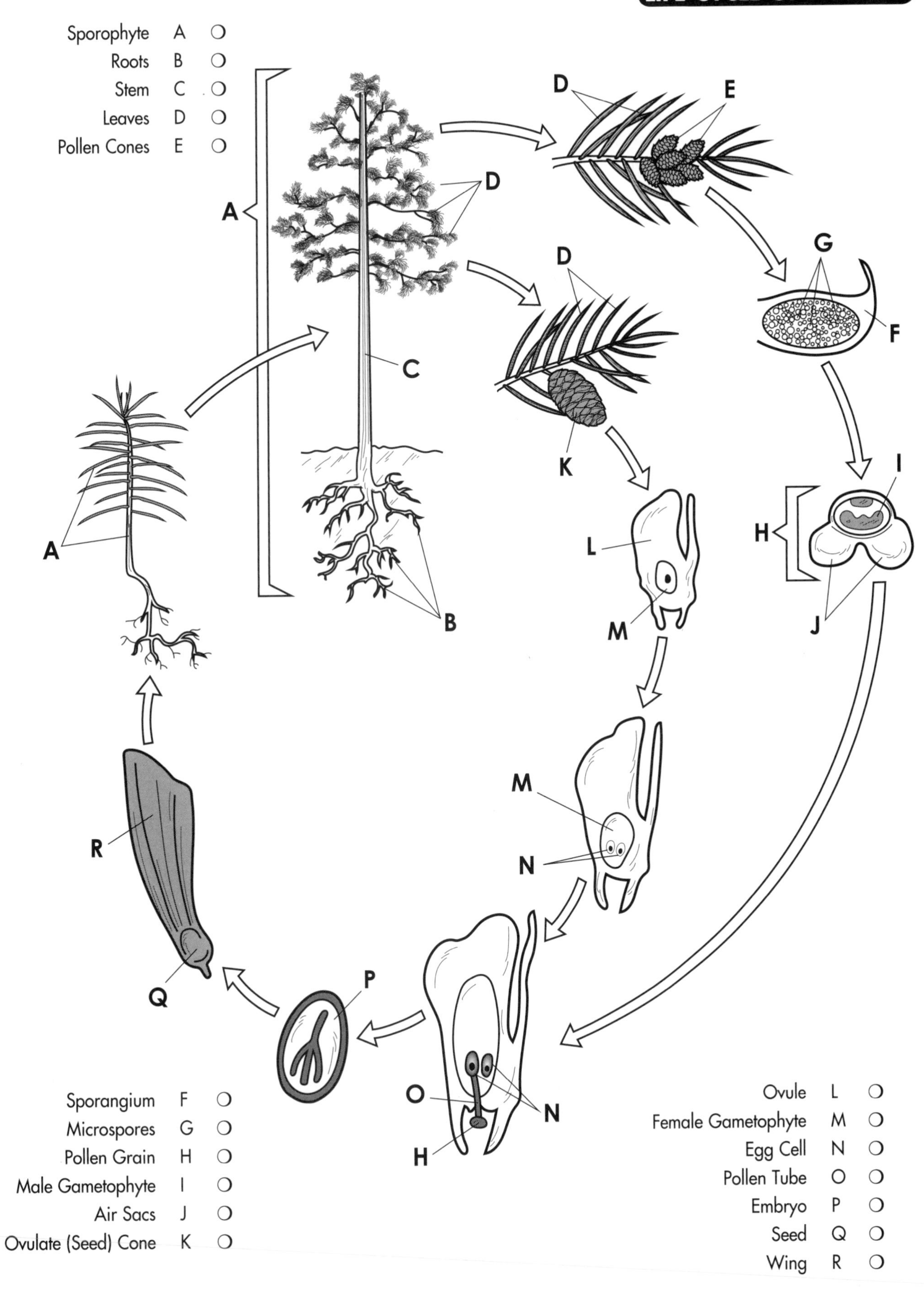

THE ROOT

The root is one of the four main organs of the plant. Its primary purpose is to provide a large surface area for the efficient absorption of water and dissolved minerals from the soil, but it also anchors plants in the soil. The root systems of plants can be taproots or fibrous, and plants can also have prop roots that provide additional support and aerial roots that absorb oxygen from the air.

Notice that this plate contains a longitudinal view and a cross section of a young root taken from a dicot plant. As you study the structures and functions of the tissues of the root, color the diagram accordingly.

The first organ to emerge from a germinated seed is the root, and the diagram at the left shows the four major regions of the root. The **root cap (A)** at the tip of the root provides protection as the root forces its way through soil particles. New cells are also produced at the root cap to replace those that are worn away.

One of the functions of the root cap is to protect the **apical meristem (D),** which is the second major region of the root. Apical meristem tissue is composed of undifferentiated cells that give rise to new cells through mitosis. In this way, the plant grows outward from the root tip. The third region is the **zone of elongation (B),** which is enclosed by a bracket. Cells from the apical meristem increase in size in this zone. Highest up is the **zone of differentiation (C),** in which cells that were once part of the zone of elongation start to become specialized cells of the epidermis, cortex, and vascular cylinder.

We have pointed out the four major regions of the primary root of a dicot plant, and will now examine the tissues found within these regions and the functions they perform. You should use light or medium colors here, since the structures tend to be small and close to one another.

Beginning at the surface of the plant, the first tissue we encounter is the **epidermis (E),** shown in both the longitudinal and cross sections. The epidermis protects the plant against invading microorganisms and environmental stresses. Some epidermal cells near the zone of differentiation become extensions called **root hairs (F).** Many root hairs are seen in the longitudinal section, and one is shown in the cross section. The root hair is an outgrowth of a single epidermal cell, and might contain its nucleus, as shown in the cross section. Root hairs increase the surface area that is in contact with the soil, which facilitates the absorption of water and minerals.

Inside the epidermis is a region called the **cortex (G).** The cortex is composed primarily of **cortex parenchyma cells (H),** which are thin-walled cells that take up mineral ions after they have passed through the epidermis. The cells may also be modified to store water or starch. Notice that many **air spaces (I)** exist between the parenchyma cells; they facilitate the free exchange of ions, fluids, and gases. The cortex is the largest area of the young root.

Having examined the epidermis and cortex of the primary root of the dicot plant, we are now ready to focus on the internal vascular cylinder, through which tubules transport materials from the soil to other parts of the plant. Read the paragraphs below and locate the appropriate structures in the diagram as you proceed.

The central core of the root is occupied by a complex group of tissues known collectively as the **vascular cylinder (L).** The outermost tissue of the vascular cylinder is the **endodermis (J),** which is a single layer of cells that completely encloses the rest of the vascular tissue. Parts of the cell walls of these endodermal cells are impregnated with a waxy material that regulates the flow of water and minerals from the cortex to the inner tissues of the vascular cylinder. This material forms a collar that is called the **Casparian strip (K).**

Inside the endodermis is the **pericycle (M),** which is made up of a layer of parenchyma cells that undergo cell division to produce lateral roots. Cells of the pericycle are also responsible for the secondary growth of roots, which causes an increase in root diameter.

Some of the tissue that extends from the roots through the stem of the plant to its leaves is called **xylem (N).** Xylem transports water and dissolved minerals from the soil to distant parts of the plant. Two types of cells, tracheids and vessel elements, make up the xylem, and we will discuss these in a future plate. The job of transporting sugar and other organic compounds in the plant falls to the **phloem (O).** The phloem transports nutrients produced during photosynthesis (in the leaves) to other parts of the plant.

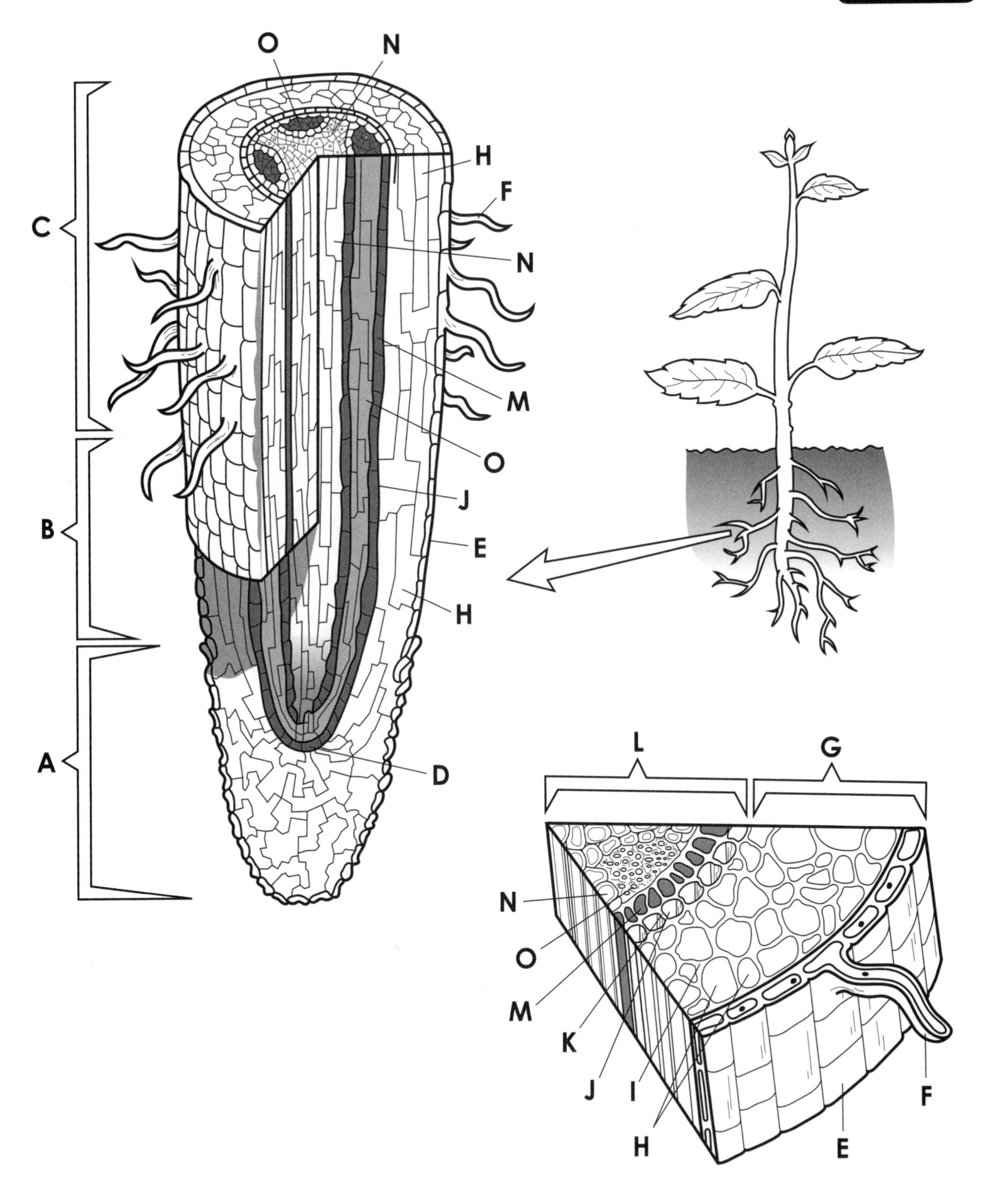

Root Cap	A	❍
Zone of Elongation	B	❍
Zone of Differentiation	C	❍
Apical Meristem	D	❍
Epidermis	E	❍
Root Hair	F	❍
Cortex	G	❍
Cortex Parenchyma Cells	H	❍
Air Spaces	I	❍
Endodermis	J	❍
Casparian Strip	K	❍
Vascular Cylinder	L	❍
Pericycle	M	❍
Xylem	N	❍
Phloem	O	❍

THE STEM

The stem is the main axis of the plant. Together with the lateral branches, the stem produces leaves and supports them so they are exposed to the sunlight.

The stem contains vascular tissue (xylem and phloem) that transports water and minerals to the leaves and transports the products of photosynthesis throughout the plant. In some plants, the stem also functions as a storage organ. For example, the tuber of a potato is a stem that stores starch. This plate examines the structure and growth pattern of a plant stem.

The stem shown in this plate is the woody twig of a forest chestnut plant. We will use this stem to discuss the structure and growth pattern that occurs within stems in general. Begin your work by focusing on the twig as you read the paragraphs below.

We will begin our discussion of the stem by focusing on its upper region, which is called the **apical meristem (A).** The apical meristem is the section that's enclosed by a circle, and it should be colored in a medium or dark color. Here, at the tip of the stem, cell division occurs, causing elongation. Also at the apical meristem is the **terminal bud (F),** which is the beginning of a new set of leaves. Other leaves arise from the **lateral buds (B)** along the margin of the stem.

The production of new leaves occurs at specific points called **nodes (C),** and you should use a bold color for the arrows that point to them. The areas between nodes are called **internodes (D).** Nodes and internodes can be seen easily at the tips of woody plants such as maples, oaks, and hickory trees. These trees all shed their leaves in the autumn.

After a leaf has fallen from a plant, a scar remains and is called a **leaf scar (E).** You should use spots of color to designate these areas. In the art, you can also see the **terminal bud scar (G)** left behind from the previous year's terminal bud (F). Each terminal bud scar represents a year's growth, so the age of the stem can be determined by counting the number of terminal bud scars. The terminal bud prevents lateral buds from sprouting in a phenomenon called apical dominance.

Because the stem contains living tissue, it must be supplied with carbon dioxide for photosynthesis and oxygen for respiration. These gases enter the plant through openings in the stem called **lenticels (H).** Spots of color should be used to increase their visibility.

We have examined the major structures of the woody stem, and will now examine its growth pattern and internal structures. The stem we are examining is a dicot, and the organization of the structure is characteristic of dicots. Continue reading below as you color the plate.

In a woody dicot plant, primary growth begins at the terminal bud; primary growth in a plant increases its length. The **present year's growth (I)** is indicated by the bracket, which should be colored in a darker color; **last year's growth (J)** is indicated by a second bracket, as is **two year's previous growth (K).** Notice that the terminal bud scars divide these periods of growth.

Take a look at the longitudinal section of the stem that shows its internal structure. At the center of the stem is the **pith (L)**; the pith is derived from parenchyma tissue and contains a lot of intercellular space. Another layer of tissue called **vascular cambium (M),** can be seen in this diagram; both **primary xylem (N)** and **primary phloem (O)** come from the division of cells of the vascular cambium.

Secondary growth of the plant increases the plant's width (girth). Once again, the vascular cambium is responsible for producing the vascular tissue that leads to this increase in width. In the longitudinal section, we see **secondary xylem (P)** and **secondary phloem (Q).**

To further explain the development of xylem and phloem, we demonstrate the pattern of secondary growth at the bottom of this plate. A **cambium cell (R)** is the original cell, and it undergoes division by mitosis to yield a **xylem cell (S).** When another division of the cambium cell takes place, the result is a **phloem cell (T),** which is deposited toward the outside of the plant.

In the fourth row, division of the cambium cell (R) has taken place again, and has resulted in a xylem cell (S). The previous xylem cell is forced to the inside. Division occurs once more, and a new phloem cell (T) pushes the previous phloem cell toward the outside. Eventually, the phloem tissue will disintegrate with each new layer deposited and the xylem cells will form annual rings, except for the newest layer of xylem cells, which functions in transport. The annual rings (wood) are the remains of the previous year's active xylem tissue.

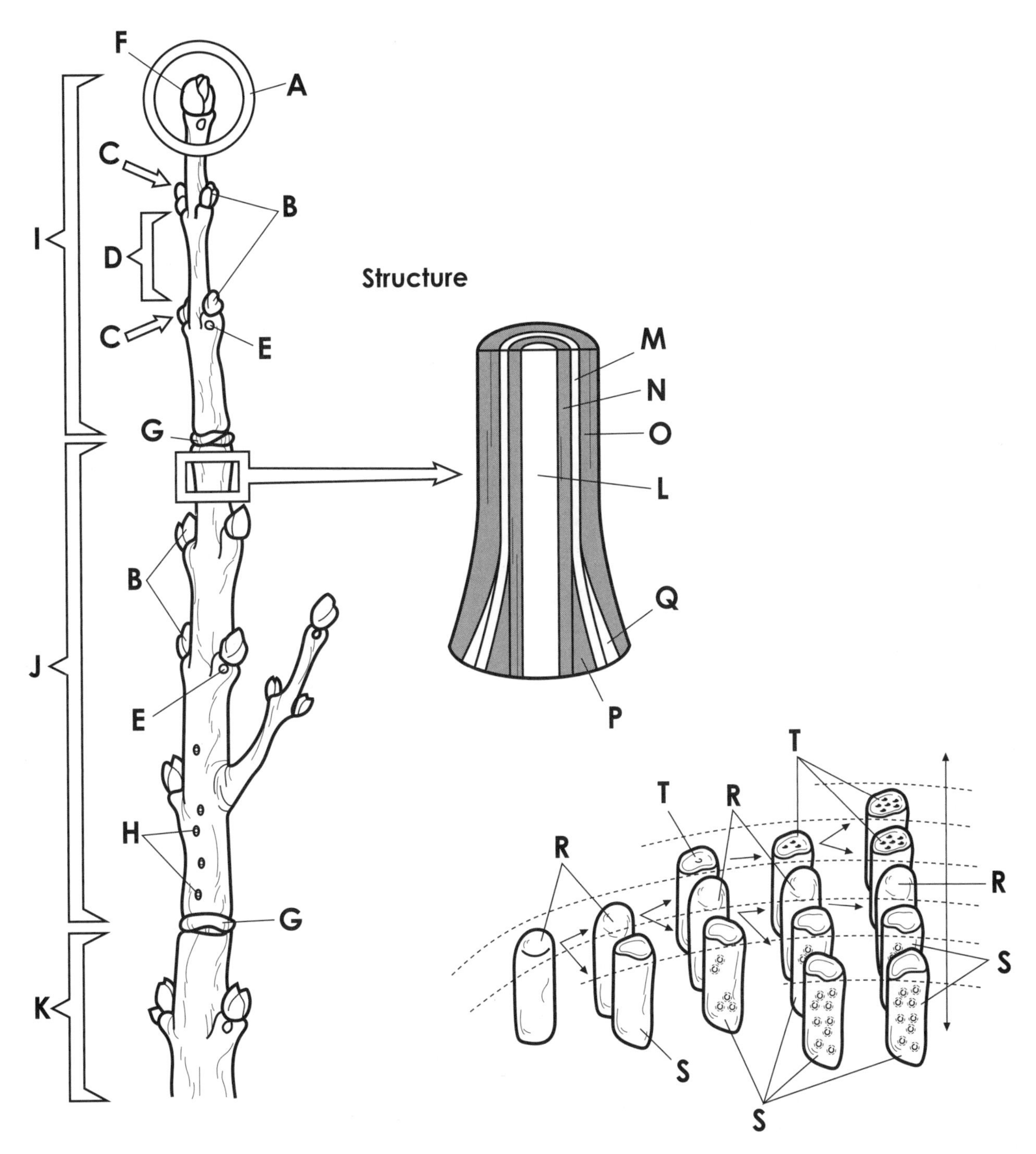

Apical Meristem	A	❍	Lenticels	H	❍	Primary Phloem	O	❍
Lateral Buds	B	❍	Present Year's Growth	I	❍	Secondary Xylem	P	❍
Node	C	❍	Last Year's Growth	J	❍	Secondary Phloem	Q	❍
Internode	D	❍	Two Year's Previous Growth	K	❍	Cambium Cell	R	❍
Leaf Scar	E	❍	Pith	L	❍	Xylem Cell	S	❍
Terminal Bud	F	❍	Vascular Cambium	M	❍	Phloem Cell	T	❍
Terminal Bud Scar	G	❍	Primary Xylem	N	❍			

THE LEAF

The ability of plants to photosynthesize depends on the exposure of much of their surface to sunlight. Leaves provide this necessary exposure. In gymnosperms, leaves tend to be round and needlelike, while in flowering plants (angiosperms), leaves have flat, thin blades that maximize light absorption and control gas exchange.

The leaf is the major photosynthetic organ of plants. In this plate, we will examine some of the leaf's structural characteristics.

This plate contains a cross section of a leaf, and we will examine its internal photosynthetic structures. Bold colors should be used throughout the diagram since the structures can be easily differentiated.

The typical leaf of a flowering plant consists of a thin, flattened area called the **leaf blade (A),** and the blade is joined to the stem of the plant by a stalk-like extension called the **petiole (B).** Vascular tissue extends from the stem through the petiole and into the blade to form a branched system of veins in dicots, and parallel veins in monocots. Look at the details of the leaf's interior as seen in the cross section.

The typical leaf blade is covered on top and bottom by a protective **cuticle (C).** The cuticle is a layer of waxes that helps prevent moisture loss from the internal portion of the leaf. The outermost layer of cells of the leaf is the epidermis; the epidermis is generally only one cell in thickness, and in the diagram you can see an **upper epidermis (D)** and a **lower epidermis (E).** The epidermal layers protect the inner tissues of the leaves and secrete the waxy cuticle. Epidermal cells interact with the environment and are permanent tissues in the plant; they vary in structure among plants and are the source of root hairs on growing roots.

At the lower surface of the leaf, we show several openings known as **stomates (F),** which are also called stomata. On either side of each stomate are two cells, the **guard cells (G).** These cells are formed from epidermal cells, and you should use bold colors for them. Stomates allow for the diffusion of carbon dioxide, oxygen, and water vapor into and out of the leaf, and these gases must be exchanged continually in order for photosynthesis to occur. The opening and closing of the stomates is controlled by a complex series of chemical reactions that involve ions and water. Stomates are generally open during the day and closed at night, and an enlarged stomate is shown at the lower right. Notice that stomates are present only on the lower surface of the leaf.

We began our examination of the leaf by focusing on the structures found on its upper and lower surfaces, and we will now examine the cells found in its interior. Continue your coloring as you read the paragraphs below.

Sandwiched between the upper and lower epidermis are the cells of the **mesophyll (H)** layer, which include the photosynthetic cells of the plant. This internal region is made up of two types of chloroplast-containing parenchyma cells arranged in layers.

The first layer of mesophyll contains the **cells of the palisade layer (I),** which contain numerous **chloroplasts (K).** Spots of green should be used for the chloroplasts, and a light green is recommended for the cells. These columnar cells are situated just below the upper epidermis. Most plant photosynthesis takes place within the palisade cells, whose shape and arrangement maximize the number of chloroplasts that are exposed to sunlight.

The next layer of mesophyll cells is the **cells of the spongy layer (J).** Also filled with chloroplasts (K), these parenchyma cells are irregular in shape and are suspended in a system of interconnected **air spaces (L)** that permit all of the cells to come in contact with air. The loose arrangement of these spongy cells permits the rapid exchange of carbon dioxide and oxygen during photosynthesis and respiration. Air passes into and out of the stomates and among the spaces of the spongy layer within the mesophyll.

The final tissues that we will examine are xylem and phloem, which comprise the **vascular bundle (M). Xylem (N)** is the vascular tissue that conducts water through the plant and transports dissolved minerals from the soil, while **phloem (O)** transports the sugars that are produced during photosynthesis (in the mesophyll), taking them to the nonphotosynthetic areas of the plant, such as the stem and root.

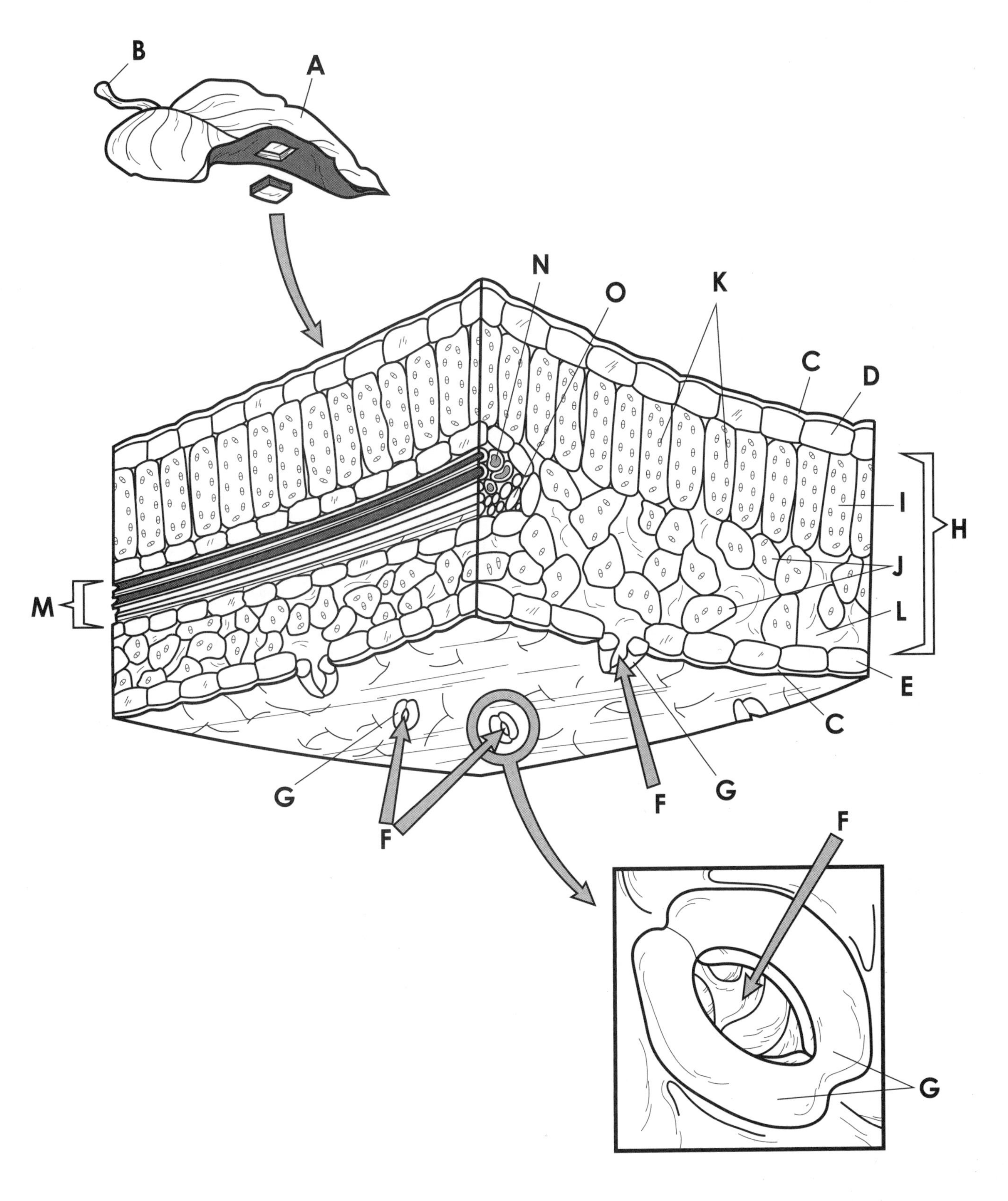

Leaf Blade A ❍
Petiole B ❍
Cuticle C ❍
Upper Epidermis D ❍
Lower Epidermis E ❍

Stomate F ❍
Guard Cells G ❍
Mesophyll H ❍
Cells of Palisade Layer I ❍
Cells of Spongy Layer J ❍

Chloroplasts K ❍
Air Space L ❍
Vascular Bundle M ❍
Xylem N ❍
Phloem O ❍

XYLEM AND PHLOEM

The major plant organs are the root, stem, and leaf, and these organs are composed of specialized tissues. Some of these tissues are epidermal tissue, which provides protective covering; ground tissue, which fills the plants interior; and vascular tissue, which is responsible for the transport of water and nutrients throughout the plant.

There are two kinds of vascular tissue in plants. The first type, xylem, transports water throughout the plant, and the second type, phloem, transports sugars and the other products of photosynthesis to nonphotosynthesizing regions of the plant. In monocot plant stems, the xylem and phloem are arranged in scattered vascular bundles, while in dicots, xylem and phloem are gathered into vascular bundles that are arranged in rings. In this plate, we will examine the structural compositions of xylem and phloem.

In the art to the right, you can see diagrams of xylem and phloem tissues. Begin your work by reading the text below as you concentrate on the diagram of xylem tissue.

Like animals, plants require material to be transported throughout their structures, but while animals have a circulatory system and a heart, plants have a vascular system that consists of xylem and phloem. As you can see in the upper portion of the plate, xylem contains two types of conducting cells. The first type is the **tracheid (A),** which is shown in the xylem tube, and the second type is the **vessel element (B);** a long tube of connected vessels elements is shown in the diagram on the left. The middle diagram shows three vessel elements before xylem formation. You should use light colors for both the tracheids and vessel elements.

Both tracheids and vessel elements are hollow, nonliving structures, and the tracheids are the smaller of the two cell types. As the diagram shows, tracheids are elongated and have **tapered ends (C);** water moves across the ends of the cells by passing through **pits (D)** in their end walls.

Vessel elements are larger than tracheids. Fully mature vessel elements have no end walls and form a continuous pipeline for the transport of water and minerals. Preliminary vessel elements are shown in the center diagram, and the complete vessel element is shown in the art to the left. Notice that the walls of the vessel elements contain pits; these pits allow water and minerals to move across the side walls of the xylem and supply the surrounding cells.

Outside the xylem is a series of **parenchyma cells (E),** which are the major ground tissue of the plant. They form the bulk of the plant and are considered typical plant cells, because they are found in all organs of the plant and are not specialized for any specific function.

We have examined the structure of the xylem tissue, and will now look at phloem. Notice several of the structural differences in these cells that are related to their function. Continue your reading as you color the bottom half of the plate.

As you are aware, among the functions performed by plants is photosynthesis, which results in the production of glucose and other nutrients that must be transported throughout the plant. Phloem is the vascular tissue that participates in this transport. As the lower part of the plate shows, phloem consists of stacked **sieve tube cells (F).** The sieve tube cells stand next to a series of **companion cells (G),** which have nuclei as well as the general structures found in most plant cells. The companion cells provide the sieve tube cells with nutrients and support them. They can be seen in the diagram on the right as well.

Returning to the sieve tube cells, notice that they contain cytoplasm, but do not have nuclei; they are living cells that rely on companion cells for support. At the end of the sieve tube cells are **sieve plates (H),** which have channels through which nutrients flow. The sieve tube cells provide the area through which nutrients move up and down in the plant. The cell membrane remains functional in the sieve tube cell and allows the movement of solutions through the phloem. The sieve tube cells are usually connected to one another by strands of cytoplasm called **plasmodesmata (I).** These pores and channels in the end walls create a system of tubes that extends throughout the plant.

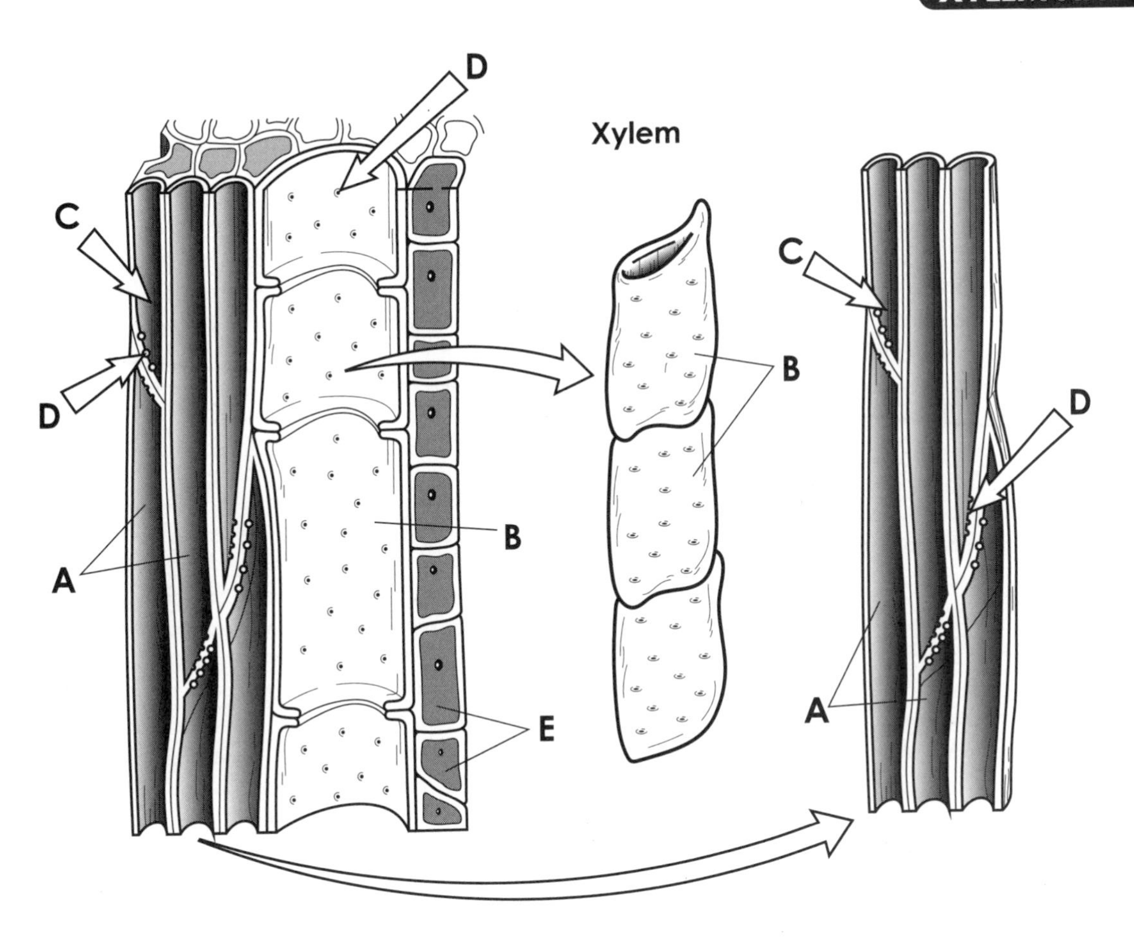

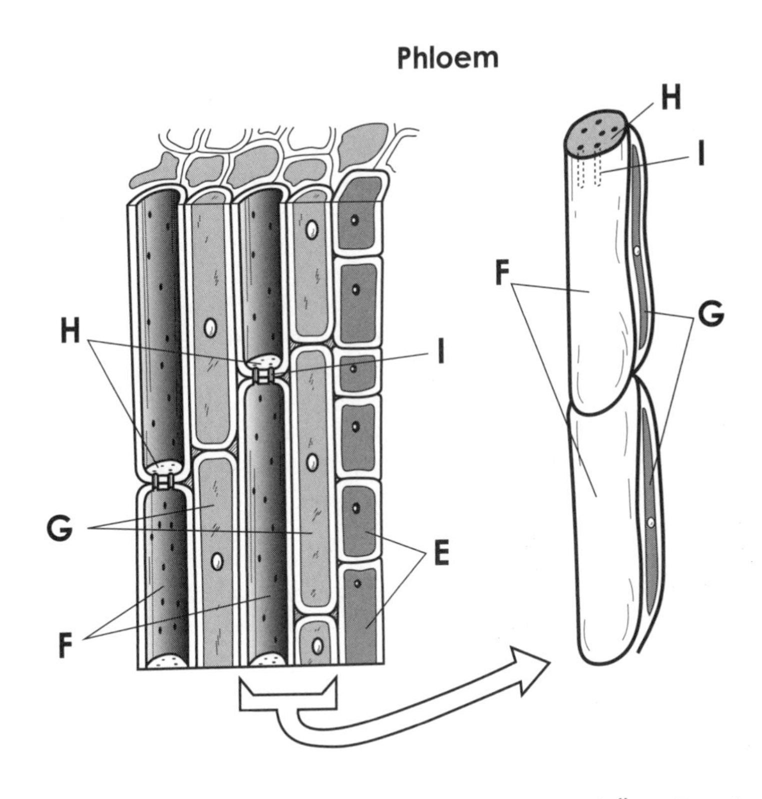

Tracheids	A	❍
Vessel Elements	B	❍
Tapered Ends	C	❍
Pits	D	❍
Parenchyma Cells	E	❍
Sieve Tube Cells	F	❍
Companion Cells	G	❍
Sieve Plate	H	❍
Plasmodesmata	I	❍

TRANSPORT IN PLANTS

The advantages gained by plants because of their adaptation to life on land are numerous—on land there is more light for photosynthesis, and more oxygen is available for the plant's cellular processes. But in order to live on land, plants had to develop a system for transporting water, minerals, and nutrients throughout their tissues. Minerals and water that are obtained from the soil must be transported upward, and the nutrients that are synthesized through photosynthesis in the leaves must be transported throughout the plant.

As you learned in earlier plates, xylem and phloem are the two components of the plant's vascular system. In this plate, we will discuss how water and nutrients reach the vascular tissues and how they are transported within them.

In this plate, the transport of two substances is discussed: water and nutrients. These are depicted in the two main diagrams in the plate. Begin your work by focusing on the diagram entitled Water Transport as you begin reading below.

In the first diagram, we show the mechanism by which water reaches the xylem tissue in the roots. Two pathways exist. One of them involves the passage of water between porous cells, and the second involves the passage of water from cell to cell. We will begin with extensions of epidermal cells known as **root hairs (A).**

The **first pathway (B)** is the extracellular route. The water passes through the **epidermis (D)** and courses between the cell walls (intracellular spaces) of the **epidermal tissue (E)** to then enter the **cortex (F).** Here the water passes around the **parenchyma cells (G),** which have thin walls and irregular shapes, and are loosely packed.

Next, water comes upon the **endodermis (H),** which is a single layer of rectangular **endodermal cells (I)** bordered on two sides by a layer of waxy material called the **Casparian strip (J).** The two cell sides contacting the cortex (F) and **vascular tissue (K)** do not have the Casparian strip, but at this point, water is forced to pass through these adjacent cells of the endodermis. The only way that water can enter the vascular tissue is through endodermal cells.

The water then enters the vascular tissue (K) and passes toward the **xylem (L),** through which it is transported up the step of the plant through a cohesion-tension process. The water molecules cling together (cohesion) because of hydrogen bonding and a water column extends up the step, while the water molecules adhere to the walls of the xylem vessels. At the top of the plant, water evaporates from leaf cells through transpiration, and this loss creates tension that pulls water up through the xylem tissue.

We now focus on the **second pathway (C),** which is an intracellular route that takes water across plasma membranes and through epidermal cells, cortex cells, and endodermal cells. The cells along this route are all interconnected by channels known as plasmodesmata, and water and minerals pass through these channels to reach the xylem. The water is then pulled upward by the cohesion-tension process.

We now turn our attention to the flow of sugars and nutrients throughout the plant. We will see a process that's very different from the one-way flow of water we saw in xylem tissue. Continue your reading below as you focus your attention on the second portion of the plate, entitled Sugar/Nutrient Transport.

Since photosynthesis takes place primarily in plant leaves, organic materials such as sugars, glucose, and other carbohydrates must be transported from them to other parts of the plant. This transport process is shown in the second diagram. We begin with **water (M),** represented by the arrow on the right side of the diagram. Water flows up through the xylem (L) through the process we discussed previously.

Within the plant leaves, water enters and then exits cells known as **source cells (N).** Its borders should be colored green, and its interior should be left pale to show that water and nutrients accumulate there. Sugars and nutrients that are produced in the source cells are moved by active transport into the **phloem (O).**

As the sugar molecules accumulate in the phloem, osmosis forces water into the phloem tubes to dilute the concentrated sugar and nutrient molecules. This additional water entering the phloem forces water through successive **sieve plates (P)** and throughout the plant.

The **sugar/nutrient solution (Q)** will eventually reach a part of the plant in which photosynthesis is not taking place, such as the root. Here the cells are called **sink cells (R)** and in them, nutrients are removed and converted to storage carbohydrates. Water then leaves the sink cells and moves back into the xylem tube, completing the circuit.

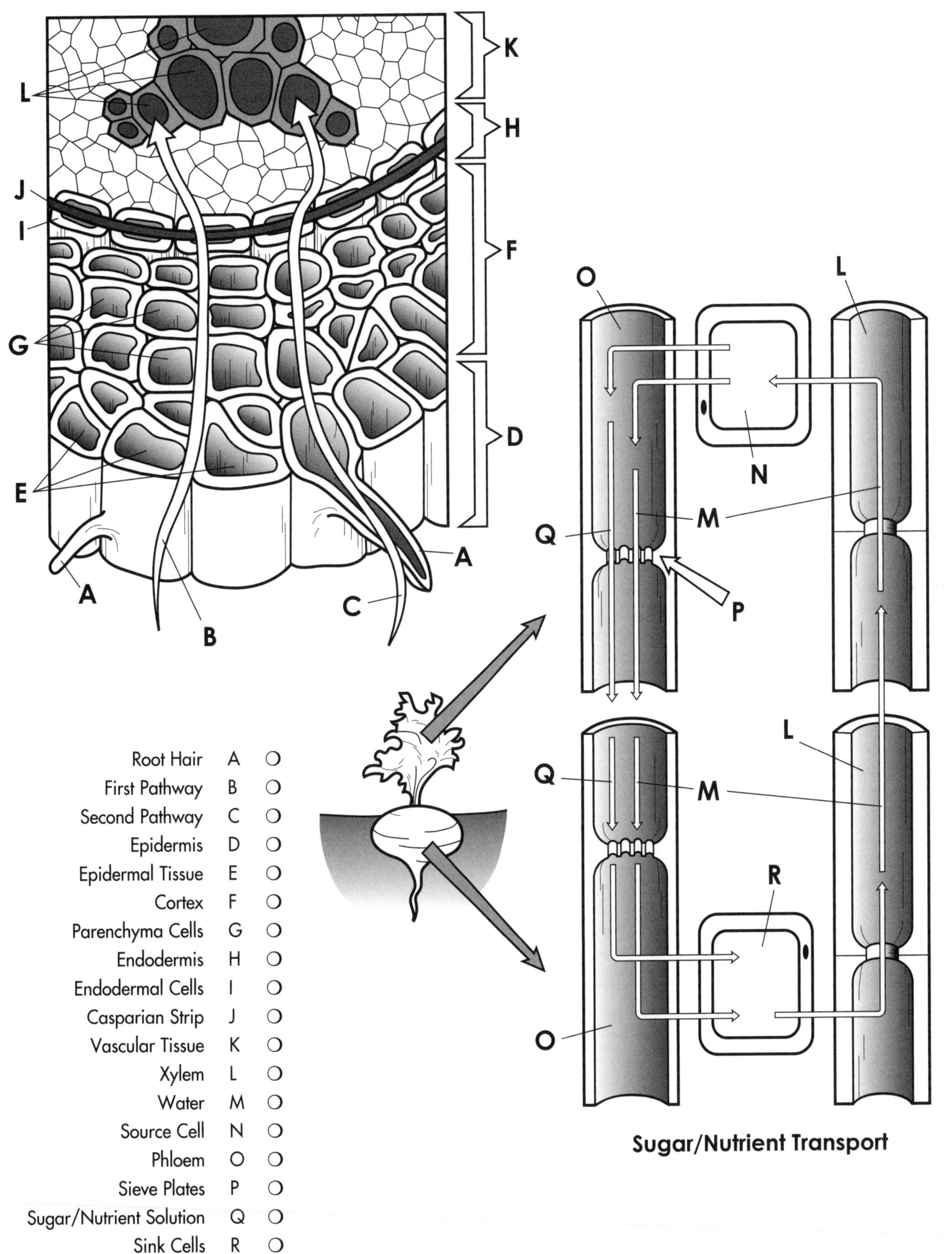

Root Hair A ❍
First Pathway B ❍
Second Pathway C ❍
Epidermis D ❍
Epidermal Tissue E ❍
Cortex F ❍
Parenchyma Cells G ❍
Endodermis H ❍
Endodermal Cells I ❍
Casparian Strip J ❍
Vascular Tissue K ❍
Xylem L ❍
Water M ❍
Source Cell N ❍
Phloem O ❍
Sieve Plates P ❍
Sugar/Nutrient Solution Q ❍
Sink Cells R ❍

PLANT HORMONES

Hormones are chemical substances that regulate the activities of living things. In humans, hormones are produced by glands such as the thyroid, pituitary, and adrenal. Plants also produce hormones, but they do not possess specific hormone-producing organs.

In plant tissues, hormones regulate growth and development. The cells affected by hormones are known as target cells. Hormones bring about physiological changes by inhibiting or stimulating the metabolic activities of target cells. This plate will discuss five plant hormones.

This plate contains an illustration of a typical plant, which includes roots, stems, leaves, and fruit. You should color the plant with light and medium colors. The main focus of this plate will be on the arrows that designate where the various hormones act. These arrows should be colored in bold reds, greens, and blues to indicate their importance.

The effect of plant hormones on their target cells is slow in contrast to the speed of many animal hormones. Most hormones are manufactured in meristematic tissues and are transported by the phloem.

The first group of hormones we will discuss are the **auxins (A);** you should color the arrow for auxin a bold color. Auxins determine to what degree a plant will bend toward the light (this phenomenon is called phototropism), and they also stimulate cell elongation. Several auxins are known to exist, including indoleacetic acid and phenylacetic acid, both of which are relatively simple compounds.

Auxins also act to delay abscission, which is the process through which leaves fall from a plant. When auxins are in short supply, leaves drop more easily because the petiole weakens where the leaf joins the stem. A decrease in auxins also causes fruit to drop from the plant.

The second hormones we will discuss are the **gibberellins (B),** hormones that promote the elongation of the stem as well as the process of cell division. There are 65 known gibberellins.

We have now examined two hormones produced by plants and have described their function in plant development. In the next section, we will turn to a third group of hormones. Continue reading and using bold colors for the arrows.

The third group of plant hormones presented here is the **cytokinins (C),** which also induce cell division. In the plate we show cytokinins in the fruit of the plant, because they also stimulate fruit development and play a part in maintaining a balance between the development of the stem and root.

Cytokinins work with auxins to regulate plant growth and development. An overabundance of either hormone produces a plant that's out of balance because unregulated cell differentiation and development occur. One example of a cytokinin is kinetin.

We will now look at the fourth hormone, **abscisic acid (D),** which is responsible for the closing of stomata on the undersides of leaves. Water passes through the stomata during transpiration, and abscisic acid works to control the opening and closing of the guard cells surrounding the stomata.

Abscisic acid also stimulates the formation of winter buds by converting primitive leaves into bud scales, which puts the plant into its dormant state. The effects of abscisic acid are counterbalanced by the effects of gibberellins.

We have now surveyed four plant hormones. Notice that hormones work broadly to influence physiological activities such as cell division, growth, and elongation. We will conclude this plate by focusing on the fifth hormone.

The last hormone we will discuss is **ethylene (E),** which functions during the ripening of fruit, as the arrow indicates. It is a simple chemical compound that regulates the growth of fruit cells to stimulate ripening. The hormone causes spoiling fruit to liberate ethylene in abundance, causing fruit situated close to it to ripen.

Ethylene also influences sex determination in some plants. In certain species, ethylene increases the number of female flowers, to make fertilization more likely. In this way, ethylene regulates and stimulates many developmental and metabolic activities in plants.

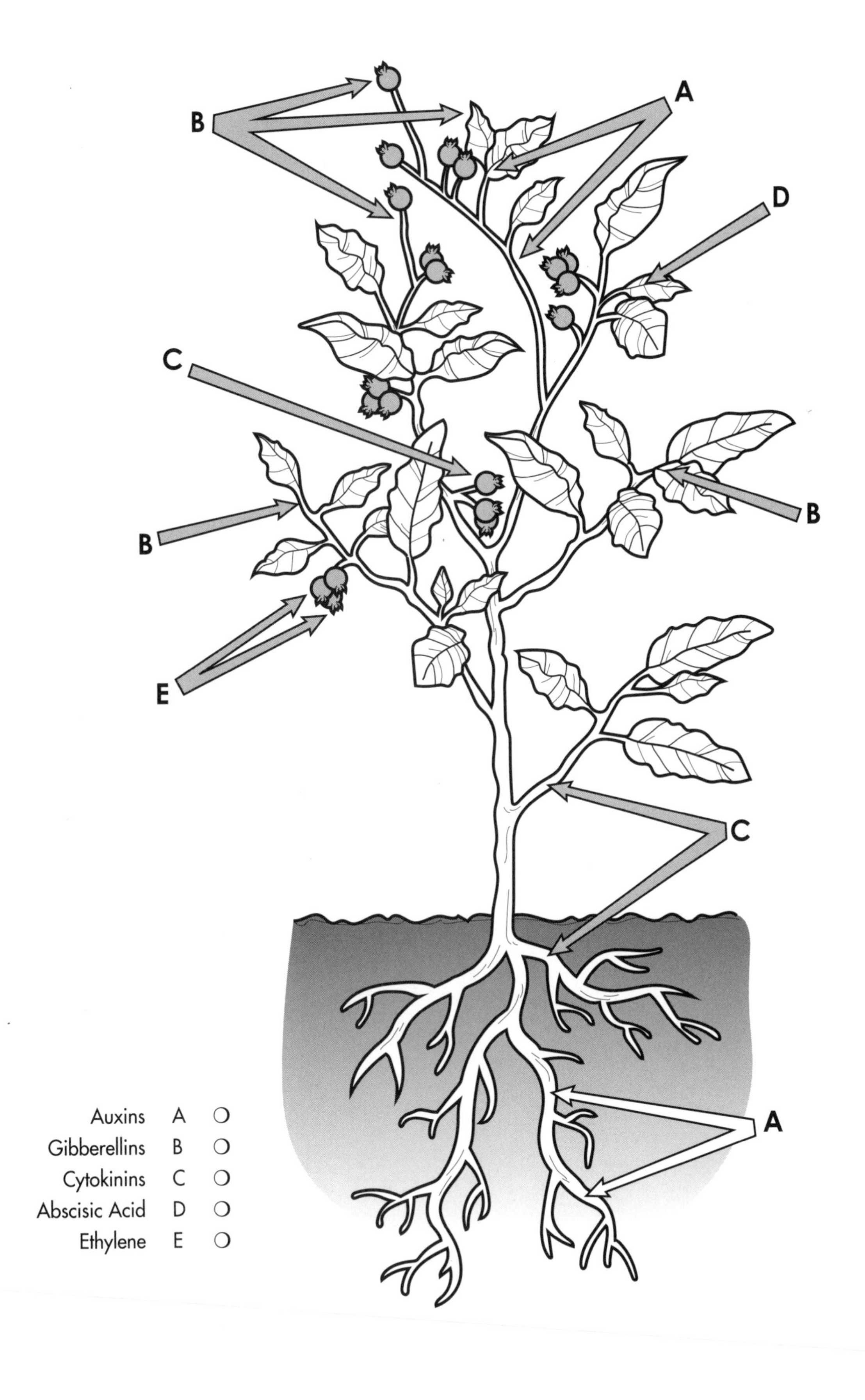
A
B
D
C
B
B
E
C
A
Auxins A
Gibberellins B
Cytokinins C
Abscisic Acid D
Ethylene E

CHAPTER 8:

BIOLOGY of the ANIMALS

PHYLUM PORIFERA

Many animals have several levels of anatomical organization: the cellular level, tissue level, organ level, and organ system level. One exception to this rule is the sponges, found in the phylum Proifera. Essentially, sponges consist of a few types of cells loosely arranged into a tissue. Sponges are the simplest animals.

In this plate, we show a colony of sponges as well as a detailed view of a single sponge and a more detailed view of its parts. As you color the plate, locate the parts in the four diagrams and relate them to one another. You should use light colors for the various cells of the sponge, but dark colors for the arrows and the whole view of the sponge.

Sponges are mainly ocean dwellers, but several species live in fresh water. They fasten to the ocean bottom where they filter tiny food particles from the water. The simplest sponges are vase sponges, which have hollow, vase-like bodies that are a few cell layers thick. A small colony of these **vase-like sponges (A)** is shown in the first diagram.

Sponges feed by filtering bacteria and other microscopic organisms from the surrounding water. They are thus called filter-feeders (or suspension feeders) because they feed by straining suspended matter and food particles from water, using a filter. This plays an important role in water clarification. The **flow of water (B)** through the sponge is shown in the major diagram as a set of arrows that should be colored in a dark color. Water is brought into the sponge through **incurrent pores (C)** and enters a main chamber called the **spongocoel (D).** It flows through the spongocoel and exits through excurrent pores, which are also called **osculum (D_1).**

After water enters the spongocoel, it flows by a series of cells called collar cells, or **choanocytes (E),** which are shown in the main diagram as well as diagrams 3 and 4. As diagram 4 shows, the color cells have a **membranous collar (E_1)** that traps food particles and ingests them by phagocytosis. Extending from the collar is a **flagellum (E_2),** which is surrounded by a ring of stiff cilia. The flagellum generates a water current that forces water by the collar cells. Once food has been extracted from the water and taken into the collar cells, it is transferred to wandering **amoebocytes (G).** These cells creep among the tissues and digest and distribute food throughout the body. They can be seen in diagrams 2 and 3.

We will continue our study of the sponges by pointing out some of their other cells and structures. Notice that sponges have no muscles or nerves and that their cells show only a limited degree of organization. Continue your reading below and locate and color the structures in the diagram.

Sponges are able to maintain their shape because they possess fibers called **spicules (F),** which can be seen in diagrams 2 and 3 and should be colored with a dark color. Spicules are composed of calcium carbonate, silica, or a flexible protein called sponging, depending upon the class of sponge. For instance, the natural bath sponge, which is now rare, is composed of protein-based spicules.

Sponges reproduce by the asexual process of budding as well as a sexual process. During sexual reproduction, sperm cells released by other sponges enter the spongocoel through the incurrent pores, where they fertilize **egg cells (H).** The fertilized egg cells grow into larvae that eventually attach to the ocean bottom and grow into adult sponges. Most sponges are hermaphroditic, meaning they possess both male and female reproductive parts.

The final cell of the sponge that we will look at is the **epidermal cell (I),** which is shown in diagram 3. The epidermal cells lie at the outer surface of the sponge and provide it with support. In the diagram, they can be seen at the base of the collar cells and amoebocytes. The spicules (F) are seen passing through the epidermal cells in diagram 3. Because these epidermal cells are organized into a covering layer, the sponge is said to show a tissue level of organization.

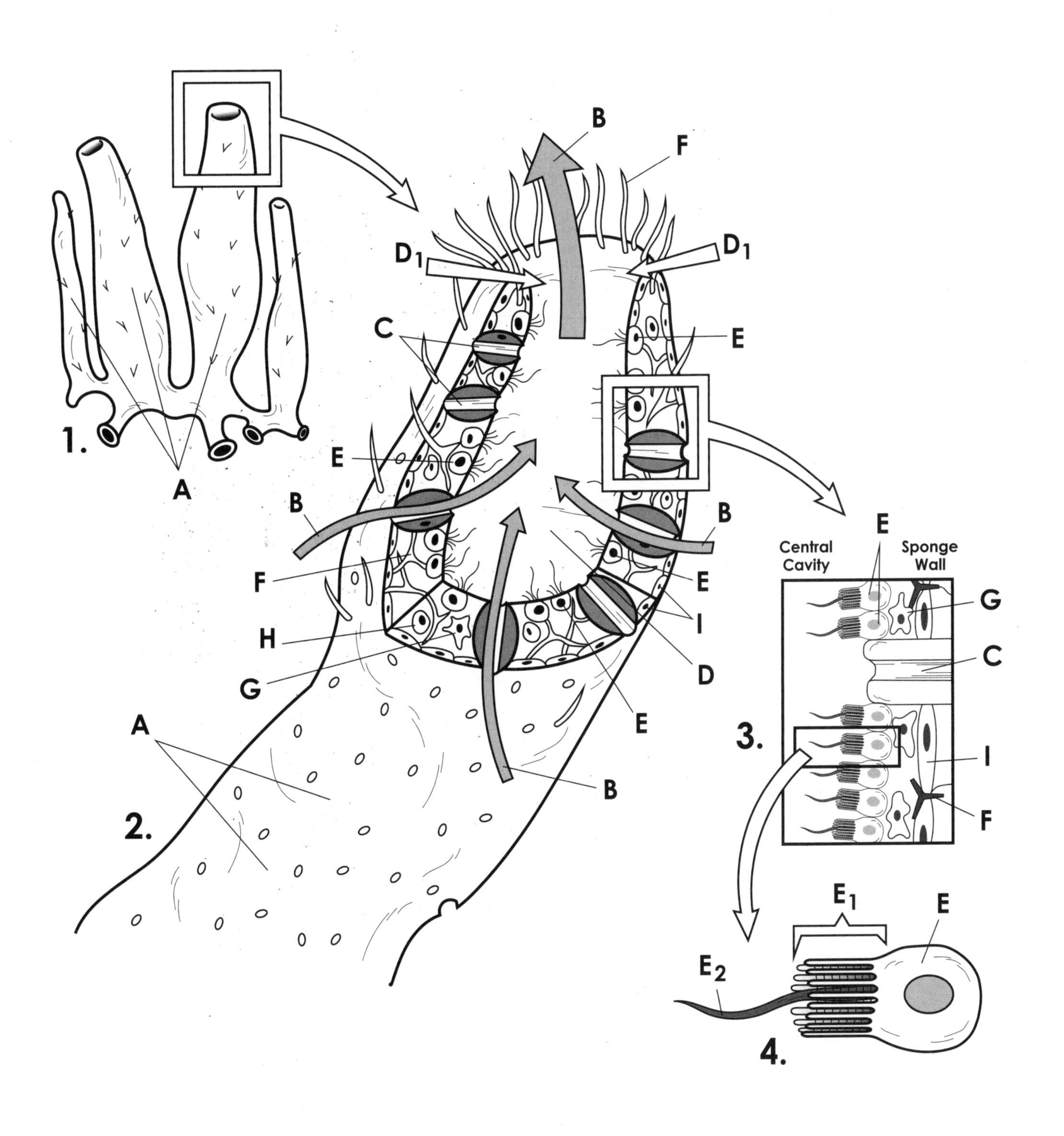

Sponge Colony	A	❍
Water Flow	B	❍
Incurrent Pore	C	❍
Spongocoel	D	❍
Excurrent Pore (Osculum)	D_1	❍
Collar Cell (Choanocyte)	E	❍
Membranous Collar	E_1	❍
Flagellum	E_2	❍
Spicule	F	❍
Amoebocyte	G	❍
Egg Cell	H	❍
Epidermal	I	❍

PHYLUM CNIDARIA

Members of the phylum Cnidaria exist predominantly in marine habitats, but some species are found in fresh water. Most are sessile or slow-moving animals but are still efficient predators, because they possess special stinging cells called cnidocytes. Each cnidocyte contains a coiled thread called a nematocyst, which is discharged when stimulated and is used for trapping and paralyzing prey.

Cnidarians include branching, plantlike sea anemones as well as jellyfishes, corals, and the common laboratory specimen, the hydra.

First direct your attention to the first diagram in the plate and read about the two types of Cnidarians in the paragraphs below. Color the species as you read.

Like the sponges, the Cnidarians are very simple, aquatic animals. Their bodies are radially symmetric, somewhat like a wheel with spokes, and contain a single opening that's lined with tentacles.

In the adult, the Cnidarian sac may take either of two forms, which are similar in composition. Take a look at the hydra in the diagram. This is an example of the first type of sac—it is an erect form known as the polyp that has a **body (A_1)** and a number of **tentacles (A_2).** The second form of Cnidarian body is called the medusa, and our example is *Physalia*, or the Portuguese man-of-war. This free-floating, umbrella-shaped type of Cnidarian possess a body (A_1) and a number of tenacles (A_2). The medusa somewhat resembles a flattened polyp tuned upside down. Hydras and sea anemones are polyps, and jellyfish have the medusa body form. Notice that both forms of Cnidarians are radially symmetrical, and that neither form as a distinct head. The medusa usually undergoes sexual reproduction. Polyps often form large colonies and undergo budding, a form of asexual reproduction.

We will now turn our attention to some of the details of the body forms of Cnidarians. The diagrams show schematic views of the polyp and medusa, and we can see the two distinctive body layers not found in sponges. As you read about these layers below, color them in the appropriate diagrams.

The bodies of all members of the phylum Cnidaria consist of two layers of cells with a jellylike mass between them. The saclike, hollow cylinder is seen in the diagrams. The **mouth (A_3)** is the only body opening, and tentacles (A_2) extend from the region of the mouth. Both polyp and medusa have these structures. Cnidarians are said to have an incomplete digestive system because there is only one opening to their pouch-like cavity. This opening must thus serve as both mouth and anus.

The central body cavity of the Cnidarian is the **gastrovascular cavity (A_4),** which can be seen in both diagrams. The arrow pointing to the cavity should be colored in a light color. Food enters the mouth and travels to the gastrovascular cavity for digestion. This cavity is also called a coelenteron, and for many years, the name of the phylum was Coelenterata.

Lining the gastrovascular cavity is a layer of cells called the **gastrodermis (A_5).** Next comes a jellylike layer called the **mesoglea (A_6),** which is very thick in the medusa and makes up the bulk of the animal's body and renders it buoyant. Because of this jellylike mass of mesoglea, meduasae are commonly called jellyfish. Outside the meoglia is a third layer called the **epidermis (A_7);** both forms have this layer.

We finish by examining some of the specialized cells in this animal. Notice that the cells have different shapes depending on their function. Special emphasis will be given to the nematocyst since this is the characteristic feature of the phylum.

You can see a cross section of the hydra in the lower part of the plate. In the epidermis (A_7), there are special cells called **epithelial cells (A_8)** that provide support. Within the meoglia (A_6), there is a diffuse nervous system that is made up of a **sensory network (A_9).** Nerve processes extend into the epidermis and throughout the mesoglea to transmit nerve impulses among the animal cells. **Gland cells (A_{10})** are also present, and **nutritive cells (A_{11})** secrete enzymes that digest food.

Within the epidermis is a specialized collection of cells called **interstitial cells (A_{12}),** which give rise to the cnidocytes. A **nematocyst (A_{13})** can be seen in the view on the lower left side of the page, both in its undischarged and discharged form. As we mentioned above, this is the stinging organ of the cnidarian. Cnidocytes and nematocysts are prevalent on the tentacles. The cnidocyte contains a trigger-like **cnidocil (A_{14})** that, when stimulated, discharges the nematocyst. An arrow points to this structure. (The **nucleus (A_{17})** of this specialized cell can also be seen in the diagram.) Once he nematocyst is discharged, its **barbs (A_{15})** hold fast to the prey. The **filament (A_{16})** shoots out to contact the prey before the barb takes hold, and rapid recoiling brings the prey to the cnidarian for consumption.

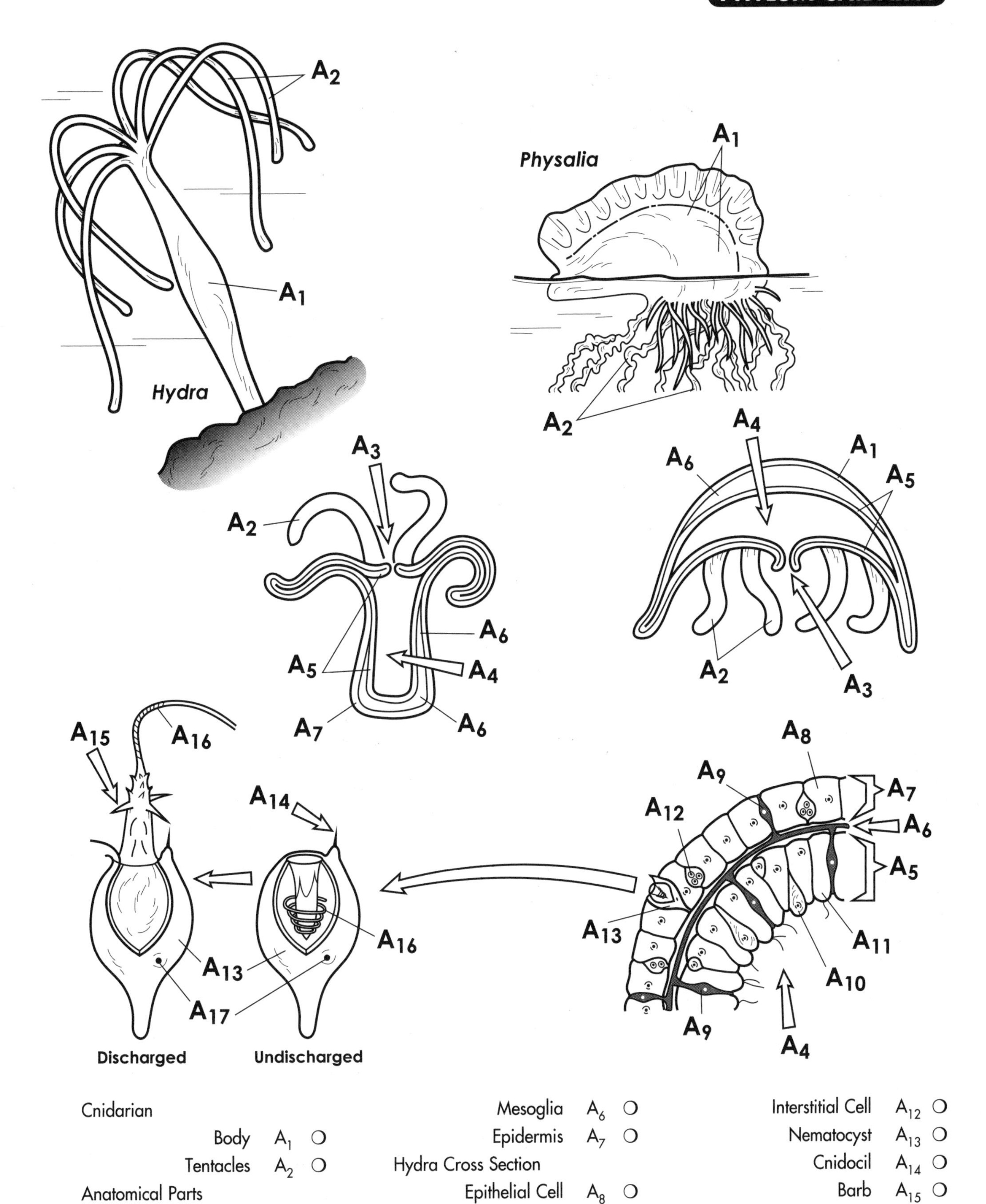

Cnidarian

Body A_1 ❍

Tentacles A_2 ❍

Anatomical Parts

Mouth A_3 ❍

Gastrovascular Cavity A_4 ❍

Gastrodermis A_5 ❍

Mesoglia A_6 ❍

Epidermis A_7 ❍

Hydra Cross Section

Epithelial Cell A_8 ❍

Sensory Network A_9 ❍

Gland Cell A_{10} ❍

Nutritive Cell A_{11} ❍

Interstitial Cell A_{12} ❍

Nematocyst A_{13} ❍

Cnidocil A_{14} ❍

Barb A_{15} ❍

Filament A_{16} ❍

Nucleus A_{17} ❍

PHYLUM PLATYHELMINTHES

There are approximately 20,000 species in the phylum Platyhelminthes. Platyhelminthes are flatworms that inhabit marine and freshwater environments and include free-living forms as well as parasitic flukes and tapeworms. They range in size from the nearly microscopic species to tapeworms that are over 20 meters in length. Not surprisingly, their bodies are flattened.

One of the most familiar flatworms is the planarian of the genus *Dugesia*. This animal is used extensively in introductory biology laboratories and will be a major topic of this plate.

As you look over this plate, notice that we present information on three types of flatworms, including the planarian. A mixture of light and dark colors is recommended; you should use darker colors for the anatomical features. Light colors should be used on the larger specimens and the background areas. Begin your work by focusing on the anatomy of the flatworm, a representative of the planarian *Dugesia*.

The planarian is a free-living flatworm that moves along rock surfaces by gliding or rhythmic muscle waves, or over a slime track secreted by its adhesive glands. The animal has a **gastrovascular cavity (A)** that is similar to that of the cnidarian discussed in the previous plate. Flatworms also have incomplete digestive systems. There is only one opening to the gastrovascular cavity at the **pharynx (B),** but there are several dead-end sacs along the branches of the cavity. This animal consumes small worms, protozoa, and insects.

The nervous system of the flatworm is organized into two **nerve cords (C)** that can be seen running along the sides of the animal. The nerve cords meet at two primitive brains called **ganglia (D),** which are masses of nerve cells. The animal also has two **eyespots (E)** that detect light.

We will continue our study of flatworm anatomy by focusing on its symmetry and germ layers. Continue your reading below as you focus on the remaining whole section and cross section of *Dugesia*.

The excretory system of the flatworm consists of a number of canals and tubules that come together to form a series of **protonephridia (F).** This system is shown in the whole section and should be colored in a dark color; as you can see, the **left side (G)** is a mirror image of the **right side (H).** This type of symmetry is called bilateral symmetry. Some flatworms can eliminate waste by performing diffusion across outer cells, and others use flame cells (specialized excretory cells that serve a function similar to mammalian kidneys).

An important evolutionary feature of the cnidarians is their three germ layers, seen here in cross section. They exist in the embryonic stage and are the layers from which all organs are formed. Surrounding the gastrovascular cavity (A) are the **endoderm (I), the mesoderm (J),** and the **ectoderm (K).** As you may recall, cnidarians, possess only two germ layers. Flatworms also show the rudimentary beginning of a head region. In other words, most of these worms have moderate cephalization; this means nervous system tissue is concentrated at one end of the organism.

We will now turn our attention to another type of flatworm call the fluke. As a representative fluke, we examine *Schistosome mansoni*. This flatworm is a parasite. Focus your attention on the section called fluke life cycle, and begin reading below.

Notice the similarity of the **fluke (L)** to *Dugesia* described previously; it possesses many of the same organs. The fluke's life cycle begins when workers in irrigated fields are infected with human feces contaminated with the larvae of flukes. They can enter the body of the worker and travel to the liver of the **human (M),** where they deposit eggs. These pass out of the human intestine and hatch to become miracidium, the larval form that infects the intermediate host, the **snail (N).** In the snail, the miracidium develops into the cercaria, which is the larval form that infects the final host—the human. This type of life cycle depends upon a primary host (human) and an intermediary host (snail).

We close the plate with a brief examination of a third flatworm, the tapeworm. Like flukes, they are parasites. Continue your reading below and focus on the diagram of the tapeworm in the plate.

Tapeworms have long, flat bodies and little internal detail. They lack digestive cavities and attach to the inner walls of their host animals, absorbing food through their skin. The attachment organ of the tapeworm is called the **scolex (O).** It has several suckers, and in some cases it also has hooks. Behind the scolex is a short neck, and then a series of repetitive segments called **proglottids (P).** At the end of the body, the gravid proglottids produce eggs, which provide the next generation of tapeworms. Flatworms have no respiratory system but do gas exchange via diffusion through skin; this is called cutaneous respiration. They can reproduce either sexually (most are hermaphroditic) or asexually (via fragmentation).

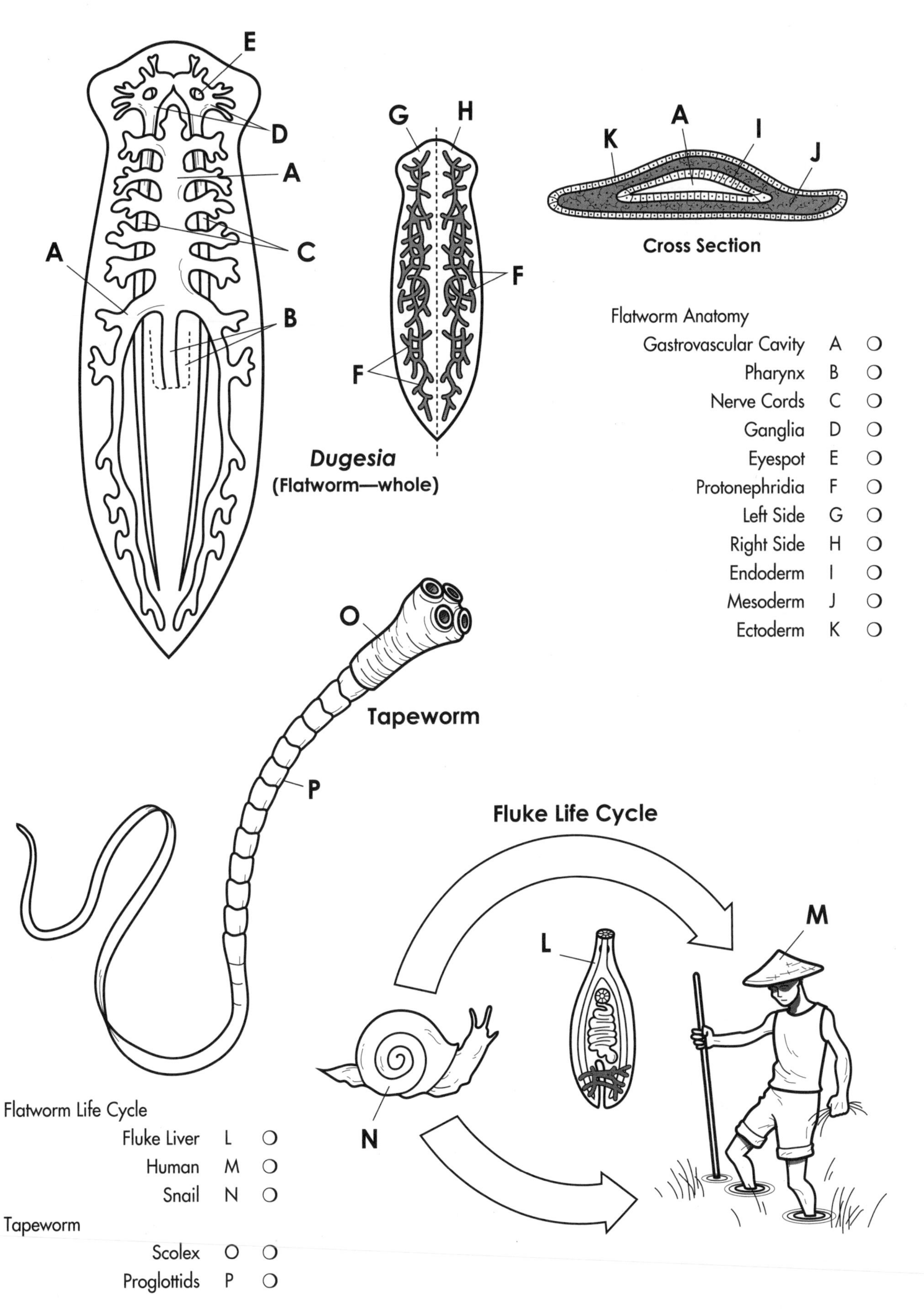

Flatworm Anatomy

Gastrovascular Cavity	A	❍
Pharynx	B	❍
Nerve Cords	C	❍
Ganglia	D	❍
Eyespot	E	❍
Protonephridia	F	❍
Left Side	G	❍
Right Side	H	❍
Endoderm	I	❍
Mesoderm	J	❍
Ectoderm	K	❍

Flatworm Life Cycle

Fluke Liver	L	❍
Human	M	❍
Snail	N	❍

Tapeworm

Scolex	O	❍
Proglottids	P	❍

PHYLUM NEMATODA

The phylum Nematoda is made up of roundworms, which are also know as nematodes. Roundworms are elongated, cylindrical individuals that may have evolved from flatworms. They display two evolutionary advances over the flatworms: They have a complete, one-way digestive tract that runs from mouth to anus, and a cavity between their digestive tract and body wall. Within this cavity there are specialized organs.

Nematodes live in fresh water and saltwater as well as in soil. Most are small, but some may reach lengths of a foot or longer. Most resemble threads, and some are parasites of animals and humans.

Among the parasitic nematodes are the hookworm and roundworm, sometimes transmitted to humans through undercooked or raw pork. One such worm is called *Trichonella spoalis.* When this nematode enters human muscle tissue, it causes a disease called trichinosis. Other parasitic roundworms include pinworms and whipworms, both of which spend time in the human intestine. In this plate, we will consider a typical roundworm known as Ascaris, which is a parasite sometimes found in dogs.

Looking over the plate, you will notice that we show diagrams of two worms, the female and male Ascaris. We also show a cross section of the female roundworm to display some internal parts. As you read about the worms in the paragraphs below, locate and color the structures in the plate.

As we mentioned earlier, roundworms have a complete digestive tract, which contrasts with the digestive cavities of the two previous phyla. The roundworm's digestive tract begins with the **mouth (A),** and continues with an enlarged opening called the **pharynx (B).** Next comes an extremely long **intestine (C)** that extends along the entire length of the roundworm. You should use a light color for this tube. The digestive tract ends at the opening to the exterior called the **anus (D).**

The intestinal tract of the roundworm is separated from the **body wall (E),** and in the cross section you can see the body wall encircling the roundworm, while the digestive tract is shown as an open tube. You should use a light color for the body wall (E). Notice that, in the cross section, a space exists between the digestive tract and body wall. This space is the **pseudocoelom (F).** The pseudocoelom is not a true body cavity, because it is not surrounded by mesoderm-derived tissue.

We will now focus on some of the internal organs of the roundworms. Continue your coloring as you read about these organs below. Notice the relatively complicated reproductive organs as we continue.

Excretion in the roundworms is accomplished by a series of **excretory tubes (G),** which run along the left and right side of the animal and can be seen in all three figures. Waste products from surrounding cells accumulate in these tubes and pass to the **excretory pore (H),** where they exit the body. Many roundworms have excretory flame cells to help process waste, similar to flatworms. Most species of nematodes are dioecious, meaning that separate male and female individuals exist. In the female Ascaris, egg cells are produced within the very thin, coiled **ovary (I),** then enter the **oviduct (J),** and finally pass to the **uterus (K),** where fertilization takes place.

The male organ of reproduction in nematodes is the **testis (L),** which, like the uterus, is a long, think tube. Sperm cells are produced here and are stored in the coiled **vas deferens (M).** When reproduction takes place, sperm cells enter the **seminal vesicle (N).** They pass out of the male tract during copulation, and a **spicule (O)** holds the female reproductive organ in place while the sperm cells pass into the female. Fertilization takes place within the female, and the fertilized eggs are stored in the uterus until they are deposited in the soil or another environment. The eggs are surrounded by a thick shell and are deposited in enormous numbers; a single female worm may deposit up to 200,000 fertilized eggs in the course of one day. Pinworms, whipworms, hookworms, and animal roundworms also multiply in this fashion. Ascaris eggs are very resistant to environmental change, and may live for years under adverse conditions.

The final structures we will consider are the two nerve cords that transmit impulses to various cells and tissues of the animal's body. A **dorsal nerve cord (P)** is shown as a spot in the cross section above the digestive tract. Below the digestive tract is the second nerve cord, which is called the **ventral nerve cord (Q).** The presence of two nerve cords is another specialized development in the roundworms that distinguishes them from the more simple flatworms. Roundworms have no circulatory or respiratory systems. Instead, nutrients are transported by diffusion, and cutaneous respiration occurs via diffusion through skin.

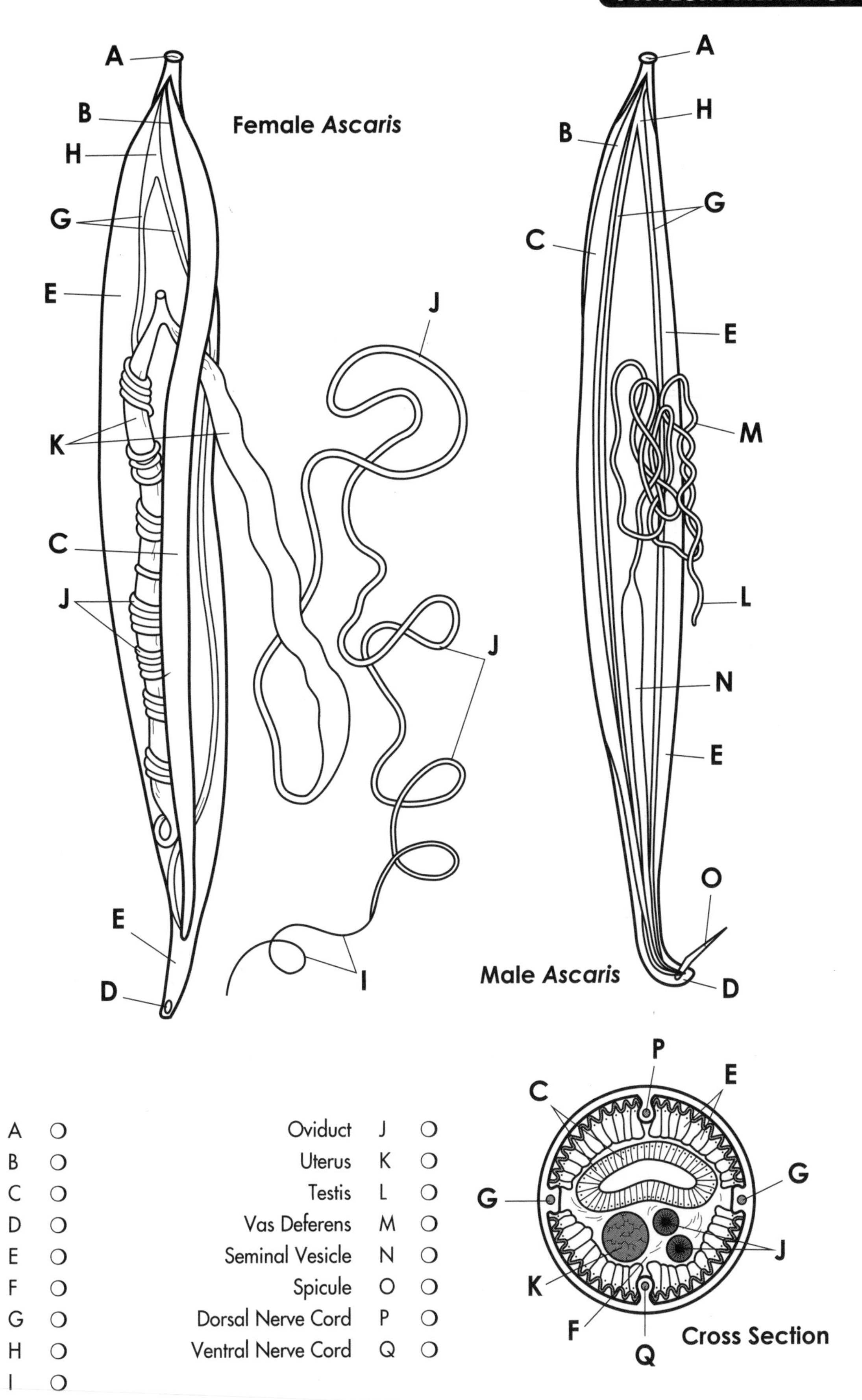

Mouth	A	❍
Pharynx	B	❍
Intestine	C	❍
Anus	D	❍
Body Wall	E	❍
Pseudocoelom	F	❍
Excretory Tube	G	❍
Excretory Pore	H	❍
Ovary	I	❍
Oviduct	J	❍
Uterus	K	❍
Testis	L	❍
Vas Deferens	M	❍
Seminal Vesicle	N	❍
Spicule	O	❍
Dorsal Nerve Cord	P	❍
Ventral Nerve Cord	Q	❍

PHYLUM ANNELIDA

Members of the phylum Annelida are segmented worms, often called annelids. There are approximately 9,000 species in this phylum, and representative animals include earthworms, leeches, and marine worms, including the clamworms, scaleworms, and lugworms. The most distinctive evolutionary feature of annelids is their segmented body, which made possible the development of structural specializations characteristic of more complex animals. Annelids have a true coelom, as compared to the nematodes, which have a pseudocoelom.

Three views of a representative annelid are presented in this plate. We show a whole worm, a somewhat detailed view of its anatomy, and a cross section. The genus we'll look at is *Lumbricus,* which is the common earthworm studied in school laboratories.

The typical annelid possesses a head region, a segmented body, and a terminal portion with an anus. As you can see in the whole mouth, the body is divided into rings called **segments (A)** and along the body is a saddle-like enlargement called a **clitellum (B),** which is involved in the worm's reproduction. (Not all annelids have a clitellum.)

The worm has a long digestive tract that runs the length of its body; it begins at the **mouth (C),** and ends at the **anus (D).** Food is drawn into the body by the sucking action of the muscular **pharynx (E)** and, after is passes through a short tube called the **esophagus (E_1),** the food passes into a thin-walled region called the **crop (F).** The food is stored in the crop before it passes into the **gizzard (G),** where grinding action breaks it into small pieces. It then passes into the long **intestine (H)** for final digestion and absorption; the wall of the intestine can be seen in the cross section of the animal. In the cross section, you can see an infolding in the side of the intestine, called the **typhlosole (H_1);** this infolding increases the intestine's digestive and absorptive area. Most of the food the earthworm ingests consists of decaying organic matter and bits of leaves and other vegetation.

We will continue our study of the annelid by focusing on some of its internal details, many of which we have not seen in the animals we've studied thus far. Continue reading and coloring the structures as you encounter them. You should use lighter colors for the next parts because some of them are small.

As we mentioned before, annelid worms have a true **coelom (I),** which is a cavity between the body wall and the intestinal tract. A light color should be used to color the coelom in the anatomical view and in the cross section. The lining of the coelom is derived from mesoderm tissue.

Lining the coelom are two types of muscle. **Longitudinal muscles (J)** run lengthwise, while **circular muscles (K)** run around the body of the animal—both muscle types are seen in Lumbricus. A layer of **epidermis (L)** encloses the circular muscle, and a thin **cuticle (L_1)** surrounds the epidermis. The longitudinal and circular muscles are used for swimming, crawling, and burrowing. The epidermis surrounds and protects these muscles, and the cuticle secretes fluid to keep the body surface moist. Inside the muscles, the fluid in the coelom bathes the internal organs and insulates them from the stresses of body movement.

Annelids possess complex, closed circulatory systems. In a closed circulatory system, blood is contained within vessels and is distinct from the fluid in the body cavity. One or more pumping hearts are required to move blood through the blood vessels. Surrounding the esophagus is a series of muscular vessels referred to as **hearts (M),** or aortic archaes. They deliver blood to a **ventral vessel (M_1)** (made up of segmented vessels) that runs along the lower surface of the body and delivers blood to the brain and much of the body. The **dorsal vessel (M_2)** runs along the top of the digestive tract and delivers blood to the hearts; you can see both of the vessels in the cross section view.

The nervous system of the annelid worm consists of a primitive **brain (N),** which is basically an accumulation of brain cells, and a ventral **nerve cord (N_1)** that connects the brain to other areas of the body. For excretory purposes, the annelids have **nephridia (O)** in each segment on both sides of the body. Each nephridium contains several loops that accumulate waste fluid and release it to the outside through a pore called the nephridiopore. Nephridiopores are located near tiny bristles called **setae (P),** which help anchor the worm as it moves and prevent backward slipping.

Segmented worms have no formal respiratory system, as they perform cutaneous respiration (gas exchange via diffusion through the skin). They can perform sexual or asexual reproduction via fragmentation. Some species are also capable of regeneration, which is a wound-healing mechanism, where lost tissues, organs or limbs can be regrown.

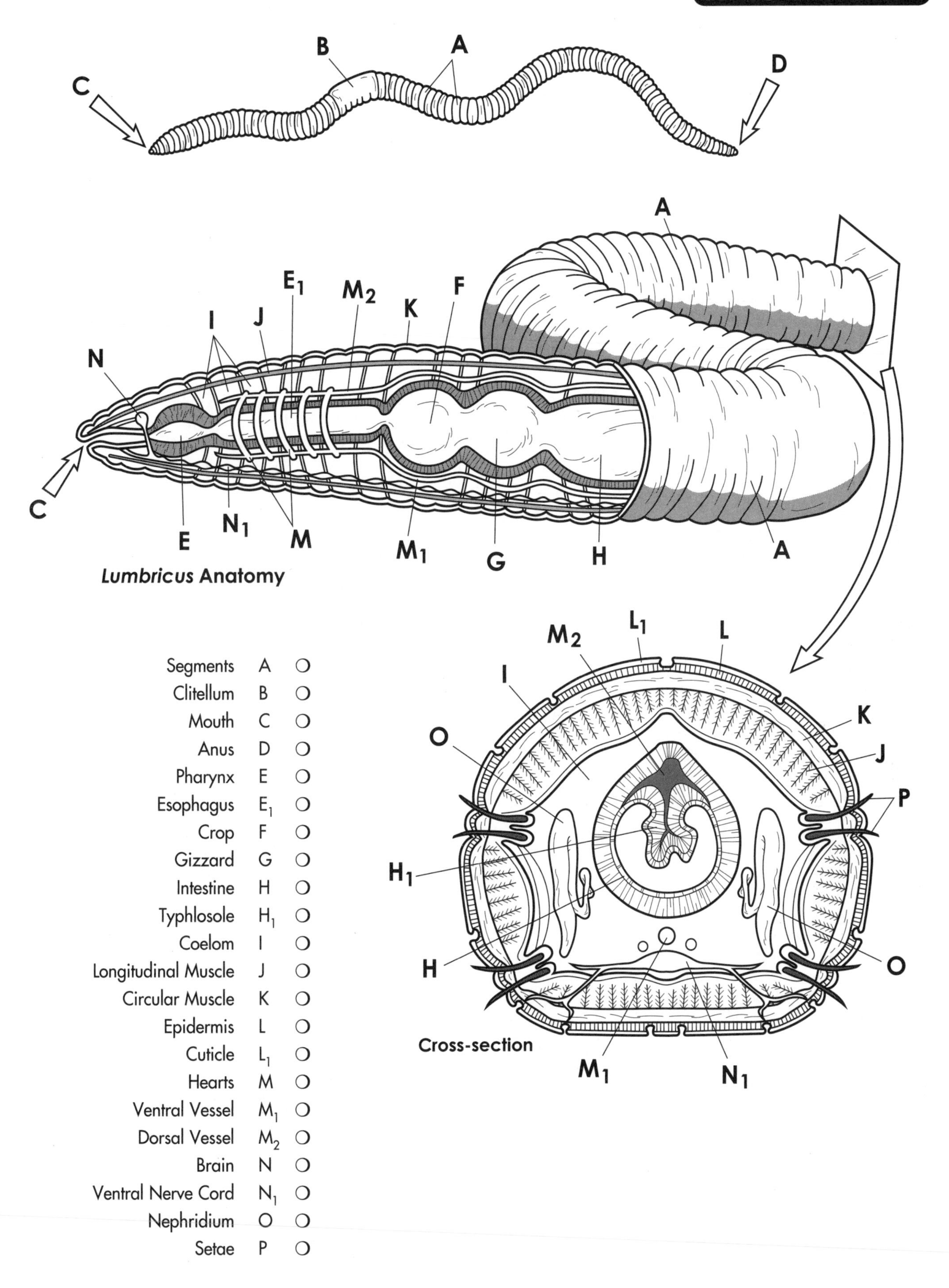

***Lumbricus* Anatomy**

Cross-section

PHYLUM MOLLUSCA

There are approximately 50,000 species in the phylum Mollusca, and this phylum contains clams, snails, and squid. These animals are unsegmented coelomates with soft bodies and other important distinguishing characteristics that we will study in this plate.

Examine the art and notice that we are studying the anatomical features of a typical mollusk, the clam. This animal is significantly more complex than the worm studied in the last plate. The second part of the plate briefly compares the anatomical structures of four different kinds of mollusks. Begin your work by concentrating on the first diagram as you read the paragraphs below.

In addition to soft bodies, all mollusks possess a thick, muscular **foot (A),** which is different in different types of mollusks. It is used for creeping, holding fast, catching prey, and swimming.

Another distinguishing characteristic of many mollusks is their external **shell (B),** seen in the clam diagram. The shell is usually made of calcium carbonate. The clam is called a bivalve because it has two hinged shells. This shell is produced by a fleshy covering of the internal mollusk body called the **mantle (C),** but both shelled and nonshelled mollusks have a mantle.

Two powerful muscles hold the hinged shells together and open and close them; these muscles are the **anterior adductor muscle (D)** and the **posterior adductor muscle (E).** In aquatic mollusks, a set of feathery structures called **gills (F)** are used for respiration, much as they are in fish. Other mollusks use lungs for respiration.

We will now study the digestive and other systems of the clam. Continue to read as you color the structures. Lighter colors such as yellows and grays are recommended since these structures tend to be small.

Bivalves like the clam obtain particles of food by filtering them from the water. Water-bearing food particles are first drawn into the cavity through an opening called the **incurrent siphon (G),** and are then trapped in mucous secretions of the **mouth (H).** As the food moves along the **digestive gland (I),** it is broken into smaller particles. In the **intestine (J),** the molecules are absorbed, and the waste flows out through the **anus (K).** Notice that the intestine twists several times through the body of the animal. Excretion in the clam is performed in part through a specialized organ called the **nephridium (N),** which collects wastes and transports them out of the body cavity. After leaving the anus, the residue is suspended in water that leaves the animal through the **excurrent siphon (L).**

Most mollusks have an open circulatory system that includes a **heart (M)** that pumps blood through their body. In an open circulatory system, there is no difference between blood and interstitial fluid. This fluid is called hemolymph, and it bathes organs and tissues. Hemolymph is composed of water, inorganic salts (such as Na^+ and K^+), and organic compounds (mostly carbohydrates, proteins, and lipids). The primary oxygen transporter molecule is hemocyanin, and free-floating cells (called hemocytes) function in immune protection. Most mollusks have a three-chambered heart.

Although clams have no nervous system, they possess a number of concentrated nerve cells known as **ganglia (O)** at different locations within the body. The reproductive structure of the clam is quite large and is known as a **gonad (P).**

We will now focus on four different kinds of mollusks. As you color the structures in the diagrams, read the paragraphs below. Medium or dark colors should be used for these features because they are large and distinct. Remember that the objective is to draw comparisons between the various mollusks.

The chiton, in the first diagram, has a flattened body, an elongated foot, and a plated shell, and is commonly found at the bottom of tidal pools. Snails and their relatives are gastropods, clams are bivalves, and squid, octopuses, and nautiluses are cephalopods.

In the four representative mollusks, notice the variation in the foot (A). The chiton has an elongated foot, while the octopus and squid have feet that are modified into eight tentacles. The chiton has a shell (B) that consists of eight plates, while the snail and clam both have a single shell, and the squid and other cephalopods have no shells. A **gut (Q)** exists in all four animals, but varies in structure and appearance.

In the octopus and squid, the mantle is modified into a propulsive device. The space within the mantle is the **mantle cavity (R);** it has a different shape in these four animals. In terrestrial snails and slugs, it is modified to extract oxygen from the air. With the exception of bivalves, mollusks also have what is called a **radula (S),** which consists of rows of teeth that are used for breaking large food particles into small ones.

Most mollusks perform sexual reproduction, and some species are hermaphroditic. Most mollusks live in marine or fresh water, and some live on land. It is worth emphasizing that cephalopods (such as squids and octopuses) have some differences from the rest of the mollusks. Almost all members of this class have a reduced or missing shell. Cephalopods are carnivores that move much faster than the other mollusks. They are also the only members of this phylum that have a closed circulatory system.

Clam Anatomy

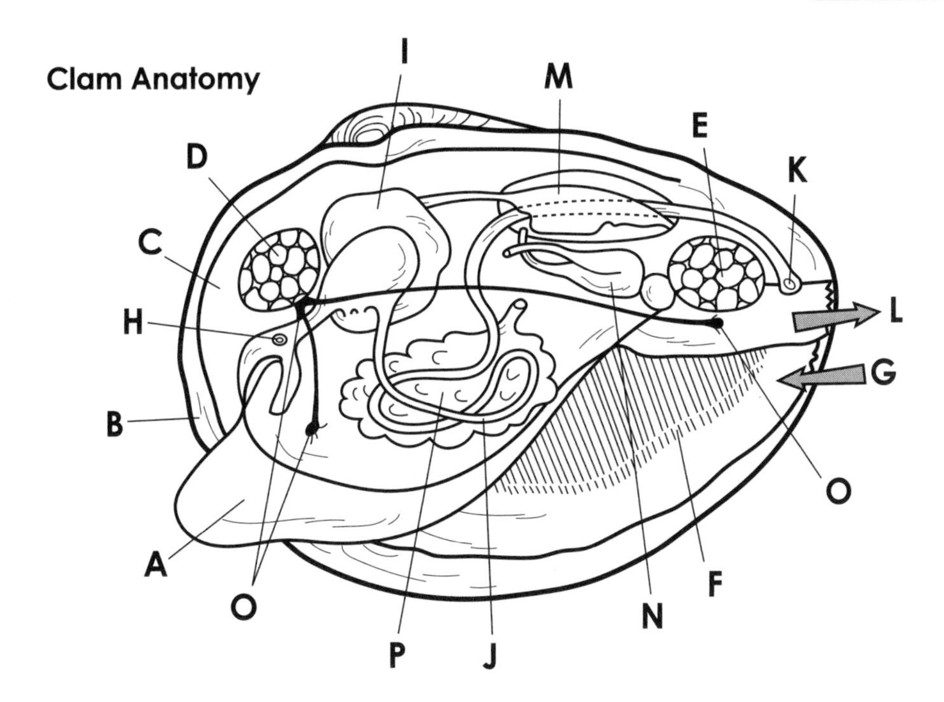

Comparison of Mollusks

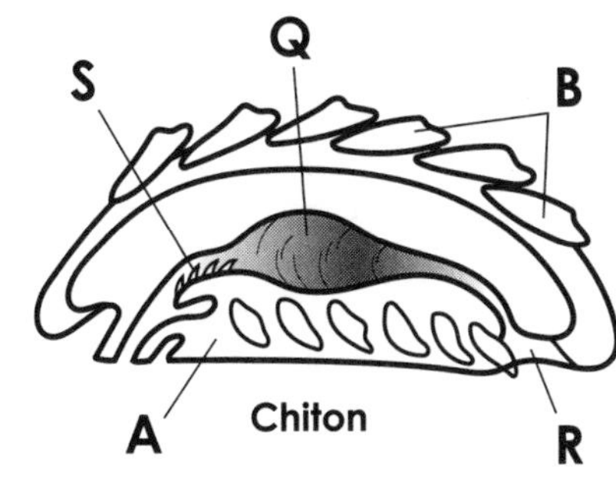

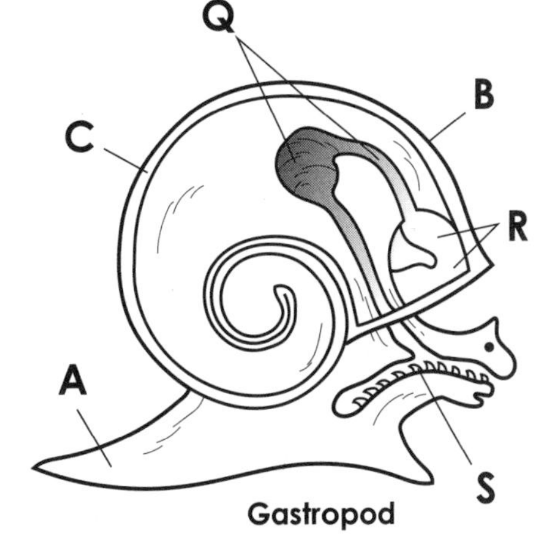

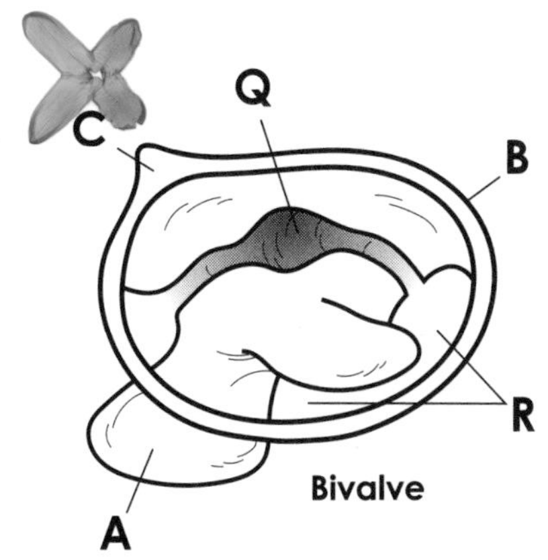

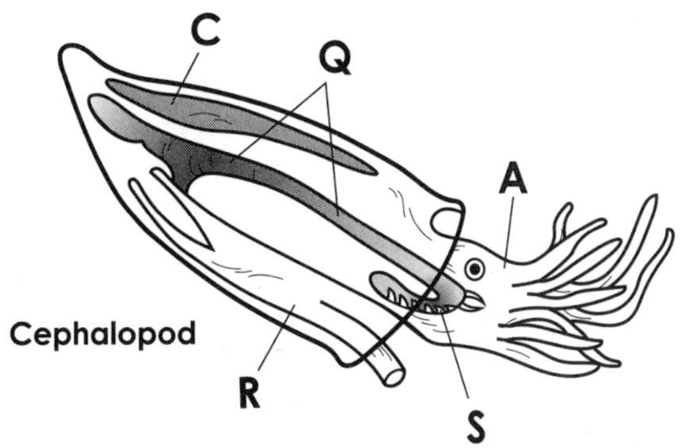

Clam Anatomy

Foot	A	❍
Shell	B	❍
Mantle	C	❍
Anterior Adductor Muscle	D	❍
Posterior Adductor Muscle	E	❍
Gills	F	❍
Incurrent Siphon	G	❍
Mouth	H	❍
Digestive Gland	I	❍
Intestine	J	❍
Anus	K	❍
Excurrent Siphon	L	❍
Heart	M	❍
Nephridium	N	❍
Ganglion	O	❍
Gonad	P	❍

Comparison of Mollusks

Gut	Q	❍
Mantle Cavity	R	❍
Radula	S	❍

PHYLUM ARTHROPODA

Most animals are arthropods; the species in this phylum outnumber all of the other species of animals and plants together. Over a million species of arthropods in the phylum Arthropoda have been described and it is generally accepted that, from an evolutionary standpoint, they are the most successful organisms on Earth.

All members of the phylum Arthropoda have jointed appendages and segmented body plans, but their segments do not simply repeat as they do in earthworms. They are highly specialized for tasks such as feeding, walking, sensing, and reproducing, In addition, arthropods have an outer skeleton called an exoskeleton. This exoskeleton is made of chitin (like the fungal cell wall) and is periodically shed in a process called molting, which enables the arthropod to grow.

In the first part of this plate, we show the anatomy of a typical arthropod, the grasshopper. Many of the typical arthropod features are found in this animal. In the second section of the plate, representative arthropods are described. Begin your reading below as you focus on the grasshopper.

The grasshopper is a member of the class Insecta. The body of the grasshopper consists of three main parts: the **head (A),** the **thorax (B),** and the **abdomen (C),** in which segmentation is most apparent. In the grasshopper, the thorax and head are fused.

Some of the distinguishing characteristics of insects are their **antennae (D),** their **compound eyes (E),** and their feeding appendages, which include a **mandible (F)** and two pairs of **maxillae (G).** The maxillae are fused together to form a sort of lower lip.

Insects vary in shape and size, but many insects have two pairs of **walking legs (H)** and a third pair called **jumping legs (I).** Along the wall of the abdomen are openings called **spiracles (J),** which lead to tubes that make up the respiratory system of the insect. Two pairs of wings are also present: the **fore wings (K)** and the **hind wings (L).** The last segment of the abdomen, called the **ovipositor (M),** is specialized for the deposit of the insect's eggs.

Having described some of the typical features of an arthropod, we will now turn to some representative arthropods from the various classes of this phylum. We will use the common names of these different arthropods as we discuss them. Focus on the representative arthropods and use light colors such as yellows, grays, and pastels in order to avoid obscuring the important features of the animals.

Among the arthropods are thousands of species of **millipedes (N),** which have cylindrical bodies and two pairs of legs per segment. Most millipedes feed on decaying plant matter. A similar group in appearance is the **centipedes (O),** which have flattened bodies and one pair of legs per segment. Centipedes are predators and scavengers and their first pair of legs is modified into claws that are used for injecting poison into their prey.

Certain arthropods are referred to as crustaceans. These animals live in marine and fresh water, and include the **crab (P)** and the **crayfish (Q).** These are the only arthropods that have two pairs of antennae as well as walking legs on their thorax and abdomen. Shrimp and lobster are also included in this group.

Arthropods that have eight legs, such as **ticks (R)** and **horseshoe crabs (S),** are arachnids. Spiders, scorpions, and mites are also included in the group, in which the head and thorax are usually fused to one another. Many species of arachnids kill their prey by injecting them with venom. Ticks are bloodsucking parasites that are vectors for various diseases including Rocky Mountain spotted fever and Lyme disease.

Over 700,000 species of insects have been identified, all of which have at least three pairs of legs, and several of the typical parts shown in the grasshopper. The diagram shows the **housefly (T)** and a common **beetle (U)** as further representatives of the insect group. Some other familiar members include termites, crickets, lice, fleas, moths, butterflies, bees, and ants.

The circulatory system in arthropods is open and is filled with hemolymph. A complete digestive system is present. Waste removal is accomplished by malpighian tubules in most arthropods; these tubules are outpocketings of the gut. They absorb water and waste from the circulating hemolymph, and dump it into the alimentary canal. Wastes are then secreted as solid from the hindgut. However, some arthropods use metanephridia instead (similar to mollusks and annelids discussed on previous plates). Arthropods typically perform sexual reproduction.

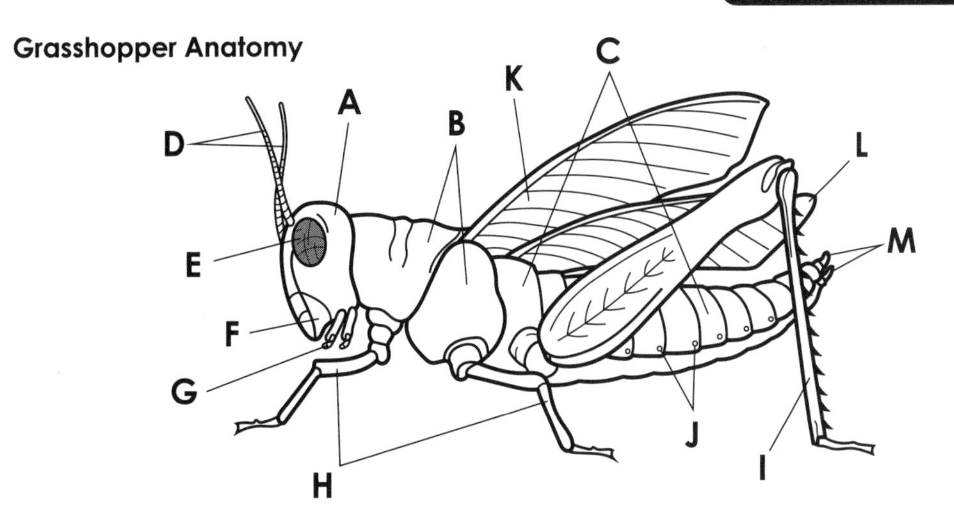

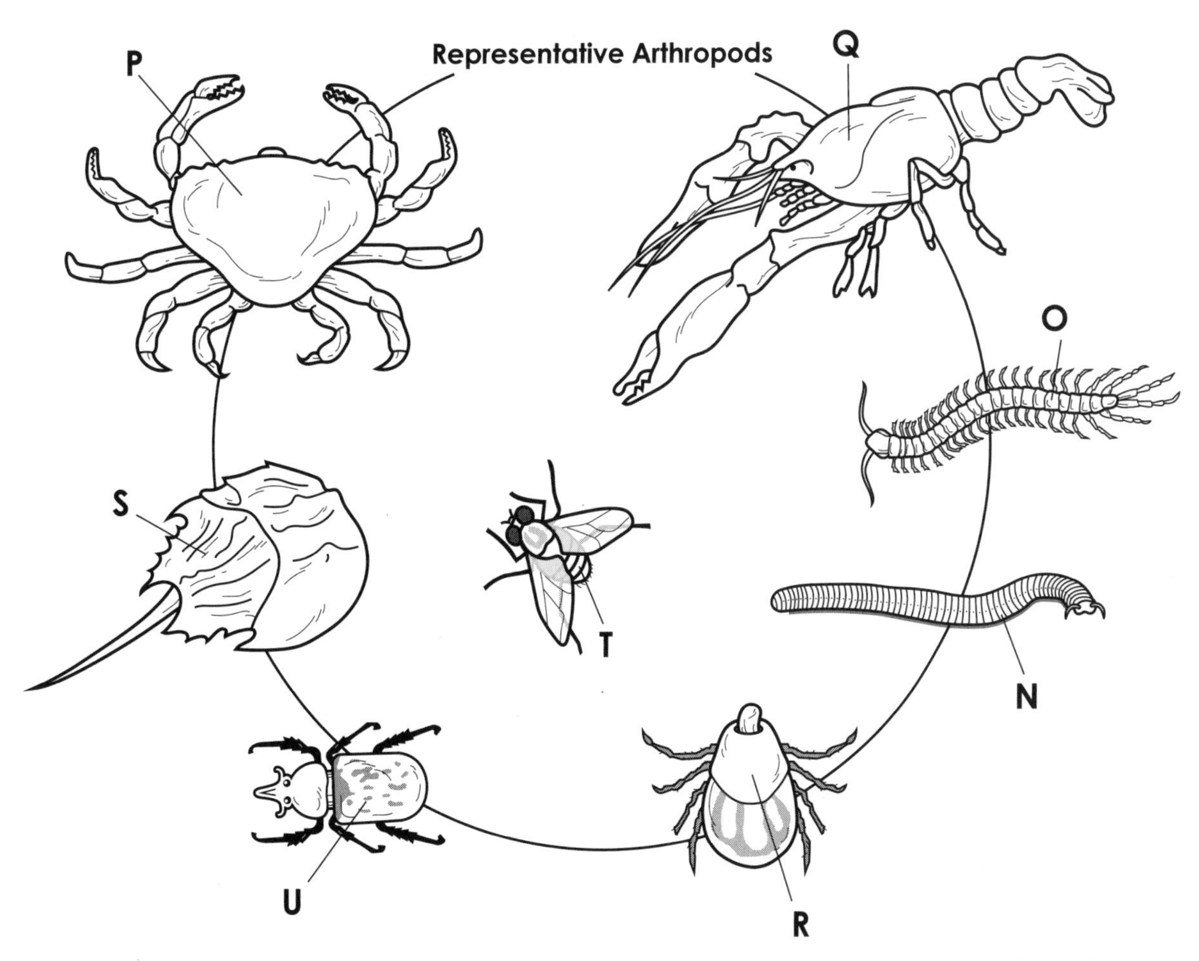

Grasshopper Anatomy

Head	A	❍
Thorax	B	❍
Abdomen	C	❍
Antennae	D	❍
Compound Eye	E	❍
Mandible	F	❍
Maxillae	G	❍
Walking Legs	H	❍
Jumping Legs	I	❍
Spiracles	J	❍
Fore Wing	K	❍
Hind Wing	L	❍
Ovipositor	M	❍

Representative Anthropods

Millipede	N	❍
Centipede	O	❍
Crab	P	❍
Crayfish	Q	❍
Tick	R	❍
Horseshoe Crab	S	❍
Housefly	T	❍
Beetle	U	❍

PHYLUM ECHINODERMATA

Animals of the phylum Echinodermata have spiny skins that are made up of hard, calcified plates that form a type of skeleton just under their epidermis. They include the starfishes, sea cucumbers, and other familiar seashore animals. A unique feature of the group is that they possess a derivative of the coelom, known as the water vascular system.

This plate contains a diagram of a representative echinoderm, the starfish. The anatomy of this animal is typical of the rest of the group. We will also highlight its water vascular system and show various representatives of the phylum. You should begin your work by focusing on the anatomy of the starfish.

Echinoderms are exclusively marine and brackish water animals and many species are found in the shallow sections of bodies of water. They are radially symmetrical animals, as the view of the starfish shows; the starfish may be cut equally in five different directions, beginning at its midpoint. This type of symmetry is also found in the phylum Cnidaria, but it is different from the bilateral symmetry found in most other phyla thus far discussed. However, it is worth noting that echinoderm larvae have bilateral symmetry. Because of this, the adults are said to have secondary radial symmetry. Echinoderms have no segmentation in their bodies and no head region.

The prefix "echino-" means spiny, and all echinoderms have skins with **spines (A),** which you should color a light color. These are formed by a spiny endoskeleton usually made of calcium carbonate and covered by thick skin.

Another unique feature is the water vascular system, which can be seen in the starfish and is extracted in the diagram below it. At the surface of the starfish, the entry to the water vascular system is called the **madreporite (B).** This structure is sometimes called the sieve plate. Through minute openings in the madreporite, water enters a tube that leads to a **ring canal (C)** that circles this opening. Water then enters any of five **radial canals (D),** which can be seen in both the starfish and the lower left diagram. Each radial canal carries water to a pair of **tube feet (E)** at the surface of the echinoderm, and above each tube foot is a rounded muscular sac called an **ampulla (F).** As the ampulla contracts, the water within it is forced into the tube foot, and the tube foot extends and attaches to an object with its sucker. The suckers permit the echinoderms to cling tightly to rocks and prevent them from being swept away by strong waves. They also allow the starfish to grip tightly to the shells of mollusks upon which they feed.

We will now concentrate on the other anatomical features of the typical echinoderm. Light colors should be used for this section. Continue your reading in the paragraphs below.

When an echinoderm feeds, its gut extends out of its body. The **stomach (G)** is located close to the opening of the starfish and it is extended to contact food. The food is digested and stored in five pairs of **digestive glands (H),** located in the five arms of the starfish, and waste is eliminated through the **anus (I),** which opens to the exterior near to the madreporite. Echinoderms have a complete digestive system.

Echinoderms are either male or female, and can perform sexual or asexual reproduction. Sperm cells and egg cells are produced within the testes or ovaries, which are located in each arm, and released into the water through openings in the arms. The reproductive organs are called **gonads (J)** and can be seen in the arm of the starfish. Echinoderms have no centralized nervous system, but nerves are concentrated near the mouth and branches extend into the arm.

We will close the plate with a brief examination of three other representative echinoderms. Light yellows and grays should be used to avoid obscuring the details of these animals. Notice that they are radially symmetric like the starfish.

The phylum Echinodermata consists of five classes, one of which contains the **sea urchins (K).** These are flattened organisms without arms that are covered by solid shells bearing jointed, movable spines. Other representative echinoderms are the **brittle stars (L),** which are similar to starfishes, but have long arms set off from the central disk. The skeleton of the arms is highly specialized, jointed, and extremely flexible, and allows the animal to move with great speed and agility.

A final representative is the **sea cucumber (M).** Sea cucumbers are soft-bodied creatures that burrow in the ocean bottom. Their tube feet are very small and their skeleton consists of isolated plates and spicules inside their leathery body wall. These marine animals have a closed circulatory system. They use skin gills for respiration and excretion; podocytes (specialized epithelial cells with excretory function) also help with excretion. Many echinoderms can regenerate lost tissues.

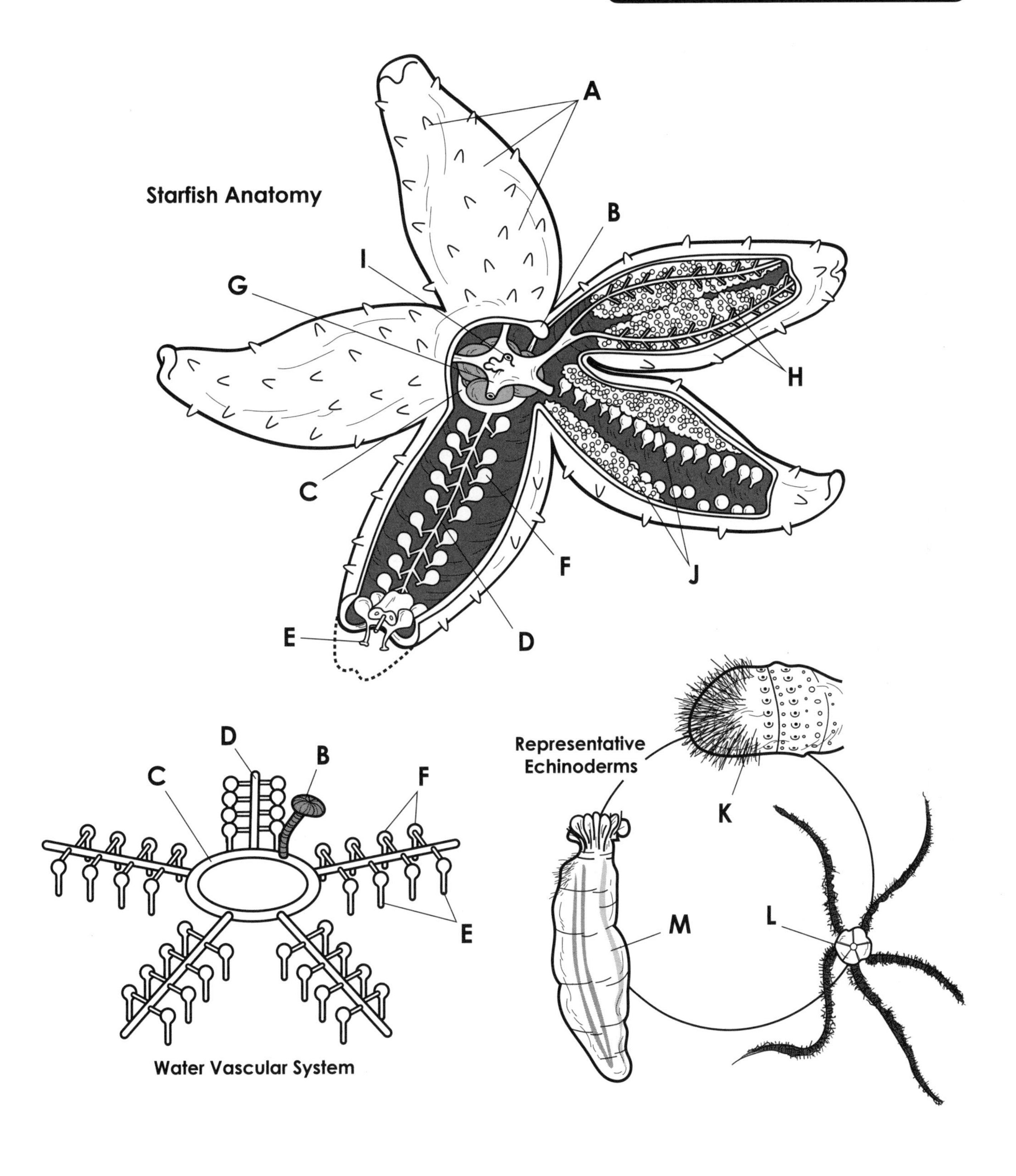

Starfish Anatomy

Skin with Spines	A	❍
Madreporite	B	❍
Ring Canal	C	❍
Radial Canal	D	❍
Tube Feet	E	❍
Ampulla	F	❍
Stomach	G	❍
Digestive Glands	H	❍
Anus	I	❍
Gonad	J	❍

Representative Echinoderms

Sea Urchin	K	❍
Brittle Star	L	❍
Sea Cucumber	M	❍

PHYLUM CHORDATA

Members of the phylum Chordata include approximately 45,000 different species of fish, birds, reptiles, amphibians, and mammals. Collectively, the organisms are known as chordates, and humans are included in this phylum. In this plate, we discuss the principal features of all chordates and survey the main properties that make up its five groups.

This plate contains a view of a typical chordate as well as representations of the five major groups of chordates. Begin your work on the plate by concentrating on the typical chordate as you read the paragraphs below.

Chordates are a fascinating and diverse group of animals that all have certain fundamental characteristics in common. A few chordates lack backbones and are therefore known as invertebrates. Tunicates and lancelets are invertebrate chordates. However, the vast majority of chordates have backbones and are vertebrates. This includes fish, amphibians, reptiles, birds, and mammals (each of which will be discussed below). In vertebrates, the notochord is replaced by a vertebral column. These animals have a well-formed head with a cranium to protect the brain, as well as neural crest cells (unique embryonic cells). Vertebrates have internal organs (such as liver, kidneys, and endocrine glands) and physiology to support increased energy demands. They also have a bone or cartilage endoskeleton, a closed circulatory system, and a complete digestive tract. Many vertebrates are tetrapods, meaning they have four limbs.

Several traits distinguish all chordates, and light colors should be used for the first chordate shown. The first characteristics we will point out are **pharyngeal gill slits (A),** which are perforations in the wall of the pharynx near the **digestive tract (A_1).** As fish develop, these slits give rise to gills, while in land-based vertebrates such as humans, they close and disappear.

A second characteristic is the **dorsal nerve cord (B).** A bold color should be used for this structure and its expanded portion, the **brain (C).** The nerve cord is tubular and is located along the back portion of the chordate; the human spinal cord is an expansion of the dorsal nerve cord.

The third characteristic feature is the **notochord (D).** This flexible rod of tissue is found in the embryo, beneath the nerve cord. It lends support and consists of large cells encased in a tight covering of tissue, and disappears as the vertebral column appears.

A final characteristic of chordates is a **tail (E).** The bracket that encloses this area should be colored. The tail is found beyond the anus, at the far end of the digestive tract, and is a muscular projection. It disappears in many vertebrates (e.g., humans) as the embryo develops.

Despite the presence of common characteristics, there is a wealth of diversity among the chordates. We will see examples of that diversity as we survey more animals. Begin your study by focusing on the fish as you read the paragraphs below.

The first group of vertebrates we will survey are the **fish (F),** which include the ray, the herring, and the shark. The shark and ray have skeletons of cartilage (class Chondrichthyes), while the herring has a bony skeleton (class Osteichthyes). Some primitive fish such as the lamprey and hagfish have no jaws. Fish require a constant stream of water forced over the gills in order to obtain oxygen and have two-chambered hearts.

The second group of chordates is the **amphibians (G)** of the class Amphibia, here represented by the frog and newt. Amphibians were the first land vertebrates, but they return to the water to lay their eggs. They breathe through their lungs and skin, and they constantly lose water through their skin, so they must remain in moist habitats. They have bone endoskeletons and three-chambered hearts.

The third group of chordates is the **reptiles (H)** of the class Reptilia, here represented by snakes and alligators. Reptiles have dry skin that's covered with scales that prevent water loss. They lay eggs on land, and their eggs contain stored food and water for the developing embryo. Reptiles are mostly terrestrial, have bone endoskeletons and three-chambered hearts. Most breathe through lungs, but some can perform cutaneous respiration.

We have examined three of the groups of chordates, and we will conclude our discussion with the final two groups, birds and mammals. Continue your reading below as you color the members of these two groups.

Perhaps the most distinctive characteristic of the members of the class Aves, or **birds (I),** is their feathers. Feathers are used for flight and as insulation to maintain constant elevated body temperatures. Their lungs have outpockets in which additional air can accumulate, and their heart rate is extremely high, providing large amounts of blood to their tissues. The eggs produced by birds have hard shells and are highly resistant to dehydration. Birds have a lightweight honeycomb bone skeleton to help with flight and a four-chambered heart. They breathe through lungs.

The final group we will consider is the **mammals (J),** in the class Mammalia. Included in this group are humans, dogs, and whales. Mammals nourish their young with milk produced by mammary glands, and have elevated body temperatures and four-chambered hearts. Mammals have hair and teeth and the largest brains. They respire using lungs.

In most mammals, the young develop internally in the uterus. Exceptions are the monotremes, or egg-laying mammals; and the marsupials, whose young develop in pouches outside the body. Mammalian limbs are generally directed downward to raise the body off the ground, which allows mammals to run, burrow, leap, and swim.

Animals can use different heat sources. Ectothermic animals are cold blooded. They do not produce enough metabolic heat to control body temperature and so get heat from an external source. Ectothermic organisms are called poikilothermic because their body temperature fluctuates with the environment. Fish, amphibians, reptiles, and invertebrates are cold blooded. Endothermic organisms are warm blooded, in that their bodies are warmed by internal metabolic heat. These animals are homeothermic, as their body temperature is constant, and must be maintained at a certain level to maintain life. Birds and mammals are warm blooded.

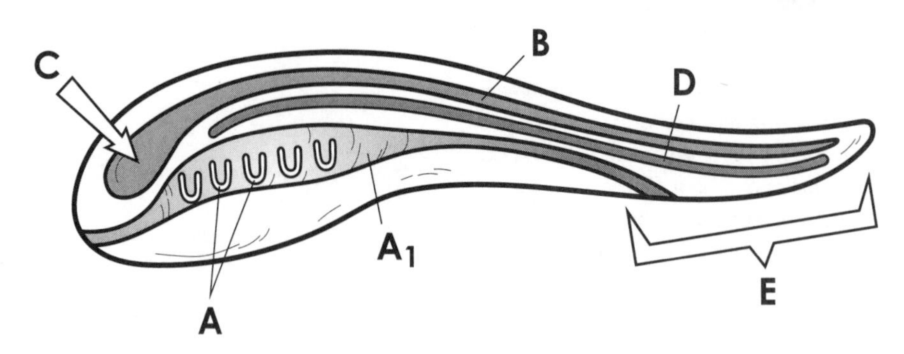

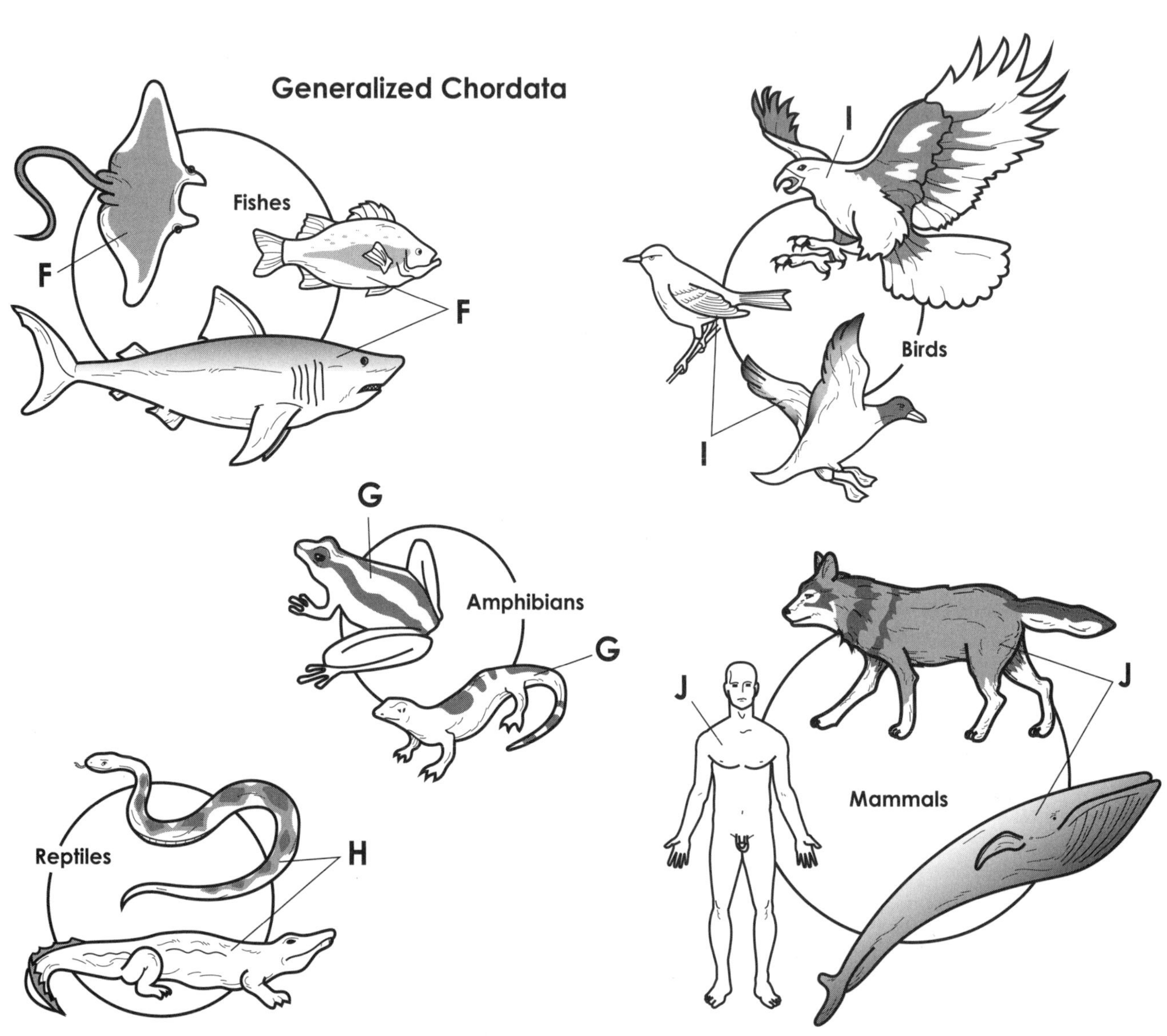

Pharyngeal Gill Slits	A	❍
Digestive Tract	A_1	❍
Dorsal Nerve Cord	B	❍
Brain	C	❍
Notochord	D	❍
Tail	E	❍
Fish	F	❍
Amphibians	G	❍
Reptiles	H	❍
Birds	I	❍
Mammals	J	❍

NOTES

ANIMAL EVOLUTION

Animals evolved from a common ancestor. This organism lived on Earth about 600 million years ago and is no longer present. No one knows for sure what it looked like, but it was probably a eukaryotic multicellular blob of some sort **(A).** All the animal species in the world today evolved from this common ancestor. The huge amount of diversity in animals results from modification by natural selection operating over millions of years in different environments. Animals vary greatly in what they look like, but there are patterns in morphological and developmental traits. These patterns evolved over time, and are the focus of this plate. In some cases, a particular characteristic evolved very early on and hasn't really changed since. Other traits evolved later. The evolution of key animal traits is denoted in grey boxes on the plate.

Early on, animals evolved tissues. There are four main types of tissues in humans: connective, muscle, epithelial, and nervous. **Parazoa** refers to animals without tissues, and **metazoa** includes animals with tissues. Porifera (or sponges) are parazoans **(B).** Only metazoans evolved symmetry, body cavities, and segmentation (although not all of them have all of these features).

Metazoans evolved two different types of symmetry. Animals with **radial symmetry (C)** resemble a pie, where each cross section of a pie piece looks the same. An example is an apple or Cnidarians such as jelly fish **(D).** Most animals with radial symmetry have less efficient locomotion and did not evolve body cavities or segmentation. **Bilateral symmetry (E)** is the same as plane symmetry, in which the organism has an internal line of symmetry and two halves that are mirror images. Humans, turtles, cats, and butterflies are all examples. Bilateral symmetry is associated with more efficient locomotion.

Some organisms with bilateral symmetry evolved a **body cavity,** a fluid- or air-filled space located between the digestive tract and the outer body wall. These animals are called **coelomates.** For example, humans have a **cranial cavity (F),** a **vertebral cavity (G),** a **thoracic cavity (H),** an **abdominal cavity (I),** and a **pelvic cavity (J).** A body cavity allows for more specialized organs and structures, and better storage of food, eggs, and sperm. **Acoelomates** do not have a body cavity and instead have a solid body. Platyhelminthes (or flatworms) are acoelomates **(K)** and so are parazoans and metazoans with radial symmetry. This is because the presence of a body cavity did not evolve in parazoans and metazoans with radial symmetry. **Pseudocoelomates** have a body cavity, but it develops differently from the body cavity of coelomates. Nematodes (or roundworms) are pseudocoelomates **(L).**

Coelomates can be divided into two distinct lines of evolution: protostomes and deuterostomes. Protostomes and deuterostomes develop differently when the single-celled zygote (formed right after fertilization) is becoming a morula (a solid ball of up to 16 cells). **Protostomes (M)** undergo spiral and determinate cleavage. This means that as the morula is forming, cell fate is decided early on and the cells in the morula become different from their neighbors (small vs. large). **Deuterostomes (N)** undergo radial and indeterminate cleavage, where the cells don't become specialized and instead all stay the same.

Segmentation refers to the division of some animal body plans into a series of repetitive segments. For example, vertebrates have a segmented vertebral column. Segmentation of the body plan is important because it allows different regions of the body to develop for different uses, and allows for better locomotion. Some protostomes, such as **annelids** (or segmented worms, **O**) and **arthropods (P)** (including arachnids, millipedes, centipedes, insects and crustaceans, evolved a segmented body plan. For example, in earthworms, each ring is a distinct segment. Lobsters and insects have developed specialized appendages on many segments. Other protostomes, such as **mollusks (Q),** do not have segmented bodies. **Chordates (R)** (including vertebrates such as fish, amphibians, reptiles, birds and mammals, evolved segmented body plans, whereas **echinoderms (S)** did not. For example, fishes exhibit segmentation in the muscles and backbone.

Animal Ancestor A ❍
Porifera B ❍
Radial Symmetry C ❍
Cnidaria D ❍
Bilateral Symmetry E ❍
Cranial Cavity F ❍
Vertebral Cavity G ❍
Thoracic Cavity H ❍
Abdominal Cavity I ❍
Pelvic Cavity J ❍
Platyhelminthes K ❍
Nematoda L ❍
Protostomes M ❍
Deuterostomes N ❍
Annelida O ❍
Arthropoda P ❍
Mollusca Q ❍
Chordata R ❍
Echinoderms S ❍

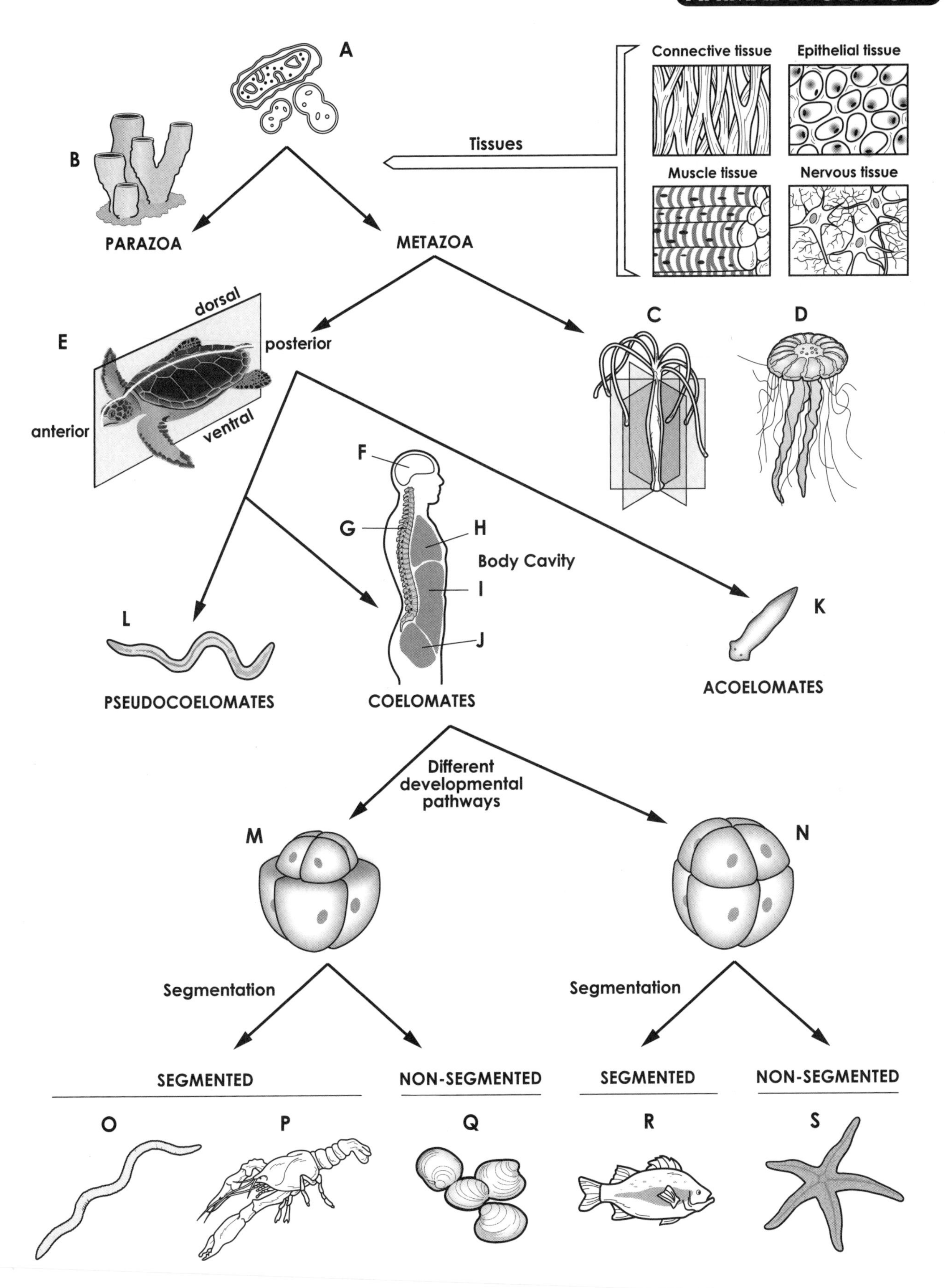
A
B
PARAZOA
METAZOA
Tissues
Connective tissue
Epithelial tissue
Muscle tissue
Nervous tissue
dorsal
E
posterior
anterior
ventral
C
D
F
G
H
Body Cavity
I
J
K
L
PSEUDOCOELOMATES
COELOMATES
ACOELOMATES
Different developmental pathways
M
N
Segmentation
Segmentation
SEGMENTED
NON-SEGMENTED
SEGMENTED
NON-SEGMENTED
O
P
Q
R
S

CHAPTER 9:

HUMAN BIOLOGY

THE INTEGUMENT (SKIN) AND DERIVATIVES

The integumentary system is composed of the integument (the skin) and its derivatives, including the hairs, sweat glands, and oil glands. The skin provides protection to the body is the largest body organ.

In looking over the plate, you will notice that we present a section of skin that includes hairs, glands, and other structures. As you begin your study of the integumentary system, prepare to use light and pale colors, because many tissues are detailed.

Structurally, the skin is composed of three parts. At its surface is the **epidermis (A),** which is outlined by a bracket. The next layer is made up of connective tissue and is called the **dermis (B),** and there is an even deeper layer called the **hypodermis (C).**

Now take a look at the detailed view of the epidermis, in the lower right-hand corner. The most superficial layer of the skin is the **stratum corneum (A_1).** This is a layer of flat, dead cells that are filled with the protein keratin. This layer protects against heat, pathogenic microorganisms, chemicals, and light. The next layer down is the **stratum lucidum (A_2).** Clear, flat cells that contain a prekeratin substance called eleidin are found here. The layer exists primarily in the palms of the hands and soles of the feet.

The next layer of the epidermis is the **stratum granulosum (A_3),** which is made up of cells that contain the substance keratohyalin. Later, this material will become keratin. The next layer is very deep, and is called the **stratum spinosum (A_4).** Keratin is produced in many of the cells of the stratum spinosum.

The deepest layer is the **stratum basale (A_5).** It is a single mixed layer of cube-shaped and tall cells, which undergo mitosis to become the cells of the more superficial layers. The layer is also called the statum germinativum.

We will now focus on the dermis and point out some of its important structures. The tissues in this layer provide protection, sensation, and immunity to disease. Continue your coloring as you read below.

The dermis contains collagen fibers as well as various types of cells. The most superficial region of dermis is the papillary region, and its shallow projections can be seen projecting into the epidermis. The remainder of the dermis is called the dermal layer.

Within the dermal layer are a number of **sebaceous glands (D).** These glands secrete an oily substance called sebum and are generally connected to hair follicles, as the plate indicates. Other glands in the dermis are the **sweat glands (E),** which are also called the sudoriferous glands. These glands deliver watery secretions (sweat) to **sweat gland ducts (E_1),** which lead to **sweat gland pores (E_2).** A spot of color should be used to trace the pores. Sweat delivers metabolic waste products to the skin surface for removal, and also helps regulate body temperature.

We now focus on the hair. Hair provides the skin with protection and decreases loss of body heat. Its color is primarily due to the pigment melanin. As you read about the hair fibers below, locate and color their parts in the plate.

Hairs are epidermal growths that vary in amount and texture throughout the body surface. The **hairs (F)** in the plate should be colored at the skin's surface.

The part of the hair that projects above the body surface is called the **shaft (F_1).** The portion that penetrates into the dermis is the **root (F_2).** The root of the hair is covered by the **root sheath (F_3),** which is a continuation of the epidermis, as the plate indicates. At the base of the hair follicle is the enlarged hair **bulb (F_4).** An indentation called the **papilla (F_5)** contains connective tissues and blood vessels that provide nourishment to the hair. At the side of the hair follicle is a section of specialized smooth muscle called the **arrector pili (F_6).** This muscle contracts during stress and pulls the hair into an upright position.

The plate closes with a brief look at the nerve receptors in the dermis and structures of the hypodermis. Complete your coloring as you read the paragraphs below.

Many different types of nerve receptors are located within the dermis; one is the **Pacinian corpuscle (G_1).** This nerve receptor detects vibrations and heavy touch sensations and sends impulses to the brain, while another type of receptor, called **Meissner's corpuscle (G_2),** detects light touch sensations.

There are also a number of nerves in the hypodermis, which is the site of the blood supply of the integumentary system. An **artery (H)** carries blood to the skin, and a **vein (I)** carries blood away. Red and blue colors may be used for these structures, respectively. Finally, the **fat tissue (J)** in the hypodermis provides support and cushioning to the skin.

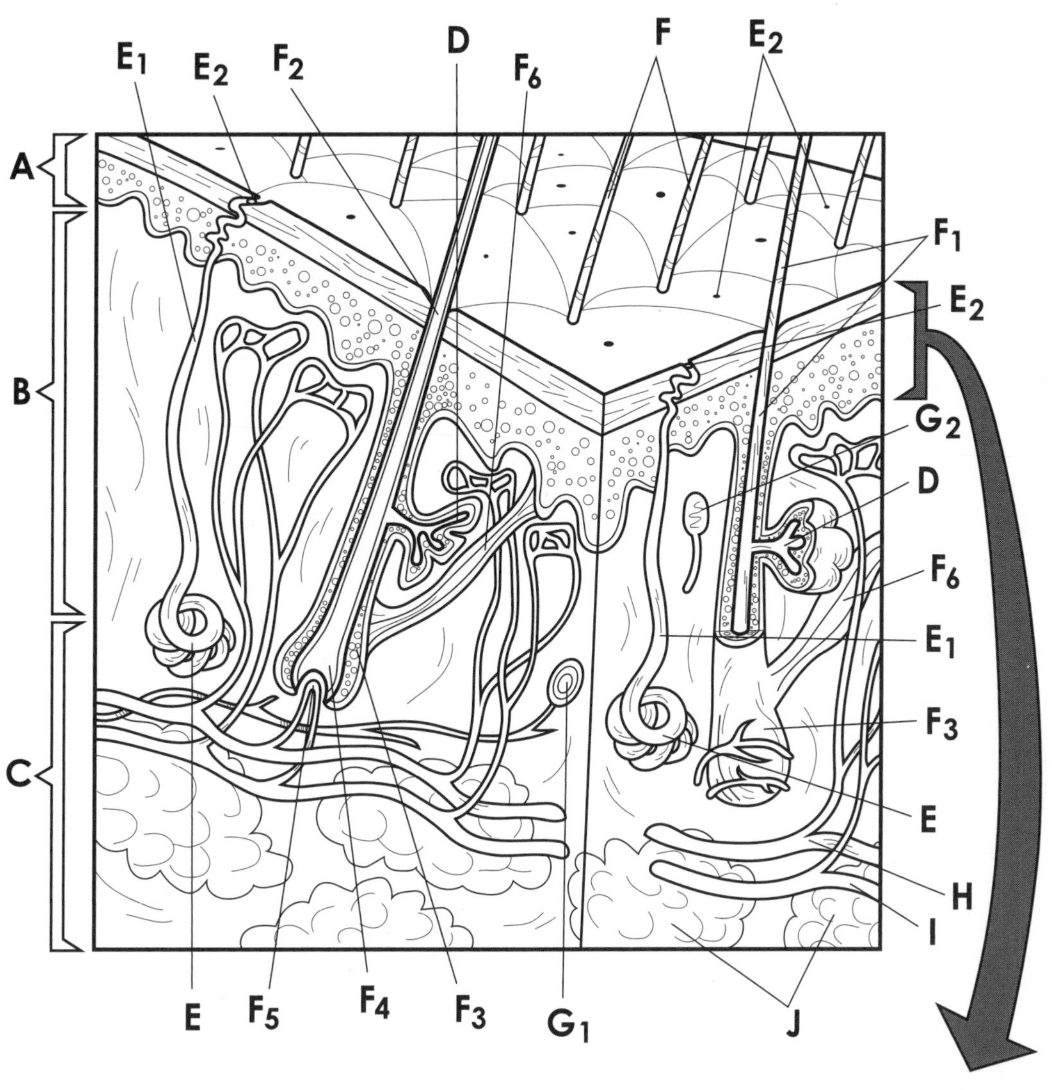

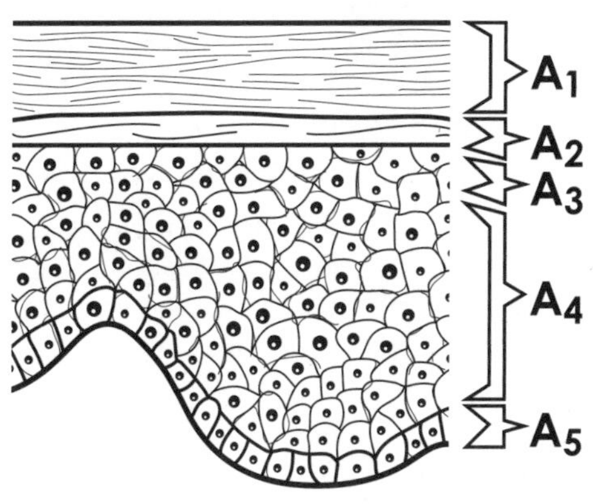

Epidermis	A	❍
Stratum Corneum	A_1	❍
Stratum Lucidum	A_2	❍
Stratum Granulosum	A_3	❍
Stratum Spinosum	A_4	❍
Stratum Basale	A_5	❍
Dermis	B	❍
Hypodermis	C	❍
Sebaceous Glands	D	❍
Sweat Glands	E	❍
Sweat Gland Ducts	E_1	❍
Sweat Gland Pores	E_2	❍
Hair	F	❍
Hair Shaft	F_1	❍
Root	F_2	❍
Root Sheath	F_3	❍
Bulb	F_4	❍
Papilla	F_5	❍
Arrector Pilius	F_6	❍
Pacinian Corpuscle	G_1	❍
Meissner's Corpuscle	G_2	❍
Artery	H	❍
Vein	I	❍
Fat Tissue	J	❍

THE SKELETAL SYSTEM

The human skeleton consists of 206 bones that differ in size, shape, weight, and composition. This diversity is related to the myriad structural and mechanical functions of the skeleton, which include supporting the body, protecting the body cavities, acting as levers for muscle activity, and providing a site for blood cell development.

The skeletal system is divided into two major parts: the axial skeleton, which is composed of the skull, vertebral column, sternum, and ribs; and the appendicular skeleton, which is composed of the upper and lower extremities and the supporting girdles.

Looking over the plate, you will note that it contains an anterior (front) view of the skeleton with the palms facing forward. As you read about the skeletal system, color the appropriate bones in the plate. There may be some overlapping, and pale colors are suggested for these areas.

The first structure of the axial skeleton that we'll talk about is the skull. This structure houses the brain and is the location of many sensory organs. The two main features of the skull are the **cranium (A)** and **face (B).** The skull contains 14 of these bones. The only bone that's not attached directly to the other bones of the skull is the lower jaw bone, which is called the **mandible (C).**

The skull and upper torso of the body are supported by another main component of the axial skeleton, the **vertebral column (G).** There are 31 bones in this column, which extends along the back of the body and connects to the thoracic cage. At the front, the thoracic cage is made up of a three-part bone called the **sternum (E$_1$)** and a set of 12 pairs of **ribs (E$_2$)** that connect the sternum to the vertebral column.

Having examined the axial skeleton, we now move to the appendicular skeleton and look at some of its bones. Continue reading, and as you encounter the bones in the text, color them in the plate.

The **upper extremity (F)** of the axial skeleton is composed of the pectoral girdle and arm bones. The **pectoral girdle (D)** is outlined by a bracket, which you should color. It contains two bones: the collar bone or **clavicle (D$_1$)** in the front of the body, and a flat, triangular bone called the **scapula (D$_2$),** which is located at the back of the body.

Connecting to the pectoral girdle is the upper arm bone, the **humerus (F$_1$).** The two lower arm bones that connect with the humerus are the **radius (F$_2$)** and the **ulna (F$_3$).** The wrist bones are called **carpals (F$_4$),** the hand bones are **metacarpals (F$_5$),** and the finger bones are **phalanges (F$_6$).**

At the lower portion of the body is the **pelvic girdle (H),** which is indicated by the bracket. This bone appears to be singular, but it is actually made up of three fused bones called the ilium, ischium, and pubis. Connecting to the pelvic girdle is the lower extremity. It consists of the thigh bone, called the **femur (I$_1$),** the kneecap, or **patella (I$_2$),** and two lower leg bones, the **tibia (I$_3$)** and the **fibula (I$_4$).** The ankle contains the **tarsals (I$_5$),** and the foot bones are **metatarsals (I$_6$).** The toe bones are the **phalanges (I$_7$).** This completes our brief sketch of the appendicular skeleton.

We will conclude by mentioning the five different types of bones that make up the skeletal system. Take a look at the detailed views on the right. Dark colors should be used for these bones.

Bones are classified according to their function in the body as well as their shape. For instance, one example of a **flat bone (J)** would be a bone of the skull. These bones are thin and serve to protect the brain; the scapula and ribs are other examples of flat bones. An **irregular bone (K)** is typified by the vertebra of the spinal column. Irregular bones are characterized by numerous extensions, and muscles often attach to them.

One example of a **sesamoid bone (L)** is the patella. Sesamoid bones are small and are usually embedded in tendons. They are used to protect the integrity of the tendon. **Long bones (M)** are used for movement. In the leg, for example, the femur acts as an attachment point for the muscles, and as the muscles contract, the bone moves.

The last bone we will consider is the **short bone (N).** Short bones have similar dimensions but irregular shapes and are found in the wrists and ankles (for example, carpals and tarsals).

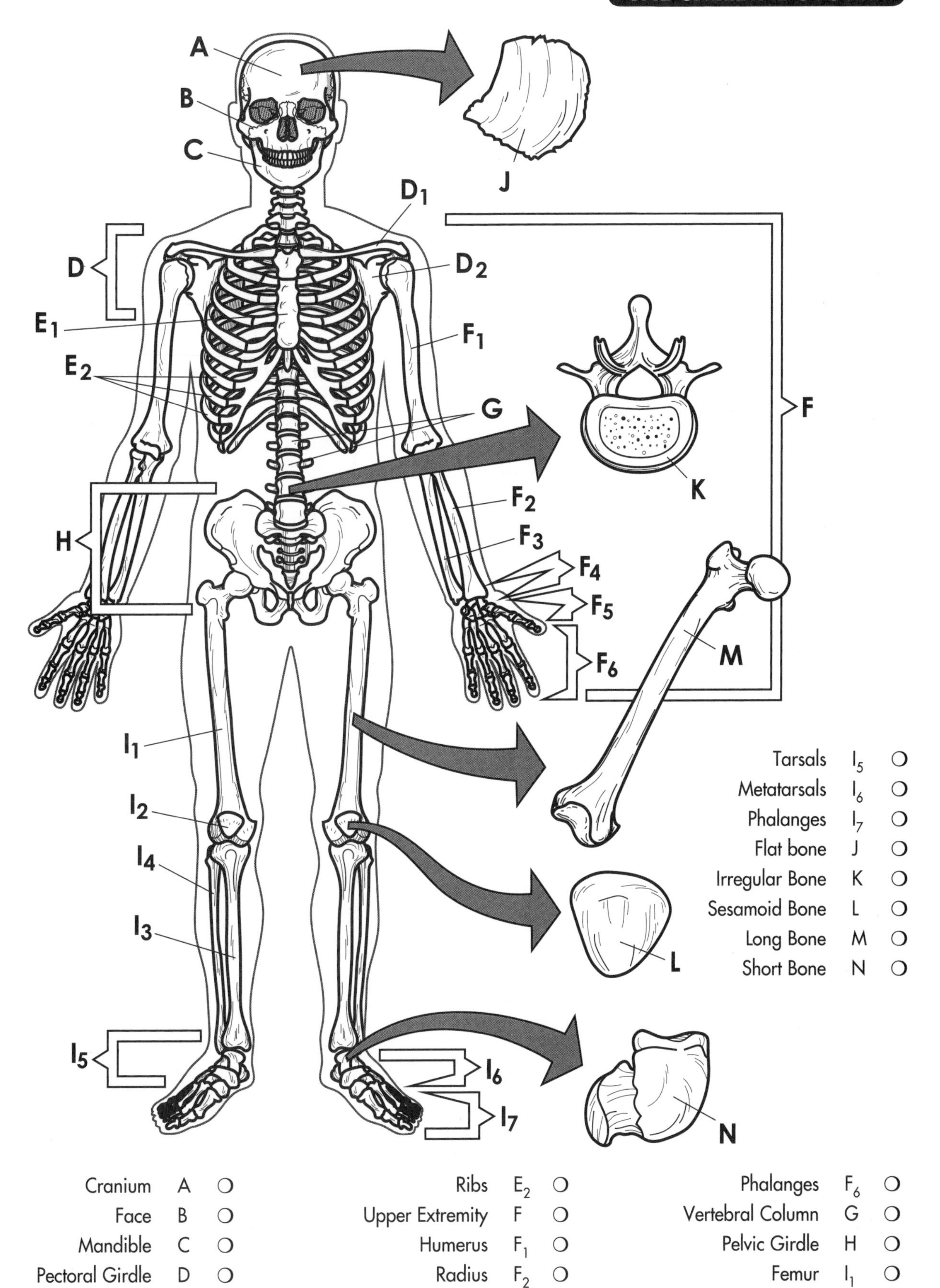

Tarsals	I_5	❍
Metatarsals	I_6	❍
Phalanges	I_7	❍
Flat bone	J	❍
Irregular Bone	K	❍
Sesamoid Bone	L	❍
Long Bone	M	❍
Short Bone	N	❍

Cranium	A	❍
Face	B	❍
Mandible	C	❍
Pectoral Girdle	D	❍
Clavicle	D_1	❍
Scapula	D_2	❍
Sternum	E_1	❍
Ribs	E_2	❍
Upper Extremity	F	❍
Humerus	F_1	❍
Radius	F_2	❍
Ulna	F_3	❍
Carpals	F_4	❍
Metacarpals	F_5	❍
Phalanges	F_6	❍
Vertebral Column	G	❍
Pelvic Girdle	H	❍
Femur	I_1	❍
Patella	I_2	❍
Tibia	I_3	❍
Fibula	I_4	❍

THE NERVOUS SYSTEM

The nervous system enables the body to adjust to changes in the outside environment and within the body. Sensing stimuli and conveying them to the brain and spinal cord, the nervous system makes analysis and coordination by the body possible. Messages from the brain are then conveyed via the nerves to the glands and muscles.

This plate lays out the general pattern of the nervous system. We will point out the major divisions and subdivisions of the system and explain their various functions.

Notice that we show the nervous system and several organ systems that are associated with it, and affected by it. We use brackets to set apart the major divisions of the nervous system. We also use subscript numbers to indicate related parts of the system and small letters to indicate organs that are associated with, but not an integral part of, the nervous system.

The nervous system is a single, unified network of communications, but on an anatomical basis, it is divided into two primary portions. The first portion we'll mention is the **central nervous system (A),** or CNS, and the second is the **peripheral nervous system (B),** or PNS. The brackets should be colored in bold colors.

The two key components of the central nervous system are the **brain (C)** and the **spinal cord (D).** The spinal cord is a continuation of the stem of the brain, so you should use the same color for them.

The brain and spinal cord constitute the central control system of the body. The tissue of these organs receives and interprets stimuli and then dispatches impulses to glands and muscles. Higher mental faculties are centered in the brain, while many automatic reflex actions take place in the spinal cord.

We now focus on the second portion of the nervous system, the peripheral nervous system, which has two major divisions and several subdivisions. Bold colors may be used to indicate the pathway of nerves, and the organs should be colored in light colors to preserve their detail.

The nerves associated with the brain and spinal cord make up the peripheral nervous system; they allow the brain and spinal cord to communicate with the remainder of the body.

There are two major divisions of the peripheral nervous system. The first is the sensory division. Nerves in this division transmit impulses from the various organs and from the surface of the body. Nerve impulses that carry impulses from body organs are called **visceral sensory nerves (E_1);** you should now color the visceral sensory nerve that leads from the **heart (a)** to the CNS. Another component of the sensory division is the **somatic sensory nerves (E_2),** which transmit nerve impulses from the body surface. A **skin section (b)** is also shown in the plate, and we recommend that you color it a bold color.

The second major division of the peripheral nervous system is the motor division, which is a system of nerve fibers that carry impulses from the CNS to the skeletal muscle. A **somatic motor nerve (F_1)** is shown carrying impulses to the **skeletal muscle (c).**

The second subdivision of the motor division is the autonomic subdivision. This is sometimes called the autonomic nervous system, and it has two parts. The sympathetic nervous system transmits impulses that stimulate organs. A **sympathetic nerve (F_2)** is shown extending to the heart from the spinal cord. The second part is called the parasympathetic nervous system. We show a **parasympathetic nerve (F_3)** that extends to the heart from the spinal cord. The activity of parasympathetic nerves opposes that of sympathetic nerves.

Finally, let's take a brief look at the tissue of the brain and the organization of the nerve cells. As you continue to read, locate the structures in the plate and color them.

The functional cells of the nervous system are called neurons, or nerve cells. In the plate, we show a small section of brain tissue. Toward the surface of the brain is the area made up of **gray matter (G),** and within this gray matter are **cell bodies (I_1)** of neurons. The next layer in from the gray matter is **white matter (H).** White matter consists primarily of extensions of the cell bodies called **axons (I_2).** We will talk a little about how nerve impulses are transmitted in the next plate.

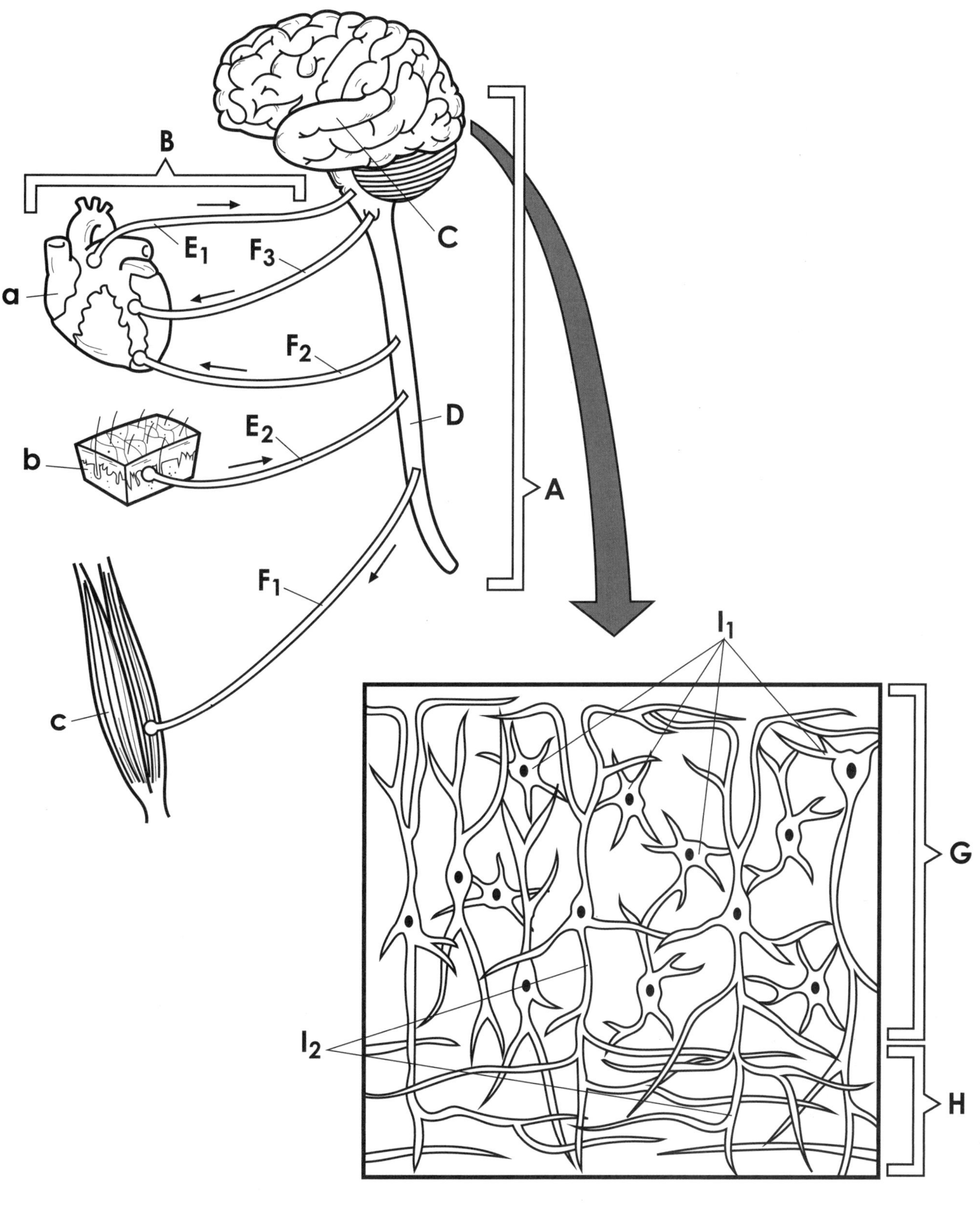

Central Nervous System	A	❍	Somatic Motor Nerve	F_1	❍	Axons	I_2	❍
Peripheral Nervous System	B	❍	Sympathetic Nerve	F_2	❍	Heart	a	❍
Brain	C	❍	Parasympathetic Nerve	F_3	❍	Skin Section	b	❍
Spinal Cord	D	❍	Gray Matter	G	❍	Skeletal Muscle	c	❍
Visceral Sensory Nerves	E_1	❍	White Matter	H	❍			
Somatic Sensory Nerves	E_2	❍	Cell Bodies	I_1	❍			

THE NERVE IMPULSES

The nervous system is responsible for directing the complex processes that take place in the body. It also links the body to the external environment and permits us to see, hear, taste, feel, and respond to stimuli.

Neurons are the cells of the nervous system and are specialized to receive and transmit information. As the structural and functional units of the nervous system, these cells are distinguished by their unique structure. In this plate, we study the structure of the neuron and the mechanism by which a nerve impulse is generated and propagated.

The plate shows three diagrams: a diagram of a nerve cell, a diagram of a neuron at rest, and a diagram of the neuron during propagation of the action potential.

We begin the plate with our diagram of the neuron. This specialized cell bears several distinguishing characteristics. The **cell body (A)** is the main portion of the cell, and it has a **nucleus (B)** and other cellular organelles.

At one end of the neuron is a collection of extensions called **dendrites (C).** In a multipolar neuron, nerve impulses enter the cell body through the dendrites. Extending away from the cell body is a long extension called the **axon (D).** The nerve impulse sweeps down the axon from the cell body, as the arrows indicate. At the end of the axon are thousands of microscopic branches called **axon terminals (E).** From these terminals, neurotransmitters are released, and transmit the nerve impulse to an adjoining neuron, muscle, or gland.

In many neurons, the axon is surrounded by a **myelin sheath (G),** made from either **Schwann cells (F)** in the peripheral nervous system, or oligodendrocytes in the central nervous system. These cells wrap around axons. This insulates the axon, which speeds up nerve impulses. Between myelin sheath cells are gaps called **Nodes of Ranvier (H).** Schwann cells and oligodendrocytes are examples of glial cells. Glial cells are specialized, non-neuronal cells that typically provide structural and metabolic support to neurons. Glia maintain a resting membrane potential but do not generate action potentials.

Having discussed the structure of the neuron, we are ready to examine its activity. We begin by focusing on the diagram entitled Resting Potential. Continue coloring as before.

The nervous system coordinates stimulus and response through neural impulses. A neuron at rest is not transmitting an impulse and is said to have a resting potential. In neurons at rest, imbalances in electrical charges exist on either side of the cell membrane. **Sodium ions (I)** are actively pumped to the exterior by the process of active transport, and the excess of sodium ions outside the cell gives it a positive charge with respect to the cell cytosol. The number of **potassium ions (J)** in a resting cell is approximately equal inside and outside, but other negatively charged ions accumulate in the cytoplasm, so that the outside of the cell bears a more positive charge than the inside. There is polarity across the cell membrane.

We now return to the axon to see what happens when a nerve impulse arises. Focus on the diagram entitled Action Potential. Continue to color the diagram as you read below.

Nerve impulses are also called action potentials. Action potentials begin when an electrical, chemical, or mechanical stimulus alters the structure of the cell membrane, which allows sodium ions to enter. The sodium gates, which are ion channels, open, and as sodium ions rush into the cytoplasm, the membrane loses its polarity; it undergoes depolarization.

In the first diagram, we show a group of **inrushing sodium ions (M),** indicated by the arrows. A stimulus has arrived and altered the membrane structure, which causes the propagation of an **action potential (nerve impulse) (L).** The horizontal arrow shows the direction of this action potential. A momentary reversal of polarity around the cell membrane takes place, as the inside of the axon becomes more positively charged.

The action potential now depolarizes the adjacent area of the membrane, and in the second diagram the action potential is seen further to the right. In a chain reaction, the next area undergoes depolarization, as the third diagram shows. Sodium ions continue to rush into the cytoplasm of the axon at adjacent areas, and a wave of depolarization sweeps down the axon of the neuron.

After the action potential has passed, the membrane repolarizes, the sodium gates close, and the potassium gates open, which allows potassium ions to move out of the cytoplasm. In the third diagram we see **outrushing potassium ions (N),** which return the external area to a positive state (repolarizing the membrane) so that another nerve impulse can occur. If no nerve impulse is immediately forthcoming, the cell pumps sodium to the exterior to reestablish the conditions seen in the resting potential. The entire depolarization and repolarization of the neuron occurs in less than a millisecond.

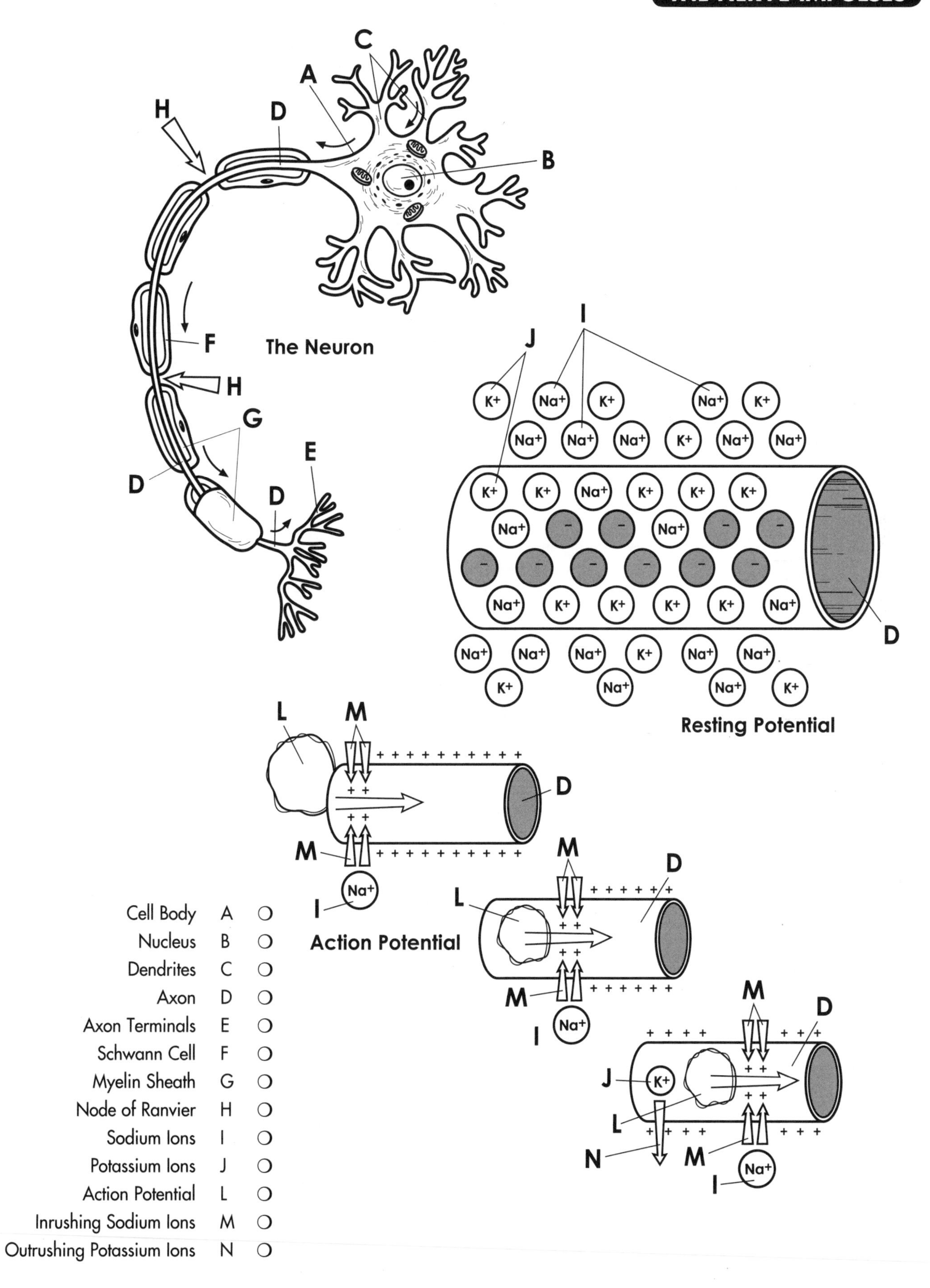
The Neuron
Resting Potential
Action Potential
Cell Body A
Nucleus B
Dendrites C
Axon D
Axon Terminals E
Schwann Cell F
Myelin Sheath G
Node of Ranvier H
Sodium Ions I
Potassium Ions J
Action Potential L
Inrushing Sodium Ions M
Outrushing Potassium Ions N

THE BRAIN

The brain is the center of human behavior and the main organ of the central nervous system. It is responsible for the functions of memory and understanding, among many other things. In this plate we will examine some of the characteristics of the human brain.

Note that this plate presents three views of the brain. We see the entire brain in the skull; we see a cross section of the brain in the skull; and we see the brain from below. You should use lighter shades when coloring to avoid obscuring the folds and details. As you read about the brain, locate the structures in all three views and color them.

We learn about our environment by means of signals that are received in the brain. Our responses are in the form of neural signals that allow us to conduct responsive activities such as talking and moving. The brain is the center of nervous activity in the body and is an exceedingly complex organ with many components.

By far the largest part of the human brain is the **cerebrum (A).** All conscious processes occur in the cerebrum, which is displayed in our diagram in the whole, sectioned, and inferior (from below) views. The surface of the cerebrum is greatly folded; the upward folds are called gyri and the downward grooves are called sulci. A series of fissures divides the cerebrum into two hemispheres, as you can see in the view from below.

A second major part of the brain is the **diencephalon (B),** which is indicated by a bracket. This portion of the brain surrounds an enlarged, fluid-filled space called the third ventricle. The diencephalon consists of the **thalamus (B_1)** and the **hypothalamus (B_2).** The thalamus contains paired masses of gray matter that are organized into bodies called nuclei, and it acts as a relay station for sensory impulses traveling to the cerebral cortex. The physiological equilibrium of the body (homeostasis) is regulated by the hypothalamus. In addition, hormones that are secreted by the pituitary gland are produced in the hypothalamus.

We will now discuss the two remaining regions of the brain and examine them briefly. Our review will be limited to the general functions that they perform. You may wish to select variations of the same color to indicate where areas of one structure begins and another ends.

The second largest portion of the human brain is the **cerebellum (C).** Seen in each of the three diagrams, the cerebellum has two hemispheres and many small surface folds. The cerebellum helps coordinate and control movements initiated by the cerebrum.

The last region we will mention is the **brain stem (D),** which is outlined by a bracket. The bracket may be colored in a bold color. The brain stem is continuous with the **spinal cord (E)** and variations of the same color should be used for them. The **midbrain (D_1)** is important because it contains fibers that transmit sensory impulses from the cerebral cortex back to the spinal cord. The **pons (D_2)** is also in the brain stem. The pons is a bridge-like structure that contains numerous fibers that carry signals between regions of the brain.

The portion of the brain that is continuous with the spinal cord is the **medulla oblongata (D_3).** This structure contains the gray matter that receives signals from the spinal cord. The cardiac center is located in the medulla; it consists of a mass of neurons that regulate heart rate. The medulla also contains other vasomotor centers that regulate the diameter of blood vessels. The depth and rate of breathing are regulated in the respiratory center, which is also in the medulla.

Various other structures may be seen in the plate. For example, the **pituitary gland (F)** is seen at the base of the brain. Also seen are the **corpus callosum (G)** and the **olfactory bulbs (H).** The corpus callosum is a mass of fibers that carry signals between the two cerebral hemispheres, and the olfactory bulb is associated with our understanding of smell. Various **cranial nerves (I)** may be seen at the base of the brain. These nerves bring nerve impulses from the senses for interpretation and return impulses to provoke appropriate responses.

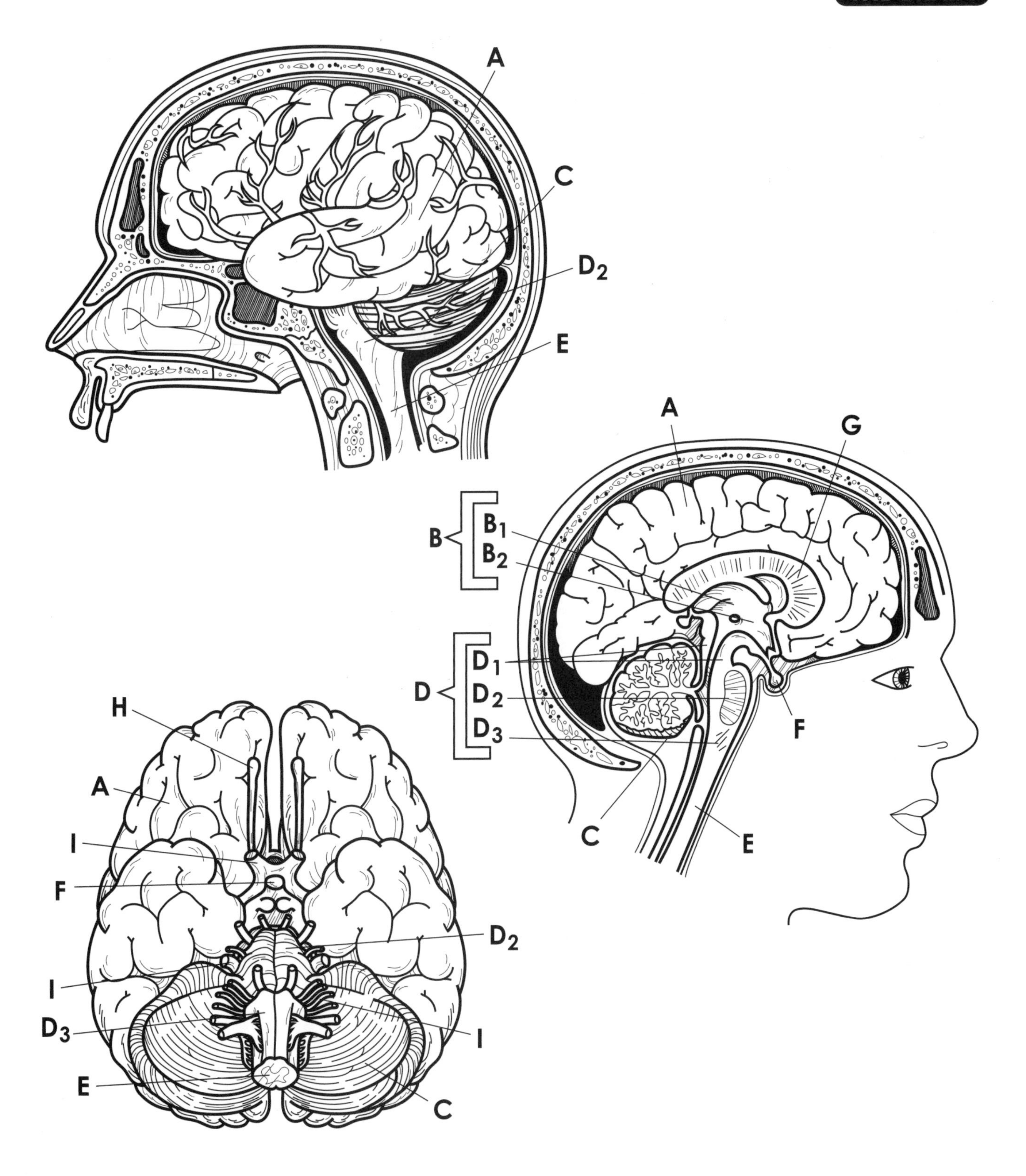

Cerebrum A ❍
Diencephalon B ❍
Thalamus B_1 ❍
Hypothalamus B_2 ❍
Cerebellum C ❍

Brain Stem D ❍
Midbrain D_1 ❍
Pons D_2 ❍
Medulla Oblongata D_3 ❍
Spinal Cord E ❍

Pituitary Gland F ❍
Corpus Callosum G ❍
Olfactory Bulbs H ❍
Cranial Nerves I ❍

THE EYE

Sight is one of the most dominant of the human senses. Over 70% of the body's receptors are the photosensitive cells of the eyes, and it has been estimated that a third of all the fibers that carry impulses to the central nervous system come from the eye.

In this plate, we will discuss the anatomy of the eye and show how some of its parts function. The structures diagramed in this plate contribute to vision directly or serve accessory purposes.

Notice that this is a diagram of a cross section through the eye. As you encounter the structures in the reading below, color them.

The wall of the eyeball consists of three layers that are often referred to as tunics. The first is the fibrous tunic, which contains the **sclera (A),** also known as the "white of the eye." The sclera gives the eyeball its shape, and can be seen around most of the eyeball's surface. A continuation of the sclera is the **cornea (B).** The cornea is a transparent structure that bulges out and contains no blood vessels; it helps focus light on the retina. The **limbus (C)** is the area where the cornea meets the sclera.

The middle layer of the eyeball is the vascular tunic, which contains many blood vessels and includes the **iris (D),** which is the colored part of the eye that can be seen through the cornea. Muscular movements within the iris cause it to open and close, increasing and decreasing the size of the opening to the eye, called the **pupil (E).** In this way, the iris regulates the amount of light entering the pupil.

Another portion of the vascular tunic is the **ciliary body (F).** This structure is continuous with the iris, and the same color or a close variation should be used for them both. The ciliary body contains the ciliary muscle, which controls eye movements. Surrounding the eyeball is the **choroid (G),** which contains an extensive capillary network that supplies blood to the retina. In the diagram, you can also see the **lens (H).**

We will now examine the chambers of the eyeball. We will also see the nerve layer of the eye and how it relates to the other structures. Continue your coloring as you read about these structures below.

The eye has three main chambers; the first is the **anterior chamber (I).** The anterior chamber is the space between the iris and the cornea. It contains a fluid material called aqueous humor. The **posterior chamber (J)** exists between suspensory ligaments and the iris; this chamber also contains aqueous humor.

The **vitreous chamber (K)** is quite large and contains a clear gelatinous mass known as the vitreous body. The vitreous body helps maintain the shape of the eye and gives support to the retina, and is sometimes called the vitreous humor.

We will finish with a brief examination of the pathway of light through the eye.

In order for vision to occur, a light image must be formed on the **retina (L),** and the light image must be converted into an action potential to be interpreted by the brain. The lens changes shape according to the distance of the object being viewed. This is because as the image is formed on the retina, the lens bends the light rays and focuses them behind itself, at a specific point called the **fovea (M).** The fovea contains photoreceptor cells called cones. Transduction of the light signal is performed by two types of photoreceptor cells, rods and cones; rods are found primarily toward the edge of the retina, and are absent in the fovea. Rods and cones are associated with bipolar cells, which in turn stimulate ganglion cells.

The cells of the retina form a network that comes together at the **optic disk (N),** which contains a blind spot, where there are no photoreceptors. The optic disk penetrates the wall of the eye and forms the **optic nerve (O),** which carries impulses to the brain.

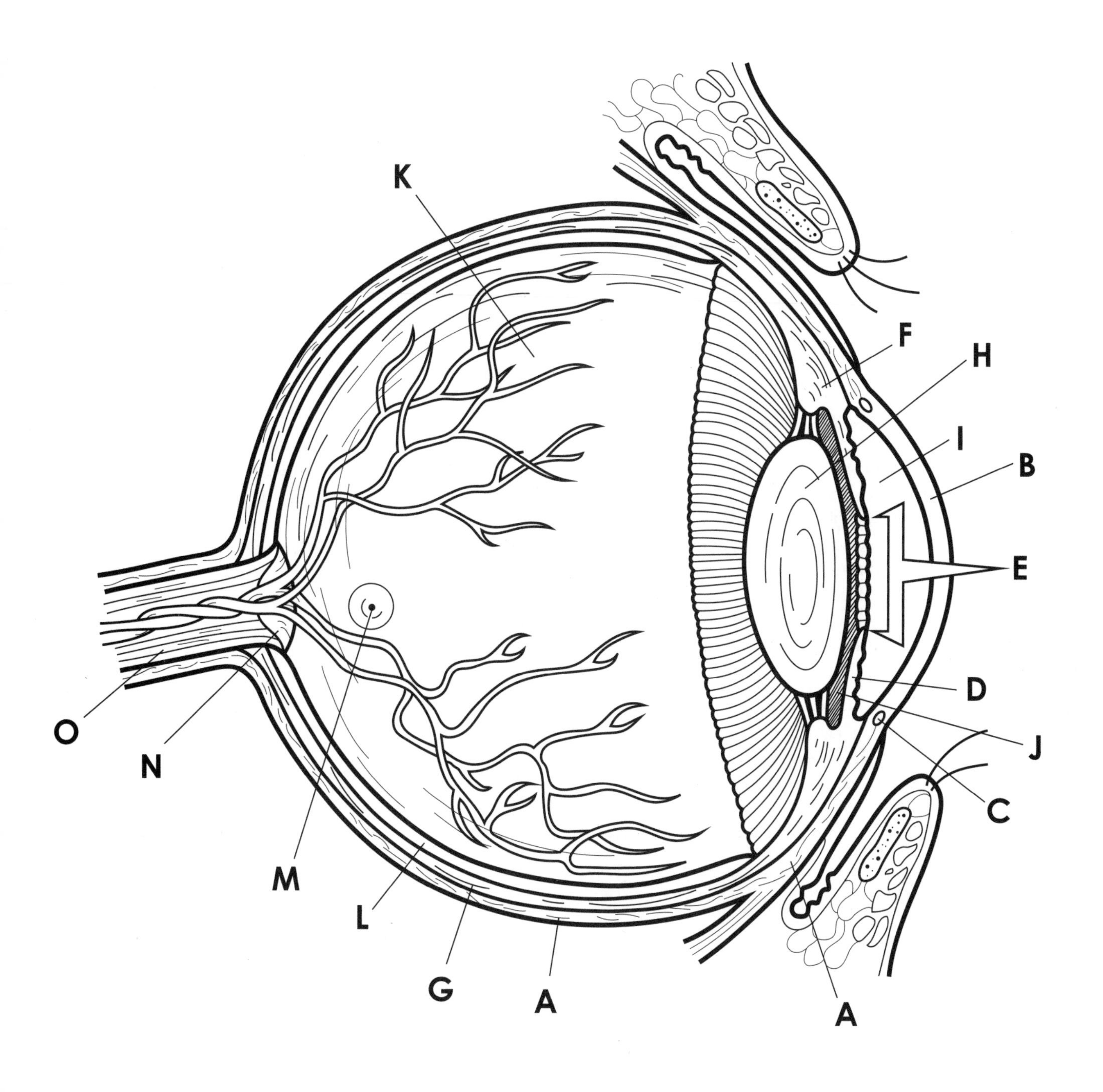

Sclera	A	❍	Ciliary Body	F	❍	Vitreous Chamber	K	❍
Cornea	B	❍	Choroid	G	❍	Retina	L	❍
Limbus	C	❍	Lens	H	❍	Fovea	M	❍
Iris	D	❍	Anterior Chamber	I	❍	Optic Disk	N	❍
Pupil	E	❍	Posterior Chamber	J	❍	Optic Nerve	O	❍

THE EAR

The ear is the organ of hearing and balance in the body. The structures of the ear translate vibrations in the air into vibrations in fluid, and then sensory impulses. The sensory impulses are interpreted at the cerebral cortex of the brain. The ear also contains receptors that act to maintain equilibrium; these are located at a different site than are the receptors for hearing.

This plate describes the general structure of the ear. Both internal and external structures are discussed, and we provide some insights into the hearing process.

Notice that this plate shows a cross section of the ear. The ear and its structures make up a complex organ system. As you read about the structures below, locate and color them in the diagram. We use subscript numbers to indicate relationships between various main structures.

The ear is generally discussed according to its three main regions: the **external ear (A),** the **middle ear (B),** and the **internal ear (C).** The brackets that encompass these regions should be colored in bold colors.

The outer ear consists of the visible portion known as the **pinna (D),** which you should color with a single, light color. Both the pinna and the **external auditory canal (D_1)** collect sound waves and channel them toward the middle ear. A light color is recommended for the canal. At the lowermost portion of the pinna is the familiar **earlobe (D_2)** known as the lobule. Its fatty tissue may be seen within.

We now move to the middle ear and see which structures are responsible for moving air vibrations along to the inner ear. Continue your coloring below as you read.

The partition between the external ear and the inner ear is the **tympanic membrane (E),** which is also known as the eardrum. Sound vibrations cause the membrane to vibrate, and this membrane also marks the opening to an air-filled space within the **temporal bone (F),** called the **tympanic cavity (F_1).** This is the site of the **Eustachian (auditory) tube (G).** The Eustachian tube leads to the nasopharynx and ensures that equal air pressures exist on both sides of the tympanic membrane (in the ear and atmosphere).

Three important landmarks of the middle ear are the **auditory ossicles (H).** They include the **malleus (H_1),** the **incus (H_2),** and the **stapes (H_3).** They vibrate in unison with the tympanic membrane and transmit vibrations to the membrane at the opening of the inner ear.

We will now discuss the inner ear. In the inner ear vibrations are converted into sensory impulses, and equilibrium is maintained as well.

The inner ear contains an intricate network of interconnecting chambers and passages. Its two main parts are the **bony labyrinth (I),** which is outlined by a bracket, and the membranous labyrinth inside it. Between the bony and membranous labyrinths, there is a vibration-conducting fluid called perilymph.

The bony labyrinth consists of three main parts. The first is the **vestibule (I_1),** the central chamber that contains fluid-filled sacs that are associated with the sense of equilibrium. As the fluid moves, equilibrium is established. The second part is made up of the three **semicircular canals (I_2),** which are also filled. The semicircular canals contain tiny hairs that are stimulated by the movement of the head. The brain interprets the information it receives from these canals to determine the rate and direction of movement.

The third structure is the spiral, **coiled cochlea (I_3).** The cochlea is firmly connected to the stapes (H_3) by a membranous partition called the **oval window (I_4).** When the auditory ossicles (H) vibrate, the vibrations are transferred to perilymph within the cochlea and as this fluid moves about, the local nerve fibers are stimulated. Sensory impulses arise from this stimulation, and these are sent to the brain. The **round window (I_5)** relieves the pressure exerted at the oval window.

The sensory impulses are transferred through the **vestibulocochlear nerve (J)** to the brain for interpretation. The nerve thus detects sensations for both equilibrium and hearing.

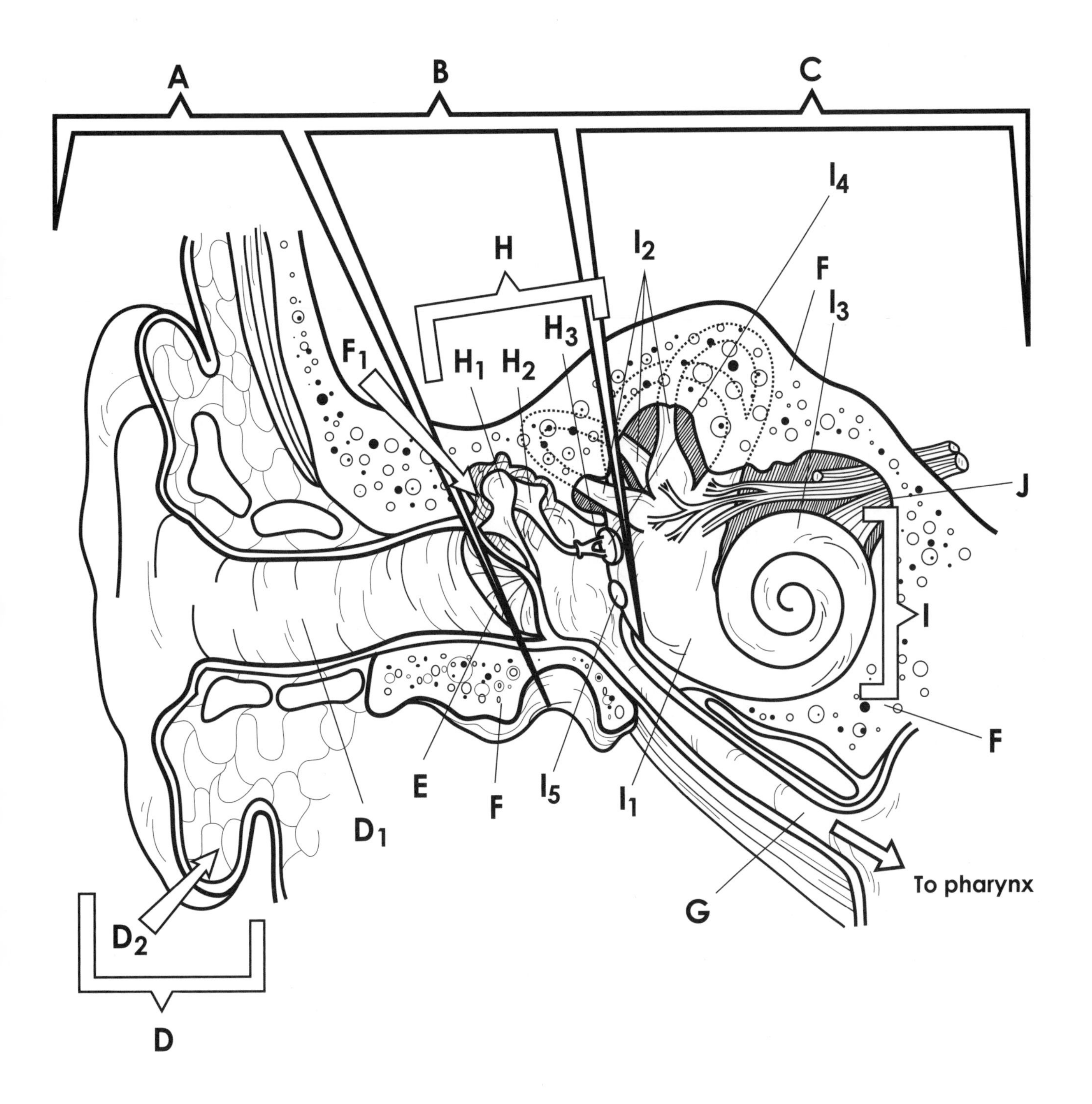

External Ear	A	❍
Middle Ear	B	❍
Internal Ear	C	❍
Pinna	D	❍
External Auditory Canal	D_1	❍
Earlobe (Lobule)	D_2	❍
Tympanic Membrane	E	❍

Temporal Bone	F	❍
Tympanic Cavity	F_1	❍
Eustachian (auditory) Tube	G	❍
Auditory Ossicles	H	❍
Malleus	H_1	❍
Incus	H_2	❍
Stapes	H_3	❍

Bony Labyrinth	I	❍
Vestibule	I_1	❍
Semicircular Canals	I_2	❍
Coiled Cochlea	I_3	❍
Oval Window	I_4	❍
Round Window	I_5	❍
Vestibulocochlear Nerve	J	❍

MUSCLE TYPES IN THE BODY

Muscle is the most abundant tissue in most animals. It allows for the movement of body parts by exerting force upon tendons, lends support to the body, and provides protection to the organs that lie beneath them. As you will see in the following plate, muscle tissue is made up of discrete cells that contract as a unit when stimulated. There are three types of muscle in animals: skeletal muscle, cardiac muscle, and smooth, or visceral, muscle. In this plate, we will review the location and composition of these three types.

Take a look at the plate, titled Muscle Types in the Body. You should focus your attention on the first detailed muscle type, skeletal muscle.

Skeletal muscle (A) makes the voluntary movement of body parts possible. These muscles are attached to bones by tendons, which are fibrous connective tissues. Skeletal muscle is also called striated muscle, because its overlapping filaments give it a striped appearance under the microscope. In the diagram, you can see the important elements of this section of the skeletal muscle: the **sarcomere (B),** the **myofibrils (C),** and the **nuclei (D).** We will discuss these structures in detail in the next plate.

The next type of muscle that we will discuss is cardiac muscle. Focus your attention on the detailed section of cardiac muscle shown in the plate.

Cardiac muscle (E) is what makes up the walls of the heart, which contracts involuntarily. One thing that differentiates cardiac muscle cells from other muscle cells is the presence of **intercalated discs (F).** Intercalated discs are found at the ends of cardiac muscle cells and, in effect, join them. They relay the nerve impulses that cause contraction from cell to cell. Cardiac muscle is striated, like skeletal muscle. In the art, you can also see the cells' **nuclei (G).**

The final muscle type that we will discuss is smooth, or visceral, muscle. Focus your attention on the third close-up in the art, and color the structures as you read the text.

Smooth, or **visceral, muscle (H)** lacks striations, and is found in the walls of the internal organs, namely the digestive tract, bladder, and arteries. These cells contract more slowly than do skeletal muscle cells, but they can contract for longer periods of time. Smooth muscle movements are involuntary, meaning that you cannot consciously control them, as you can control skeletal muscle. In the art, we have pointed out the **nucleus (I)** of one of these cells.

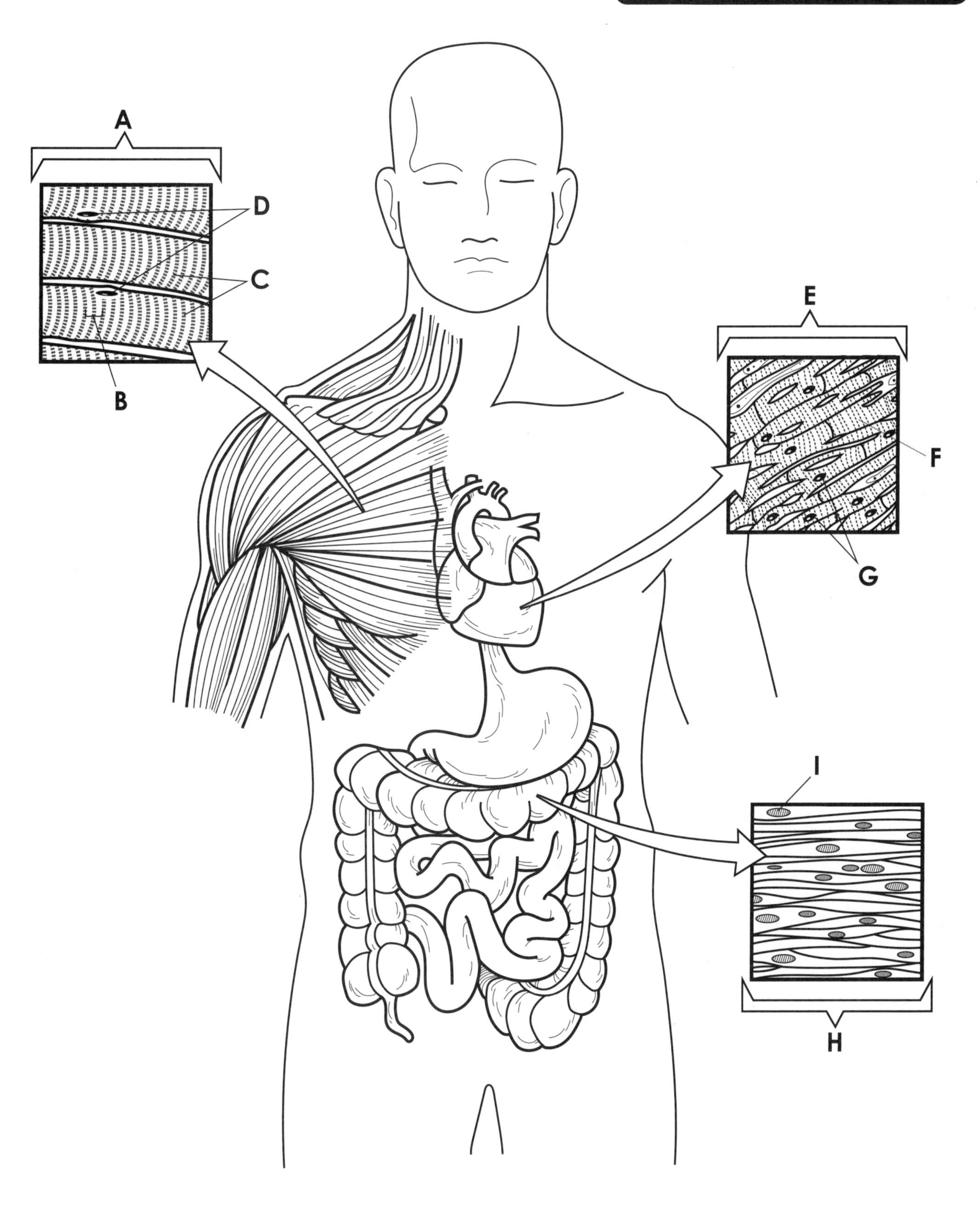

Skeletal Muscle	A ❍	Nuclei	D ❍	Nuclei	G ❍
Sarcomere	B ❍	Cardiac Muscle	E ❍	Visceral Muscle	H ❍
Myofibrils	C ❍	Intercalated Discs	F ❍	Nucleus	I ❍

MUSCLE CONTRACTION

The human body contains three types of muscle: skeletal muscle, which is mainly associated with the body trunk and movement; smooth muscle which is associated with organs; and cardiac muscle, which is associated with the heart. The movement of the body depends on the activity of skeletal muscle, and we consider that type of muscle here. Muscle tissue is distinguished from other tissues by its ability to contract and perform mechanical work. Now we will examine how contractions occur.

> In this plate, we see three diagrams that detail muscle fiber structure, the sliding filament, and the smaller details of muscle contraction.

The most abundant type of muscle is skeletal muscle, which is also called striated muscle because it has visible light and dark bands. In the first diagram, entitled Muscle Structure, we see a **bone (A)** with an attached muscle. With very few exceptions, all skeletal muscle is connected to some part of the skeleton, and **tendons (B),** which are composed of fibrous protein, connect muscle to bone.

Each **muscle (C)** is set of hundreds of fused cells called muscle fibers. Spots of color may be used to denote the **muscle fiber (D)** in the top diagram. In the next diagram, we single out one muscle fiber.

As we said, a muscle fiber is a muscle cell. Muscle cells are very long and have **nuclei (E)** that contain their genetic material. The muscle fiber (cell) is enclosed by a plasma membrane known as the **sarcolemma (F),** and the cytoplasm of a muscle cell is called the sarcoplasm.

In the diagram, you can see a number of rod-like filaments known as **myofibrils (G).** Each muscle cell contains up to 20 of these filaments in its sarcoplasm. A single myofibril (G) is extracted so that we can study it, and as you will see, a smaller unit called the **sarcomere (H)** is the unit of muscle responsible for contraction.

> We will now study the sarcomere and the process of muscle contraction, focusing on two types of filaments within the myofibril. Continue your reading while you color the plate, focusing on the third diagram entitled The Sliding Filament.

An ultramicroscopic view of the sarcomere reveals that each one is composed of two types of filaments. One type is **thin actin filaments (I),** and the other is **thick myosin filaments (J).** Bold colors should be used to trace over these filaments in the sarcomere shown in diagram 1. The filaments run parallel to one another and are composed of the proteins actin and myosin, respectively. Where the adjacent sarcomeres interweave, there is a dense black line called the **Z line (K).** A central zone called the **H zone (L)** exists between myosin filaments; there are no actin filaments in this zone, as the diagram shows. The thin actin filaments are anchored to the Z line, but the myosin filaments are not.

During muscle contraction, nerve impulses enter the muscle cell, stimulating actin filaments to slide toward one another along the myosin filaments. Arrows show this sliding in diagram 2. Notice that the distance between the Z lines has decreased and that the H zone is beginning to disappear. In diagram 3, the actin filaments continue to slide and soon overlap one another, obliterating the H zone. The attached Z lines are drawn toward one another, and the distance between them shortens while the filaments remain the same length. The muscle fiber has contracted.

When this process occurs simultaneously in thousands of sarcomeres in numerous muscle fibers, an entire muscle shortens and contracts. The contracting muscle pulls a bone, which creates movement.

> We now focus briefly on some of the details of contraction. Continue your reading below.

The action of sliding filaments was further elucidated by recent discoveries that explain how actin and myosin filaments operate. The actin filaments (I) consists of chains of globular proteins, which are indicated by the circular globules. Located alongside these chains of proteins is a protein called **tropomyosin (M).** At rest, tropomyosin prevents myosin from binding to actin by masking the **actin binding site (N)** on the globular actin molecules. Another protein called **troponin (O)** lies near the actin filament and helps stabilize the tropomyosin molecule.

Calcium ions (P) are released in the muscle fiber as a result of an action potential and the muscle contracts. Calcium ions unite with the troponin molecules, as the diagram shows, and this union causes the troponin molecules to shift position. The movement reveals the actin binding sites (N), and they unite with the **myosin heads (Q).** The myosin head then curves and pulls the actin filament. This is the power stroke that moves the actin filament across the stationary myosin filament. When nerve impulses cease, the bound calcium ions are released and the binding sites are once again covered; the muscle is now at rest.

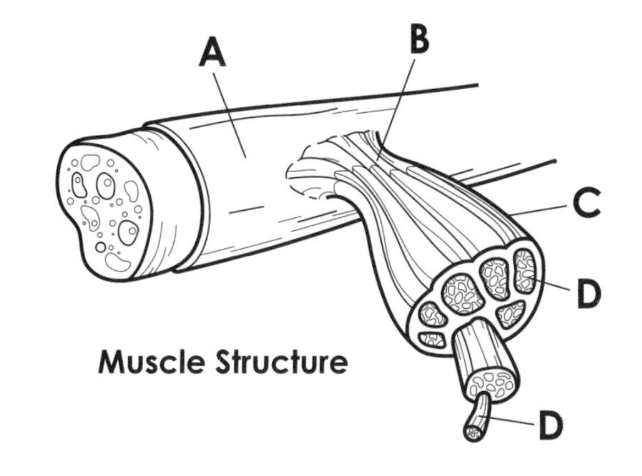

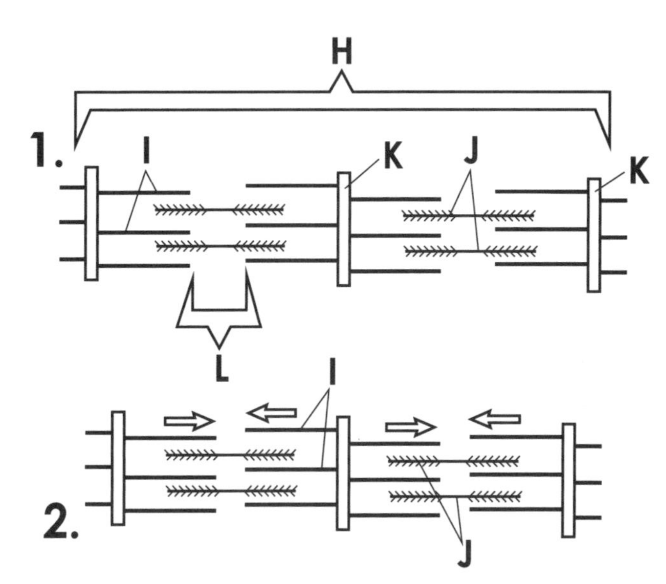

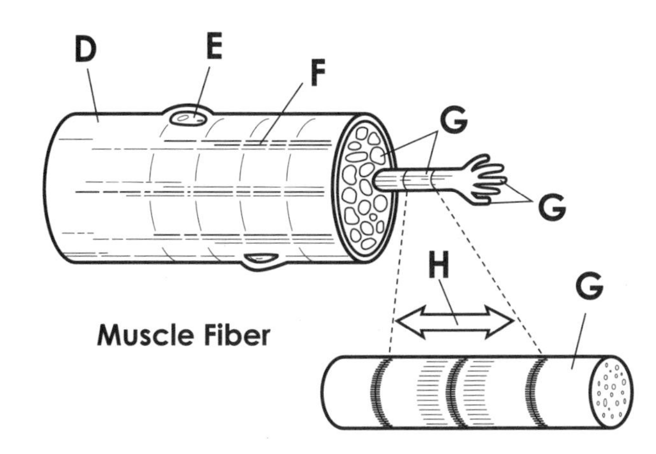

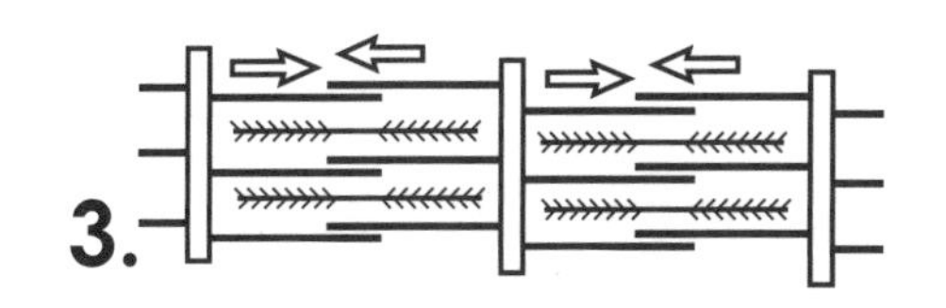

The Sliding Filament

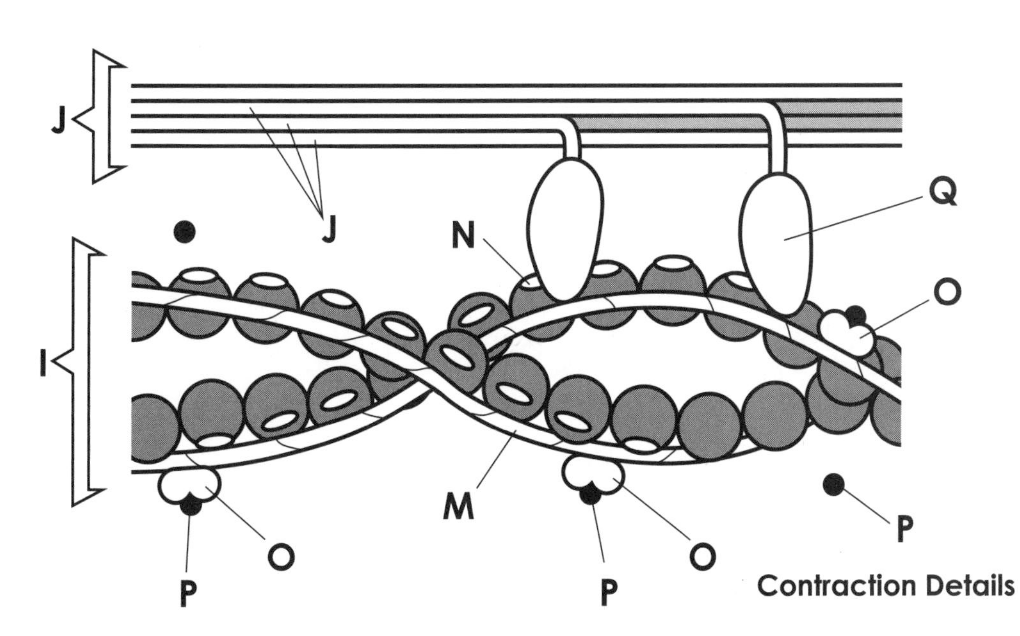

Contraction Details

Bone	A	❍	Myofibrils	G	❍	Tropomyosin	M	❍
Tendon	B	❍	Sarcomere	H	❍	Actin Binding Site	N	❍
Muscle	C	❍	Thin Actin Filaments	I	❍	Troponin	O	❍
Muscle Fiber	D	❍	Thick Myosin Filaments	J	❍	Calcium Ion	P	❍
Nucleus	E	❍	Z Line	K	❍	Myosin Head	Q	❍
Sarcolemma	F	❍	H Zone	L	❍			

THE ENDOCRINE SYSTEM

The endocrine glands secrete substances called hormones directly into the fluids of the body. (Exocrine glands secrete their products into tubes or ducts.) As a group, glands of the endocrine system help regulate metabolic processes, including rate of chemical reactions, the transport of substances across membranes, and water concentrations in the body. They also affect body development and growth. A hormone is a substance that is secreted by a cell and has an effect on a distant metabolic cell/tissue. Two mechanisms for hormone action are discussed in the next plate.

> Begin by looking over the plate and the various endocrine glands of the body. Most of the glands are large enough to be distinct, and we recommend that you use dark colors to color them. Start in the head and then work your way down the body.

We will begin our study of the endocrine glands with a brief examination of a pea-sized gland called the **pituitary gland (A).** The pituitary gland has anterior and posterior divisions, each of which secretes a number of hormones. Lying above the pituitary gland and connected to it by a stalk is the **hypothalamus (B).** The hormone-releasing cells in the hypothalamus are actually specialized neurons that are different from both other secretory cells and other nerve cells.

The third endocrine gland in this area is the **pineal gland (C).** The pineal gland is a small, oval structure that lies deep within the cerebral hemisphere, which is why it cannot be seen clearly in the plate. We suggest a spot of light color to indicate its general location. The pineal gland secretes melatonin. Varying light conditions outside the body appear to regulate its activity.

> We will now move to the neck and thorax region and briefly examine four endocrine glands.

Located just below the larynx on either side and in front of the trachea is the **thyroid gland (D).** As the plate shows, this gland consists of two large lobes that are connected by a broad isthmus. It secretes a number of hormones that affect metabolism in body cells. Located among the tissues of the thyroid gland are four tiny **parathyroid glands (E).** (These glands are located on the dorsal surface, but we show them here on the lateral surface to indicate their general location.) Hormones from these glands regulate calcium metabolism in the body.

The plate shows a rather large and prominent **thymus gland (F).** In the mature adult, the gland shown in the plate is atrophied. However, in very young individuals, the thymus is quite large. Hormones called thymosins that participate in immunity are believed to be synthesized by this gland.

The function of the **heart (G)** in blood circulation is well known, but its endocrine function is less appreciated. Fibers of cardiac muscle in the right atrium produce a hormone called atrial natriuretic peptide, which controls the release of a hormone from the posterior pituitary gland and is involved in the regulation of water levels in the body.

> In the final portion of this plate, we will examine the endocrine glands of the abdominal and pelvic cavity. In some cases, the organs have other functions beside endocrine functions. Continue your coloring as before.

Many of the digestive processes are controlled by hormones, such as gastrin and secretin, that are released by cells of the **digestive organs (H).** The **pancreas (I),** which is a digestive organ as well as an endocrine organ, has specialized cells that produce insulin and glucagon, both of which regulate the levels of glucose in the blood.

Lying on top of the kidneys are the **adrenal glands (J).** These pyramid-shaped glands are also known as suprarenal glands. Each gland has distinctive cortex and medulla regions, and numerous hormones are produced in both of these regions. Excretions of the **kidneys (K)** have a urinary function as well as an endocrine function. The endocrine cells of the kidney produce the hormones erythropoietin and renin, which are a part of the angiotensin system that regulates water balance in the body.

The reproductive organs produce numerous hormones as well as sex cells. The **testes (L)** produce hormones such as testosterone that regulate sperm production and induce the development of secondary male characteristics. The **ovaries (M)** produce hormones that induce the maturation of eggs and the growth of the reproductive structures. These hormones are called estrogens.

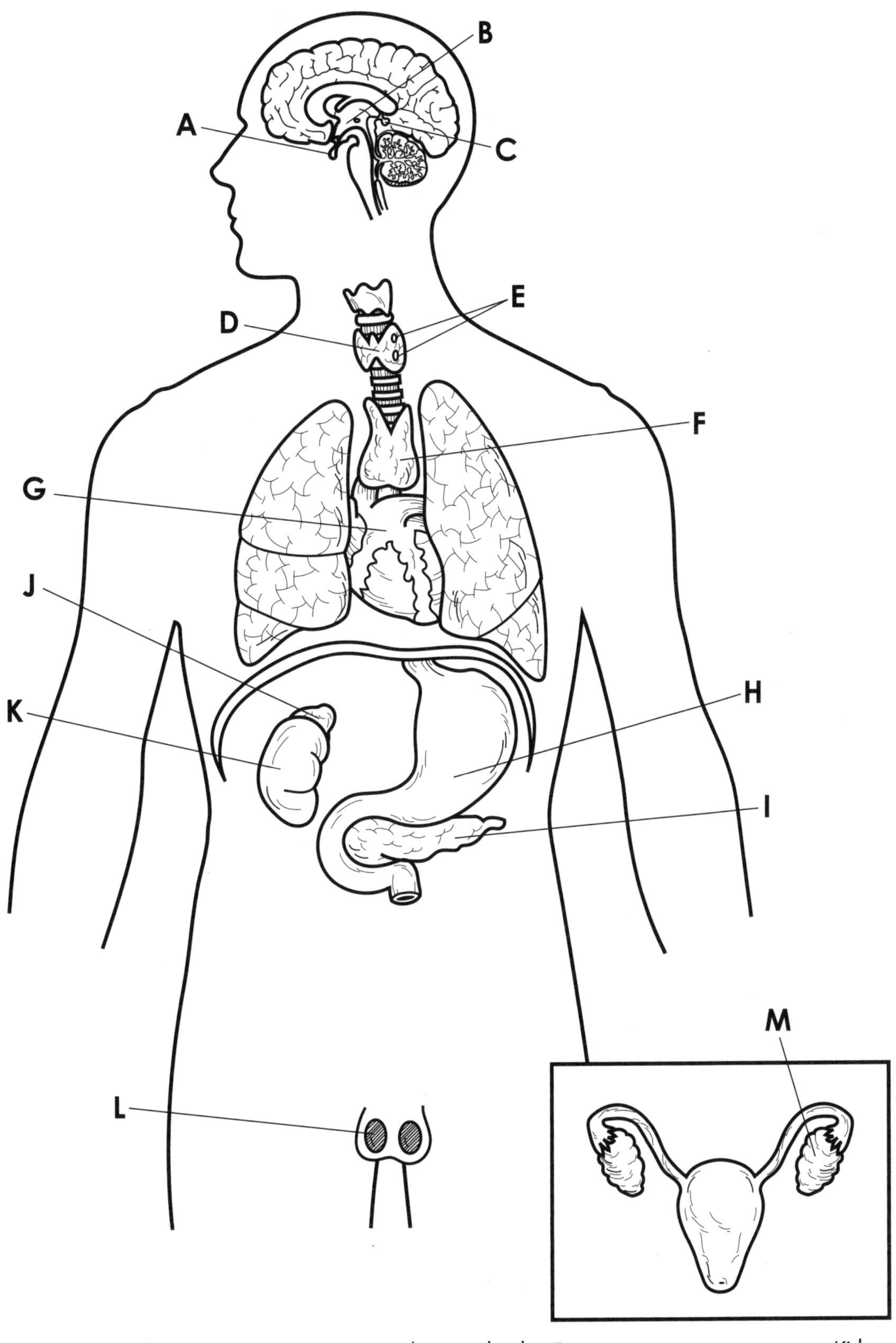

Pituitary Gland	A	❍
Hypothalamus	B	❍
Pineal Gland	C	❍
Thyroid Gland	D	❍
Parathyroid Gland	E	❍
Thymus Gland	F	❍
Heart	G	❍
Digestive Organs	H	❍
Pancreas	I	❍
Adrenal Glands	J	❍
Kidney	K	❍
Testis	L	❍
Ovary	M	❍

THE ACTION OF HORMONES

Hormones bring about chemical changes in the body by modifying the metabolic activities of their target cells and tissues. Although there is still much to be learned about the mechanisms of hormone activity, scientists do know that different kinds of hormones act in different ways. For instance, steroid hormones are lipid-soluble and enter the cytoplasm of the target cell, where they bind to receptors, but protein and amine hormones cannot diffuse through the cell membrane, so they combine with receptors on the surface of the cell membrane of their target cells.

In this plate, we will examine both of these mechanisms and study other aspects of the methods of hormone action. Bold colors such as reds, greens, and blues should be used throughout the plate.

Look over the plate and note that there are two diagrams that reflect the two pathways of hormones. Read about the mobile receptor mechanism below as you locate the appropriate structures on the diagram.

Steroid hormones act on their target cells by a mechanism currently known as the mobile receptor mechanism (in older texts, the process is called the gene-activated mechanism). This mechanism is so-named because the receptor for the hormone molecule becomes mobile after the hormone binds to it.

The process begins when a **steroid hormone (A)** leaves the bloodstream and approaches the **cell membrane (B)** of the target cell. The hormone passes easily through the membrane (because steroids are lipids) to enter the **cytoplasm (C)** of the cell, where it combines with the **receptor (D).** This receptor is a protein, and the union of hormone and receptor creates the **hormone-receptor complex (E).**

The hormone-receptor complex then enters the **nucleus (F)** of the cell. Here the complex binds to a **DNA molecule (G)** in the nucleus and activates one or more genes. (This gene activation is the reason that the alternate name of this mechanism is sometimes used.) The activation induces transcription: DNA is used as a template to made molecules of **messenger RNA (mRNA) (H).** Next, mRNA is transported from the nucleus to the cytoplasm and is then translated by ribosomes either in the cytoplasm or associated with the **endoplasmic reticulum (I).** Overall, this causes the synthesis of a **protein (J).** This protein brings about the desired effect in the cell. For example, the protein may act as an enzyme that alters the rate of specific cellular processes, or it may act as part of a membrane transport system to alter membrane permeability.

We will now study the second mechanism of hormone activity in the body. This mechanism is called the fixed-membrane receptor mechanism.

Many protein and amine hormones act by combining with specific receptors that are present on the target cell membranes, and after the union has occurred, the receptors remain on the membrane.

The first step in this mechanism occurs when a **protein hormone (K)** approaches the **cell membrane (L),** where it unites with a specific **receptor (M)** in the membrane. The union of hormone and receptor has an effect on downstream proteins and enzymes. For example, the enzyme adenylyl cyclase is typically bound to the cytoplasmic side of the membrane and is normally **inactive (N).** In response to epinephrine (a hormone) binding its receptor on the plasma membrane, G proteins are activated and this causes adenylyl cyclase **activation (O).** The activated adenylyl cyclase enzyme catalyzes the breakdown of an **ATP molecule (P)** into **cyclic AMP (Q),** which is also called cAMP. Cyclic AMP acts as a "second messenger" in this series of reactions. (The hormone is considered the "first messenger.") It triggers the chain of events in the **cytoplasm (R)** that constitutes the cellular response.

The mechanism continues with cAMP. cAMP activates an enzyme called **protein kinase (S),** which in turn incites a specific event. For example, protein kinase activates enzymes in the liver to alter the rate of glucose metabolism and also acts to reduce tension in smooth muscles, increase the rate of contractions in cardiac muscle, and stimulate the secretion of thyroid hormone.

If cyclic AMP remained active too long in the cell, the chemical changes could be disastrous. Therefore the cAMP must be changed back to an **inactive cAMP molecule (U).** This change is brought about by **phosphodiesterase enzymes (T).** The inactive cAMP then passes out of the target cell and into the **bloodstream (V).** Now the receptors at the membrane are available once again.

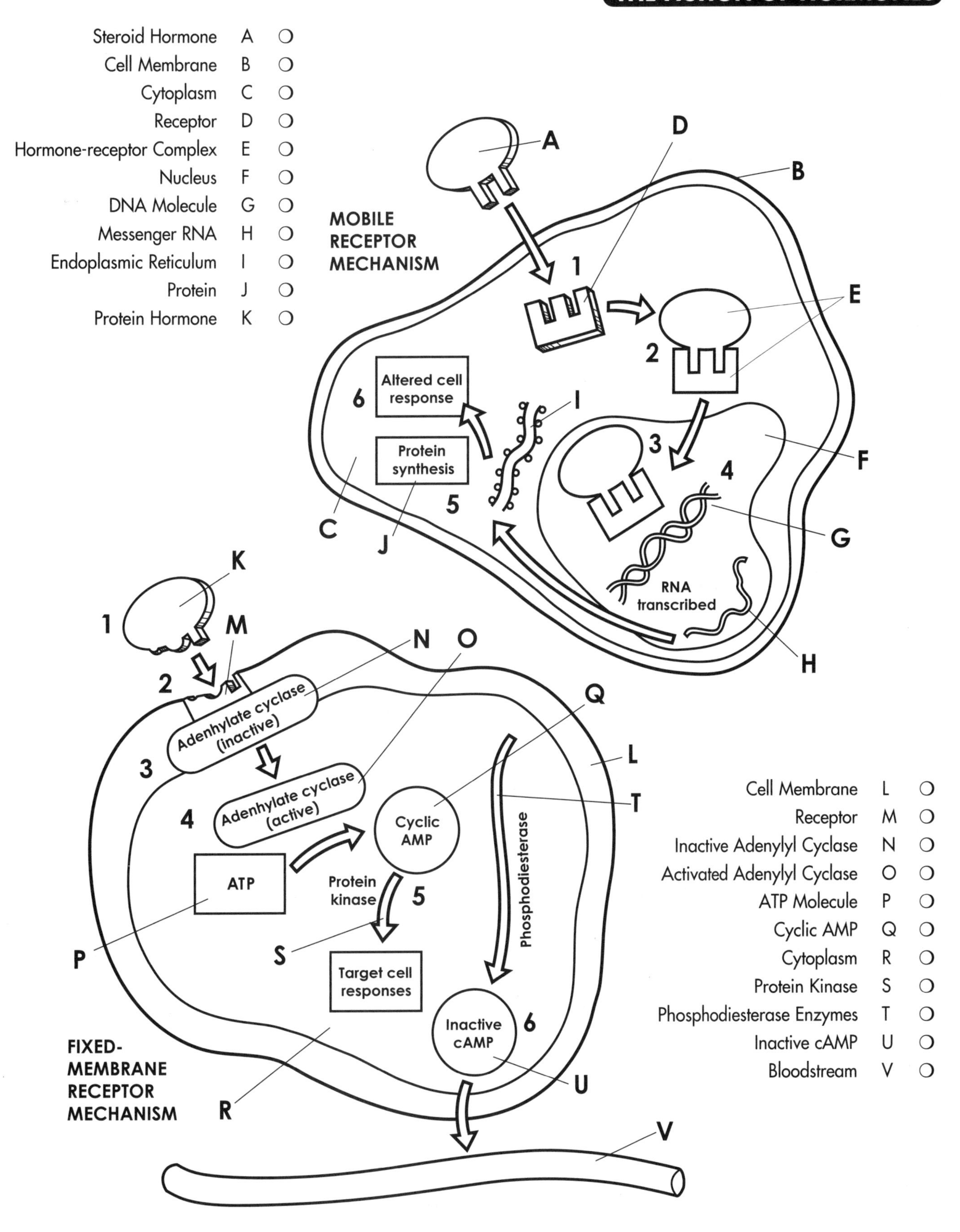
Steroid Hormone A
Cell Membrane B
Cytoplasm C
Receptor D
Hormone-receptor Complex E
Nucleus F
DNA Molecule G
Messenger RNA H
Endoplasmic Reticulum I
Protein J
Protein Hormone K
MOBILE RECEPTOR MECHANISM
Altered cell response
Protein synthesis
RNA transcribed
1
2
3
4
5
6
Adenhylate cyclase (inactive)
Adenhylate cyclase (active)
Cyclic AMP
ATP
Protein kinase
Phosphodiesterase
Target cell responses
Inactive cAMP
FIXED-MEMBRANE RECEPTOR MECHANISM
Cell Membrane L
Receptor M
Inactive Adenylyl Cyclase N
Activated Adenylyl Cyclase O
ATP Molecule P
Cyclic AMP Q
Cytoplasm R
Protein Kinase S
Phosphodiesterase Enzymes T
Inactive cAMP U
Bloodstream V

BLOOD CELLS

Blood is the medium for the transportation of oxygen from the lungs to the body's cells, as well as carbon dioxide from the cells to the lungs. It also transports nutrients and chemical waste products and participates in the immunological defense of the body. The blood is composed of a pale, somewhat yellow fluid known as plasma in which three basic types of blood cells are suspended. In this plate, we will describe the three types and their function.

This plate shows the various types of blood cells, divided into three major groups. We will discuss each type in turn, emphasizing their functions.

The first type of cell we will concentrate on is the **red blood cell (RBC) (A),** also known as the erythrocyte. The bracket enclosing these cells may be colored a dark color, but a pale shade should be used for the cells themselves to avoid obscuring their details. Technically, red blood cells are not true cells because they have little internal organization, no nucleus, and no organelles. They are membranous sacs filled with hemoglobin and used for the transportation of oxygen.

There are approximately 5.4 million RBCs per cubic millimeter in an adult male and about 4.8 RBCs per cubic millimeter in a female. Red blood cells are biconcave discs (meaning that they're thinner at the center than at the edge), and are approximately 7.8 μm in diameter. They are produced in the red marrow of bones, and they contain hemoglobin, which complexes with oxygen outside the cell to form oxyhemoglobin. This complex diffuses into the cell. Red blood cells circulate for approximately 120 days and then are destroyed in the spleen and other organs.

We now focus on the second type of blood cells: white blood cells. Five types are located in this group, as the diagram indicates. Continue your reading as you continue to color.

The second type of blood cells we will discuss are the **white blood cells (WBCs) (B),** also known as leukocytes. Within this group are the **neutrophils (C),** indicated by the brackets. Neutrophils comprise about 60% of the total white blood cell count. They have **granular cytoplasm (D).** Their **nuclei (E)** have between two and five lobes, as the diagram shows. Neutrophils are responsible for phagocytosis at infection sites.

Another type of white blood cell is the **eosinophil (F).** These cells also have granular cytoplasm (D), and each cell is approximately 10 to 14 μm in diameter. Eosinophils make up approximately 1% of the total white blood cells and are believed to play a role in allergic reactions.

The third type of white blood cell is the **basophil (G).** Basophils also have granular cytoplasm (D) and a multilobed nuclei. Basophils constitute about 1% of the total white blood cells and are believed to function in allergic reactions, clotting, and inflammation.

The fourth type of white blood cell is the **monocyte (H).** This cell, about 15 to 20 μm in diameter, is the largest type of white blood cell. Monocytes make up 6 to 8% of the white blood cells and have large nuclei that are indented along one wall. Their nuclei (E) are surrounded by **nongranular cytoplasm (I).** Monocytes squeeze through the walls of capillaries to enter the tissues, where they phagocytosize microorganisms. They also transform into large phagocytic cells called macrophages.

The final type of white blood cell is the **lymphocyte (J),** which measures about 8 to 10 μm in diameter. Here we see a large, circular nucleus (E) that takes up almost the entire nongranular cytoplasm (I). Lymphocytes account for about 30% of all white blood cells. They include B-lymphocytes and T-lymphocytes, both of which are important in the immune process.

We conclude the study of blood cells by focusing on platelets.

The last type of blood cell shown in the plate is the **platelet (K).** Platelets are cell fragments that are approximately 1 to 2 μm in diameter. They are also known as thrombocytes and consist of small amounts of cytoplasm enclosed by membranes. They form from large cells called megakaryocytes. Approximately 300 thousand platelets exist per cubic millimeter of blood. They function in platelet plugs (which form at several parts of a blood vessel), where they react with collagen fibers, and are also present in blood clots.

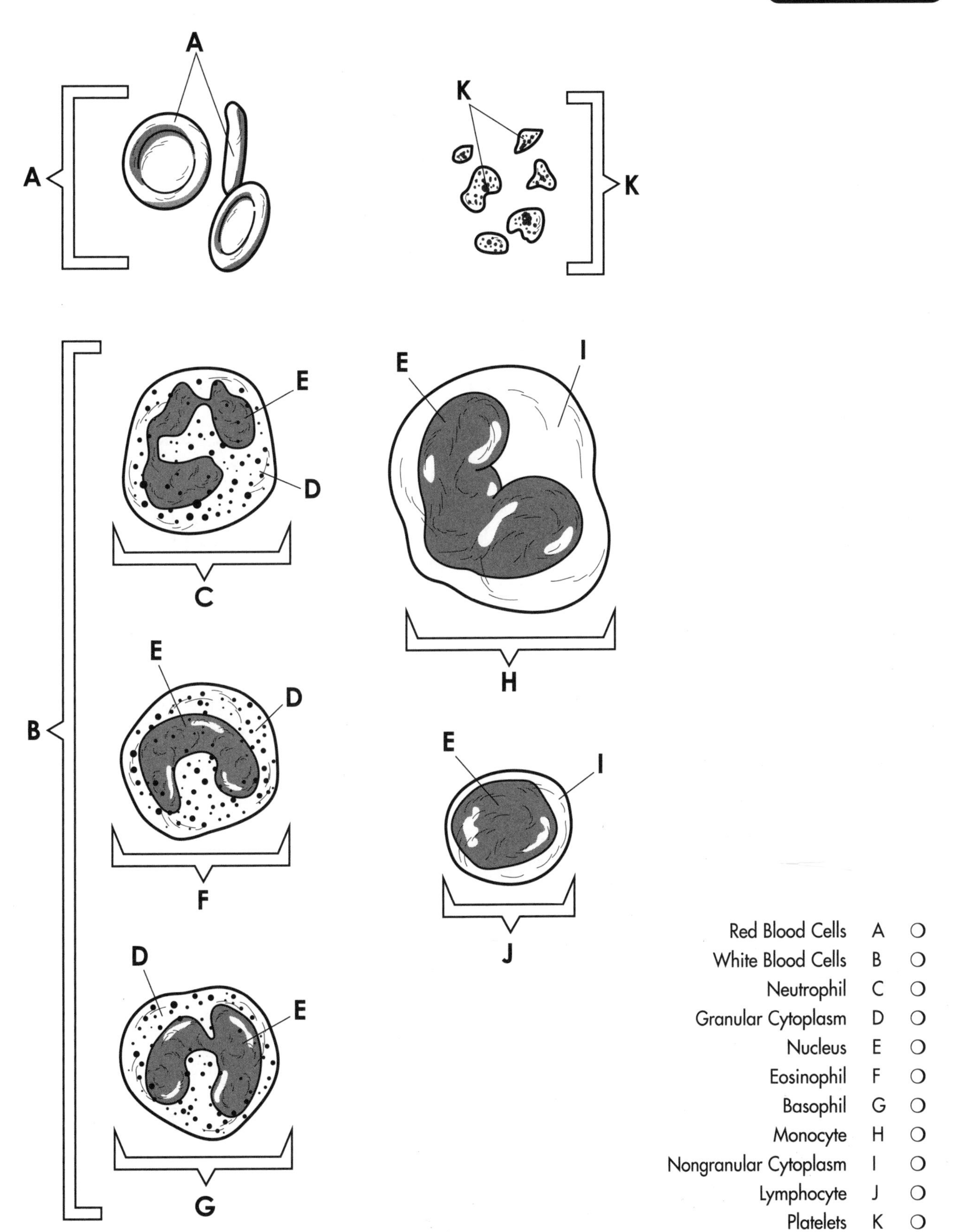
A
A
K
K
E
I
E
D
C
H
E
D
B
E
I
F
J
D
E
G
Red Blood Cells A
White Blood Cells B
Neutrophil C
Granular Cytoplasm D
Nucleus E
Eosinophil F
Basophil G
Monocyte H
Nongranular Cytoplasm I
Lymphocyte J
Platelets K

THE HEART

The functions of the cardiovascular system depend primarily on the activity of the heart, which distributes blood to the lungs and the rest of the body. Each day, the heart beats approximately 100,000 times, at a rate of about 70 beats per minute.

To help you follow the pathway of blood, we have included a number of arrows that you should color as you read.

The heart pumps blood into two closed circuits of the body: the systemic circulation, which supplies the body cells, tissues, and organs with blood; and the pulmonary circulation, which carries blood to the lungs. After completing the systemic circuit, blood returns to the heart through two veins, the **superior vena cava (A),** which comes from the head, and the **inferior vena cava (B),** which comes from the lower body. The vena cavae meet at the **right atrium (C).**

We now follow the flow of blood from the right atrium into the right ventricle and out to the lungs. Continue your coloring as before.

From the right atrium, the blood flows down through the **tricuspid valve (D),** which is also called the right atrioventricular valve. This valve has three flaps, or cusps, and one cusp is indicated. Strands of connective tissues called **chordae tendinae (F)** support the valve and prevent the cusps from flapping back into the right atrium, and the **papillary muscles (G)** hold the chordae tendinae in position.

Now the blood enters the **right ventricle (H),** which is the smaller of the two ventricles; note that the muscle wall here is thinner. Blood flows into this ventricle and, when it contracts, the blood is forced upward, as the arrows show. Note the substantial size of the **interventricular septum (I)** that separates the right and left ventricles. The blood is forced out through the **pulmonary semilunar valve (J),** and then into the pulmonary trunk. The semilunar valve prevents the blood from flowing back into the ventricle.

The **pulmonary trunk (K)** now divides to become the **left pulmonary artery (L)** and the **right pulmonary artery (M),** which lead to the two lungs. This begins the pulmonary circuit. Note the direction of the arrows and color them blue.

Blood is circulated to the lungs for oxygenation and then returns to the heart for distribution to the rest of the body. We will now follow the path of blood through the left side of the heart.

Blood returns to the heart by means of the pulmonary veins. Since the blood is oxygenated, the arrows should be colored in red. The blood now enters the **left atrium (N),** which is the second receiving chamber. The left atrium is separated from the right atrium by the **intra-atrial septum (O).**

Blood now flows through the **left atrioventricular valve (P),** which is also called the mitral valve, and enters the **left ventricle (Q),** which is the larger of the two ventricles. When the ventricle undergoes contraction, the blood is forced up to the aorta, passing through the **aortic semilunar valve (R),** which cannot be seen because it lies behind the pulmonary trunk.

On passing through the valve, oxygenated blood enters the **arch of the aorta (S).** The aorta turns to the posterior region and flows behind the heart. It can be seen emerging as the **descending aorta (T).** Arteries that arise from the aorta travel to the thorax, abdomen, pelvic cavity, and lower extremities.

We will now briefly discuss electrical control of the heart and the heart cycle.

The cells of the heart are self-excitable, which means that they can contract without first receiving a signal from the nervous system. Their contraction is initially caused by a region of the heart called the **sinoatrial (SA) node (U),** which is also sometimes called the pacemaker. As you can see in the art, the SA node is in the wall of the right atrium. When the SA node contracts, a wave of excitation travels through the heart wall, causing the two atria to contract in unison. At the boundary of the atria is the atrioventricular node, or **AV node (V).** The impulse travels to the AV node, where it is delayed for a fraction of a second in order to allow the atria to empty completely. It then spreads through the ventricles via Purkinje fibers, which causes them to contract.

The heart cycle refers to the sequence of events that occur during the course of a heartbeat. There are two phases of this cycle: systole, in which the heart contracts and blood is pumped; and diastole, during which the heart is relaxed. These two phases are approximately equal in duration. The flow of blood is as follows: During the first part of systole, the atria contract and blood is squeezed into the ventricles; then toward the end of systole the ventricles pump blood into the arteries. Diastole, which comes next, allows the ventricles to refill with blood from the atria.

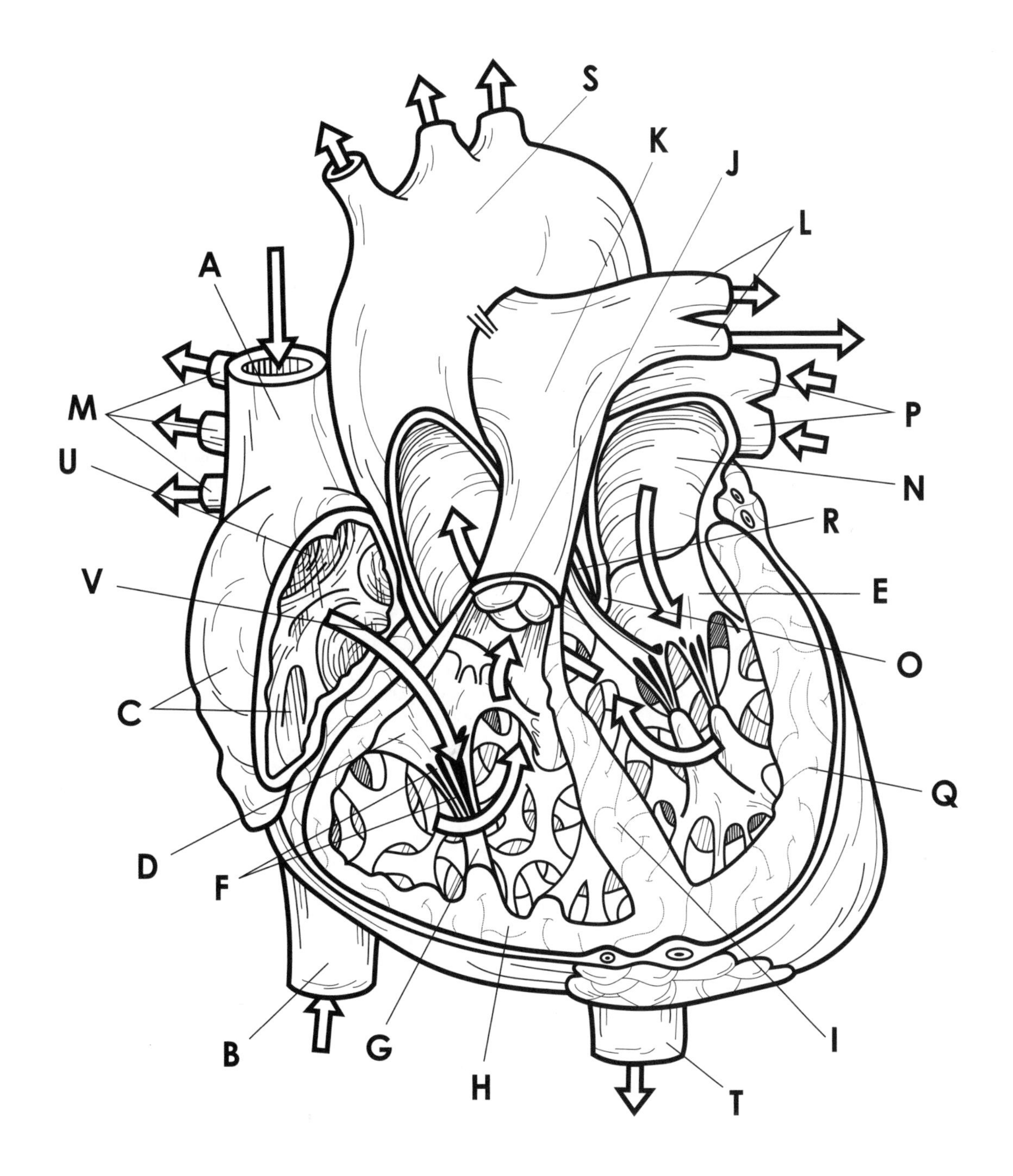

Superior Vena Cava	A	❍
Inferior Vena Cava	B	❍
Right Atrium	C	❍
Tricuspid Valve	D	❍
Cusp of the Valve	E	❍
Chordae Tendinae	F	❍
Papillary Muscles	G	❍
Right Ventricle	H	❍
Interventricular Septum	I	❍
Pulmonary Semilunar Valve	J	❍
Pulmonary Trunk	K	❍
Left Pulmonary Artery	L	❍
Right Pulmonary Artery	M	❍
Left Atrium	N	❍
Intra-atrial Septum	O	❍
Left Atrioventricular Valve	P	❍
Left Ventricle	Q	❍
Aortic Semilunar Valve	R	❍
Arch of the Aorta	S	❍
Descending Aorta	T	❍
SA Node	U	❍
AV Node	V	❍

PRINCIPAL ARTERIES OF THE BODY

The arteries of the circulatory system transport blood away from the heart. Their main purpose is to carry oxygen and nutrients to body tissues, but they also transport hormones, elements of the body's immune system, and metabolic waste. All of the arteries of the body branch from the aorta. Veins return the blood to the heart through the vena cava.

Note that many letters bear a subscript 1 or 2. Arteries that have the subscript 1 lie on the anatomical left side of the body (your visual right), while the arteries labeled with 2 lie on the anatomical right side (your visual left). It is often difficult to distinguish where one artery begins and another ends, so we have indicated the boundaries with short lines. When coloring, darker colors should be used for the large arteries, but as you reach smaller ones, light colors are recommended.

Arising from the left ventricle of the heart is the largest artery of the body, the **aorta (A).** In the plate, the artery is seen curving to the right to become the **thoracic aorta (A_1).** The thoracic aorta passes down along the spine and through the diaphragm to become the **abdominal aorta (A_2),** which in turn splits to become the common iliac arteries.

A branch of the aorta at its first major arch is the **brachiocephalic trunk (B),** which is also called the innominate artery. It branches into the common carotid artery, which in turn branches into the **left common carotid (C_1)** and the **right common carotid artery (C_2).** The carotid arteries supply the neck and head with blood.

The third branch from the brachiocephalic trunk is the **right subclavian artery (E_2).** On the right side of the plate, the **left subclavian artery (E_1)** arises from the arch of the aorta. The subclavian arteries supply the upper limbs with blood. Arising from the right subclavian artery is the **vertebral artery (D),** which supplies the vertebrae, deep muscles of the neck, and spinal cord with blood.

Also arising from the subclavian arteries are the **left and right axillary arteries (F_1 and F_2).** Axillary arteries supply the muscles of the shoulder and the thoracic muscles, and give rise to the **brachial arteries (G_1 and G_2),** which service the arm. The **radial arteries (H_1 and H_2)** and **ulnar arteries (I_1 and I_2)** arise from the brachial arteries and carry blood to muscles of the forearm.

To this point we have briefly surveyed the principal arteries of the head, neck, and upper extremity. Now we will return to the thoracic and abdominal regions and locate the other branches from the aorta. Continue your coloring, and continue to locate the arteries. Be careful to note the beginnings and endings of the arteries.

Arising from the aorta just as it leaves the left ventricle, the **coronary arteries (J)** pass into the heart muscle, where they supply this organ with oxygen and nutrients.

Another trunk emerges from the aorta after it has passed through the diaphragm. This is an unpaired artery called the **celiac trunk (K).** Arteries from the celiac trunk branch to the liver, stomach, and spleen, as well as other regions of the upper abdomen. The **hepatic artery (L)** branches from the celiac trunk and extends to the liver. From the abdominal aorta, the **gastric artery (M)** supplies the stomach, and the **splenic artery (N)** moves in the direction of the spleen.

Also extending from the celiac trunk are the paired, renal arteries. The **left renal artery (O_1)** supplies the left kidney, and the **right renal artery (O_2)** extends to the right kidney. Next we see the unpaired **superior mesenteric artery (P).** This artery carries blood to the small intestine, pancreas, and portions of the large intestine. The **gonadal artery (Q)** leads to arteries that supply the ovaries in females and the testes in males. Beyond the gonadal artery is the **inferior mesenteric artery (R).** The diagram shows its numerous branches as it services portions of the transverse colon, descending colon, sigmoid colon, and rectum.

Further down the celiac trunk, two major arteries arise: the **common iliac arteries (S_1 and S_2).** They soon split to form the internal and external iliac arteries. Only the **external iliac arteries (T_1 and T_2)** are shown. These arteries lead to the **left and right femoral arteries (U_1 and U_2).** Blood from these arteries services the leg muscles.

Aorta A ❍
Thoracic Aorta A_1 ❍
Abdominal Aorta A_2 ❍
Brachiocephalic Trunk B ❍
Left Common Carotid Artery C_1 ❍
Right Common Carotid Artery C_2 ❍
Vertebral Artery D ❍
Left Subclavian Artery E_1 ❍
Right Subclavian Artery E_2 ❍
Left Axillary Artery F_1 ❍
Right Axillary Artery F_2 ❍
Left Brachial Artery G_1 ❍
Right Brachial Artery G_2 ❍
Left Radial Artery H_1 ❍
Right Radial Artery H_2 ❍
Left Ulnar Artery I_1 ❍
Right Ulnar Artery I_2 ❍
Coronary Arteries J ❍
Celiac Trunk K ❍
Hepatic Artery L ❍

PRINCIPAL ARTERIES OF THE BODY

Gastric Artery M ❍
Splenic Artery N ❍
Left Renal Artery O_1 ❍
Right Renal Artery O_2 ❍
Superior Mesenteric Artery P ❍
Gonadal Artery Q ❍
Inferior Mesenteric Artery R ❍

Left Common Iliac Artery S_1 ❍
Right Common Iliac Artery S_2 ❍
Left External Iliac Artery T_1 ❍
Right External Iliac Artery T_2 ❍
Left Femoral Artery U_1 ❍
Right Femoral Artery U_2 ❍

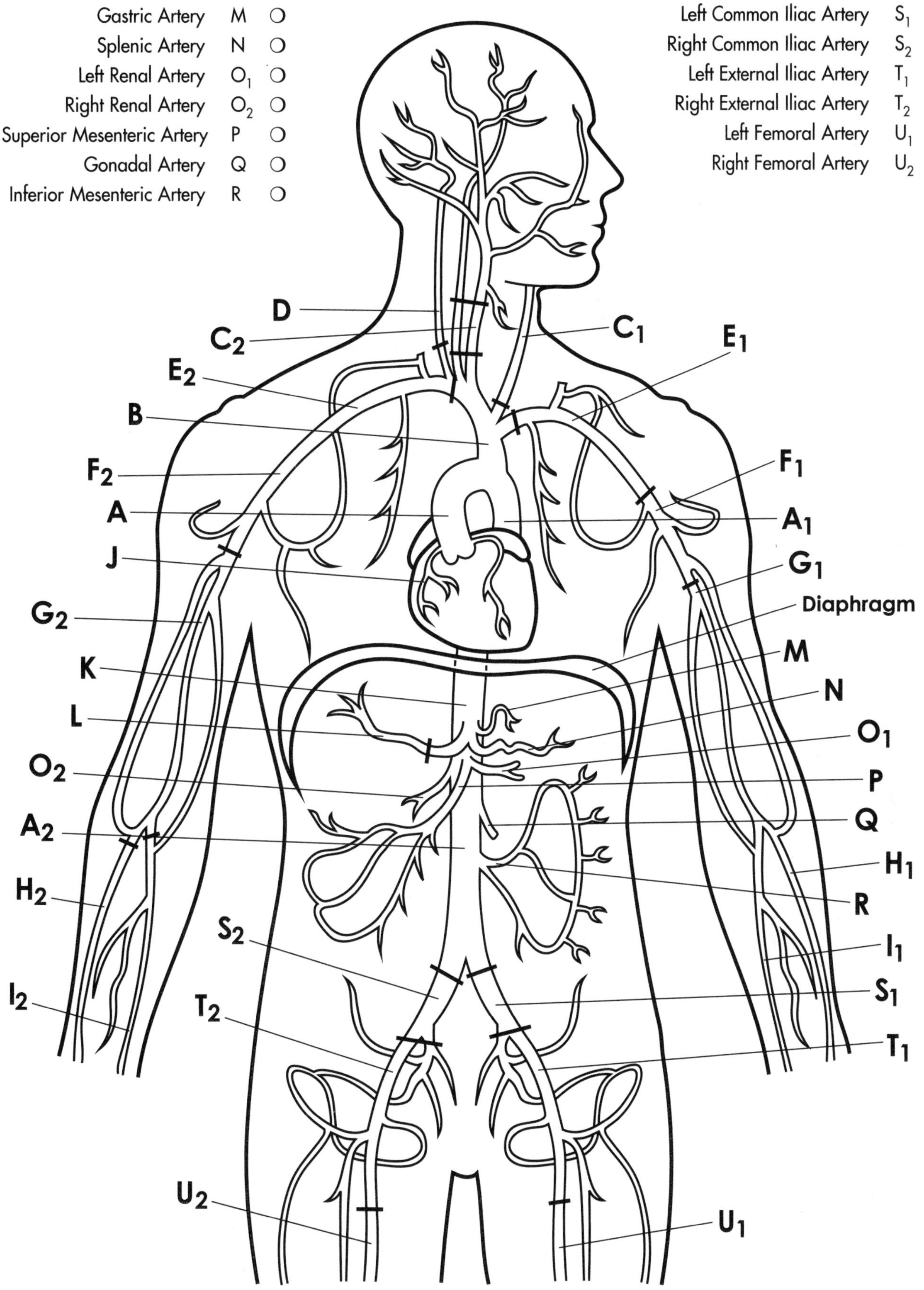

D
C_2
C_1
E_1
E_2
B
F_2
F_1
A
A_1
J
G_1
G_2
Diaphragm
K
M
N
L
O_1
O_2
P
Q
A_2
H_1
H_2
R
S_2
I_1
I_2
T_2
S_1
T_1
U_2
U_1

THE LYMPHATIC SYSTEM

The lymphatic system is a series of vessels, structures, and organs that collect proteins and fluid that is lost from the blood and return it to the main circulation. This system contains cells known as lymphocytes that function in the immune process. In this plate, we examine the structure of the lymphatic system. This plate serves as prelude to the following one, which discusses the immune system.

Looking over the plate, you will note that it displays the layout of the lymphatic system. We also illustrate the two body regions drained by the two major vessels of the lymphatic system. Spots of color should be used, since lymph nodes are pockets of tissue. Begin reading below, and as you encounter anatomical structures, locate them on the diagram. Certain structures are designed with lowercase letters because they are associated with other systems. The uppercase letters are reserved for parts of the lymphatic system.

The fluid drained by the lymphatic system is called lymph. It is a clear fluid that is similar to the plasma portion of blood but contains no large proteins. The lymphatic system returns lymph to the blood by means of two major vessels. The first is the left lymphatic duct, which is also known as the **thoracic duct (A).** The left lymphatic duct begins as a dilation called the **cisterna chyli (A_1)** and throughout its course it receives blood from the left side of the head, the left portion of the neck and chest, the left upper limb, and the entire body below the level of the ribs. Color this portion in the small diagram.

The second major duct of the lymphatic system is the **right lymphatic duct (B).** It can be seen on the visual left (anatomical right) of the plate, where it empties its contents into the **right subclavian vein (a_1).** In the same way, the left lymphatic duct empties lymph into the **left subclavian vein (a_2).** Color the drainage area in the small diagram for the right lymphatic duct using the color you used for the duct.

Having noted the main drainage areas for the two main lymphatic vessels, we now focus on the lymph nodes, which are pockets of lymphatic tissue. Note that these organs are found throughout the body. Continue your reading below, and use spots of color to denote the various lymph nodes.

The vessels of the lymphatic system pass through small lymphatic structures known as lymph nodes. These approximately oval organs contain cells of the immune system, including phagocytes, which engulf foreign organisms and debris found in the lymph. Some lymph nodes you might be familiar with are the tonsils, which are in the pharyngeal area. The **palatine tonsil (C)** is shown in the plate, and the other tonsils are the lingual tonsils.

Many lymph nodes can be found along lymphatic vessels; we will mention only a few. The **submandibular lymph nodes (D_1)** are located beneath the mandible, and the **cervical lymph nodes (D_2)** are found in the neck region—these nodes drain the head area. **Axillary lymph nodes (D_3)** are situated in the region of the armpits, and the **mammary lymph nodes (D_4)** are close to the mammary glands in females.

Many thoracic lymph nodes are found close to the thoracic duct, and a collection of lymph nodes called **Peyer's patch (D_5)** is located at the surface of the **small intestine (b).** Also in the abdomen, near the major blood vessels, are the **iliac lymph nodes (D_6),** which drain lymph that comes from the legs. The **inguinal lymph nodes (D_7)** are near the groin, and **intestinal lymph nodes (D_8)** are found near the **large intestine (c).** These nodes receive lymph from the numerous **lymphatic vessels (E)** that are shown in the leg and arm.

There are many other organs that are associated with the lymphatic system, and as the organs are mentioned, color them.

The **thymus gland (F)** is prominent in young children, and it is within this group that the T-lymphocytes of the immune system mature before they move to the lymph nodes. The thymus diminishes in size in the teenage years and is quite small in the adult.

Near the stomach and pancreas, on the left side of the body is the **spleen (G).** Also a lymphatic organ, the spleen contains the B-lymphocytes and T-lymphocytes of the immune system, which we will discuss in the next plate. The **appendix (H)** is also associated with the lymphatic system because it contains many phagocytic white blood cells, which engulf debris in the digestive contents. Lastly, **bone marrow (I)** is associated with the lymphatic system because lymphocytes originate here.

THE LYMPHATIC SYSTEM

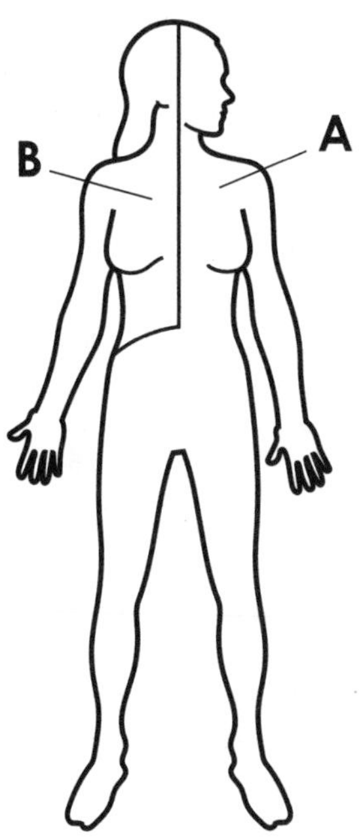

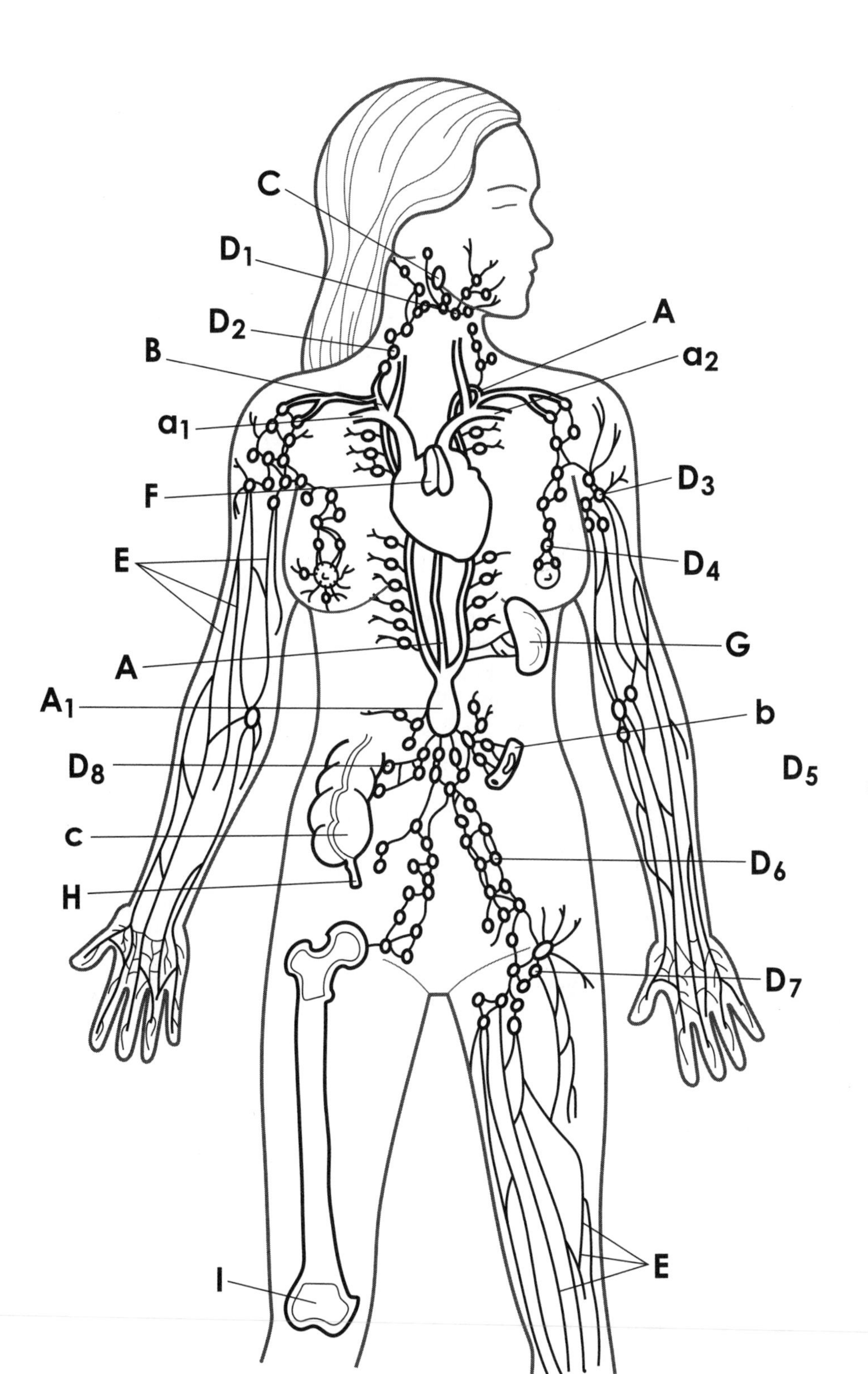

Thoracic Duct A ❍
Cisterna Chyli A_1 ❍
Right Lymphatic Duct B ❍
Palatine Tonsil C ❍
Submandibular Lymph Nodes D_1 ❍
Cervical Lymph Nodes D_2 ❍
Axillary Lymph Nodes D_3 ❍
Mammary Lymph Nodes D_4 ❍
Peyer's Patch D_5 ❍
Iliac Lymph Nodes D_6 ❍
Inguinal Lymph Nodes D_7 ❍
Intestinal Lymph Nodes D_8 ❍
Lymphatic Vessels E ❍
Thymus Gland F ❍
Spleen G ❍
Appendix H ❍
Bond Marrow I ❍
Right Subclavian Vein a_1 ❍
Left Subclavian Vein a_2 ❍
Small Intestine b ❍
Large Intestine c ❍

THE IMMUNE SYSTEM

The immune system is responsible for defending the body against bacterial and viral invasion and also provides long-term immunity to disease. The cells and other agents of the immune system are located in the lymphatic system, throughout the lymph nodes, spleen, tonsils, and other lymph-associated organs. In this plate, we present a brief overview of the immune process, and show how it relates to the lymphatic system. This is the domain of immunology and is much more complex than this plate depicts.

As you review the plate, note that we are presenting a series of events that take place in the course of an immune response. Here anatomical features are of less consequence than is the process itself. Medium light colors would be best for most of the cells and structures shown, as well as the arrows used to indicate passage from one cell group to another.

The immune system is made up of a complex series of cells, chemical factors, and organs. The cells of the immune system originate in the **bone (A).** During the fetal stage of human development, a series of primitive cells called stem cells emerge within **bone marrow (A_1).** These stem cells develop into the cells of the immune system.

In the fetus, some of the stem cells develop into **immature T-lymphocytes (B),** which migrate to the **thymus gland (C),** where they mature, eventually leaving as one of two cell types. **Cytotoxic T-lymphocytes (D_1)** are also called T-killers, and express a glycoprotein on their plasma membrane called CD8. Other emerging cells become **helper T-lymphocytes (D_2).** These cells are sometimes called T-helpers and express CD4. Both types of lymphocytes have a number of protein complexes called **T-cell receptors (D_3)** on their membranes. The T-lymphocytes gather in the **lymphatic tissue of the lymph nodes (a).** You should color the bracket that indicates this tissue.

Some of the stem cells follow a different route, maturing in the fetus in the bone marrow, liver, and other areas of the body to become **mature B-lymphocytes (H).** The B-lymphocytes also migrate to the lymphatic tissue and replicate, forming clusters of cells. The lymphatic tissues are pointed out in the previous plate on the lymphatic system, and you should refer to it if you feel you need to review. The T-lymphocytes and B-lymphocytes are the fundamental participants in the immune process.

Having explored the origin of the immune system, we now turn to its activity. We will discuss the major aspects of its activity in sequence. Our initial focus will be on T-lymphocytes. Color the appropriate sections as you read along.

A T cell recognizes a bad cell by "examining" (binding to) proteins on its surface. Most of our cells have major histocompatibility complex (MHC) proteins on their surfaces so that the immune system can keep an eye on what is going on inside every cell. Their role is to randomly pick up peptides from the inside of the cell and display them on the cell surface. This allows T cells to monitor cellular contents. For example, if a cell is infected with a virus, one of its MHC complexes will display a piece of a virus-specific protein. When a T-killer cell detects the viral protein (by binding to it) displayed on the cell's MHC, it becomes activated. T-killer cells can also detect transplant tissue and cancer cells.

The T-cell receptor (D3) binds an MHC complex that is displaying a **foreign molecule (E).** A reaction takes place on the surface of the cell, and this **activates the T-killer cell (E_1).** The **active cytotoxic T-lymphocyte (F)** moves to the **circulation (b)** and will attack the foreign cell and help clear it from the body. This activation is encouraged by the helper T-lymphocyte through a process called **helper assistance (G_1).** T-lymphocytes unite with and destroy cells in a process known as cell-mediated immunity or CMI.

In the second part of the plate, we take a look at the second process of immunity, known as antibody-mediated immunity. As you will see, there is no cell-to-cell interaction here. Continue coloring as you read.

An antigen is a foreign substance that induces an immune response in the body, and it is able to bind an antibody. Antigens in the form of bacteria, viruses, and foreign chemicals enter the body and stimulate the mature B-lymphocytes (H) in the lymphatic tissue. The **antigen (I)** does this by binding to a specific **B-cell receptor (H_3)** on a subset of B cells. Note that the B-cell receptor complements the antigen. The cells then undergo **activation of the B-lymphocyte (I_1).**

When the B-lymphocytes are activated, they transform into protein-secreting cells called **plasma cells (J).** Plasma cells secrete enormous numbers of **antibody molecules (K),** which are strands of protein that enter the circulation (b). They travel to the sites of the virus, bacteria, and other antigen, and their thick strands bind to them, rendering them inactive. Soon, large clumps of organisms accumulate and phagocytes come along and engulf and destroy them. Specific defense to disease is provided by this process, known as antibody-mediated immunity, or AMI. Antibodies remain in the circulation for years, which brings about long-term immunity.

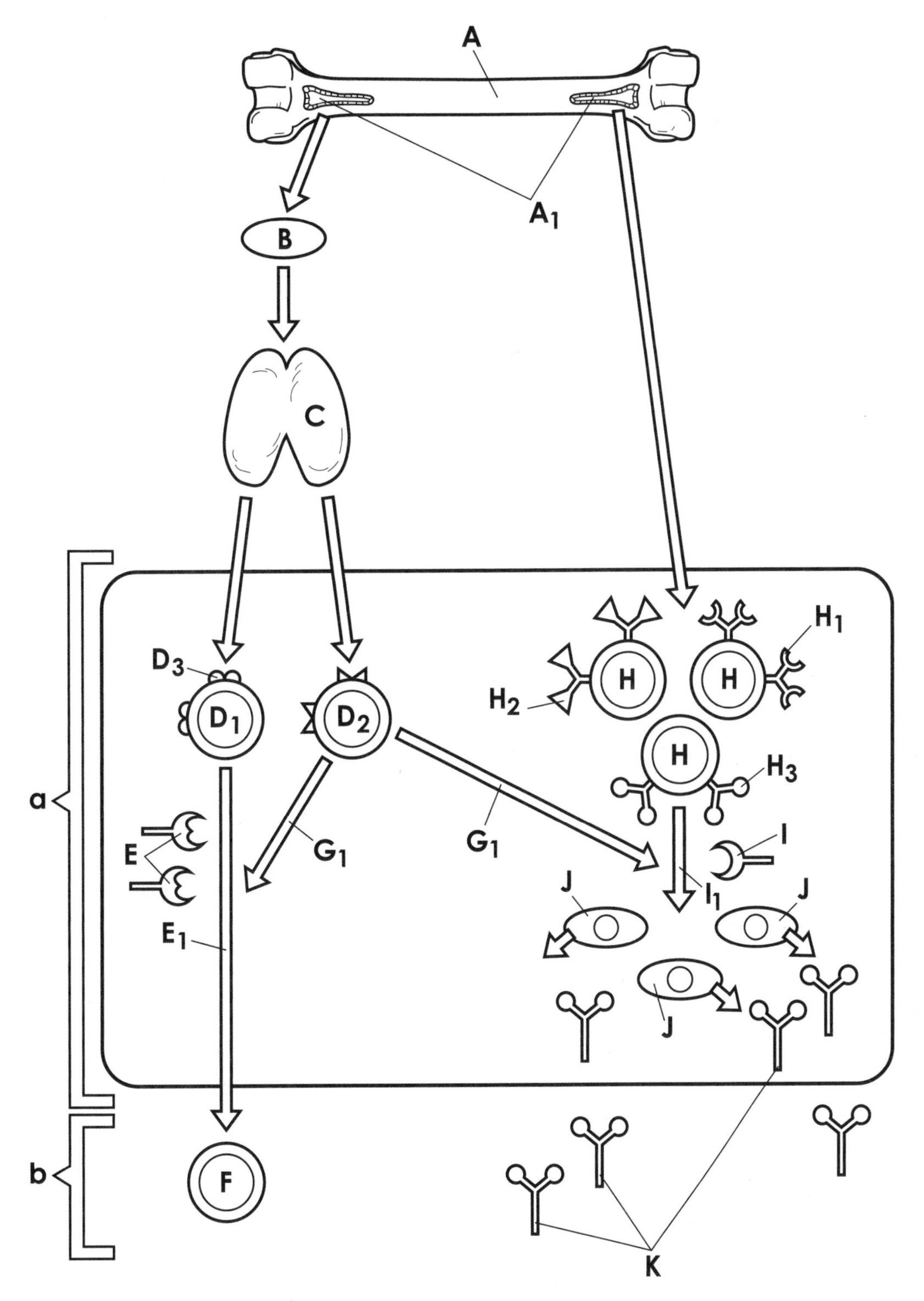

Bone A ❍
Bone Marrow A_1 ❍
Immature T-lymphocytes B ❍
Thymus Gland C ❍
Cytotoxic T-lymphocytes, or T-killers D_1 ❍
Helper T-lymphocytes, or T-helpers D_2 ❍

T-cell receptor D_3 ❍
Foreign molecule expressed on a cell E ❍
T-killer activation E_1 ❍
Active T-killer F ❍
Helper Assistance G_1 ❍
Mature B-lymphocytes H ❍
B-lymphocyte Receptor H_1 ❍

B-lymphocyte Receptor H_2 ❍
B-cell receptor H_3 ❍
Antigen for B-lymphocyte I ❍
Activation of B-lymphocyte I_1 ❍
Plasma Cells J ❍
Antibody Molecules K ❍
Lymphatic Tissue a ❍
Circulation b ❍

THE DIGESTIVE SYSTEM

The function of the digestive system is to break down large food particles into smaller ones that can be absorbed into the membranes or cells. Two main groups of organs comprise the digestive system. The first group is made up the organs of the gastrointestinal (GI) tract, also known as the alimentary canal. This is a tube that extends from mouth to anus and is open at each end. The second group is made up of accessory structures such as the teeth, tongue, and glands that line the GI tract. These aid in the mechanical and chemical breakdown of foods.

This plate gives a view of the digestive system and focuses on the two groups mentioned above. Be prepared to use bold colors in this plate, since the structures and regions are easy to see and distinguish.

Note that we are presenting the complete digestive system, including the organs of the GI tract and their accessory organs. Dark colors such as reds, oranges, greens, and blues may be used. As you read about the organs, locate and color them in the plate.

The process of digestion begins when food enters the digestive system at the **oral cavity (A).** Here the food is broken down mechanically and moistened with secretions. It is shaped into a football known as the bolus by the tongue, and then enters the next part of the GI tract, the **pharynx (B),** which starts at the rear of the mouth. A color that blends with the one used for the oral cavity should be used.

The organ that transports food to the stomach is the muscular **esophagus (C).** Transport of food through the alimentary canal occurs via **peristalsis,** a series of wave-like muscle contractions. The process of peristalsis begins in the esophagus when a bolus of food is swallowed. The esophagus passes through the thoracic cavity and pierces the diaphragm before entering the pouch-like **stomach (D).** Here food mixes with acid and protein-digesting enzymes. Passing from the stomach, food enters the 20-foot long **small intestine (E).** A great portion of the abdominal cavity is taken up by the numerous folds and twists of this organ, and the main processes of digestion and absorption occur here.

At the lower left portion of the plate (the anatomical right), the small intestine leads into the **large intestine (F).** This tube can be seen ascending along the anatomical right side, passing along the midline and then turning and descending. Undigested material is dehydrated and compacted in this organ. Prior to defecation, indigestible waste is stored in the **rectum (G).** After a certain amount of feces has accumulated, it passes through the **anus (H).** A spot of color should be used to designate this opening.

The digestive process is aided by several organs that lie along the GI tract and contribute secretions. Three of these organs are mentioned briefly in this section, and we recommend the use of dark colors to indicate their location.

Three sets of **salivary glands (I)** supply enzymes that digest food molecules in the oral cavity. Saliva also contains mucus for lubrication, and an enzyme called lysozyme, which helps break down bacteria that may be in our food. In the abdominal cavity, the **liver (J)** secretes bile that participates in fat digestion. The liver also processes most of the products of digestion before sending them to cells of the body, stores glucose, and helps our bodies clear some toxins. Beneath the liver is the **gallbladder (K).** Bile from the liver is stored here before it is delivered to the intestine.

An important contributor of enzymes to the digestive process is the **pancreas (L).** Exocrine cells of this gland make many digestive enzymes, and also bicarbonate, which neutralizes stomach acid in the intestine. Exocrine cells deliver their secretions into the first part of the small intestine, the duodenum. The pancreas also contains endocrine cells, which make hormones and secrete them into the blood stream. You learned about this process in the plate called "The Endocrine System."

In the final portion of this plate, we will mention the important digestive enzymes, in the order in which they are encountered by food passing through the digestive system.

Present in salivary secretions is the enzyme salivary amylase. This enzyme breaks down starch through the process of hydrolysis, producing the disaccharide maltose. The next digestive enzymes encountered by the food are present in the stomach. The stomach contains gastric juice, which has a pH of about 2, and this gastric juice contains the enzyme pepsin, which breaks down proteins. Chemical breakdown of food continues in the small intestine, where pancreatic amylase continues to hydrolyze starch. Other enzymes called disaccharidases (such as lactase, maltase, and sucrase) are secreted by the cells that made up the duodenum, and also help break down sugars. (Based on their names, can you guess what each of these enzymes does?)

The small intestine also contains trypsin, chymotrypsin, carboxypeptidase, aminopeptidase, and other peptidases, which are all responsible for breaking down proteins. Some of these enzymes are made by the pancreas and others are made by cells in the duodenum. Bile is secreted from the gallbladder into the duodenum, and it helps fat molecules (or lipids) mix with digestive juices. This process is called emulsification and provides a larger surface area on which the enzyme pancreatic lipase can act to digest the fats into fatty acids and glycerol. Finally, the pancreas also secretes nucleases (such as DNase and RNase) into the duodenum. These enzymes break down DNA and RNA in the food we eat.

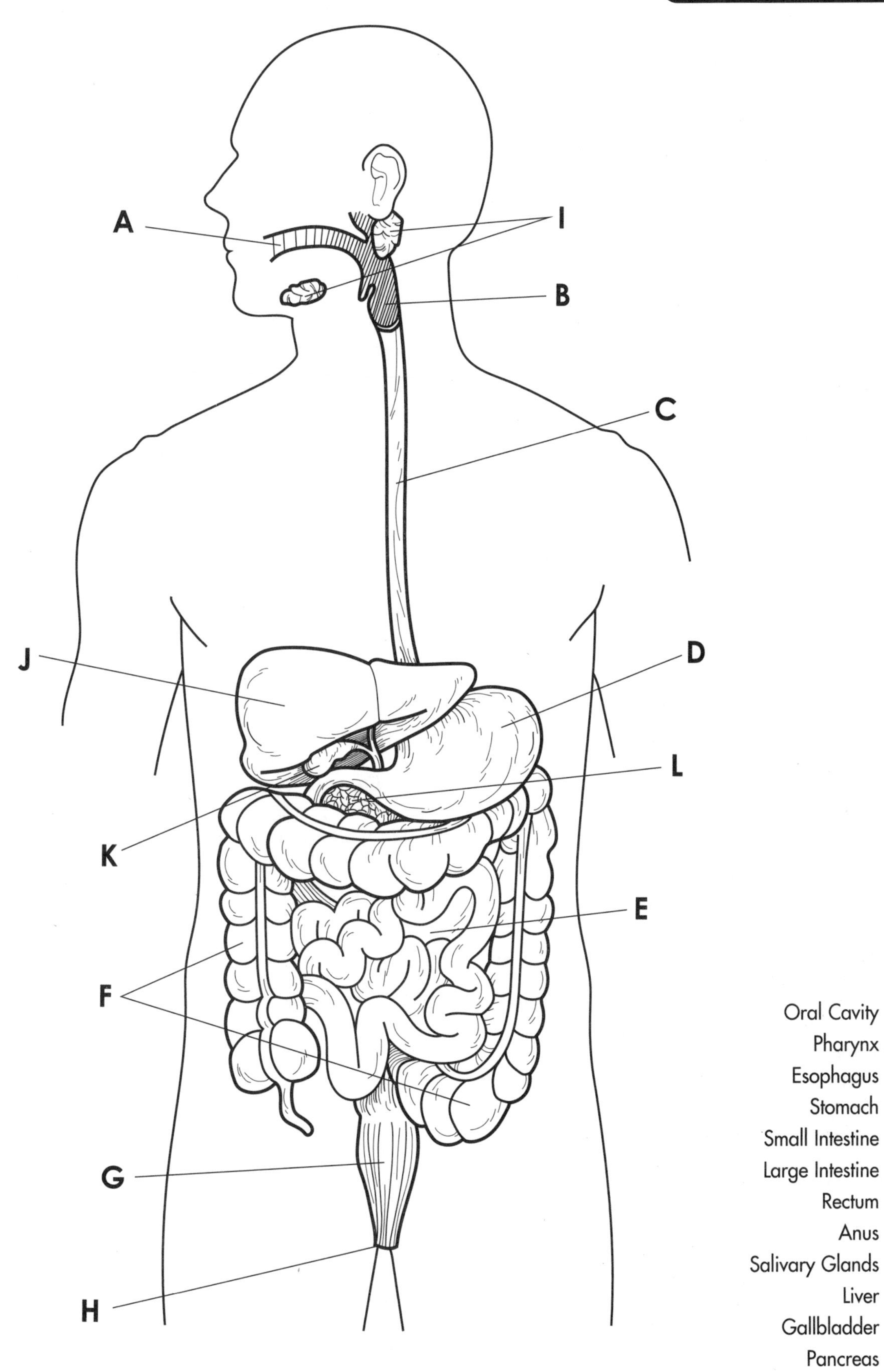

Oral Cavity A ○
Pharynx B ○
Esophagus C ○
Stomach D ○
Small Intestine E ○
Large Intestine F ○
Rectum G ○
Anus H ○
Salivary Glands I ○
Liver J ○
Gallbladder K ○
Pancreas L ○

THE ORAL CAVITY

The digestive system begins at the oral cavity, which is the subject of this plate. The respiratory system also uses this cavity for the passage of air, and elements of the lymphatic system are found here as well.

> Many of the structures discussed will be seen in more than one of the diagrams of the oral cavity, and these structures should be colored the same color in both views. The areas can be shaded in with a pale or light gray, tan, or other color, but be careful to avoid obscuring the arrows.

The oral cavity is also known as the buccal cavity. It is outlined by the **cheek (A).** The arrow should be colored in lieu of this structure. The cavity is also defined by the **lips (B),** which are also called the labia. The region between the lips and the teeth is the **vestibule (C).**

The mechanical breakdown of food is accomplished by the **teeth (D).** Note in the cross section that the upper and lower teeth are encased in bone. A ridge of flesh called the gum or **gingiva (E)** lies at the base of each tooth and surrounds it.

> We now turn our attention to the back of the oral cavity. As before, many areas will be designated with arrows, and the arrows should be colored with dark colors.

The roof of the mouth is made up of two palates, the **hard palate (F)** and the **soft palate (G).** The hard palate is supported by bones, while the soft palate is composed primarily of skeletal muscle. The last portion of the soft palate is a dangling structure called the **uvula (H),** which helps prevent food from leaving the oral cavity unintentionally.

The soft palate joins with the tongue at two pairs of arches. The arch that is closer to the back of the mouth is the **palatopharyngeal arch (I);** it marks the entry to the pharynx. The **palatoglossal arch (J)** is closer to the front of the mouth.

> We will now study the tongue. This organ can be seen in all three views, and a light color such as pink or tan is recommended. Darker colors can then be used to indicate areas along the surface of the tongue.

Most of the region at the base of the oral cavity is occupied by the **tongue (K).** This thick, muscular organ is covered by mucous membranes and anchored in the midline of the floor of the mouth by a fold of membrane called the **lingual frenulum (L).** The tongue is partly anchored to the **hypoid bone (M)** and is controlled by several muscles.

The tongue is divided into two sections: the **body (N)** and the **root (O),** both of which are outlined by brackets. On its superior surface, the tongue contains numerous **papillae (P).** There are three types of papillae, two of which act as taste buds, and at the base of the tongue are the openings of the two **sublingual salivary glands (Q).**

> We conclude the plate by pointing out structures that are related to the oral cavity but not necessarily associated with the digestive system. As you encounter these structures in the reading below, locate and color them in the plate. Darker colors may be used for the structures and arrows.

The oral cavity contains elements of other important body systems. For instance, the two sets of tonsils are parts of the immune system and are composed of lymphoid tissue. At the back of the tongue are the **lingual tonsils (R_1)** and lateral to them, along the wall of the oral cavity, are the **palatine tonsils (R_2).** the **pharyngeal tonsil (R_3)** lies in the pharynx at the back of the oral cavity; these tissues can also be seen in the plate on the lymphatic system.

The oral cavity continues as the pharynx, and the first portion of the pharynx is the **oropharynx (T_2).** Above the oropharynx is the continuation of the nasal cavity, which is known as the **nasopharynx (T_1).** The nasopharynx contains the pharyngeal tonsil (noted above), and is the site at which the **Eustachian tube (S)** opens. As we mention in the plate on the ear, this tube leads to the middle ear and helps equalize the pressure in it. The **epiglottis (U),** which is part of the respiratory system, is also visible. This flap of tissue prevents food from entering the trachea.

Cheek	A	❍
Lips	B	❍
Vestibule	C	❍
Teeth	D	❍
Gingiva	E	❍
Hard Palate	F	❍
Soft Palate	G	❍
Uvula	H	❍
Palatopharyngeal Arch	I	❍
Palatoglossal Arch	J	❍
Tongue	K	❍
Lingual Frenulum	L	❍

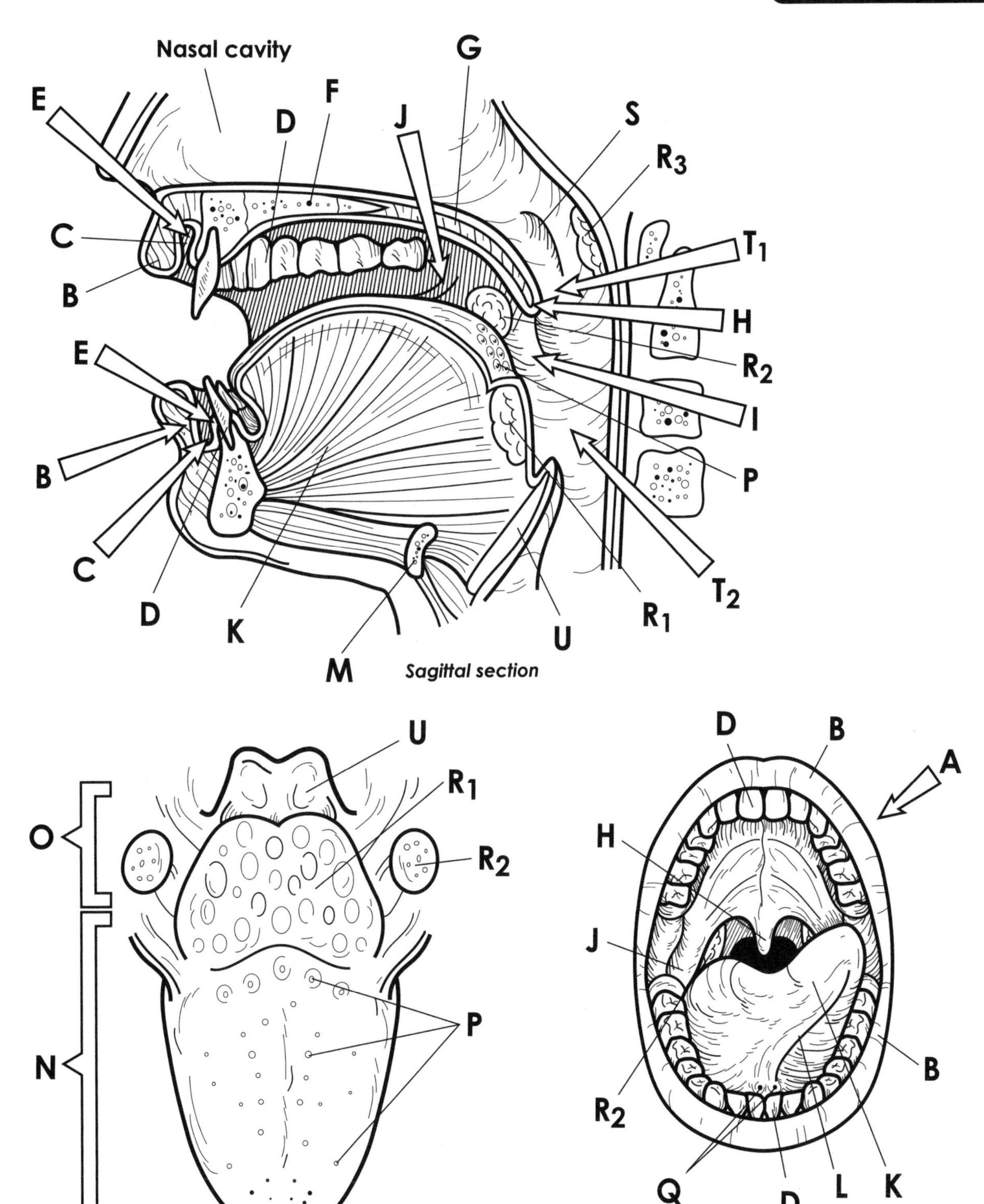

Hypoid Bone	M	❍	Sublingual Salivary Glands	Q	❍	Eustachian Tube	S	❍
Body	N	❍	Lingual Tonsils	R_1	❍	Nasopharynx	T_1	❍
Root	O	❍	Palatine Tonsils	R_2	❍	Oropharynx	T_2	❍
Papillae	P	❍	Pharyngeal Tonsil	R_3	❍	Epiglottis	U	❍

THE RESPIRATORY SYSTEM

The respiratory system consists of passageways that filter incoming air as it enters the lungs. Here, in microscopic air sacs, gas exchange takes place between the external, atmospheric air and the body environment; the next plate explores these exchanges. Some of the structures we'll see in this plate have already been encountered in previous plates.

> Certain structures are designed with lowercase letters because they are associated with other systems. The uppercase letters are reserved for parts of the respiratory system.

On entering the body, air passes through the first structure of the upper system: the **nasal passage (A).** A light color should be used for this passageway. Within the nasal passage, outcroppings of bone from the lateral wall divide the main passageway into smaller ones. These outcroppings are called **nasal conchae (A_1).** Air circulates around these ridges and becomes warm and moist before entering the lungs.

The upper respiratory system also contains a number of air-filled spaces between several of the bones of the skill, called sinuses. The diagram shows the **frontal sinus (B_1)** and the **sphenoid sinus (B_2);** air is warmed and moistened in these spaces. Cilia and mucus along the walls of the sinuses also trap foreign microorganisms and debris in the air.

In this section of the head, we also see features of the digestive system. The **tongue (a)** is the large muscle that fills most of the space of the oral cavity. It assists in the mechanical digestion of the food. The oral cavity leads to a major passageway called the **pharynx (C),** which serves both the respiratory and digestive systems. The **esophagus (b)** leads from the pharynx to the stomach.

> As we study the next area we encounter several passageways that lead to the lungs; notice how these passageways look like an inverted tree. Pale colors should be used for these structures to avoid obscuring their details.

Below the pharynx, we encounter the **larynx (D),** which is the first portion of the passageway that leads to the lungs. The bracket outlining this structure should be colored boldly, and a pale color should be used for the structure itself. The flap-like epiglottis guards the entry to the larynx, and several plates of cartilage make up the walls of the larynx.

Continuous with the larynx is the windpipe, also known as the **trachea (E).** The rings of the trachea are made up of cartilage, and you may use a dark color to highlight them.

The trachea extends in front of the esophagus into the thoracic cavity and then splits to form two passageways called bronchi. On the visual right (the anatomical left) is the **left bronchus (F_1),** and at your visual left (the anatomical right) is the **right bronchus (F_2).** The color that was used for the trachea should be used here to show the continuity of the tube, and the arrows should be colored in a bold color such as a blue or purple.

Each bronchus continues as a bronchial tree. The **left bronchial tree (G_1)** and the **right bronchial tree (G_2)** are designated with arrows that may be colored boldly, but the tubes themselves should be colored with a light color. Cartilage rings hold the tubes open through most of their length, and the tubes themselves are composed of smooth muscle. The right and left bronchial trees lead to the smaller alveolar ducts and then to the air sacs of the lungs.

> The main area of gas exchange in the body is the lungs. The lungs should be shaded with a very light color in the plate. We also show some of the bones that surround the lungs, which should be highlighted in light colors.

The **left lung (H_1)** and the **right lung (H_2)** occupy most of the space of the thoracic cavity. The lungs are soft and spongy and cone-shaped. They are separated from each other by the heart and a central area of the thorax called the mediastinum.

The expansion of the lungs depends heavily on the activity of the large dome-shaped muscle known as the **diaphragm (I).** When this muscle contracts, air enters the lungs, as the next plate explains. Enclosing the lungs are the bones of the thoracic cage; these bones protect the lungs and provide support for the thoracic cavity. These include the **ribs (c),** the **sternum (d),** and the **clavicles (e).** We suggest you outline the margins of these bones to show their proximity to the lungs. The muscles associated with the ribs are an essential factor in breathing, as the next plate demonstrates.

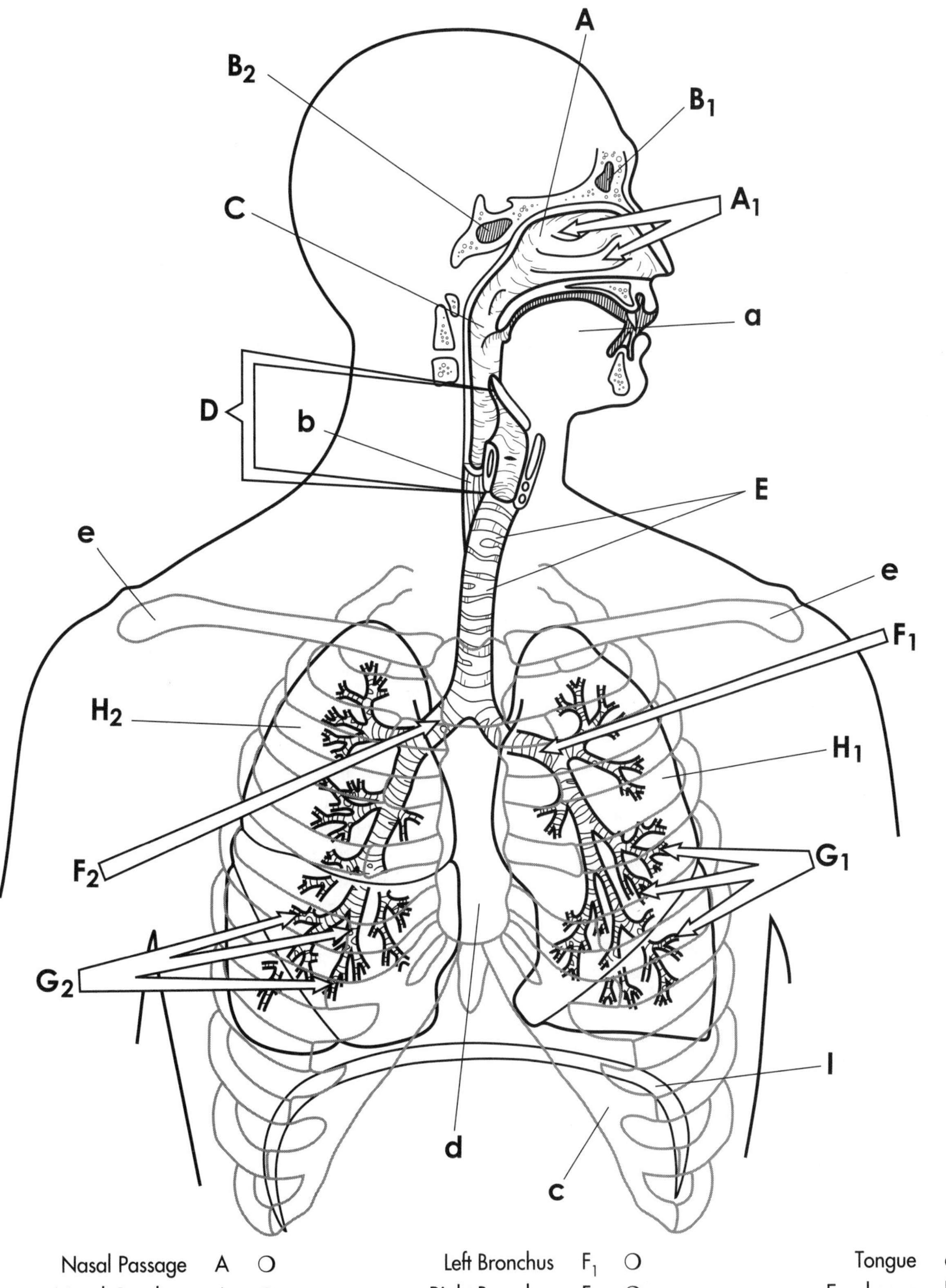

Nasal Passage	A	❍
Nasal Conchae	A_1	❍
Frontal Sinus	B_1	❍
Sphenoid Sinus	B_2	❍
Pharynx	C	❍
Larynx	D	❍
Trachea	E	❍
Left Bronchus	F_1	❍
Right Bronchus	F_2	❍
Left Bronchial Tree	G_1	❍
Right Bronchial Tree	G_2	❍
Left Lung	H_1	❍
Right Lung	H_2	❍
Diaphragm	I	❍
Tongue	a	❍
Esophagus	b	❍
Ribs	c	❍
Sternum	d	❍
Clavicles	e	❍

THE MECHANISM OF BREATHING

The respiratory system functions in the exchange of oxygen and carbon dioxide between body cells and the external environment. The system contains highly branched, hollow tubes that form air passageways and conduct air into and out of the lungs. The mechanism by which inhalation and exhalation takes place is the primary subject matter of this plate. We also discuss how gas is exchanged at clusters of microscopic air sacs called alveoli.

This plate contains two parts: In the first series of diagrams we consider the mechanism by which breathing takes place, and in the second, single diagram, we study the structure of the air sacs and their associated vessels. Focus your attention on the first set of diagrams.

Breathing takes place when air enters and leaves the lungs. We begin this study by focusing on the first diagram entitled At Rest. The **lungs (A)** are shown as paired organs that occupy most of the space of the **thoracic cavity (B).** Notice that the thoracic cavity is a closed environment, bounded along the right and left sides by the ribs and **intercostal muscles (C),** and at the bottom by a dome-shaped muscle, the **diaphragm (D).** Air enters and leaves the lungs through a set of passageways that include the windpipe, or **trachea (E).**

In the process of inhalation, the intercostal muscles receive a series of nerve impulses, which cause them to contract. When these muscles contract, the diaphragm contracts as well. An arrow shows the diaphragm (D) contracting and moving downward. These contractions increase the size of the thoracic cavity (B). When the volume of the thoracic cavity increases, the elastic lungs stretch out because they are attached to the walls of the cavity by a membrane called the pleura. The lungs expand as the thorax expands, and this lung expansion increases the volume space within them. Expansion lowers the air pressure (the air molecules have more room to move into) and, since air flows from a region of high pressure to a region of low pressure, atmospheric air flows freely into the lungs. This flow is indicated by the arrow showing **inhaled air (F).**

The inhalation of air is followed by exhalation. During exhalation, the intercostal muscles and diaphragm relax. The arrow shows the diaphragm relaxing and returning to its original dome shape. In the process of relaxation, the volume of the thoracic cavity decreases, and this activity compresses the lungs so that they return to their original size. The air within the lungs is compressed, and the **exhaled air (G)** flows out into the atmosphere, as is shown by the arrow. The flow of air is a passive process; it depends on movement of the thoracic compartment, intercostal muscles, and diaphragm.

Having discussed the breathing process, we will now turn to the process of gas exchange that occurs in the lungs, Our focus is on the bottom diagram, entitled Gas Exchange. Continue your reading as you color the plate.

The lung consists of millions of microscopic, cup-shaped sacs called alveoli. In the diagram, we point out a single **alveolus (H)** that is grouped with several others. Alveoli have extremely thin membranes across which gases pass freely. The average adult set of lungs contains about three million alveoli.

Blood reaches the alveoli by flowing through the **pulmonary artery (I),** which leads from the heart. This major artery divides and subdivides until it forms microscopic **arterioles (J).** As arterioles approach the alveoli, they become **capillaries (K),** through which red blood cells must pass single file. Blood in the capillaries carries dissolved carbon dioxide. During exhalation, the level of carbon dioxide in the alveoli is low, so it diffuses out of the blood and into the alveolus so that it can be expelled.

At the same time that carbon dioxide leaves the blood, oxygen flows into it from the alveoli, because the level of oxygen in the alveoli is high following an inhalation.

In the diagram, we show that the capillaries become **venules (L).** These venules lead blood away from the alveolus, uniting with one another as the diagram illustrates, and eventually forming the **pulmonary vein (M).** After it passes through the heart, the blood is pumped to all parts of the body.

The driving force behind gas exchange is diffusion: the movement of gas molecules from an area of high concentration to an area of low concentration. To summarize, after inhalation, the oxygen content is high in the alveolus and low in the blood. Therefore, oxygen diffuses into the blood, and carbon dioxide diffuses in the opposite direction.

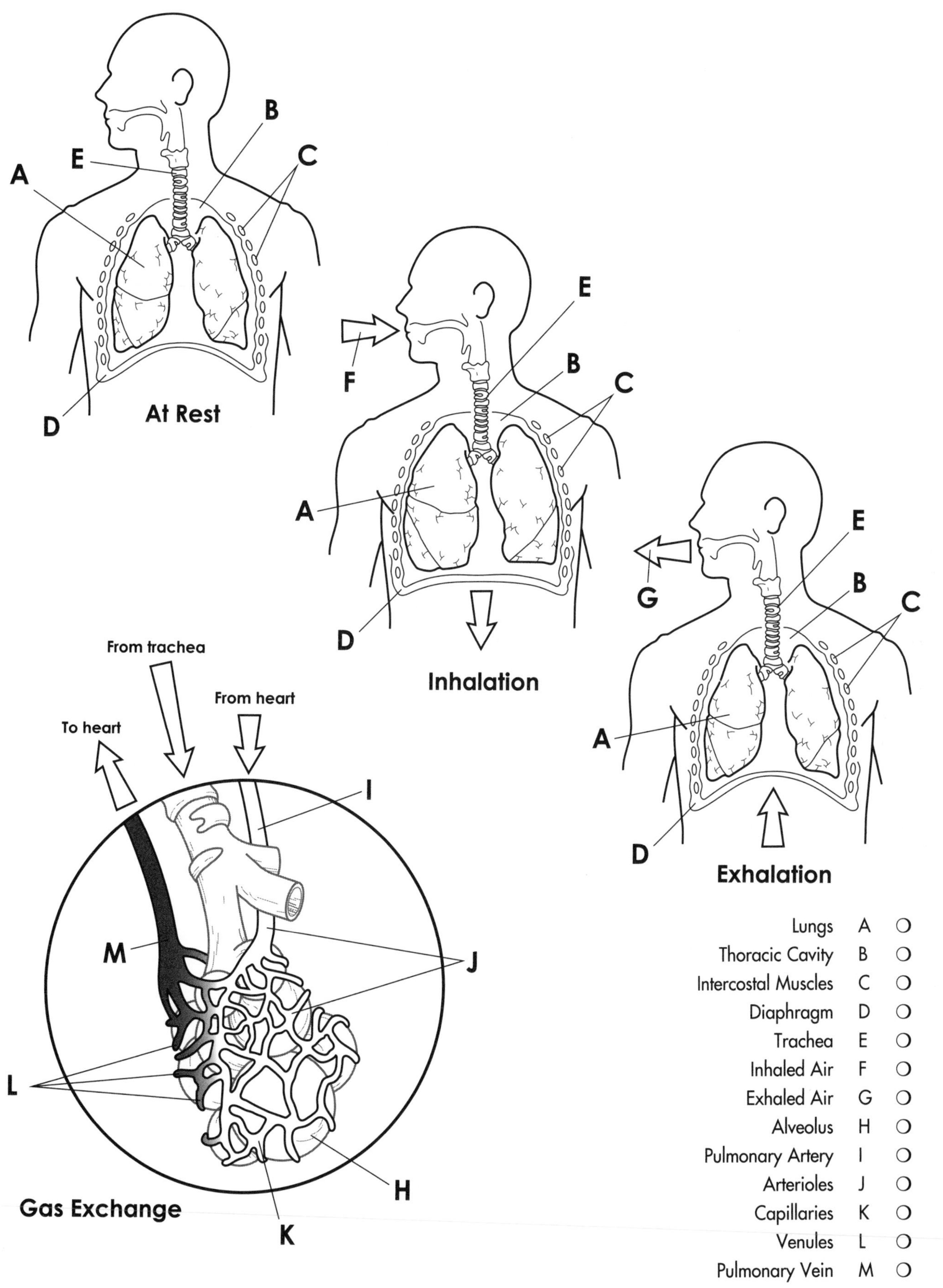
A
B
C
D
E
At Rest
F
Inhalation
G
Exhalation
From trachea
From heart
To heart
I
J
M
L
H
K
Gas Exchange
Lungs A
Thoracic Cavity B
Intercostal Muscles C
Diaphragm D
Trachea E
Inhaled Air F
Exhaled Air G
Alveolus H
Pulmonary Artery I
Arterioles J
Capillaries K
Venules L
Pulmonary Vein M

THE URINARY SYSTEM

Excretion refers to the elimination of the waste products of cellular metabolism and undigested food from the body. It also refers to the removal of surplus materials from tissues. Excretions help to regulate the water and salt content of the body. These regulatory functions are performed by the kidneys and their accessory structures in the urinary system. In this plate, we present an overview of the urinary system in two views. We point out the major organs of the system and describe their structure.

Looking over the plate, you will notice that it shows the organs of the urinary system in two views. In the front and back views, the organs are shown relative to each other and nearby structures. As you read the text, locate the structures and color them.

The main organs of waste filtration in the body are the **kidneys (A_1 and A_2).** A medium color is recommended for both kidneys; red or purple would be suitable.

The kidneys are bean-shaped organs that are roughly the size of the fist. They are located on each side of the vertebral column, and as the posterior view shows, the 12th rib partially protects them. It should be colored a light color.

Leading from the kidneys are tubes called **ureters (B).** These tubes carry urine away from the kidney after it has been formed (this topic is discussed in depth in the next plate), and the ureters lead to the main storage organ, the **urinary bladder (C).** This hollow muscular sac is located at the floor of the pelvic cavity.

The tube that leads from the bladder to the exterior of the body is the **urethra (D).** This tube of smooth muscle is about 1.5 inches long in the female and about 8 inches long in the male; in the male, the urethra passes through the penis.

We will now discuss the blood supply to the kidney. The passageways for blood were mentioned previously in the discussion of the circulatory system, and we will review them here. As you encounter the structures in the reading, color them in the anterior and posterior views.

The main circulatory vessel that transports blood to the kidney is the **renal artery (E),** seen in the posterior view. The **renal vein (F)** lies behind the renal artery in the posterior view and is difficult to see. A light color is recommended for this structure. The renal vein transports blood away from the kidney after it has been cleansed. The renal artery is supplied with blood by the **abdominal aorta (G),** while the renal vein empties its blood into the **interior vena cava (H).**

We close the plate with a transverse section through the body. We are looking down from above at a level of the stomach, transverse colon, pancreas, and other organs of the abdominal cavity.

The view in the plate on the bottom shows structures of several systems. For example, we see an outline of the **lumbar vertebra (I),** and the spinal cord is visible. Sections are also shown through the kidneys (A_1 and A_2). Note the renal arteries (E), which arise from the abdominal aorta (G). Also note the renal veins (F), which lead on both sides to the inferior vena cava (H).

The kidney is surrounded by three layers of supportive tissue. Immediately adhering to the kidney surface is the **renal capsule (J),** which provides an impenetrable barrier to infection of the kidney surface; a dark color may be used to highlight this layer. Outside the renal capsule is a layer of fat called the **adipose capsule (K).** The fat tissue helps cushion the kidney against blows. Outside the adipose capsule is the **renal fascia (L),** which is composed of fibrous connective tissue. It helps anchor the kidneys to nearby tissues.

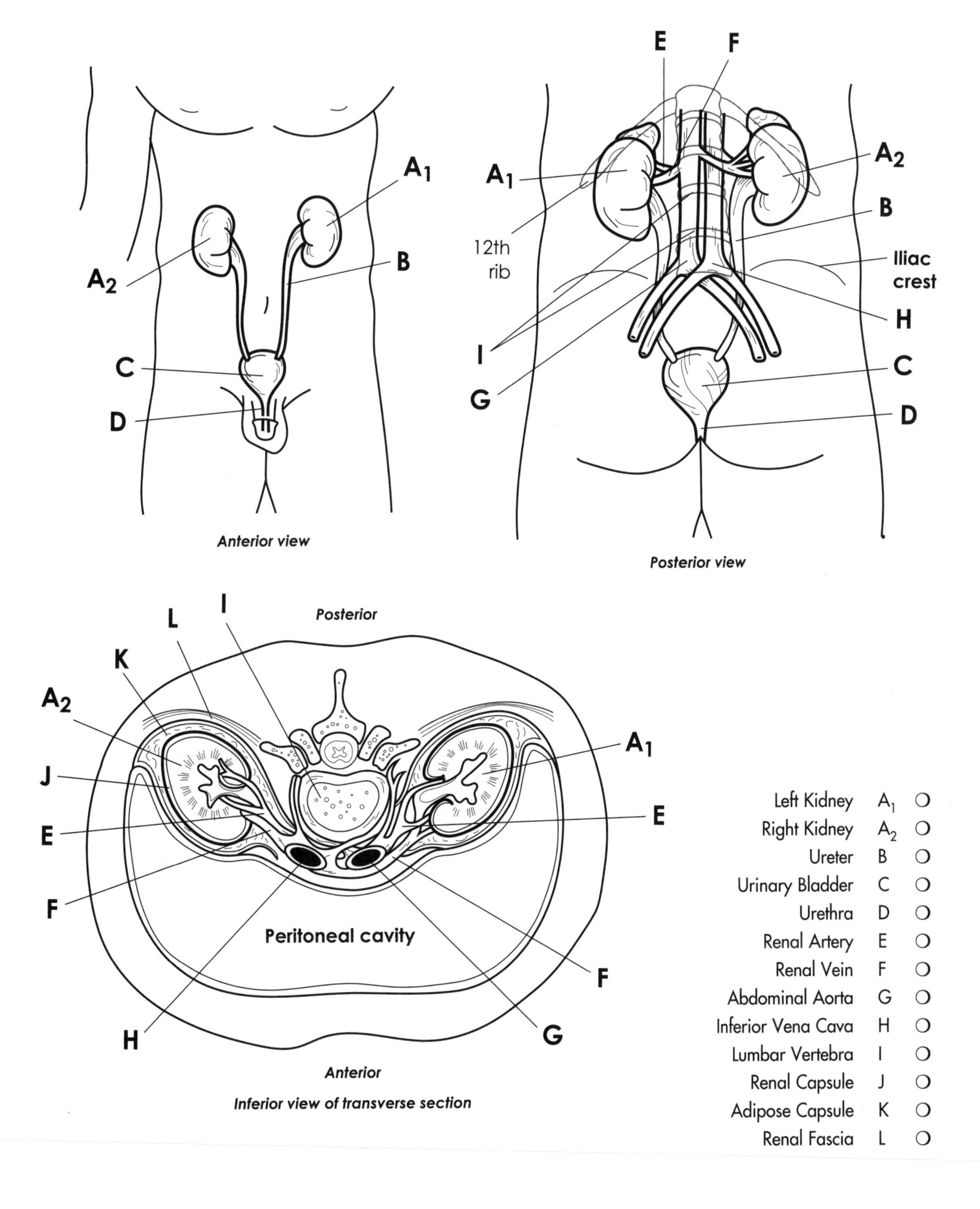
A1
A2
B
C
D
Anterior view
E
F
A1
12th
rib
A2
B
Iliac
crest
I
G
H
C
D
Posterior view
L
I
Posterior
K
A2
J
E
F
A1
E
F
Peritoneal cavity
H
G
Anterior
Inferior view of transverse section
Left Kidney A1
Right Kidney A2
Ureter B
Urinary Bladder C
Urethra D
Renal Artery E
Renal Vein F
Abdominal Aorta G
Inferior Vena Cava H
Lumbar Vertebra I
Renal Capsule J
Adipose Capsule K
Renal Fascia L

THE NEPHRON

As you learned in the preceding plate, the two major organs involved in the process of excretion are the kidneys. In this plate, we will discuss the nephron, the independent unit inside the kidney that produces urine. There are approximately one million nephrons in each human kidney. Nephrons perform the functions of filtration, reabsorption, and secretion.

Looking over the plate, you will notice that it is composed of three diagrams: the entire kidney, nephrons in their environment, and a close-up of a single nephron, which is the microscopic unit of the kidney. Color the structures as you encounter them.

In the first diagram, we see the **kidney (A),** as it was shown in the previous plate. One section of the kidney is enlarged and shown in detail, and this area contains a **nephron (B).** A pale shade is recommended to highlight it. The **renal artery (C)** delivers blood to the kidney, the **renal vein (D)** removes blood from it, and the **ureter (E)** is responsible for carrying urine away.

We now focus on the expanded view of a section of the kidney, in which we see eight nephrons. We will briefly study this perspective before going on to the detailed view of a single nephron.

The second diagram of the plate shows the **cortex (F),** which is indicated by a bracket and should be colored in a bold color; its general area can be colored in a pale hue. This diagram also shows the **medulla (G).**

Within the cortex and medulla we present the simplified views of eight nephrons. Each of the eight nephrons has a **renal corpuscle (H),** and we have circled one of these. Nephrons are oriented so that they're perpendicular to the kidney's surface. There are **cortical nephrons (H_1)** and **juxtamedullary nephrons (H_2),** both of which have capsules, proximal tubules, and distal tubules that are in the cortex, and long **loops of Henle (R)** that extend into the medulla. In this diagram, we recommend that you color in the cup-like renal corpuscles and then the tubules that lead from the corpuscle as they extend toward a collecting duct. The **collecting duct (I)** receives urine from many nephrons.

Now we come to a single nephron and study its vascular and tubular components. We recommend variations of one color for all parts of the vascular system, and variations of a different color for parts of the tubular component. Continue reading below as you study the nephron, and color the structures as you come upon them.

Blood travels from the heart through a series of arteries and finally reaches the **interlobular artery (J);** one branch of this artery is the **afferent arteriole (K).** Blood flows through this vessel into a small network of capillaries called the **glomerulus (L).** Filtration takes place here (as we describe below), and then the blood enters a vessel called the **efferent arteriole (M).** The efferent arteriole branches into a network of capillaries, called the **peritubular capillary network (N).** You should use variations of a single color for these vascular tubules.

Filtration occurs when blood passes through the glomerulus. As blood passes through the glomerulus, normal blood pressure forces fluid across the wall of the nephron tubule, which is also known as **Bowman's capsule (O).** Together the glomerulus and Bowman's capsule are called the renal corpuscle. This fluid, or filtrate, contains water, salts, glucose, nitrogenous wastes, and other small molecules. The blood then continues to flow through the tubular component, first encountering the **proximal convoluted tubule (P).** At several points along the nephron tubule, fluid is filtered out and reabsorbed back into blood that is traveling through the peritubular capillary network. Most this reabsorption occurs in the proximal convoluted tubule. Here, large amounts of ions, glucose, amino acids, and water are brought back into the body. These are useful substances, so our bodies do not want to eliminate them in the urine.

Now the proximal tubule descends toward the renal medulla. The descending tubule is called the **descending limb (Q)** and functions in water reabsorption. This tubule turns abruptly at the **loop of Henle (R)** and ascends as the **ascending limb (S),** where salt is reabsorbed. Variations of the color used for the proximal tubule should be used to color here. The peritubular capillaries surround the tubules as they dip into the medulla, and water and salts leaving the nephron tubule are reabsorbed back into the blood.

As the tubule ascends, it forms the **distal convoluted tubule (T).** Again, it intertwines with capillaries of the **peritubular capillary network (N)** and selective reabsorption takes place. This reabsorption is critical; without it the body would lose an excess of water, vitamins, and other important molecules and ions in the process of excretion.

The hormone vasopressin allows water reabsorption from the distal convoluted tubule, and the hormone aldosterone allows sodium reabsorption. This means the distal convoluted tubule plays an important role in regulating salt and water balance in the body, and thus blood pressure. In addition, both bicarbonate (HCO_3^-, a base) and protons (H^+, an acid) can be secreted in the urine. Bicarbonate can also be reabsorbed back into the body. The relative rates of these processes help regulate our body's pH over the long term.

The capillary network then moves toward the renal corpuscle and forms the **interlobular vein (U).** This vein ultimately leads to the renal vein, which removes the cleansed blood from the kidney. Following the tubule once again, you will notice that the distal convoluted tubule (T) comes to the collecting duct (I). The fluid in the collecting tubule is urine, which is sent to the ureter for discharge.

The kidney is an amazing organ! To help give you a sense of how hard your kidneys work, here are some interesting facts: An adult human has about 5 L of blood and produces 1–2 L of urine a day, but 140 L of fluid goes through the glomerulus a day. Think of how much urine you would produce if reabsorption wasn't an efficient process! Your body filters 1 L of blood every minute, and filters your entire blood pool about 30 times a day.

THE NEPHRON

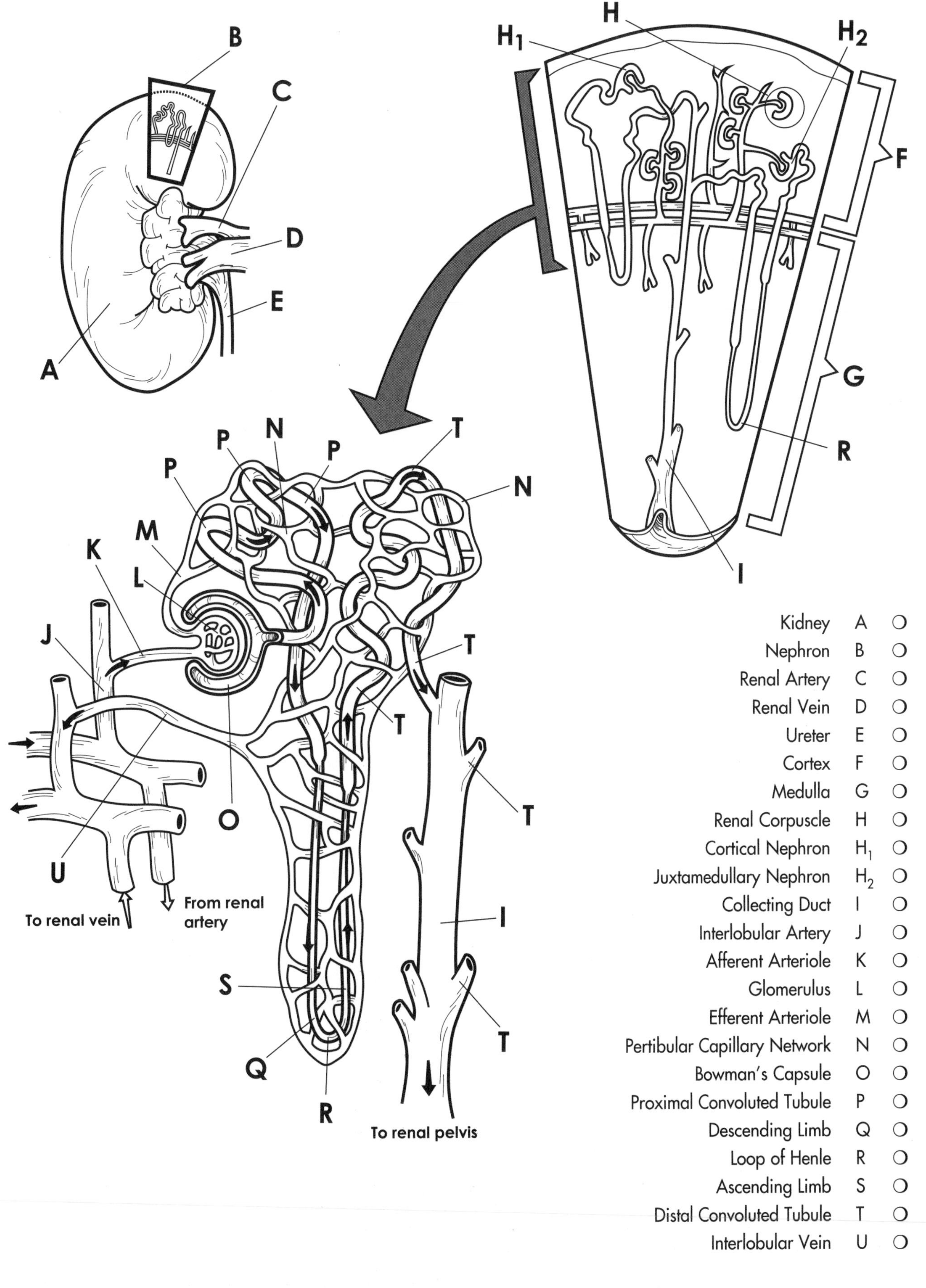

A
B
C
D
E
F
G
H
H1
H2
I
J
K
L
M
N
O
P
Q
R
S
T
U
To renal vein
From renal artery
To renal pelvis
Kidney A
Nephron B
Renal Artery C
Renal Vein D
Ureter E
Cortex F
Medulla G
Renal Corpuscle H
Cortical Nephron H1
Juxtamedullary Nephron H2
Collecting Duct I
Interlobular Artery J
Afferent Arteriole K
Glomerulus L
Efferent Arteriole M
Pertibular Capillary Network N
Bowman's Capsule O
Proximal Convoluted Tubule P
Descending Limb Q
Loop of Henle R
Ascending Limb S
Distal Convoluted Tubule T
Interlobular Vein U

CHAPTER 10:

REPRODUCTION and DEVELOPMENT

TYPES OF REPRODUCTION

Organisms can reproduce either asexually or sexually. **Asexual reproduction** is a mode of reproduction in which offspring arise from a single parent, and inherit the genes of that parent only. As a consequence, there is no variation in genetic material and the offspring are genetically identical to the parent. Asexual reproduction has short-term benefits when rapid population growth is important, in stable environments. Asexual reproduction is common in each of the major domains/kingdoms.

There are six common modes of asexual reproduction:

1. In **fission (A),** the parent organism is replaced by two daughter organisms by dividing in two. The two daughter cells are usually approximately equal in size. This is common in bacteria, Archaea, protists, and some unicellular fungi (yeasts). Diagram A shows a bacterial cell undergoing fission. Be sure to color the cell wall, cell membrane, genome, and cytoplasm consistently across the four parts of this diagram.

2. Parthenogenesis (B) does not involve a male gamete, and occurs when an unfertilized egg develops into a new individual. In other words, the female's egg is not fertilized by a sperm cell, but still grows into a new organism. Parthenogenesis occurs naturally in many plants, worms, insects, and vertebrates (particularly reptiles, amphibians, and fish). In diagram B, color the unfertilized egg at the bottom left the same color as the lizard hatching out of its shell. Color the shelled egg in the middle and the broken egg on the right the same light color.

3. Budding (C) occurs when a small offspring grows out of the body of a parent. The bud grows into a fully mature organism and breaks off from the parent. This can occur in unicellular organisms such as budding yeast and in multicellular organisms such as the hydra. Color the hydras in diagram C the same color as each other.

4. Fragmentation (D) occurs when a new organism grows from a fragment of the parent. Note that this is different from regeneration (discussed below). Each fragment develops into a mature, fully grown individual. Fragmentation is seen in animals (such as annelid worms and sea stars), fungi, and plants. Color the sea stars in diagram D the same color as each other.

5. Vegetative reproduction (E) occurs when a new plant is formed without the production of seeds or spores. For example, a new strawberry plant can grow from a runner attached to a parent plant.

6. Spore formation (F) can be done by many organisms (all plants, most protists, most fungi, some bacteria). Some spores are resistant to harsh conditions, and so are a survival mechanism for organisms when growing conditions are not ideal. When spores germinate, they grow into multicellular individuals without a fertilization event. Sporogenesis can involve meiosis (in which the ploidy number is lower in the spore compared to the parent organism) or mitosis (in which the parent and the spore have the same genetic content). Plants typically perform meiotic sporogenesis (in which a diploid parent plant generates haploid spores) and fungi typically perform mitotic sporogenesis (in which the parent and the spore are both haploid). Image F shows bread mold reproducing using asexual spores.

Regeneration is commonly mistaken as asexual reproduction, but is not a mode of reproduction. This is a wound-healing mechanism, in which lost tissues, organs, or limbs can be regrown in an organism. For example, sea stars can regrow lost arms, geckos can regenerate tails **(G),** humans can regenerate liver tissue, and the planarian flatworm can regenerate just about anything.

Sexual reproduction (H) increases genetic variation in individuals, because each offspring inherits genetic information from two parents. Genetic diversity allows adaptation to changing environments. Most multicellular organisms (fungi, plants, and animals) can perform sexual reproduction. Plants use both spores and gametes in sexual reproduction. Fungi typically use sexual spores. Animals reproduce sexually via gametes. In diagram H you can see a sperm cell (male gamete) and an egg cell (female gamete) joining in fertilization to produce a zygote. The zygote then develops into an embryo. Sexual reproduction via gametes requires two processes:

1. Genetic ploidy is decreased via meiosis, and gametes are generated.
2. Gametes join during fertilization, restoring the ploidy of the organism.

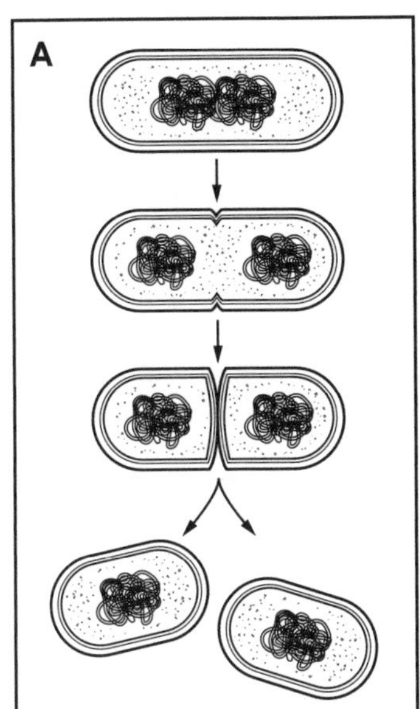

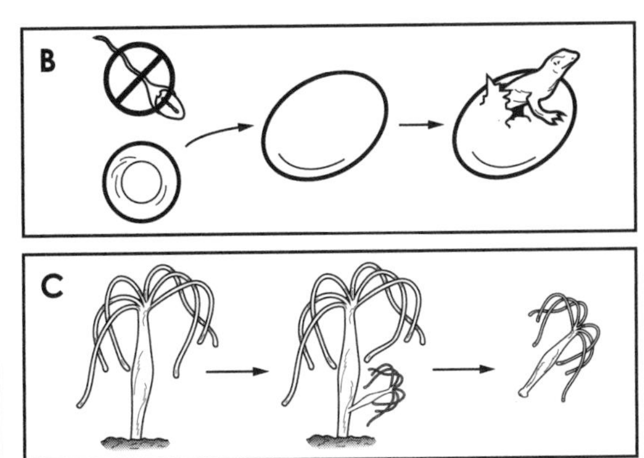

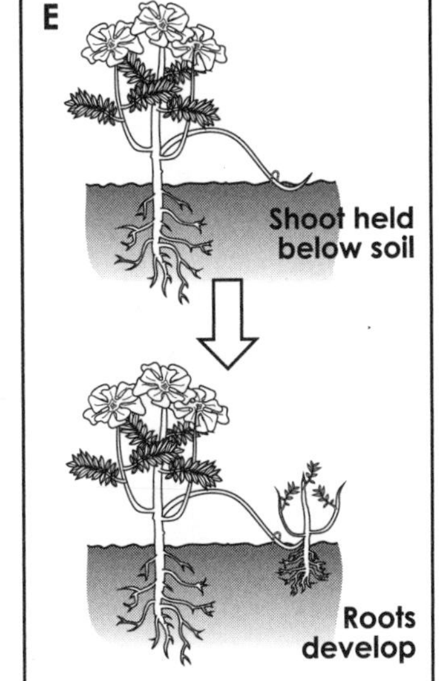

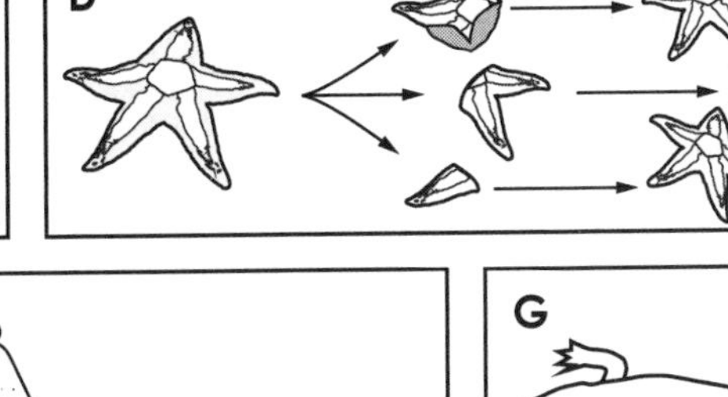

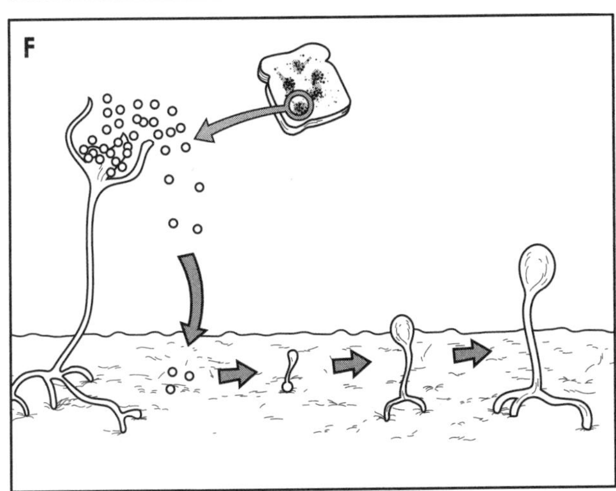

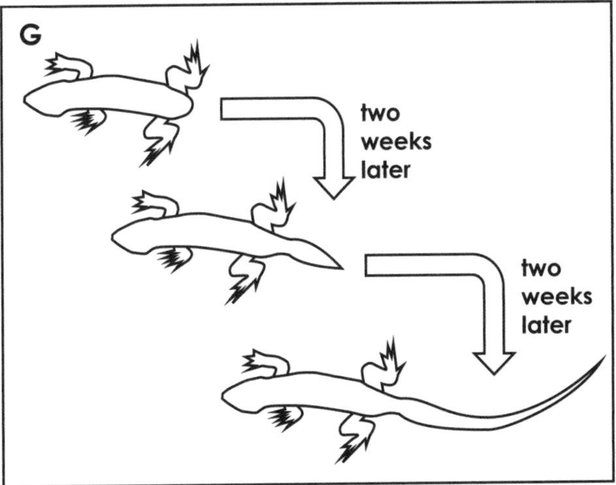

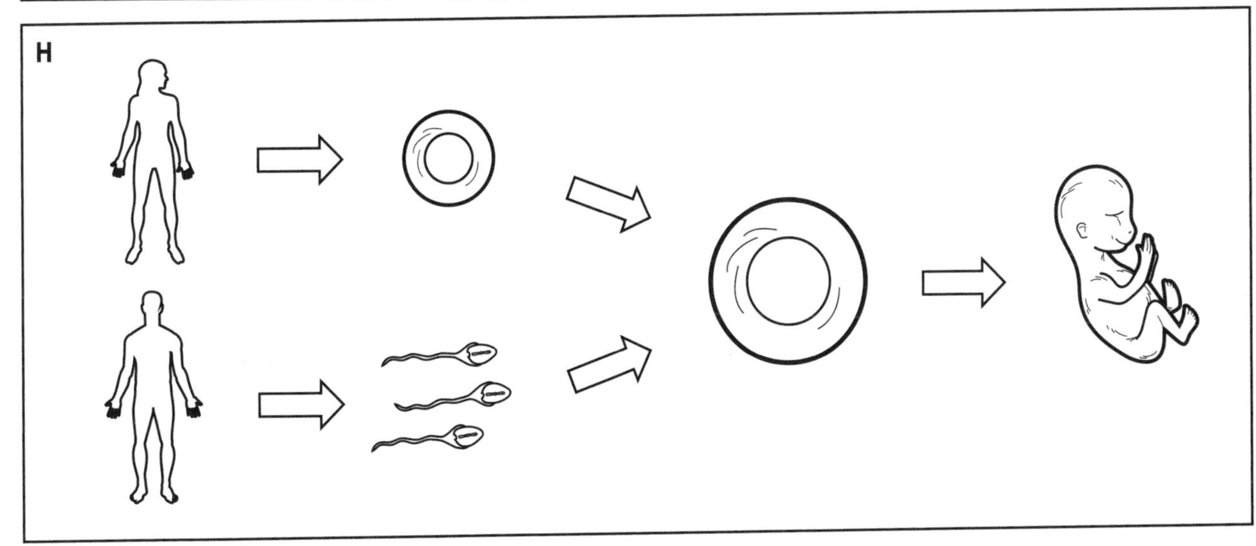

Fission A ❍
Parthenogenesis B ❍
Budding C ❍
Fragmentation D ❍
Vegetative Reproduction E ❍
Spore Formation F ❍
Regeneration G ❍
Sexual Reproduction H ❍

THE MALE REPRODUCTIVE SYSTEM

The primary function of the male reproductive system is to produce sperm cells and deliver them to the female reproductive system where they will fertilize egg cells. In this way, genetic material is passed from generation to generation and new individuals are produced. In this plate, we will present some of the structures of the system and their functions.

As you look over the plate, note that we are presenting a cross section. We will examine the male system from the left side and note its main structures. Light colors are recommended for the smaller structures.

The major organs of the male reproductive system are the two testes, where sperm cells are produced. A single **testis (A)** is noted in the diagram, and a dark color should be used for this large structure. Surrounding the testis is a comma-shaped structure called the **epididymis (B).** The tightly coiled tubules contained in this organ store sperm cells while they mature.

Arising from the epididymis is a long tube that leads sperm cells out of the body. This tube is called the **vas deferens (C).** Note the long length of this tube as it courses from the epididymis up into the body, curves to the left, and passes the **urinary bladder (O).** The vas deferens soon meets the duct that leads from the **seminal vesicle (K)** and forms the **ejaculatory duct (D).** The ejaculatory duct joins with the urethra from the urinary bladder.

We now follow the pathway of sperm cells out of the body. A common tube, the urethra, services both the reproductive and urinary systems at this point. As you continue to read below, locate and color the structures in the plate.

The **urethra (E)** is discussed in an earlier plate on the urinary system. It is quite a long tube in the male, and runs alongside the prostate gland, and then enters the **penis (F),** which is approximately 6 to 8 inches long.

One important feature of the penis (F) is the **corona (G),** which is the outer edge of the **glans penis (H).** Covering the glans in the uncircumcised penis is a portion of skin tissue called the **prepuce (I).**

The penis is suspended from the muscle above by a suspensory ligament, and is attached to the sac that contains the testis. This sac hangs from the root of the penis and is called the **scrotum (J).**

We will now consider three accessory glands that add secretions to the sperm cells to produce semen. These glands are quite small, and you should color them lightly.

The first accessory gland is the **seminal vesicle (K).** This paired gland lies near the base of the urinary bladder. Its alkaline fluid is added to the sperm cells as they enter the ejaculatory duct. The second gland is the **prostate gland (L).** This large, single gland is about the size of a walnut and surrounds the urethra. The acidic fluid that it contributes to sperm contains several enzymes.

The third gland we will mention is the **bulbourethral gland (M),** which adds alkaline secretions to the sperm. This paired gland is also known as the Cowper's glands.

For reference purposes, we have labeled the **pubis (N),** which is near the point where the urethra enters the body. The urinary bladder (O) is prominent in the diagram, and the last structures we point out are the **rectum (P),** and the temrinus of the gastrointestinal tract, the **anus (Q).**

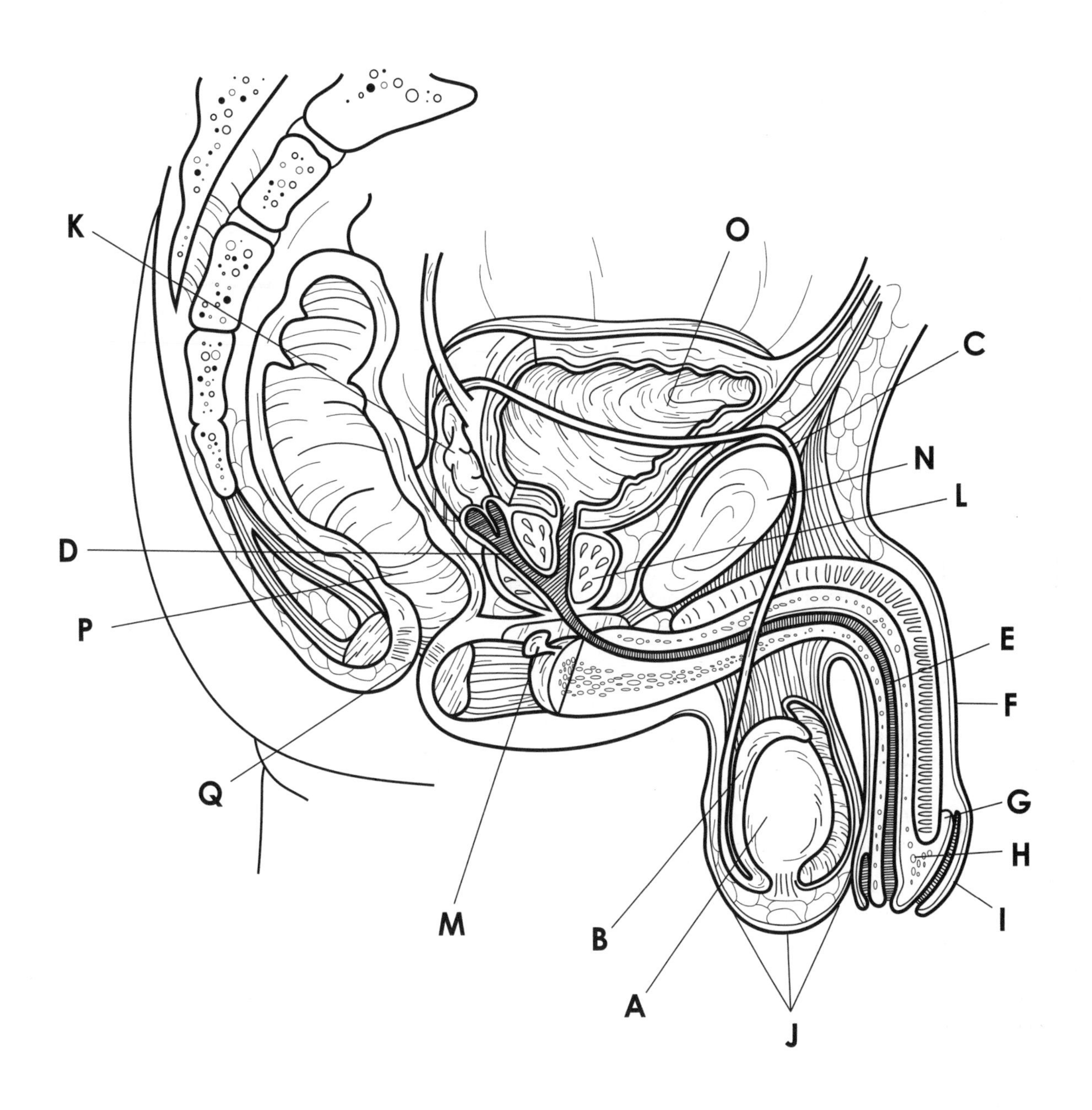

Testis A ❍
Epididymis B ❍
Vas Deferens C ❍
Ejaculatory Duct D ❍
Urethra E ❍
Penis F ❍

Corona G ❍
Glans Penis H ❍
Prepuce I ❍
Scrotum J ❍
Seminal Vesicle K ❍
Prostate Gland L ❍

Bulbourethral Gland M ❍
Pubis N ❍
Urinary Bladder O ❍
Rectum P ❍
Anus Q ❍

THE FEMALE REPRODUCTIVE SYSTEM

The female reproductive system produces sex cells that join with male sperm cells in the process of fertilization. In addition, the female system nurtures the developing embryo and fetus for approximately nine months. For this reason, the female system is more complex than the male system.

In this plate, we will present an overview of the female reproductive system as a prelude to discussions in the following plate.

> Looking over the plate, you will see that it contains a single cross section of the female reproductive tract as seen from the left side. Lighter colors are suggested for the reproductive structures.

The **ovaries (A)** are female reproductive organs in which egg cells are formed, and they are also the site of the production of the female sex hormones. The egg cells travel from the ovaries into the **Fallopian tubes (B).**

The Fallopian tubes lead the egg cell to the **uterus (C),** which is a muscular organ that stretches to hold the fetus until its birth. The uterus is found just above the **urinary bladder (N).** The thick muscle of the uterus is apparent in the plate, and the innermost lining in the uterus is the **endometrium (D),** which is rich in blood vessels. The entire area that surrounds the uterine tubes, uterus, and ovaries should be colored in gray.

> We continue with our examination of the female reproductive system by focusing on the passageways that lead from the uterus and the external genitalia. Continue to follow the reading below and locate the structures in the plate. Light colors should be used because of the complexity of the areas studied.

The narrow opening that leads from the uterus is called the **cervix (E),** and the next structure encountered in the system is the **vagina (F).** A light color is recommended for this passageway. The vagina is a tubular organ that's approximately four inches in length. It receives the penis during sexual intercourse and acts as the birth canal through which babies are expelled; the walls of the vagina are thinner than are the very muscular walls of the uterus. The vagina passes through the **urogenital diaphragm (G),** which is near to another secretory gland, the greater **vestibular gland (H).** Also called Bartholin's gland, this paired gland produces lubrication in the form of mucus. The vagina opens at the **vaginal orifice (I),** which is indicated by an arrow.

In the region of the external orifice is the vulva, which is the area that encompasses the external genitalia of the female. One component of the vulva is the **mons pubis (J).** This is an elevation of adipose tissue that is covered by skin and coarse pubic hair. It cushions the **pubis (R),** which is where the pubic bones come together.

Another structure of the vulva is the **clitoris (K),** a small mass of erectile tissue that is homologous to the penis of the male. The labia are also located in this area; the **labium minora (L)** is the smaller fold of skin tissue of the vulva, and the **labium majora (M)** is larger fold of tissue. The labia are homologous to the scrotum of the male.

> We complete the plate by noting some of the other structures in the area.

In the pelvic region of the abdominal cavity, other organs lie close to the female reproductive system. Among these is the urinary bladder (N), which can be seen in the plate. The short tube that leads from the urinary bladder is the **urethra (O).** Posterior to the uterus is the **rectum (P)** of the gastrointestinal tract. It terminates at the **anus (Q).**

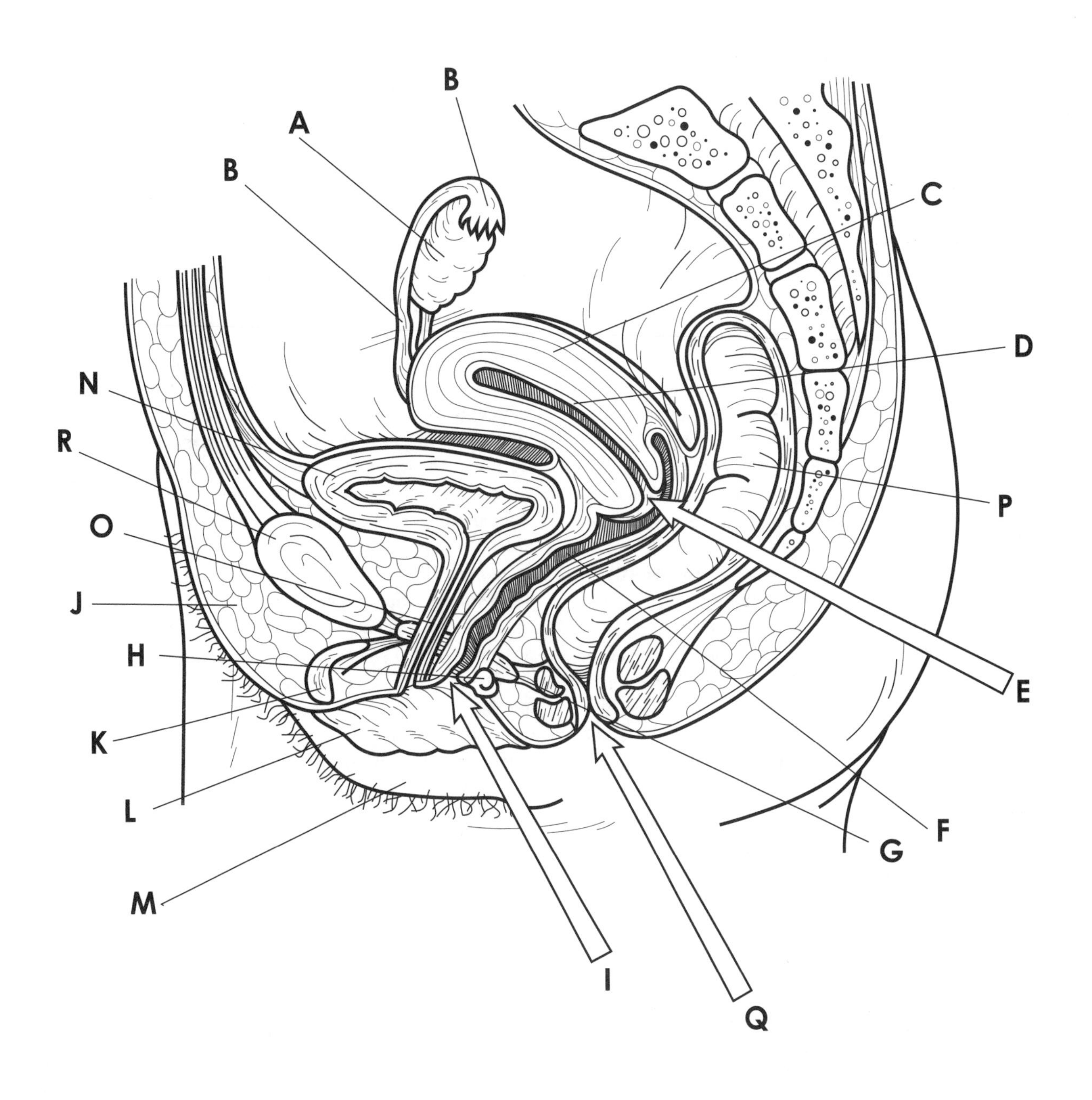

Ovary	A	❍	Urogenital Diaphragm	G	❍	Labium Majora	M	❍
Fallopian Tube	B	❍	Vestibular Gland	H	❍	Urinary Bladder	N	❍
Uterus	C	❍	Vaginal Orifice	I	❍	Urethra	O	❍
Endometrium	D	❍	Mons Pubis	J	❍	Rectum	P	❍
Cervix	E	❍	Clitoris	K	❍	Anus	Q	❍
Vagina	F	❍	Labium Minora	L	❍	Pubis	R	❍

GAMETOGENESIS

During the process of reproduction, gametes unite to form the zygote that develops into the new individual. Gametes are the sex cells: the sperm and egg cells. They are produced in the testes and ovary, respectively, in the process of gametogenesis.

This plate consists of two diagrams detailing the production of sperm cells by spermatogenesis and the production of egg cells by oogenesis. Concentrate on the left portion of the plate, where spermatogenesis is described.

Spermatogenesis takes place in the seminiferous tubules, within the testes. The process involves the production of primary spermatocytes from cells called spermatogonia. The primary spermatocytes then undergo the process of meiosis, which involves two subdivisions. The first is **meiosis I (A),** indicated by the arrow on the top half of the page. In meiosis I, primary spermatocytes become secondary spermatocytes. The secondary spermatocytes then enter **meiosis II (B),** indicated by the second arrow, and sperm cells result.

In the plate, we show the **primary spermatocyte (C),** in which you can see the **nucleus of the spermatocyte (D).** Near the nucleus is the **centrosome (E),** which is duplicated prior to meiosis, and splits at the beginning of meiosis, when it causes the growth of the microtubules used in meiosis. Three pairs of homologous chromosomes are shown: **homologous chromosomes #1 (F); homologous chromosomes #2 (G);** and **homologous chromosomes #3 (H).** In a human cell, there are a total of 23 pairs—46 chromosomes per cell.

As meiosis I proceeds, the three pairs of homologous chromosomes line up along the equator of the cell, as is shown in the second diagram. Toward the end of meiosis I, the pairs separate and one chromosome from each pair winds up in the **secondary spermatocytes (I)** that form in telophase. The process of meiosis is fully explained in the Meiosis plate early in this book. Note that at the end of meiosis I, each secondary spermatocyte has one member of each of the three original chromosome pairs. This means that each cell is haploid.

Now meiosis II begins. Both secondary spermatocytes (I) are involved. A chromosome consists of two sister chromatids, and the three chromosomes line up along the equator of the cell. Toward the end of meiosis II, the sister chromatids separate and move toward the poles of the cells, and cytokinesis occurs. Now each spermatid has three haploid chromosomes, and each **spermatid (J)** will develop into a **sperm cell (K)** with these haploid chromosomes. The original cell that begins spermatogenesis, the primary spermatocyte, has 46 chromosomes (the diploid number). Each sperm cell resulting from spermatogenesis has 23 chromosomes (the haploid number).

We have seen, in spermatogenesis, how a single cell with 46 chromosomes yields sperm cells, each with 23 chromosomes. We now turn to oogenesis, and see how a similar process results in egg cells, which also have 23 chromosomes. Continue your reading below as you color the plate.

Oogenesis is similar to spermatogenesis, but there are a few differences. During fetal development, about 7 million oogonia develop in the ovary. Before birth, about 2 million of them develop into a **primary oocyte (L)** by starting meiosis I. The oocyte arrests in prophase I until puberty. By this time, approximately half a million primary oocytes are still alive in the female's ovaries. Only about 400 oocytes are ever actually ovulated (released) in the average woman, and the remaining 99.9% will simply degenerate and die. We will now briefly review the events that occur when a primary oocyte continues through meiosis I. This happens once per month to a few primary oocytes. We show three homologous chromosomes: **homologous chromosomes #4 (M); homologous chromosomes #5 (N);** and **homologous chromosomes #6 (O).**

As meiosis I proceeds, the three pairs of homologous chromosomes line up and then separate during telophase. The result is two new cells, each of which has one of the three chromosome pairs. That means that each cell is haploid. One cell is the **secondary oocyte (P).** The second cell, which is reduced in size, is the **polar body (Q).** Although the polar body undergoes meiosis II, it will not divide to produce functional cells.

The secondary oocyte develops, deriving nutrients and growing within the follicle. It then moves toward the surface of the ovary, as is discussed in the next plate. At ovulation, the follicle erupts, and the secondary oocyte is released into the Fallopian tube. If sperm cells are present, a sperm nucleus will enter the oocyte, and meiosis II will take place.

In meiosis II, the chromosomes line up along the equator of the cell. The sister chromatids separate, and each chromatid becomes a chromosome. The cell that results is the **egg cell (R),** and it has 23 chromosomes. The other cell that results from meiosis II is a **polar body (Q).** In some animals, two other polar bodies also form from the first polar body, and these three polar bodies disintegrate. In mammals, the first polar body normally disintegrates before dividing, so only two polar bodies are produced.

Now fertilization can take place. The 23 chromosomes from the sperm cell unite with the 23 chromosomes of the egg cell to form a **zygote (S).** This union reestablishes the diploid condition of 46 chromosomes and is an essential hallmark of human reproduction. Through gametogenesis, sperm and egg cells are produced with the haploid condition, and when the union takes place to form a zygote, the diploid condition is reestablished.

GAMETOGENESIS

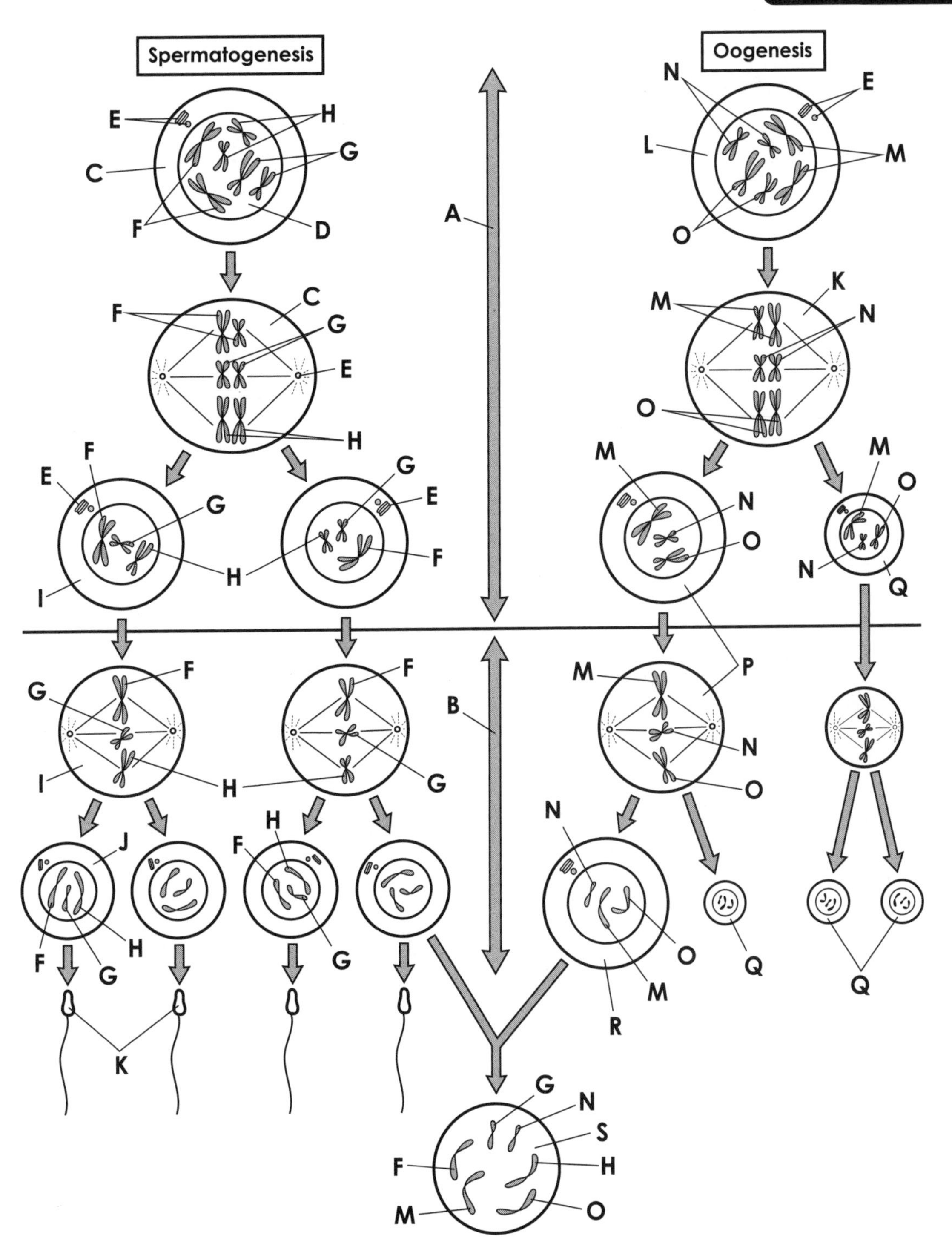

Meiosis I A ❍
Meiosis II B ❍
Primary Spermatocyte C ❍
Nucleus of Spermatocyte D ❍
Centrosome E ❍
Homologous Chromosomes #1 F ❍
Homologous Chromosomes #2 G ❍
Homologous Chromosomes #3 H ❍
Secondary Spermatocyte I ❍
Spermatid J ❍
Sperm Cells K ❍
Primary Oocyte L ❍
Homologous Chromosomes #4 M ❍
Homologous Chromosomes #5 N ❍
Homologous Chromosomes #6 O ❍
Secondary Oocyte P ❍
Polar Body Q ❍
Egg Cell R ❍
Zygote S

THE FEMALE HORMONE CYCLE

The female hormone cycle is a complex process that involves a number of hormones as well as various structures of the reproductive system. It initially involves the building up of the lining of the uterus; the lining is retained if a fertilized egg cell is present, and the lining is released if there is no fertilized egg. The cycle occurs over a period of approximately 28 days.

This is a fairly complex plate because many of the processes go on simultaneously. We have divided the plate into three areas for study. Begin your reading below.

During the female hormonal cycle, an egg cell develops within the follical in the ovary and is released. Several events accompany this development, as the plate will indicate. There are three different series of events that make up the female hormone cycle, and they are designated with brackets. The cycles are **gonadotropic hormone cycle (A),** the **ovarian cycle (B),** and the **endometrial cycle (C).** The days in the cycle are designated in the plate for your convenience.

We begin at the left side of the plate with the production of hormones by the **pituitary gland (D).** This gland, at the base of the brain, produces **luteinizing hormone,** also known as **LH (E),** and a second hormone called **follicle stimulating hormone,** also called **FSH (F).** Two light colors should be used to color the arrows and hormone levels during the cycle.

The cycle begins with the production of FSH by the pituitary gland. FSH stimulates the maturation of the egg cell from the secondary oocyte, and also triggers the production of estrogen in the ovary. In the ovarian cycle (B), we show the **follicle (G)** with the **egg cell (H)** inside. This time period is called the **follicular phase (I)** of the cycle. At the start of the follicular stage, the level of **estrogen (J)** is low, but it rises as this phase continues. A second hormone, **progesterone (K)** is also in low supply at the start of this phase.

While these events are taking place, changes are occurring in the endometrial lining of the uterus. One **menstrual phase (O)** is coming to an end, and the **endometrial lining (N)** has been almost fully released (menstruation). At about day three of the endometrial cycle (C), the endometrium begins its rebuilding process.

We examined the left side of the female hormone cycle and noted the events at the beginning of the cycle. We now continue to day 7 and day 14 of the cycle to examine some of the changes that take place. Continue your reading as you color the diagrams in the plate.

At day 7 in the hormonal cycle, we see that the level of gonadotropic hormones is still relatively low. In the ovarian cycle, the follicle is growing and the egg is developing inside. The level of estrogen (J) is starting to increase dramatically, and the level of progesterone is still relatively low. In the endometrial cycle, the endometrium continues to build up and the **proliferative phase (P)** has begun.

Day 14 is arbitrarily set at the midpoint of the cycle, but this point may vary two to three days in either direction. A surge of LH (E) occurs, and the mature follicle bursts to release the egg cell (H). A surge of FSH brings about final maturation of the egg cell. Ovulation occurs when the egg is released, and this happens around day 14.

Now the follicular phase comes to an end, and the **luteal phase (M)** begins. The estrogen level has begun to decline, and now the progesterone level rises significantly. Once the egg cell has been released from the follicle, the remaining cells revert to a body called the **corpus luteum (L),** and it is these cells that secrete progesterone. Progesterone inhibits FSH, which prevents additional ovulation and stimulates buildup of the endometrial lining. As the amount of tissue increases in the endometrium, we enter the **secretory phase (Q)** of the cycle.

We conclude the plate by examining the events that take place at the latter half of the cycle, as is shown in the diagram. Continue your reading as we discuss the events that take place when fertilization does not occur.

If fertilization takes place, the zygote develops into an embryo, as the next plate indicates; here we show the events that occur in the absence of fertilization. At day 21, levels of LH and FSH are extremely low. The corpus luteum has continued to produce progesterone (K), which is as its highest level. The endometrial lining is thick with blood and tissue in anticipation of a fertilized egg cell.

As we come to day 28 of the hormone cycle, fertilization has not taken place, and the corpus luteum begins to break down. The level of progesterone drops dramatically, and the endometrial lining begins to disintegrate; menstruation takes place. Neither estrogen nor progesterone are being secreted, and the pituitary once again begins its production of FSH, which initiates another cycle.

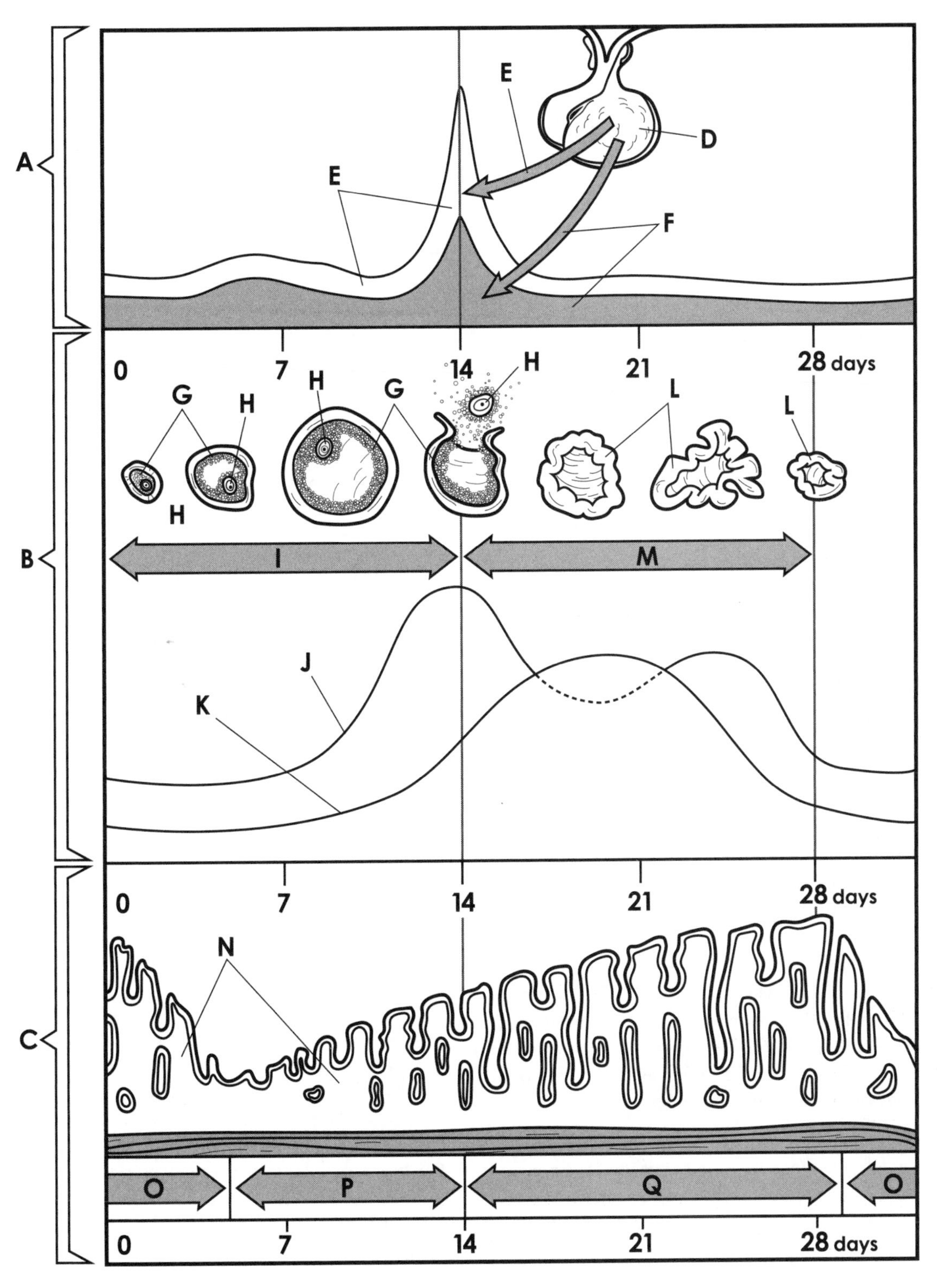

Gonadotropic Hormone Cycle A ❍
Ovarian Cycle B ❍
Endometrial Cycle C ❍
Pituitary Gland D ❍
Luteinizing Hormone (LH) E ❍

Follicle Stimulating Hormone (FSH) F ❍
Follicle G ❍
Egg Cell H ❍
Follicular Phase I ❍
Estrogen J ❍
Progesterone K ❍

Corpus Luteum L ❍
Luteal Phase M ❍
Endometrial Lining N ❍
Menstrual Phase O ❍
Proliferative Phase P ❍
Secretory Phase Q ❍

HUMAN EMBRYONIC DEVELOPMENT

The development of the human embryo involves numerous cell movements and changes that begin with the single, fertilized egg. From the zygote will arise the hundred trillion cells that comprise the adult. In this plate we will see how rapid cell divisions create a ball of cells, which differentiate into primary tissues.

In this plate, we follow the development of the embryo from the fertilized egg cell to the eighth week of life. During the embryonic stages, developments take place that set the stage for the appearance of organ systems in the fetus.

The zygote is created when a sperm cell joins with the egg cell within a Fallopian tube of the female. After about 30 hours, the zygote undergoes mitosis and two cells are formed. The first view we see is the **two-cell stage (A).** Light colors should be used to preserve the appearance of both cells. The **polar bodies (B)** that result from meiosis can also be seen associated with these cells. About 30 hours later, each cell undergoes mitosis again, and the embryo enters the **four-cell stage (C).**

Mitosis continually takes place, but the overall size of the embryo does not increase, even though the cells continue to divide. In diagram C, we see a tightly packed mass of cells called the **morula (D).** These cells are smaller than the ones in the earlier cell stages; up to 32 cells make up the morula.

Next comes the **blastocyst (E)** stage (called the blastula in other animals). The blastocyst is a ball of cells that results from mitosis of the cells of the morula. It contains between 500 and 2,000 cells. At its outer edge is the **trophoblast (E_1),** a layer of cells that develops into the membranes that surround the embryo. The blastocyst has a fluid-filled cavity known as the **blastocoel (E_2).** At one end of the blastocoel is the **inner cell mass (E_3).** These cells will become the developing embryo.

We have followed the development of the embryo through four diagrams as the cells undergo mitosis. We will continue to study their development by showing the cells along the lining of the uterus. Continue your reading below as you color.

By the third day after fertilization, the embryo is a morula. By the fifth day, the morula has become the blastocyst, and the outer cells of the trophoblast have begun the implantation process. The embryo is now about the size of the period at the end of this sentence and is called the embryonic disc.

In diagram E, we see the embryonic disc in place in the **uterine wall (F).** The **embryonic disc (G)** has two layers, the ectoderm and endoderm, as the next plate explores. Soon, the third germ layer, the middle mesoderm, will develop and the nervous and circulatory systems will appear.

In the sixth diagram, we again see the uterine wall. The embryo is slightly larger, and a **connecting stalk (I)** attaches it to the uterine wall. The embryo is now in its fourth week, and the umbilical cord has begun to form. The embryo is surrounded by a sac called the **amnion (J).** In the **four-week embryo (K),** the nerve cord can be seen and the connecting stalk is also visible. The embryo measures approximately 5 mm in length.

We will not follow the embryo's continued development through the remaining diagrams in this plate. Continue your reading as you color the views of the embryos.

In the **five-week embryo (L),** limb buds have appeared, and these later become arms and legs. The head has enlarged, and the nerve cord is fully formed. Sense organs have become prominent, and the eyes can be seen. The embryo is now about 8 mm in length. In the **six-week embryo (M),** the head continues to enlarge, and the limbs become more prominent. The digestive system in this embryo is fully formed, and is about 12 mm in length.

The **seven-week embryo (N)** is about 17 mm in length. Its back has straightened and its muscles have differentiated. Eyelids are present, and the external genital organs have begun to form.

Embryonic development ends at about eight weeks, and the **eight-week embryo (O)** is about 23 mm in length. It is recognizable as a human being, and hereafter it is called a fetus. Further development involves growth and maturation of the fetus's organs. During the next seven months, it will continue to grow as its body is defined and its organs mature.

Now that you have learned about the stages of embryonic development, we will introduce the concept of cell potency, which is a cell's ability to differentiate into other cell types. The more cell types a cell can differentiate into, the greater its potency. Both the zygote and the morula are totipotent, because they are able to develop into any cell in the blastocyst. Cells of the inner cell mass are more specialized and are referred to as pluripotent. They can differentiate into any of the three primary germ layers (ectoderm, mesoderm, or endoderm) and therefore have the capability to become any of the 220 cell types that make up an adult human. However, they cannot contribute to the trophoblast of the blastocyst.

As development continues, cells continue to specialize. After gastrulation, cells from the early embryonic germ layers are each considered multipotent. This means they can become many, but not all, cell types. For example, cells of the mesoderm can differentiate into muscle and bone cells, but not into neurons or digestive epithelium.

In other words, totipotent cells differentiate into pluripotent cells, which specialize to become multipotent cells.

Most cells in the adult have lost all potency and have become completely specialized mature cells, incapable of changing into other cell types. In humans, embryonic stem cells are the only pluripotent cells that have been found. These cells are isolated from the inner cell mass of the blastocyst.

HUMAN EMBRYONIC DEVELOPMENT

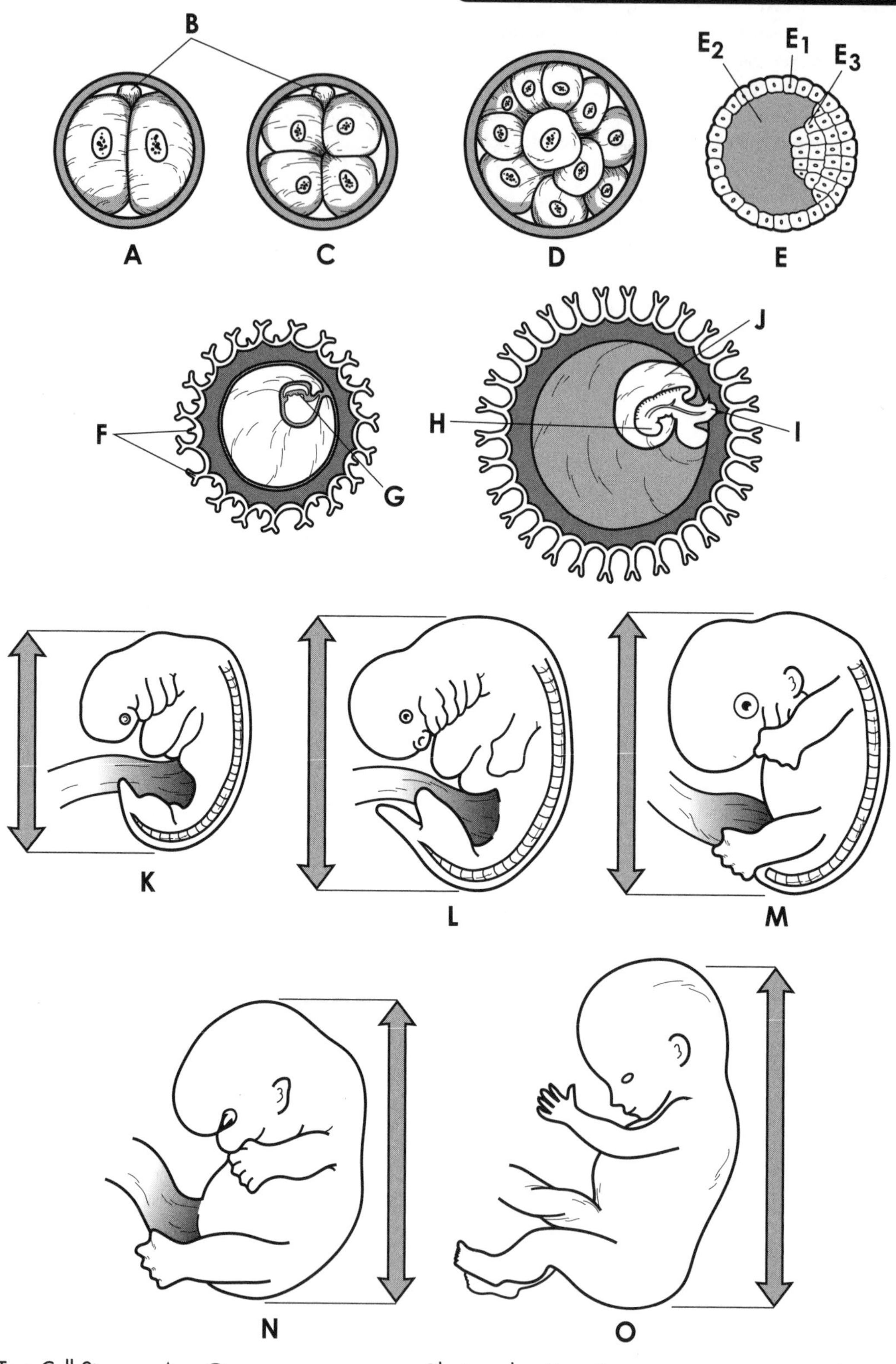

Two-Cell Stage A ❍
Polar Bodies B ❍
Four-Cell Stage C ❍
Morula D ❍
Blastocyst E ❍
Trophoblast E_1 ❍

Blastocoel E_2 ❍
Inner Cell Mass E_3 ❍
Uterine Wall F ❍
Embryonic Disk G ❍
Embryo H ❍
Connecting Stalk I ❍

Amnion J ❍
Four-Week Embryo K ❍
Five-Week Embryo L ❍
Six-Week Embryo M ❍
Seven-Week Embryo N ❍
Eight-Week Embryo O ❍

NOTES

THE PRIMORDIAL GERM LAYERS

During embryonic development, the fertilized egg forms a morula, and then a blastocyst, as we discussed in the previous plate. At the outer wall of the blastocyst, the trophoblast develops, and at one side of the blastocyst there is an inner cell mass, which becomes most of the embryonic tissues.

In the next stage, there is a special rearrangement of cells, and this phase is called gastrulation. At the completion of gastrulation, the embryo is called a gastrula. The rearrangements of cells during gastrulation results in three primordial germ layers called the ectoderm, mesoderm, and endoderm. These three germ layers will later form all of the organs and organ systems of the adult, and we examine them in this plate.

In this plate, we show a section of the embryo during the gastrula stage. Begin your reading below.

The process of gastrulation involves patterns of cell movement, and begins with the blastocyst, the hollow ball of cells. First, there is an infolding of the surface called an invagination, then an inward turning of cells called involution, and finally a flattening and spreading of the cell layer, after which cells migrate to their functional position.

In the art, we see a cross-section of the embryo in its third week of development, after gastrulation has taken place. The lining of the uterus is made up of a layer cells, the **endometrium (A).** A membrane called the **chorion (B)** has developed from the trophoblast and is surrounding the embryo. A bold color should be used to denote the cells of this membrane. A collection of cells, the **body stalk (C),** is growing, and it will become the umbilical cord. We also see a large **yolk sac (D),** which should be colored lightly. This small cavity below the embryo will eventually become the umbilical cord.

Another membrane, the **amnion (E),** lies between the trophoblast and the inner cell mass. The **amniotic cavity (F)** is developing at this stage. These membranes will be discussed in detail in the next plate. The remaining membrane outlines the **extraembryonic cavity (G)** outside the amnion. A pale or light color should be used to shade it.

We have established the position of the developing embryo in the uterus and noted its surrounding membranes. We now focus on the three germ layers of the embryo, here seen in cross section. We also mention the organs into which they will eventually develop.

In the gastrula stage, the embryonic disc has three layers of cells. The first layer of cells that we'll look at is the **ectoderm (H),** which forms the outer wall of the gastrula. The large callout leads from the ectoderm to show a fully formed fetus, in which you can see the **ectoderm derivatives (H_1).** These derivatives include the entire nervous system; including the brain and spinal cord. The outer layer of skin (epidermis) also derives from the ectoderm, and some endocrine glands such as the pituitary glands come form this layer. The inner lining of certain structures such as the nose, mouth, and anus are also derived from the ectoderm layers.

We now focus on the middle layer, the **mesoderm (I).** Several cells of this layer are seen in the cross section, and a triangle leading from the mesoderm shows its derivatives. The **mesoderm derivatives (I_2)** include the bones and cartilage of the skeleton, and the muscles and organs of the circulatory and excretory systems. Most of the reproductive organs and the inner layer of skin tissue also come from mesoderm tissue, as do certain structures of the respiratory system.

The final layer of the gastrula is the **endoderm (J).** The triangle projecting from it shows the embryo with its **endoderm derivatives (J_2).** These include the lining of the digestive tract and respiratory passages. The liver, pancreas, urinary bladder, and most glands are also derived from the endoderm.

The germ layers are so-called because they are the beginnings of tissues, organs, and organ systems (germ means beginning). In simple animals such as sponges and cnidarians, only endoderm and ectoderm layers exist.

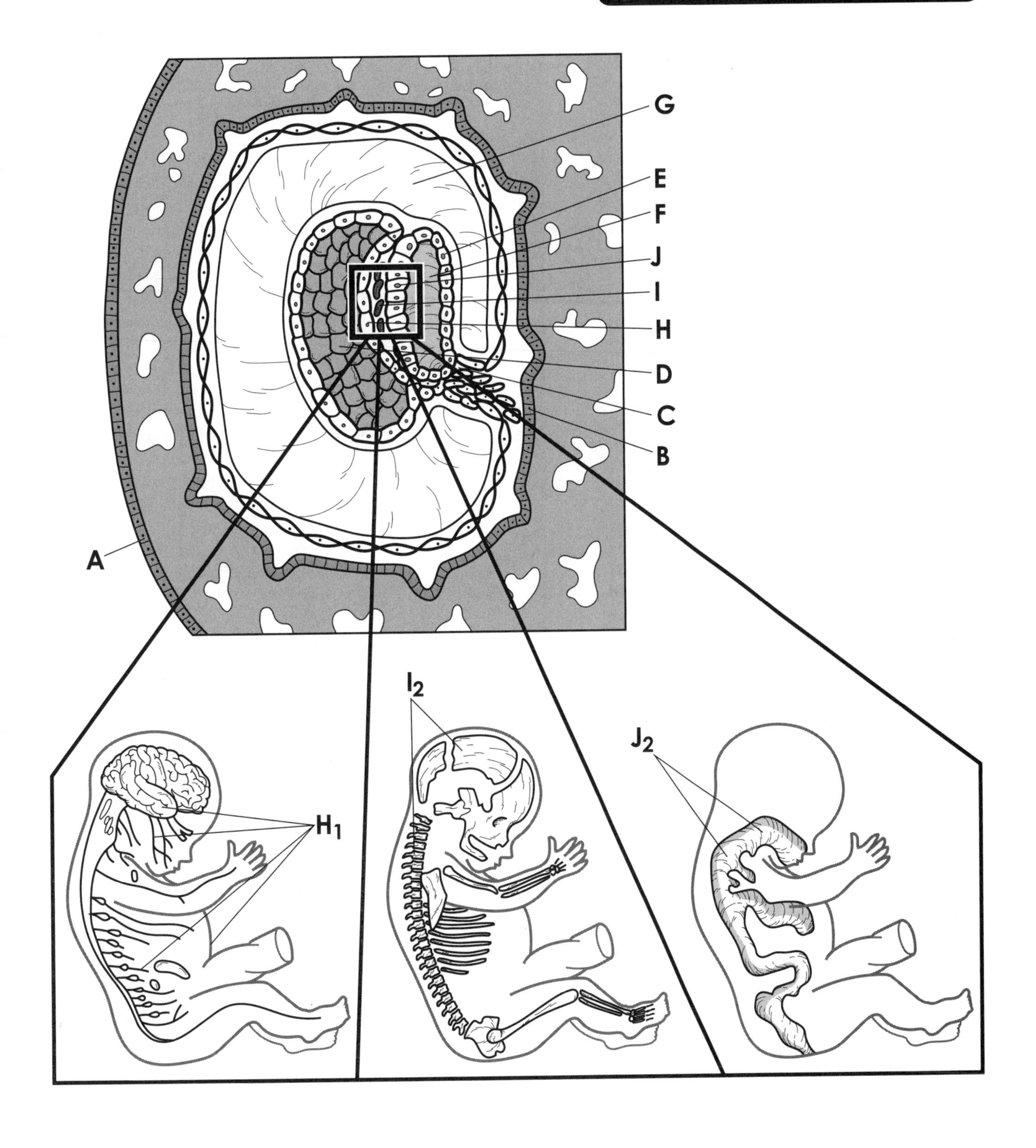

Endometrium A ❍
Chorion B ❍
Body Stalk C ❍
Yolk Sac D ❍
Amnion E ❍
Amniotic Cavity F ❍
Extraembryonic Cavity G ❍
Ectoderm H ❍
Ectoderm Derivatives H_1 ❍
Mesoderm I ❍
Mesoderm Derivatives I_2 ❍
Endoderm J ❍
Endoderm Derivatives J_2 ❍

EMBRYONIC MEMBRANES

In the course of their development, all terrestrial (land-based) animals have four membranes that surround the embryo. These membranes are often referred to as extraembryonic membranes. They derive from the three germ layers discussed in the previous plate, but are not part of the embryo itself. The embryonic membranes are adaptations that protect the terrestrial embryo, preventing it from drying out, and enabling it to obtain food and oxygen and eliminate waste. In this plate, we will discuss the four membranes of the developing embryos of a reptile and human.

This plate contains a diagram of both a human embryo and a reptile embryo. Each is surrounded by four membranes. We see differences in the structure and function of the membranes and how they correspond to differences in the organisms' physiology.

In this plate, we see a human embryo and a reptile embryo. Perhaps the most obvious difference between them is that the reptile embryo is developing within a **shell (A),** while the human embryo is developing within the **endometrium of the uterus (B).** Begin by coloring the **human embryo (H)** and the **reptilian embryo (I).** Light colors are suggested for them.

The first embryonic membrane we will mention is the chorion. In the first diagram, we see the **human chorion (C_1)** with many outgrowths that penetrate the endometrium of the uterus. The human chorion provides a surface for the exchange of gases, nutrients, and wastes between mother and the embryo.

In the reptilian egg, the **chorion (C_2)** remains in contact with the inner surface of the shell. It acts as a respiratory surface that regulates the exchange of gases and water between the embryo and the air. The chorion later fuses with the allantois (discussed later) to form the chorioallantoic membrane.

The next membrane we consider is the amnion. The **human amnion (D_1)** is a sac that surrounds the embryo, as is shown in the diagram. It is filled with amniotic fluid, and the cavity should be colored in a pale color. In the reptilian egg, the **amnion (D_2)** serves the same purpose. The amnion, like the chorion, is derived from cells of the trophoblast. The amnion cushions the embryo and enables it to maintain a constant temperature. In the human, at the end of development, the bursting of the amnion signals impending birth. It is filled with clear, watery amniotic fluids that are secreted by the membrane. The fluid prevents drying of the embryo, absorbs shock, and prevents the amniotic membrane from sticking to the embryo.

We have now examined two of the four membranes of the developing embryo and pointed out their structures and functions. We will end the plate with a discussion of the other two membranes.

The third membrane we will consider is the allantois. The **human allantois (E_1)** derives from the developing digestive tract. It appears in the third week and is a part of the umbilical cord, which contains blood vessels responsible for the exchange of nutrients and gases between the embryo and mother.

The **reptilian allantois (E_2)** is quite large in comparison with the human allantois. It is a depot for waste products, and it excretes such chemicals as uric acid, which dissolve as crystals in its cavity. Later the allantois fuses with the chorion to form the chorioallantoic membrane, which is rich in blood vessels. These vessels provide oxygen and carbon dioxide to the reptilian embryo and allow the exchanges of gases through the shell. The membrane is discarded at birth when the egg hatches.

The final embryonic membrane we consider is the yolk sac. The **human yolk sac membrane (F_1)** also forms from the developing digestive tract, and it encloses very little material. Eventually, it becomes a part of the umbilical cord. It is the source of the cells that will later give rise to gametes in the developing reproductive organs.

Within the egg, the **reptilian yolk sac membrane (F_2)** stores **reptilian yolk (G_2).** The yolk contains nutrients that are digested by the embryo. Bird and reptilian eggs are notable for the large amount of yolk they contain. Because there is no connection to the mother, the developing reptile or bird relies upon the nutrients in the yolk to sustain them until the egg hatches. By contrast, the material in the **human yolk (G_1)** contains a negligible amount of nutrients. It is believed that in mammals, the forerunners of red blood cells and other blood cells are found in the yolk sac membrane.

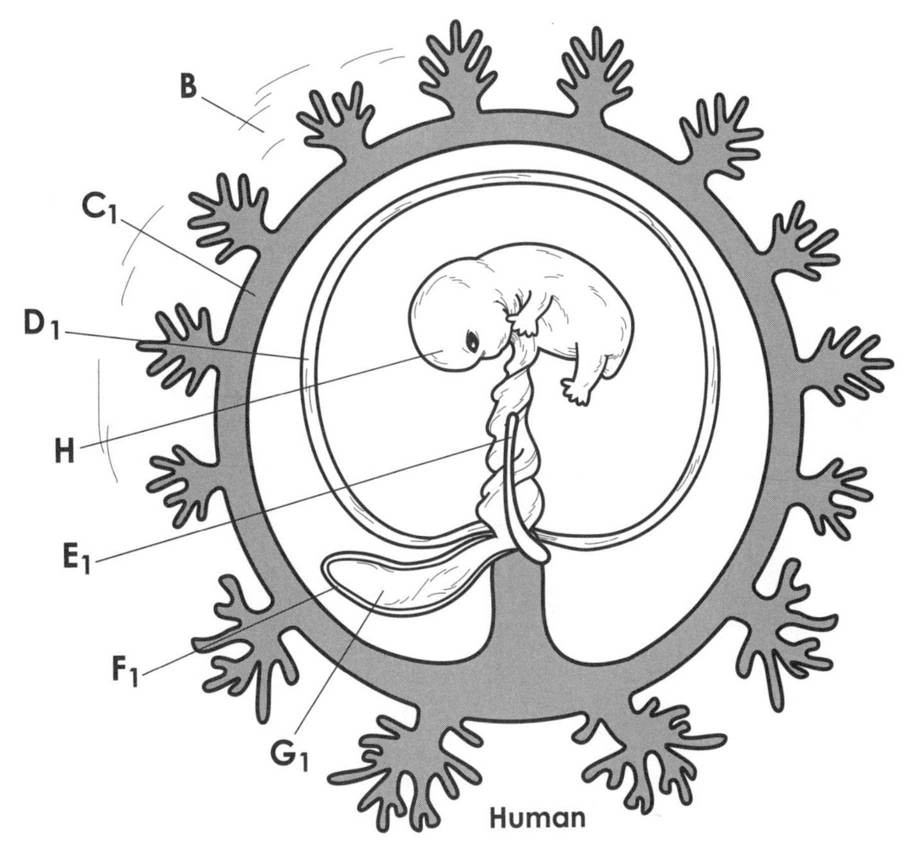

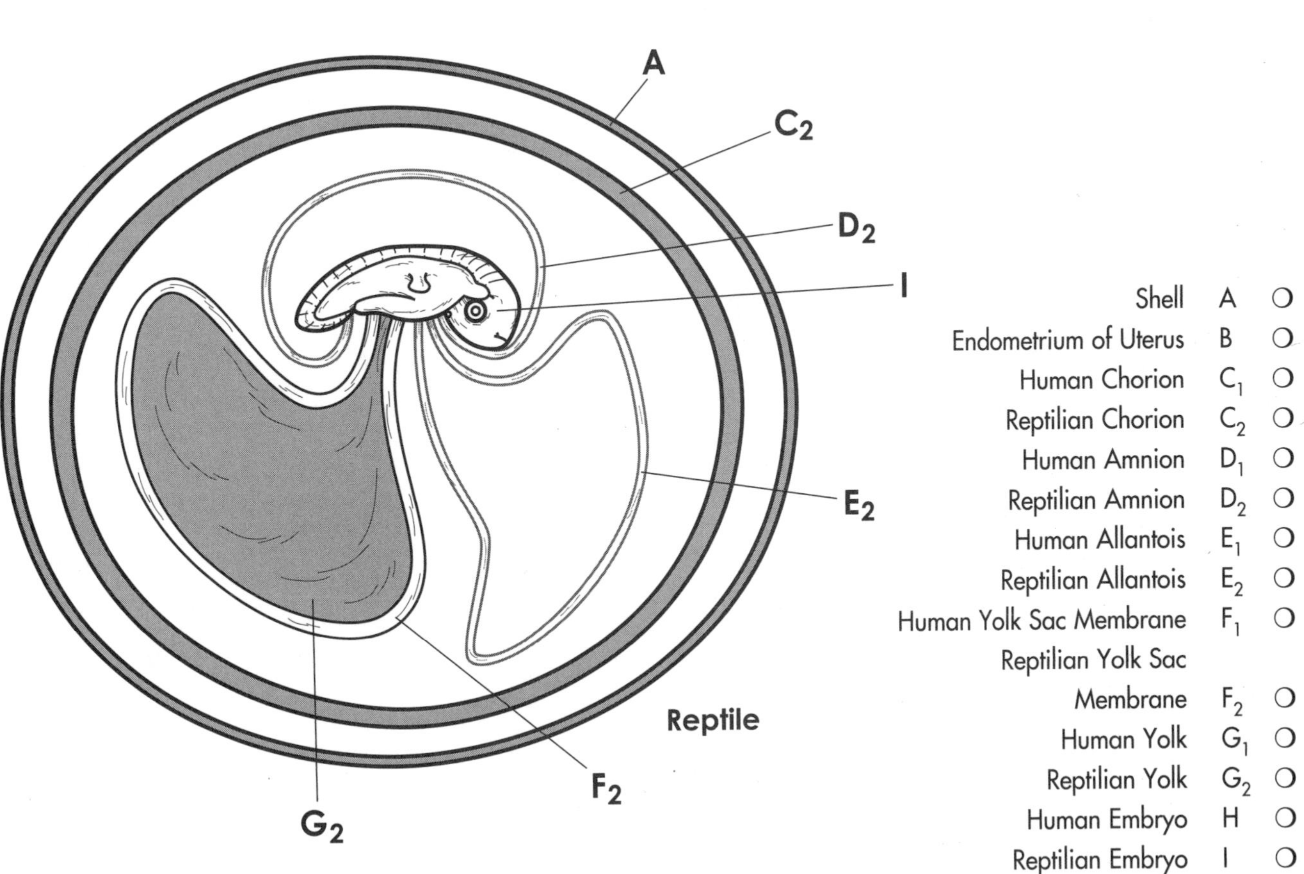

Shell A ○
Endometrium of Uterus B ○
Human Chorion C_1 ○
Reptilian Chorion C_2 ○
Human Amnion D_1 ○
Reptilian Amnion D_2 ○
Human Allantois E_1 ○
Reptilian Allantois E_2 ○
Human Yolk Sac Membrane F_1 ○
Reptilian Yolk Sac Membrane F_2 ○
Human Yolk G_1 ○
Reptilian Yolk G_2 ○
Human Embryo H ○
Reptilian Embryo I ○

CHAPTER 11:

PRINCIPLES of ECOLOGY

ECOLOGICAL COMMUNITIES

Ecology is the study of the relationships of organisms to their environment and each other. The term *ecology* is derived from the Greek word *oikis,* meaning "A house or place where one lives," and *logos,* meaning "study of."

Ecologists study assemblages of interacting organisms known as communities. They categorize organisms within communities according to their course of food. In this plate, we will examine two communities to identify their trophic levels and which organisms exist within them.

Looking over the plate, you will notice that we are examining an aquatic (water) community as well as a terrestrial (land) community. We examine four trophic levels within each to show the various organisms that make up this ecological assemblage. Examine the first trophic level, at the top of the plate.

A community is the set of all populations inhabiting a certain area. The area a community encompasses can be very small, such as a small puddle of water, or it may be very large, encompassing hundreds of square miles.

The first trophic level in the community is made up of organisms known as **producers (A).** Producers obtain their food by synthesizing it from inorganic matter through photosynthesis. In the aquatic community you see in the plate, the producers are microscopic marine organisms known as **phytoplankton (A_1).** The group of phytoplankton includes diatoms, dinoflagellates, and cyanobacteria (blue-green algae). Living in enormous numbers in the ocean communities, phytoplankton trap sunlight and produce carbohydrates for food.

In the terrestrial community, the producers are **grass plants (A_2).** Grasses of all types and climates engage in photosynthesis. The number of organisms is the highest at the producer's trophic level; for instance, if the biomasses (the entire mass of biological material) of each of the trophic levels were graphed with producers at the bottom, the resulting structure would be pyramidal in shape.

We have examined the producers in aquatic and terrestrial communities, and we now move to the trophic level of consumers. Photosynthesis does not occur in these organisms, and so they consume producers to obtain energy. Continue your coloring as you read the paragraphs below.

The next level in the aquatic and terrestrial communities includes the **primary consumers (B),** which use the producers as food. In the plate, we show **zooplankton (B_1)** in the aquatic community. Zooplankton are tiny, microscopic animals and animal-like organisms that use phytoplankton as food. In the terrestrial community, primary consumers are represented by an **insect (B_2).** This organism consumes grasses and other green plants. The biomass of primary consumers in a community is less than the biomass of producers.

The next trophic level is made up of **secondary consumers (C).** An example of a secondary consumer is a small fish, a **perch (C_1).** This fish eats zooplankton and invertebrates such as worms and tiny insects. In the terrestrial environment, the insect is consumed by a **small bird (C_2).** By consuming the insect, this secondary consumer obtains proteins, carbohydrates, and fats to fulfill its nutritional needs. The biomass of secondary consumers is generally smaller than that of the primary consumers.

We conclude our examination of the tropic levels in a community by examining the final trophic level, the tertiary consumer. You should note that the animals have become larger and less numerous at each trophic level.

The highest trophic level is occupied by the **tertiary consumer (D).** In the aquatic environment, one tertiary consumer is a large fish such as the **bass (D_1).** It consumes the perch and other small fish to fulfill its nutritional needs. In the terrestrial community, the **hawk (D_2)** preys on small birds to obtain its nourishment, The number of tertiary consumers is lowest of all trophic levels. Humans are an example of the other tertiary consumers in the terrestrial environment.

Consumers that feed only on green plants are herbivores, while those that feed on other animals are call carnivores; secondary and tertiary consumers are usually carnivores. Omnivores eat both plants and animals.

One type of consumer that was not considered here is the decomposer. Decomposers such as bacteria and fungi process and consume the remains of animals and plants and are critical to elemental cycles in the soil.

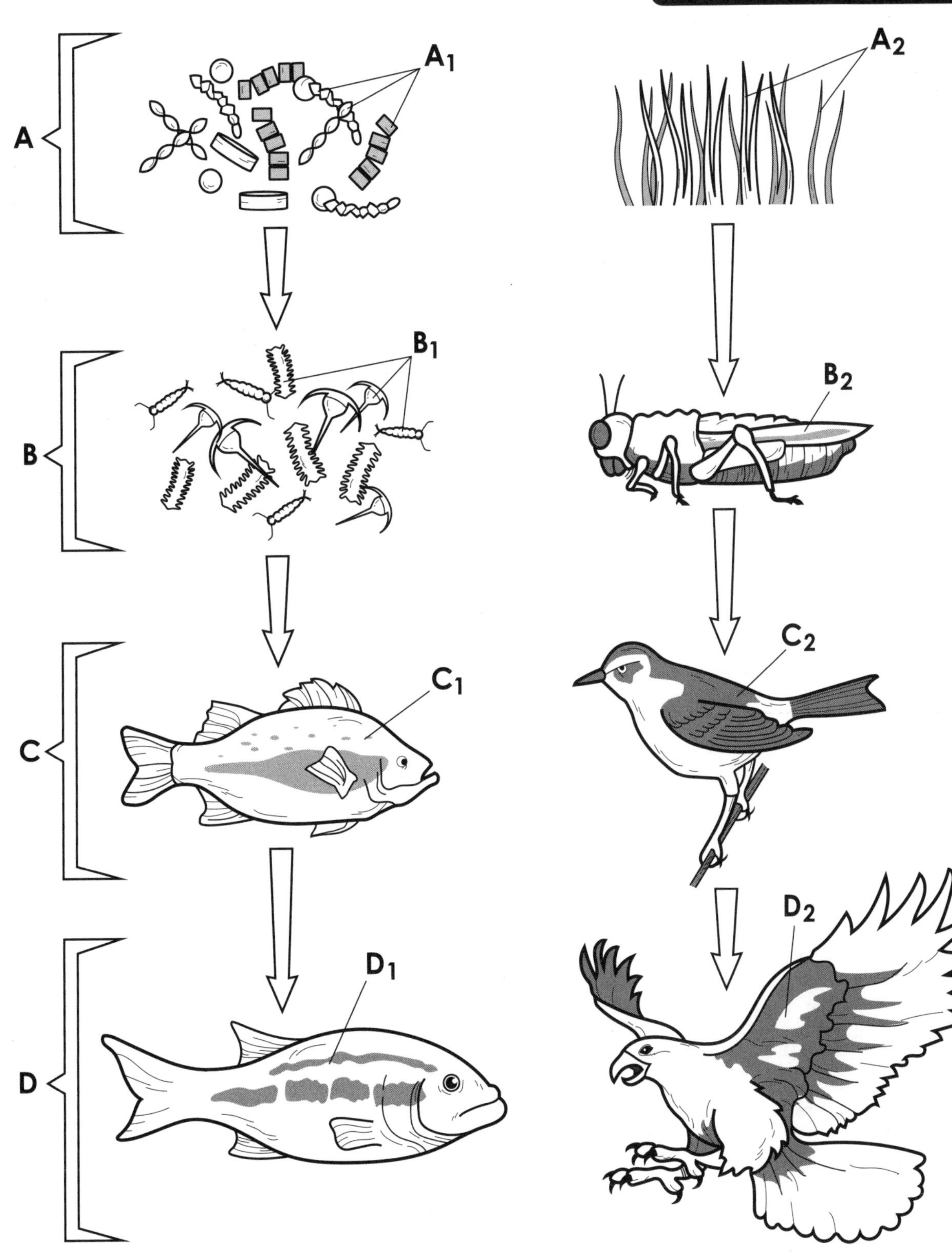

Aquatic Community

Terrestrial Community

Producers A ❍
Phytoplankton A_1 ❍
Grass Plants A_2 ❍
Primary Consumers B ❍
Zooplankton B_1 ❍
Insect B_2 ❍
Secondary Consumers C ❍
Perch C_1 ❍
Small Bird C_2 ❍
Tertiary Consumers D ❍
Bass D_1 ❍
Hawk D_2 ❍

ECOSYSTEMS

A community is a complex system of mutual dependency. Numerous interrelationships exist between the community and the physical features of the environment it inhabits. This dynamic relationship between the community and its physical environment is called an ecosystem. Ecosystems vary in size from very large forests to tiny droplets of water. In this plate, we will study some of the dynamics of a freshwater pond ecosystem.

This plate consists of a single diagram in which a collection of organisms is shown in its natural environment. Both the living organisms and the physical environment will be considered in our discussion of the ecosystem.

The first item we will mention is the **Sun (A).** The amount of solar heat that reaches the pond is one of the factors that determine the type and number of organisms that can exist in the pond. Ecologists are scientists that study ecosystems. For instance, they would study a pond's photosynthetic organisms to determine how solar energy is used. They would also study the strength of the ultraviolet light that reaches the surface of the pond, and its effect on its organisms.

Another item in a community's physical environment is the **air (B).** Ecologists might run tests to determine whether the air is clear or polluted, or whether the atmosphere is stagnant or kept circulating by the wind. They might determine how much moisture the air could hold, or how high its temperature could rise due to solar radiation. All of these factors affect the organisms living in the pond.

A third important physical factor is the **water (C).** Ecologists might investigate whether there is an outlet that refreshes the pond, or whether the water is stagnant. They might also test the water for organic nutrients that enable microorganisms to grow and compete with fish for oxygen.

The final physical component we will mention is the **soil (mud) (D).** In studying the ecosystem, ecologists determine if the soil is sandy or if it is rich in organic matter. Limited populations of organisms inhabit sandy soil, while a variety thrive in organic-rich soil. They would also determine if the soil contained dissolved oxygen to support aerobic microorganisms, or lacked it and supported anaerobic microorganisms.

Now we have named the physical factors of a pond ecosystem, and we will turn to its living members to see how they interact with each other and their physical environment.

We will begin our survey of the living members of our ecosystem with the **floating microorganisms (E)** depicted in the box. The type and variety of microorganisms present vary according to the amount of organic matter present. For example, if the pond receives outflow from a food processing plant, much organic matter will be present and will support a variety of microorganisms.

At the bottom of the pond are **mud-dwelling microorganisms (F).** As we mentioned above, conditions in the mud influence the type and number of organisms present. Mud-dwelling microorganisms serve as a source of food for **invertebrates (G),** which include tiny fish, insects, worms, mollusks, and echinoderms. These invertebrates float in the water and provide food for larger animals.

Among the plant population is photosynthetic **floating algae (H).** Its metabolic processes cause it to produce carbohydrates, which are consumed by vertebrates and other small organisms. The addition of floating algae may in turn cause fish to suffocate.

This pond also contains a variety of **insects (I)** that feed on both floating algae and invertebrates. Because of their extreme resistance to environmental fluctuations, the insect population remains fairly stable, and provides food for the **fish (J).**

Two type of plants inhabit this ecosystem. The type and number of **submerged plants (K)** depend on the conditions of the mud and water. Mud, for example, provides anchorage for submerged plants. As you can see, the second type of plant is the **aerial plant (L),** which grows at the border of the pond. When the nature of aerial plants changes, whether it is due to pollution or the effects of weather, the entire community structure of the ecosystem changes.

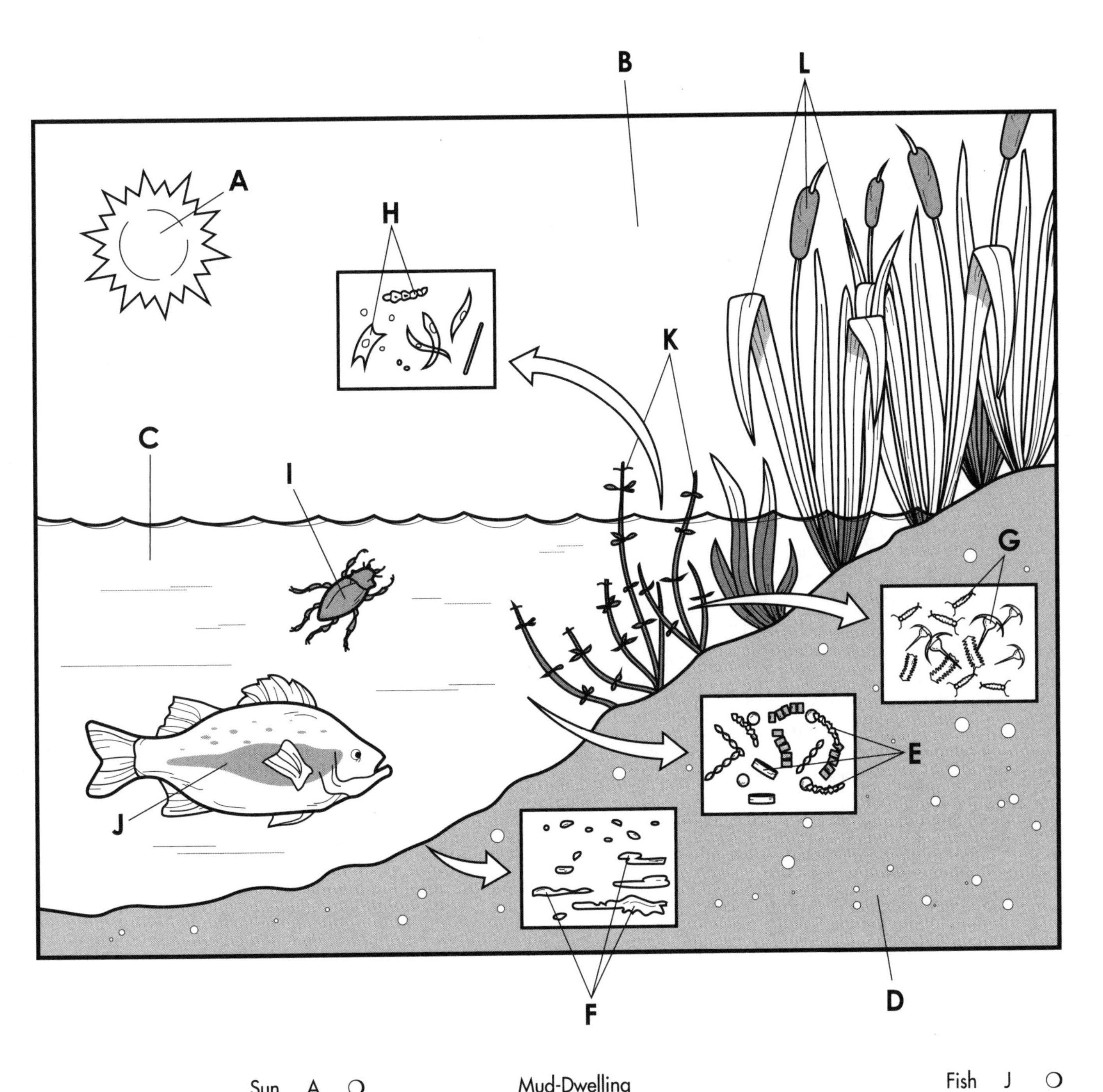

Sun A ❍
Air B ❍
Water C ❍
Soil (Mud) D ❍
Floating Microorganisms E ❍
Mud-Dwelling Microorganisms F ❍
Invertebrates G ❍
Floating Algae H ❍
Insects I ❍
Fish J ❍
Submerged Plants K ❍
Aerial Plants L ❍

ECOLOGICAL NICHES

A population is defined as a group of individuals of a single species living together within a habitat. A habitat affords variables such as space, food, climate mating conditions, behavior, and other factors. An ecological niche is the very specific place an organism occupies in its environment, and the very specific role it plays within it.

This plate discusses the ecological concept of the niche. We examine five different species of warbler and see how each fits into different niches in a very small environment. Examine the first diagram as you begin your reading below.

The noted ecologist, Robert H. MacArthur, described habitats as being subdivided so that each species comes to live where it will survive and propagate.

Begin your work with diagram 1. This diagram shows the **spruce tree (A),** which you may wish to color green. One species of warbler studied by MacArthur was the **Cape May warbler (B).** A light color may be used to outline this bird, or you may wish to color its lighter parts. The niche of the **Cape May warbler (B_1)** is shown in the top, shaded portion of the tree, A color similar to the one uses for the bird should be used for its niche. The Cape May warbler used the resources and shelter of the upper portion of the tree and, by feeding and nesting here, the bird avoids some competition with other species of warblers.

A second species of bird studied by MacArthur is the **bay-breasted warbler (C);** the **niche of the bay-breasted warbler (C_1)** is shown in diagram 2. Note that this niche lies below that of the first bird, and that it occupies areas at the exterior of the tree as well as the interior.

We proceed to diagram 3 and the **Blackburnian warbler (D).** The **niche of the Blackburnian warbler (D_1)** is very similar to that of the Cape May warbler seen in diagram 1. Although there is a slight overlap, the two species are not in direct competition because the Blackburnian warbler spends most of its time in the outer branches of the tree and feeds in an area that's lower in the tree. This bird may be in the tree at the same time as the other two warblers, but by feeding in different areas, it avoids direct competition.

We conclude our study of ecological niches by examining those occupied by two remaining species studied by MacArthur.

The species of warbler studied by MacArther belongs to the genus *Dendroica.* The fourth member of this genus is the **black-throated green warbler (E).** This bird spends at least half of its feeding time in the upper portion of the spruce tree. In diagram 4, we see that the bird occupies the exterior portion of the tree and avoids the interior area. This prevents its overlap with the bay-breasted warbler. The **Niche of the black-throated green warbler (E_1)** should be colored the same color as the bird.

A clear differentiation of niches is seen for the **Myrtle warbler (F).** The **niche of the Myrtle warbler (F_1)** includes areas at the central portion of the spruce tree as well as the bottom. This is a ground-dwelling warbler that feeds on insects from the soil and in the tree.

The principle of competitive exclusion states that evolutionary forces pull the niches of similar organisms apart so that the organisms adapt differently, resulting in niches of differentiation. No two species can occupy the same ecological niche in a community. When two or more species are found to coexist on a long-term basis, their niches will always differ, and if their niches do not differ, extinction of one of the species occurs. The divergence of feeding behavior is a factor in niche separation, and the resources exploited to meet the energy, nutrient, and survival demands of the species are all aspects of species' niches.

Spruce Tree A ❍
Cape May Warbler B ❍
Niche of Cape May Warbler B_1 ❍
Bay-Breasted Warbler C ❍
Niche of Bay-Breasted Warbler C_1 ❍
Blackburnian Warbler D ❍
Niche of Blackburnian Warbler D_1 ❍
Black-Throated Green Warbler E ❍
Niche of Black-Throated Green Warbler E_1 ❍
Myrtle Warbler F ❍
Niche of Myrtle Warbler F_1 ❍

TERRESTRIAL BIOMES

The biosphere is composed of all the ecosystems on Earth. Within the biosphere are a number of large and somewhat distinct geographical regions known as biomes. The biosphere contains two major types of biomes: terrestrial, or land-based biomes; and aquatic, or water-based biomes. This plate discusses terrestrial biomes. Types of terrestrial biomes are distinguished by their appearance and climate and are similar, wherever they occur in the biosphere.

This plate shows an ecological map of North America and its surrounding regions. Within this region, there are eight biomes. Strictly speaking, a biome does not refer to a particular geographic area; it refers to a number of regions that have similar climates, soil conditions, and communities. Thus, a biome in North America may have a counterpart in Europe or Asia.

The first terrestrial biome we identify is the **tropical forest (A).** Located in Central America and the Caribbean islands, the tropical rain forest is the biome that is richest in the number of species. Because it is near the equator, it experiences warm, humid conditions, and there is abundant rainfall and sunlight that supports a diversity of animals. In general, its trees are tall and form a dense canopy that shades the forest floor.

The second terrestrial biome identified in the map is the **grass land (B).** Grasslands extend from the base of the Rocky Mountains to the forests of the Mississippi Valley. The prairies and the Great Plains, both of which are grasslands, dominate North America. Rainfall in these areas tends to be scarce, and the large animals that inhabit these regions, such as bison, are grazers. Grasslands are the most important biome for the production of food for humans.

A third biome of North America is the **desert (C).** Large areas of southwestern United States and Mexico are covered by desert. Rainfall in these areas is very scarce and the temperatures are alternately very hot and cold. Desert plants include cacti, yuccas, and fan palms, all of which have small leaves and thick waxy coverings to prevent the loss of water through evaporation.

We have begun our discussion of terrestrial biomes by considering three different biomes of North America; we will continue our study with five others.

At the coast of southern California is a distinctive biome called the **chaparral (D).** The chaparral is characterized by long, hot, dry summers and mild, rainy winters. Only small trees and shrubs are able to grow there because of the harsh conditions. Chaparrals are also found near the Mediterranean Sea and parts of Chile, as well as along the coast of Australia.

Most of the eastern United States is within a **temperature forest (E).** The trees in this biome are deciduous, meaning that they shed their leaves in the winter. Shrubs and herbs are numerous at ground level in this biome, and the forest floor is rich in plant life. The temperate forest climate is stable, rainfall is abundant and evenly distributed, temperatures are moderate, and there are distinct summer and winter seasons. Animal life is also varied and abundant.

North of the Great Lakes and extending far into Canada is the biome known as the **taiga (F).** Most trees in this biome are coniferous, including pine, spruce, and fir. These trees do not shed their leaves in winter and continue to photosynthesize. Their dense foliage shades the ground, so few shrubs and herbs are present. Taigas can also be found across parts of Asia, and are characterized by dry, cold conditions.

North of the coniferous forests of Canada and into the Arctic area is the **tundra (G).** The tundra is an extensive, treeless plain in which the topsoil remains frozen almost year round. Small shrubs, lichens, and grasses are found here. Plant and animal life is limited since this climate is extremely inhospitable.

The last biome we will discuss is the **mountain/ice (H).** This biome exists in the Rocky Appalachian Mountains, and in Greenland. The extremely cold climate of the mountain/ice biome limits vegetation substantially, and little life is found at these high elevations.

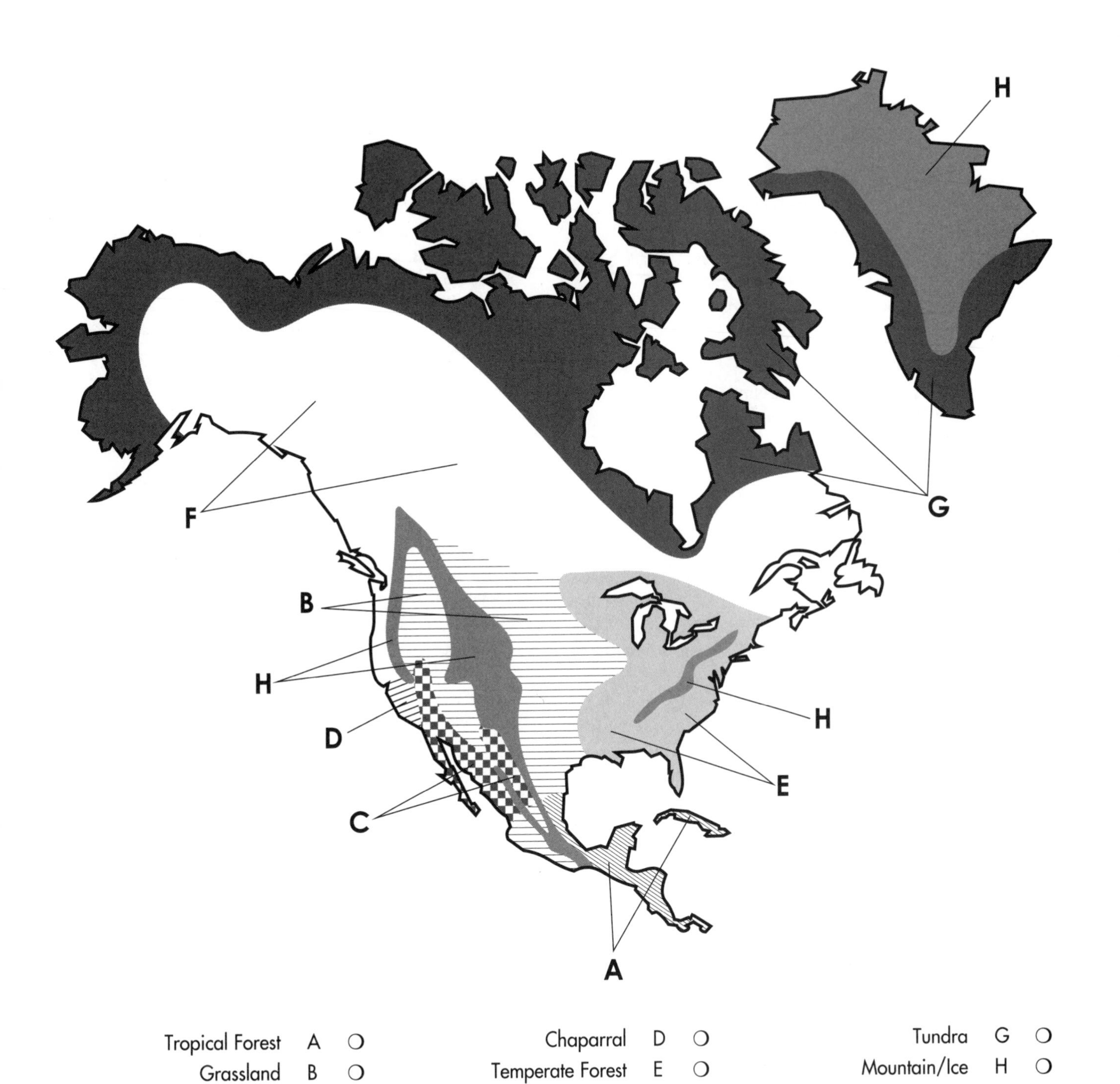

Tropical Forest A ❍
Grassland B ❍
Desert C ❍
Chaparral D ❍
Temperate Forest E ❍
Taiga F ❍
Tundra G ❍
Mountain/Ice H ❍

AQUATIC BIOMES

The biosphere is the thin layer of the Earth in which life exists. In the biosphere, communities of plants, animals, and microorganisms living within distinct geographic areas are referred to as biomes. Biomes share distinctive climatic conditions and are identified primarily by their plant populations. Several communities may exist within a biome.

Both terrestrial (land-based) and aquatic (water-based) biomes exist in the biosphere. In the previous plate, we examined terrestrial biomes. In this plate, our attention is on aquatic biomes.

This plate consists of two diagrams: One displays a freshwater biome such as is found in a lake or pond; the second diagram is of a marine biome. Your work should begin with the top diagram.

Ecologists generally recognize two types of aquatic biomes: the freshwater biome and the marine biome.

In comparison to terrestrial biomes, aquatic biomes undergo only slight temperature changes. In a lake or pond, the slow motion of water leads to stratification, in which oxygen levels decrease with depth. In this view of the lake, we see **sunlight (A)** providing energy to the biome. At the shore of the lake is the **littoral zone (B).** In this zone, light reaches all the way to the lake's bottom, and plants take root there. Many consumers such as worms, insects, and snails feed on the plants.

Further from the water's edge is the **limnetic zone (C).** A light color should be used for this area. This zone extends vertically as far down as light penetrates; turbid water has a shallower limnetic zone. Producing organisms such as algae and plankton exist here.

The area of depth that light does not reach is called the **profundal zone (D).** Photosynthesis does not take place here, but nutrients float down from the littoral and limnetic zones. The profundal zone is inhabited by primary consumers. Within the **sediment (E)** is a population of microorganisms, including bacteria and fungi. These organisms act as decomposers to break down organic matter that sinks to the bottom. Below the sediment is the **sand (E_1),** as well as the rock of the Earth's mantle.

Having examined a freshwater biome, we now move on to the marine biome. About three-quarters of the Earth's water exists in the oceans, and water temperatures vary considerably depending upon their depth. Continue your study by focusing on the marine biome at the bottom of the plate.

As is the case with ponds and lakes, there are various zones in the marine biome. At the margin of the ocean is the **intertidal zone (F).** Sand beaches and estuaries where rivers empty into the oceans are found in this zone. This environment is subject to tides and is inhabited by organisms that have adapted to wide ranges of environmental conditions. Nutrients are plentiful, and breeding grounds for many types of organisms are found here.

The next zone is the **neritic zone (G).** This zone extends from near the shoreline to where the **continental slope (H)** ends beyond the **continental shelf (I).** In the neritic zone, phytoplankton produce large amounts of carbohydrates. Animals and plants are adapted for clinging to rocks, and kelps and seaweeds form extensive beds. In many shallow areas, light reaches to the ocean bottom, and nutrients are suspended by waves, winds, and tides. The next major zone is the **oceanic zone (J).** In this area, the net productivity is comparable to that of a desert; there are few organic nutrients.

The marine biome can also be subdivided according to how far light penetrates the waters. The upper region is the **euphotic zone (K),** and reaches from sea level to a depth of approximately 200 meters. Light penetrates this zone, and the organic matter is rich.

Below this level is the **aphotic zone (L).** This area begins at the edge of the continental slope (I) and extends from 200 meters to 5,000 meters below the surface. This zone is in perpetual darkness and supports a limited variety of life. Nutrients descend from the euphotic zone and permit life to exist in this region.

The deepest part of the ocean constitutes the **abyssal zone (M)** and extending down from this zone are the **oceanic trenches (N),** some of which are five miles deep. Light cannot reach this area, but nevertheless living forms exist here; there are sparse populations of decomposers that depend on organic matter that drifts down from above. Cracks in the ocean floor also release sulfur compound that permit certain species of bacteria to thrive. Abyssal regions are characterized by tremendous pressure and numbing cold, and the organisms that live there are known as benthic scavengers.

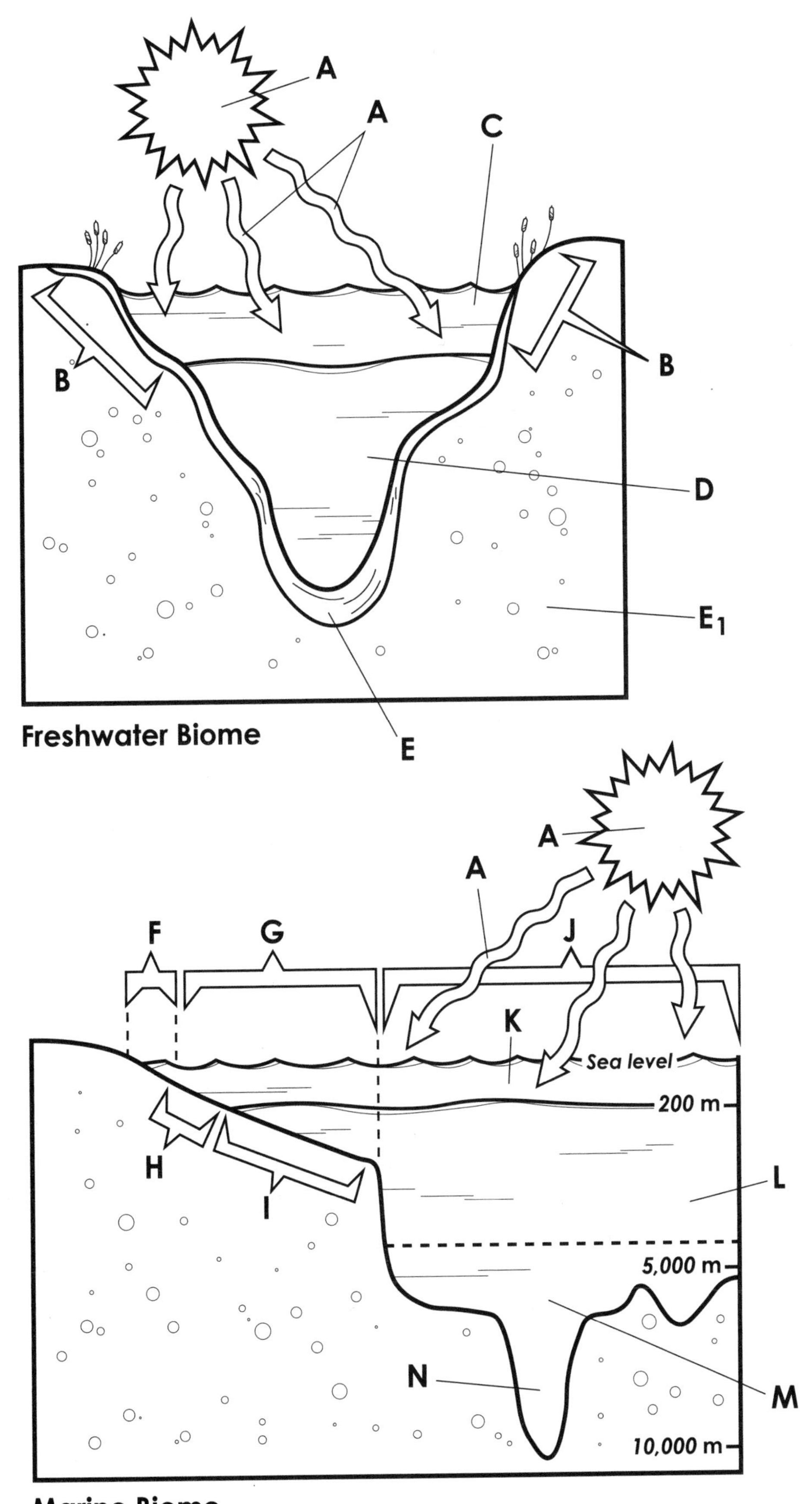

Sunlight	A	○
Littoral Zone	B	○
Limnetic Zone	C	○
Profundal Zone	D	○
Sediment	E	○
Sand	E_1	○
Intertidal Zone	F	○
Neritic Zone	G	○
Continental Slope	H	○
Continental Shelf	I	○
Oceanic Zone	J	○
Euphotic Zone	K	○
Aphotic Zone	L	○
Abyssal Zone	M	○
Oceanic Trenches	N	○

THE ENERGY PYRAMID

Although communities vary considerably in structure, certain basic processes are common to them all. Among these processes are the cycling of nitrogen, carbon, water, and phosphorus, which are discussed in future plates. Another basic process is the flow of energy through the community, which is the subject of this plate. All organisms take in energy to stay alive, and energy flows unidirectionally through the community. Feeding relationships within a community form a pyramid, and in this plate we will see how this pyramid is created in an oceanic community.

As you look over the plate, you will see that it contains a large pyramid with five levels that are designated by letter. As the energy pyramid extends upward, its levels contain a decreasing number of organisms.

All organisms must obtain nutrients and energy from their environment in order to remain alive and reproduce. Biologists categorize organisms as producers or consumers according to their means for acquiring energy. These terms are defined in the plate on ecological communities. We begin our study of the energy flow in an oceanic community by focusing on the feeding habits of members in the community. Each step of the pyramid is called a trophic level; at the bottom is the **first trophic level (A).** A light color should be used to shade this broad area.

The organisms in this first trophic level are **phytoplankton (A_1)** and the entire collection of phytoplankton is known as biomass. The box shows several of these, and they may be colored in a medium color that's similar to the one used for the trophic level. Phytoplankton are primary producers, that is, they generate nutrients that are used by all other organisms in the community. By performing photosynthesis, they trap sunlight and convert it to the chemical energy contained in carbohydrates. The pyramid is widest at its bottom because the biomass of phytoplankton is greater than the biomass at any other level.

Look at the **second trophic level (B)** next. It consists of primary consumers in the form of **small crustaceans (B_1).** These microscopic arthropods feed on the phytoplankton, obtaining energy from them. But some energy is lost in the transfer, and this leakage of energy continues throughout the pyramid. For this reason, less energy is present in the second trophic level than in the first, and the biomass of crustaceans is less than the biomass of phytoplankton.

We have examined the first two trophic levels of the energy pyramid and have seen how energy enters the ecosystem and is subsequently concentrated. Passing to the next level, even more energy is lost and the biomass is further reduced.

We now pass to the **third trophic level (C),** which should be shaded with a pale color. Within this trophic level are a number of **herring (C_1),** which feed on the crustaceans in the second trophic level. As energy is transferred from the second level to the third, some of it is lost because it has been used up in metabolic processes of the crustaceans. The herring are secondary consumers, and their total biomass is less than the crustaceans, so the area of the third trophic level is less than that of the second.

We now proceed to the **fourth trophic level (D),** and the energy flow encounters a new consumer. Here we see the **mackerel (D_1),** which feed on the herring. Notice that there are many fewer mackerel in this trophic level. The mackerel are tertiary consumers, and again, this trophic level has significantly less energy than the previous one since energy was used by the herring for their metabolism. As you can see, the biomass of mackerel is less than that of herring.

At the uppermost portion of the pyramid is the **fifth trophic level (E).** Here we find a **shark (E_1),** which consumes mackerel. About 90% of the total energy of the mackerel trophic level is lost as we move to this final level. So little energy is left at the top that relatively few sharks can be nutritionally supported in ocean communities. The amount of food at the fourth trophic level acts as a limiting factor for the number of organisms at the fifth level. Very rarely does an energy pyramid have more than five levels.

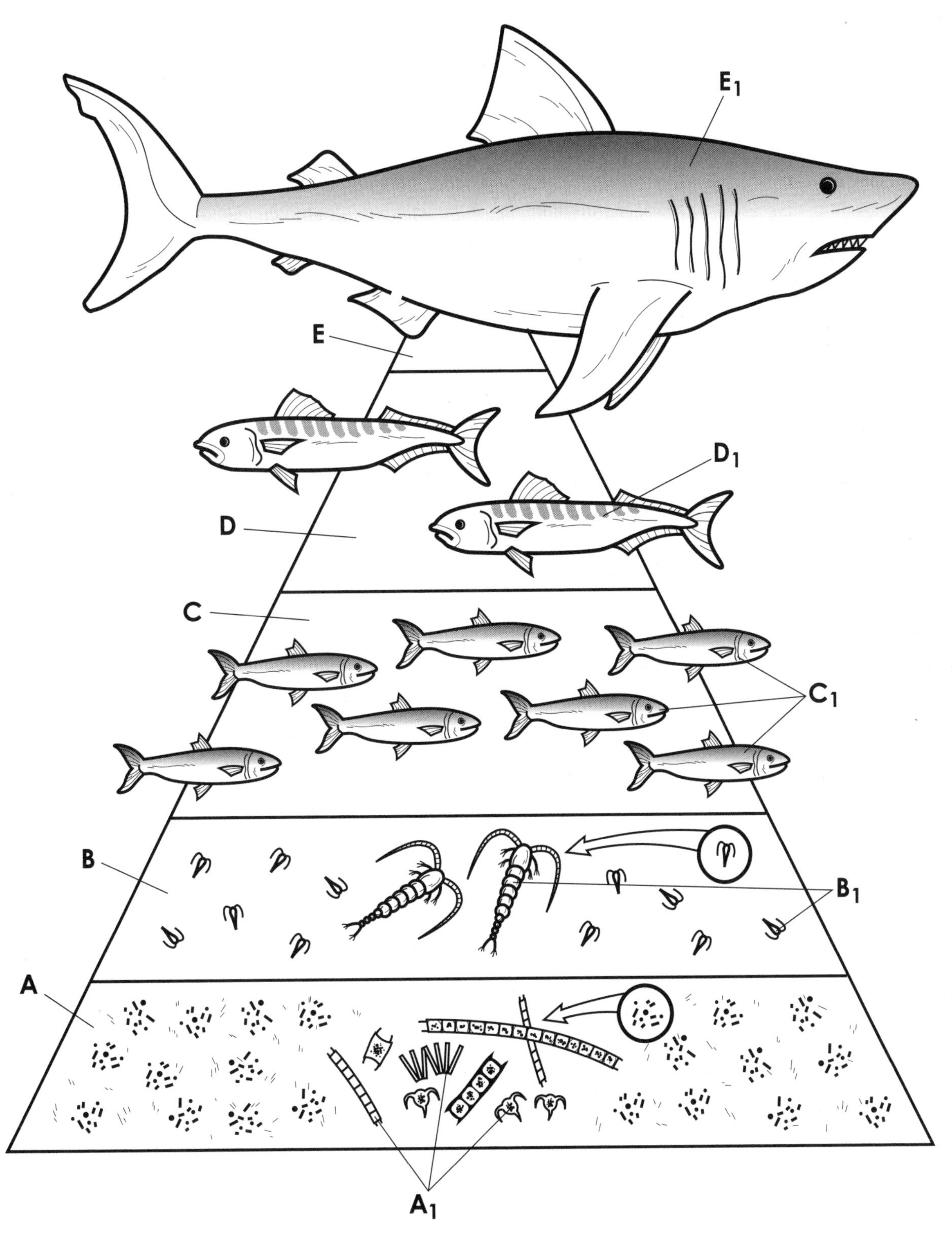

First Trophic Level A ❍
Phytoplankton A_1 ❍
Second Trophic Level B ❍
Small Crustaceans B_1 ❍
Third Trophic Level C ❍
Herring C_1 ❍
Fourth Trophic Level D ❍
Mackerel D_1 ❍
Fifth Trophic Level E ❍
Shark E_1 ❍

FOOD CHAINS

The food chain accounts for the flow of energy and the cycling of matter in an ecosystem, and is one of the processes that unites communities within ecosystems. Organisms derive energy and nutrients through the dynamics of the food chain, as we will see in this plate.

Looking over the plate, you will note that we present three different food chains. Each is a series of linkages in which we identify the individual organisms. You can compare these to the trophic levels in an energy pyramid, which are made up of groups of organisms.

In general, plants are eaten by animals, which are in turn eaten by other animals. In the first diagram, we show a highly simplified food chain that involves humans. Each link in the food chain is a trophic level.

In the first food chain, we start by looking at a **rice plant (A).** Plants are autotrophs, which means that they use photosynthesis to produce their own food. In the food chain, autotrophs are producers; they use the Sun's energy to produce carbohydrates, which are rich in energy. Rice plants also absorb minerals from the soil to produce inorganic matter for use by other organisms.

Humans (B) feed on the rice plants, and occupy the second trophic level. They are heterotrophs because they consume organic matter rather than producing it themselves, and are acting as herbivores because they consume plants. In this case, they are also the primary consumers, because they are feeding directly on the autotrophs.

The next organisms we see in diagram 1 are **microorganisms (C).** After humans die, microorganisms feed on their decaying bodies. These microorganisms, which include bacteria and fungi, are also called decomposers or detritivores, since they feed on detritus. The decomposed matter becomes nutrients in the soil that are consumed by the autotrophs, returning them to the food chain.

You should now have some idea of how the food chain operates. One organism feeds on autotrophs, and then becomes food for another organism. We will study a second food chain in diagram 2, in which humans play a different role. Continue your coloring as you read the following text.

In the second food chain, shown in diagram 2, the autotroph, or primary producer, is a **corn plant (D).** The corn plant provides food for the primary consumers, which in this case are **beef cattle (E).** The cattle, in turn, are food for humans (B). In this chain humans are acting as carnivores, since they eat meat. They are also secondary consumers in the food chain, and occupy the third trophic level. Since humans are herbivores in the first food chain and carnivores in the second, you can see that they are able to feed at several trophic levels; in other words, they are omnivores.

We now focus on a third food chain, one that exists in the ocean. Once again, humans are involved, but they occupy a high trophic level. Continue your reading below as you color the third diagram.

In the third section of art, we show a typical oceanic food chain. Here the autotrophic producers are microscopic organisms called **phytoplankton (F).** Phytoplankton are food for a number of microscopic, insect-like **crustaceans (G),** which are food for **fish (H),** which are the secondary consumers in the food chain. Humans (B) are at the fourth trophic level, and are tertiary consumers.

Much energy is lost as it travels from producers to tertiary consumers, so the fourth food chain is the least efficient—it contains the most participants. Unfortunately, humans could now subside on phytoplankton or crustaceans, which are low on the food chain.

These food chains are simplifications of the relationships inherent in ecosystems because many different organisms feed on producers and consumers along the chain. In addition, a consumer may be both a herbivore and a carnivore. The tangled mass of interconnections that evolves as a result of this is the food web.

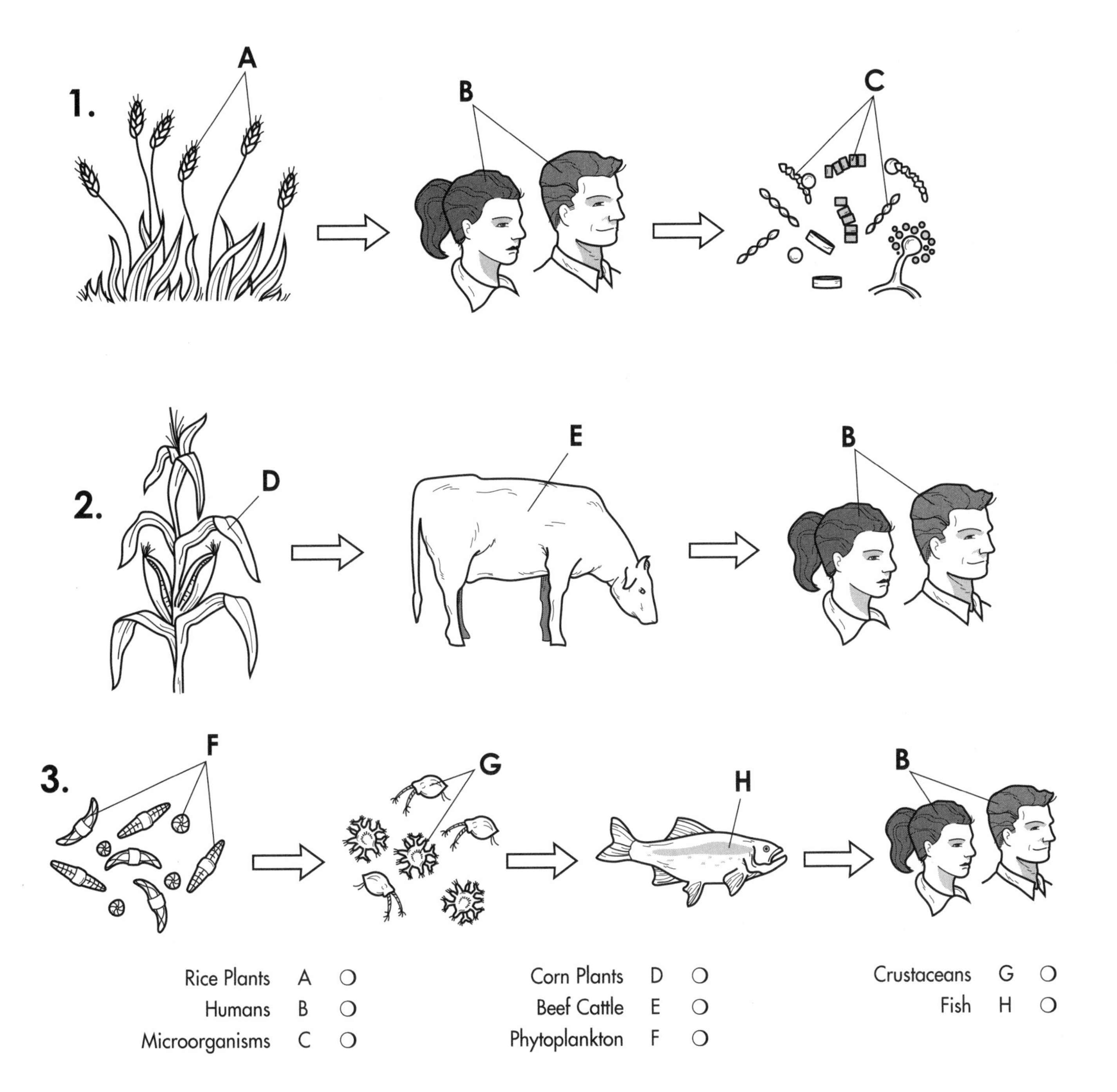
1.
A
B
C
2.
D
E
B
3.
F
G
H
B
Rice Plants A
Humans B
Microorganisms C
Corn Plants D
Beef Cattle E
Phytoplankton F
Crustaceans G
Fish H

A FOOD WEB

The sequence of relationships between predators and prey in a community manifests itself in a system known as the food chain, as we discussed in the previous plate. All of the food chains in an ecosystem form a food web. A food web includes the food sources and feeding patterns of all the organisms in a particular ecosystem.

Food chains are deceptively simple, and they do not necessarily reflect all of the interrelationships in nature. Food webs, by comparison, are quite complex, and as it may be impossible to identify all the feeding relationships that exist, they indicate only probable feeding relationships.

This plate consists of a single, large diagram that shows the relationships between numerous living organisms. Begin your work at the bottom of the diagram and work toward the top.

Communities that have complex food webs and contain large numbers of species are more subject to changes in size. In a food web, consumers rely on more than one species for sustenance, so it is unlikely that any one of these species will be eliminated through overexploitation. In simple systems, by contrast, fewer species of plants and animals live, and if one heavily-relied-upon species in the food chain disappears, the entire system might fall apart.

In this plate, we begin the food web at the bottom, where there are three different types of **producers (A).** The bracket indicating the three plants should be colored in a bold color. To the far left, we see the **grasses (A_1).** These engage in photosynthesis to produce simple and complex carbohydrates. They also extract minerals from the soil and incorporate them into organic materials.

In the center are the **water plants (A_2),** represented by a periwinkle. Living at the water's edge, these plants also produce carbohydrates. Like the grasses, they are known as autotrophs because they produce their own food. The **terrestrial plants (A_3)** are represented by goldenrods.

Having surveyed the producers in the food web, we now turn to the primary consumers. Your attention should now be directed to the next level as you continue your reading below.

Animals that feed only on plants are herbivores. In our food web, the **primary consumers (B)** are herbivores. The **field mouse (B_1)** feeds on grass, which is also consumed by the **grasshopper (B_2).** The grasshopper also feeds on periwinkles and terrestrial plants such as the goldenrod. As the diagram shows, two plants are consumed by the **butterfly (B_3),** and the goldenrod serves as a food source for both the butterfly and common **housefly (B_4).** Already we see that the flow of nutrients in a food web is more complex than that in a food chain.

We now focus on the higher levels of the food web. Notice that many of the primary consumers are prey for consumers that are higher on the food web. Your focus should be on the upper level of the plate and the animals in bracket C. Continue your reading as you examine the food web.

The animals that feed directly on the primary consumers are **secondary consumers (C).** The secondary consumers provide food for tertiary consumers; we in turn provide food for quaternary consumers.

Focus for a moment on the grasshopper. It is a primary consumer and is preyed on by the **frog (C_1),** a secondary consumer. The **snake (C_4)** is a tertiary consumer, since it eats the frog, and the **hawk (C_6)** is a quaternary consumer, since it eats the snake. The hawk (C_6) is also a secondary consumer because it feeds on the **field mouse (B_1).** Now focus on the **coyote (C_5).** This animal feeds on the field mouse, so it is a secondary consumer.

Consider the **dragonfly (C_2).** This animal is a secondary consumer since it feeds on the **housefly (B_4),** a primary consumer. Follow the food chain from the goldenrod, to the housefly, to the dragonfly, and on to the frog, snake, and hawk. Note that there are many trophic (feeding) levels in these relationships.

Conclude your work by focusing on the **bird (C_3).** It feeds on both the butterfly (B_3) and the housefly (B_4). It is clear that should the housefly disappear, the bird could continue to live on the butterfly. This is why the food web yields a stable population. If the bird relied solely on the housefly, then it would become extinct if the housefly disappeared.

We have not included all possible predator-prey relationships in this food web. You may complete the plate by drawing arrows between various animals that may feed on others. For example, the bird may be considered food for the hawk. See if you can determine any other possible relationships in this food web.

Producers	A	❍
Grasses	A_1	❍
Water Plants	A_2	❍
Terrestrial Plants	A_3	❍
Primary Consumers	B	❍
Field Mouse	B_1	❍
Grasshopper	B_2	❍
Butterfly	B_3	❍
Housefly	B_4	❍
Secondary Consumers	C	❍
Frog	C_1	❍
Dragonfly	C_2	❍
Bird	C_3	❍
Snake	C_4	❍
Coyote	C_5	❍
Hawk	C_6	❍

THE WATER CYCLE

Water is the most abundant substance in living things. The human body, for example, is composed of about 70% water, and jellyfish are 95% water. Water participates in many important biochemical mechanisms, including photosynthesis, digestion, and cellular respiration. It is also the habitat for many species of plants, animals, and microorganisms, and it participates in the cycling of all of the materials used by living things. Water is distributed through the biosphere in a cycle known as the water, or hydrologic, cycle. In this plate, we will examine some aspects of that cycle.

In this plate, we show the biosphere and several arrows that show the movement of water through it. Our primary emphasis will be on the arrows, and you should color them in darker colors than the other aspects of the biosphere.

We begin by looking at the atmosphere, which includes the clouds. When water vapor cools, it condenses and falls to Earth as rain. For instance, look at the arrow labeled **(A),** or **precipitation over land;** gravity draws the water back to Earth in the form of rain, sleet, and snow. Precipitation also occurs **over oceans (B).**

We have begun our discussion of the water cycle by showing how water reaches the Earth. We will now see how it is stored in living things before it is returned to the atmosphere. Continue your reading as you color the diagram, including its arrows.

The living on Earth are represented, in our diagram, by the trees. Water is absorbed by the roots of the trees and used in photosynthesis, but it is also lost from their leaves through the process of **transpiration (C).** Water also returns to the atmosphere through evaporation from the soil and from numerous other sources. In general, the amount of precipitation received by an area helps determine what types of plants will grow there. The nature of the vegetation, in turn, determines the types of animals that inhabit a region.

Water from the land enters the ocean through **seepage from the ground (D);** it percolates from the surface down to the water table. This water-saturated zone of soil and rock is called an aquifer, and water seeps from the aquifer to the ocean. Water also reaches the ocean as **runoff from the surface (E).** Runoff from the surface includes flow from rivers as well as melting snowfields and glaciers.

Now that we have described how water reaches the oceans, we will explore how it returns to the atmosphere, completing the hydrologic cycle. Continue reading below as you complete your coloring.

The major reservoirs of water on Earth are the oceans. Oceans cover about three-quarters of Earth's surface and contain about 97% of its water. Solar radiation causes water's **evaporation from the ocean (F).** Over 80% of the evaporated water in the hydrologic cycle enters the atmosphere in this way, and about 52% of this falls back into the oceans in the form of rain. The remainder remains in the atmosphere as clouds, ice crystals, and water vapor and then precipitates over land. On a global scale, the quantity of ocean water that evaporates each year is equivalent to a layer that's 120 cm deep and covers the entire surface of the ocean.

Precipitation over Land	A ❍		Transpiration	C ❍		Runoff from Surface	E ❍
Precipitation over Ocean	B ❍		Seepage from Ground	D ❍		Evaporation from Ocean	F ❍

THE CARBON CYCLE

Energy flows from the Sun into the biosphere, but nutrients do not enter the biosphere from an outside source. Essentially, the same pool of nutrients has circulated for the billions of years that the Earth has been in existence. Some nutrients, called macronutrients, are used by organisms in large quantities, while others, micronutrients, are used only in trace quantities. Macronutrients include carbon, hydrogen, oxygen, nitrogen, and phosphorus; micronutrients include iodine, iron, zinc, and some others.

Both macronutrients and micronutrients are recycled; they are passed back and forth between living and nonliving components of the ecosystem in processes that we call biogeochemical cycles. This plate and the ones that follow trace the pathways of several elements through biogeochemical cycles.

The prime focus of this plate is on the arrows that show how carbon travels among components of the biosphere. You should use darker colors for the arrows.

Material substances are incorporated into organic compounds by primary producers. Primary producers are then consumed by secondary consumers, and decomposers are ultimately responsible for releasing the material back into the nonliving environment.

We will begin our study of the carbon cycle with the **atmosphere (A),** which is Earth's major reservoir of carbon, in the form of carbon dioxide. Carbon enters the biotic (living) part of the excosystem through **photosynthesis (B).** We suggest a green color for the arrow. Plants of the **forest (C)** take the carbon in carbon dioxide and fix it in organic compounds such as glucose, starch, cellulose, and other carbohydrates. **Respiration in plants (D)** returns carbon dioxide to the atmosphere; an arrow shows this process.

We have seen how carbon enters the cycle of living things through photosynthesis, and we will now see how it passes through various life forms. Continue your reading below as you color.

Plants are primary producers. In the course of **plant consumption (E),** carbon passes into primary consumers, animals. When **animal consumption (F)** occurs, or when the primary consumer is eaten, carbon passes to a secondary consumer, represented by the lion in the plate. **Respiration (G)** takes place in cells of the primary and secondary consumers, and carbon is released back into the environment as carbon dioxide.

When the primary and secondary consumers die, their organic matter enters the soil through the process of **decay (H).** It is broken down by the decomposers, or **detritus feeders (I),** which are small animals and microorganisms that subsist on decaying matter such as fallen leaves, dead bodies, and animal waste. Earthworms, mites, centipedes, insects, and crustaceans are detritus feeders. Thus, **respiration in detritus feeders (J)** also returns carbon to the atmosphere.

We have seen how carbon cycles through various living things on Earth. We will now turn to a storage process for carbon in the soil. Continue your reading below as you complete the plate.

Throughout history, much carbon has been **converted to fossil fuel (K).** High pressure and temperature transform carbon-containing organic matter into coal, oil, and natural gas. **Fossil fuel processing (L)** follows. There are many **uses for fossil fuels (M).** Some power plants generate electricity using fossil fuels, and automobiles are powered by gasoline. The **products of the combustion (N)** of fossil fuels include carbon dioxide and other carbon compounds that enter the atmosphere. Carbon also enters the environment from the burning of wood and plants that occurs during **forest fires (O).**

A final aspect of the carbon cycle that we will examine is **exchange with oceans (P).** Some carbon dioxide from the air dissolves in oceans and combines with calcium to form calcium carbonate, which is incorporated into the shells of mollusks and other creatures. When these shells decay, they transform into limestone, which, over time, dissolves as it is exposed to water. Carbon is released from the limestone and may return to the atmosphere.

Atmosphere A ❍
Photosynthesis B ❍
Forest C ❍
Respiration in Plants D ❍
Plant Consumption E ❍
Animal Consumption F ❍

Respiration in Animals G ❍
Decay H ❍
Detritus Feeders I ❍
Respiration in Detritus Feeders J ❍
Conversion to Fossil Fuel K ❍

Fossil Fuel Processing L ❍
Uses for Fossil Fuel M ❍
Products of Combustion N ❍
Forest Fire O ❍
Exchange with Oceans P ❍

THE NITROGEN CYCLE

An important process in ecosystems is the recycling of nitrogen through its living (biotic) and nonliving (abiotic) components. The living components, or biota, of the ecosystem participate in the nitrogen cycle in a number of ways, as you will see in this plate.

> If you look closely at the plate, you will notice that we show the various ways in which nitrogen cycles through nature. As you color the plate, the arrows should be emphasized.

Approximately 78% of the air is composed of diatomic nitrogen. Nitrogen is essential to life because it is a key component of amino acids and nucleic acids. Even ATP, the basic energy currency of living things, contains nitrogen.

Neither plants nor animals can obtain nitrogen directly from the **atmosphere (A).** Instead, they must depend on a process called **nitrogen fixation (B).** Key players in nitrogen fixation are **legumes (C)** and the symbiotic bacteria that are associated with their root nodules. Legumes include clover, peas, alfalfa, and soybeans. The bacteria associated with their root nodules are **nitrogen-fixing bacteria (D).** These bacteria convert nitrogen in the soil to ammonia (NH_3), which can be taken up by some plants. The bacteria and the plant are in a symbiotic relationship. Cyanobacteria are also nitrogen-fixing bacteria; they are prominent in aquatic ecosystems.

> We have seen how nitrogen is brought into the biotic component of the ecosystem via nitrogen-fixing bacteria. We will now focus on how nitrogen is cycled through the living aspects of the ecosystem.

Nitrogen is fixed into the soil through the actions of free-living bacteria and, as we mentioned above, through bacteria that's associated with root nodules of legumes. Both of these methods of fixing nitrogen lead to its incorporation into ammonia (NH_3) in the process known as **ammonification (E).** The soil is a major reservoir for ammonia and other nitrogen-containing compounds. After nitrogen has been fixed, other bacteria convert it into nitrate, in a process called **nitrification (F).** In the first step of nitrification, ***Nitrosomonas*** **(G)** convert ammonia to nitrate (NO_2), and in the second step, nitrite is converted to nitrate (NO_3), by ***Nitrobacter*** **(H).** The nitrate (NO_3) is then **consumed by plants (I),** as the diagram shows.

But not all plants consume nitrate; as we mentioned before, some plants are able to use the ammonia from the soil. In both cases, nitrogen enters the primary producers in the biotic community. The plants may then be **consumed by animals (J).** Herbivores are the primary consumers, and the nitrogen of the plants is used for the synthesis of key organic compounds such as amino acids, proteins, and nucleic acids.

> We have seen how nitrogen is fixed in the soil and eventually utilized by plants and then animals. We will now complete the cycle of nitrogen by showing how it returns to the atmosphere. Continue your reading as you color the final aspects of the plate.

The final aspect of the nitrogen cycle is the process of **denitrification (K).** This process is performed by a variety of microscopic bacteria, fungi, and other organisms. Nitrates in the soil are broken down by these organisms, and nitrogen is released into the atmosphere (A). This completes the nitrogen cycle.

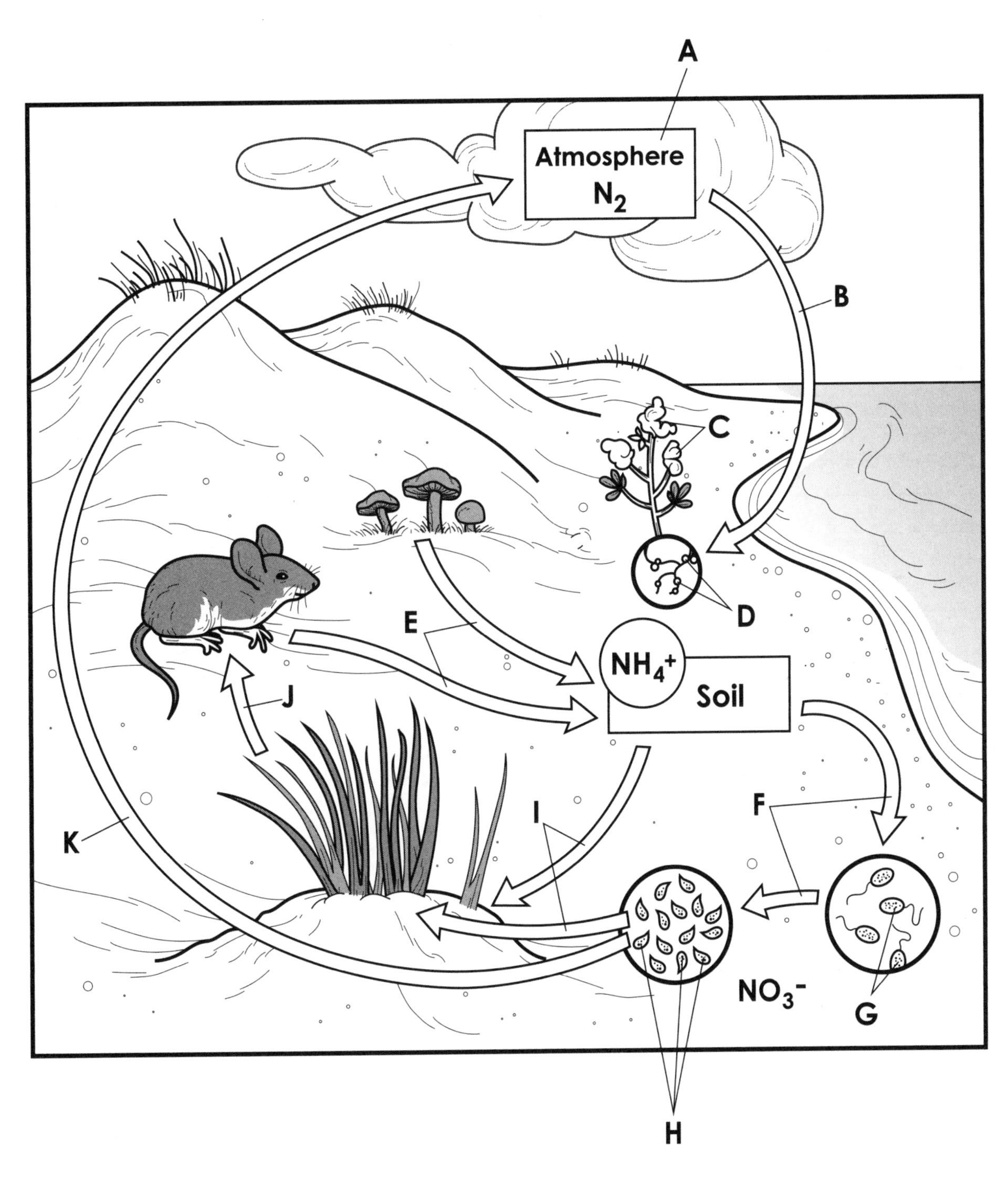

Atmosphere	A	❍	Ammonification	E	❍	Consumption by Plants	I	❍
Nitrogen Fixation	B	❍	Nitrification	F	❍	Consumption by Animals	J	❍
Legume Plant	C	❍	*Nitrosomonas*	G	❍	Denitrification	K	❍
Nitrogen-Fixing Bacteria	D	❍	*Nitrobacter*	H	❍			

THE PHOSPHORUS CYCLE

Although nitrogen and carbon exist as gases, certain elements that cycle in the biosphere do not exist in gaseous form. These elements accumulate in rocks and soil, and participate in what are called biogeochemical cycles.

Among the elements that undergo sedimentary cycles are calcium, sulfur, magnesium, and phosphorus. As you will see in this plate, phosphorus is one of the key elements in organic matter.

In this plate, we will follow the cycling of phosphorus in nature. The arrows should be the most prominent feature in the final, colored drawing.

Phosphorus is one of the critical elements in biological molecules. For example, it is a component of adenosine triphosphate (ATP) and the coenzyme $NADP^+$, which are used in important cellular processes such as photosynthesis. Phosphorus is also present in the sugar-phosphate backbone of nucleic acids, and is an essential element of phospholipids, which make up the cell membrane.

The main reservoir of phosphorus is rock and soil, so we will begin the cycle with **erosion from rocks (A).** Erosion occurs as water rushes over rock, dissolving phosphorus and washing it into rivers and streams. Phosphorus unites with oxygen to form phosphate and enters a major body of water, depicted here as a lake.

In the plate, we see plants growing along the border of the lake. Here, the water gives up its phosphates, which are **absorbed by the plants (B)** and used in the synthesis of organic molecules. Some of the phosphates also enter the soil along the margins of the lake. Dissolved phosphate is readily absorbed by the roots of plants, concentrated by cyanobacteria and protists such as *Euglena*, and then incorporated into organic molecules.

The plant is the primary producer in the phosphorus cycle. The phosphate is **concentrated in plant tissues (C),** and then the **plant is consumed (D)** by an animal, which is seen **grazing (E).**

Phosphates are returned to the lake when the plants and animals die. **Plant waste (F)** and **animal waste (G)** return phosphate to the water. Once again it may be reabsorbed by the plants that line the lake, and it enters the cycle again.

Having explored the passage of phosphorus through various aspects of the biosphere, we now turn to the marine environment to see how phosphorus cycles there. Continue your reading as you color the plate.

Large amounts of phosphorus are carried by rivers and streams as **runoff to the ocean (H).** Phosphorus exists in the form of phosphate here, as it does on land. Much of this phosphate then concentrates in **marine sediment (I).** Some of the phosphate is eventually incorporated into the bodies of marine animals such as a fish. For example, the scales and bones of bony fish contain phosphorus. As is the case on land, primary producers in the ocean incorporate phosphates into organic compounds. These primary producers are eaten by fish and other invertebrates. For instance, sea birds consume the fish and return phosphorus to the ocean in the form of excrement.

As we have seen, the atmosphere is not involved in the phosphorus cycle. In order for the phosphorus to leave the oceanic environment, geologic **upthrust (J)** must occur. Upthrust is the process through which once-submerged sedimentary rock rich in phosphorus is exposed because of the movement of the Earth's plates. This rock then enters the terrestrial ecosystem and begins to weather, participating in the phosphorus cycle.

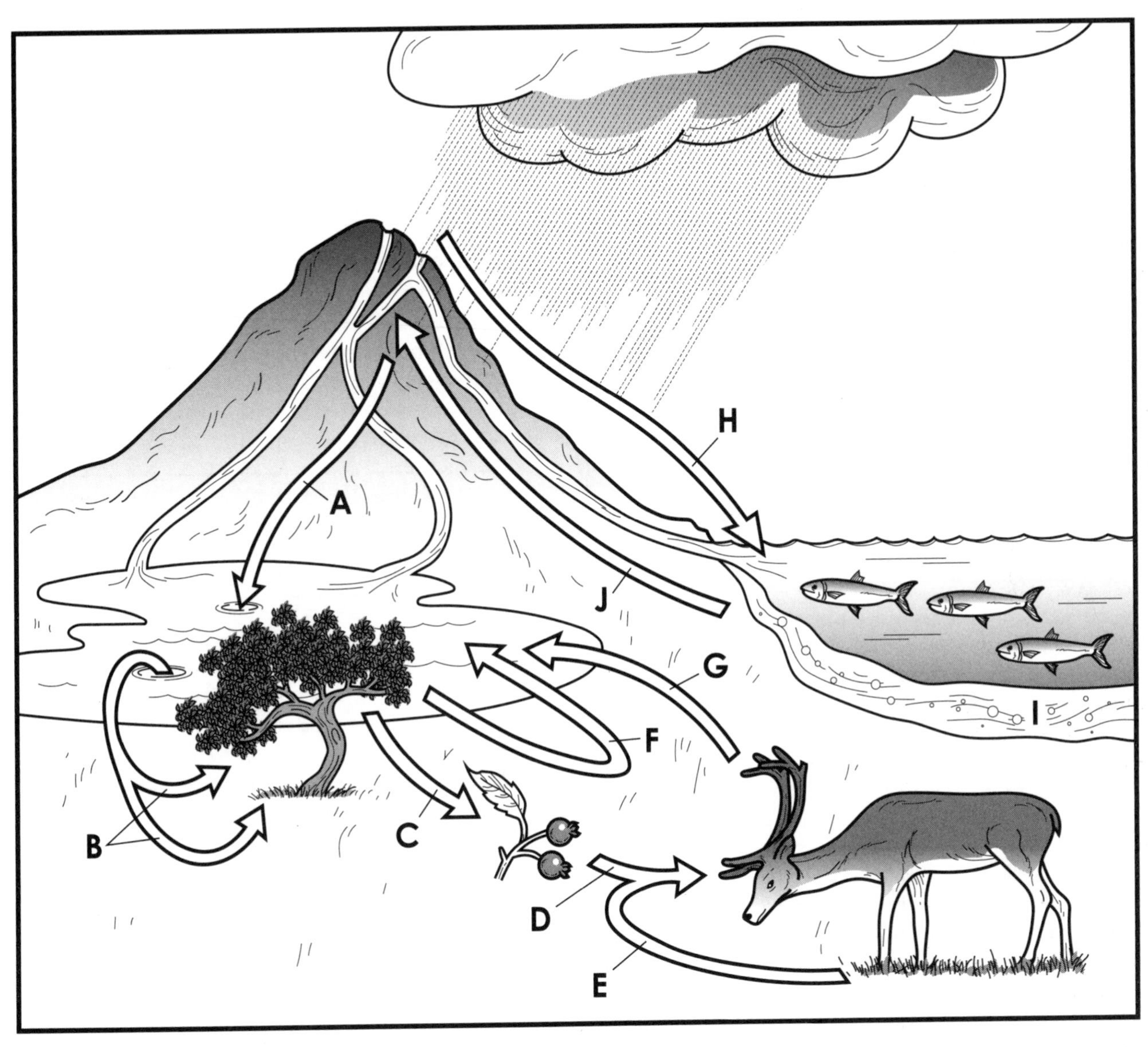

Erosion from Rock	A	❍
Absorption by Plants	B	❍
Concentration in Plant Tissues	C	❍
Plant Consumption	D	❍
Grazing	E	❍
Plant Waste	F	❍
Animal Waste	G	❍
Runoff to Ocean	H	❍
Marine Sediment	I	❍
Geologic Upthrust	J	❍

THE GREENHOUSE EFFECT

In your study of the carbon cycle in an earlier plate, you saw how carbon is cycled through many aspects of the biosphere including the atmosphere, plants, animals, and microorganisms. Under ideal circumstances, the processes of carbon dioxide uptake and release from the atmosphere would operate at equal rates. Recent decades, however, have witnessed a steady, gradual increase in the carbon dioxide content in the atmosphere. This has led to a frightening phenomenon called the greenhouse effect.

This plate describes the roots of the ecological phenomenon known as the greenhouse effect. In the art, you can see the surface of the Earth, the atmosphere, which includes a cloud layer, and the ozone layer.

We will begin our discussion of the greenhouse effect with the **Sun (A).** The Sun is the source of virtually all the energy that reaches Earth. Energy from the Sun drives photosynthesis and travels through various trophic levels, as we explained in the plate entitled The Energy Pyramid. Energy from the Sun passes through a vast expanse of **space (B)** to enter the **Earth's atmosphere (C).** Finally, the energy reaches the surface of **Earth (D).** Choose one color for the landmasses and a second, light color for the oceans.

We show the two possible fates of heat energy that comes from the Sun. Much of it is reflected from the face of the Earth; it is shown as the **Sun's reflected heat (E).** There energy that isn't reflected is absorbed as it passes through the atmosphere to reach the Earth's surface, and is shown as the **Sun's absorbed heat (F).** This energy will pass through the various trophic levels, and both forms of energy will participate in the greenhouse effect.

We will now explore some of the gases that contribute to the greenhouse effect. Your attention should be on the three types of clouds that are shown in the plate.

Approximately 300 million years ago, during the Carboniferous Period, huge quantities of carbon in the form of dead plants and animals were buried in the Earth. Time, heat, and pressure converted these carbon compounds into fossil fuels such as coal, oil, and gas. With the Industrial Revolution in the 1800s, the fossil fuels were burned in power plants, factories, and automobiles, and huge amounts of **carbon dioxide (G)** were released into the atmosphere. Clouds of carbon dioxide began to accumulate, causing the atmospheric content of carbon dioxide to increase by about 25%.

Another gas that contributes to the greenhouse effect is **methane (H).** Clouds of methane are released from marshes, during the burning of forests, and from fermentation occurring in the digestive tracts of grazing animals, which is expelled as flatulence.

A third type of gas that contributes to the greenhouse effect is **chlorofluorocarbon (I).** Chloroflurocarbons include industrial gases such as refrigerants, solvents, propellants, and plastic foams. Discarded and leaky units permit these gases to escape into the atmosphere.

Having identified the contributing factors, we will now see how they create the greenhouse effect. Continue your coloring as you read the text below.

In a greenhouse, glass enclosures allow sunlight to enter, and once it has entered, its energy is converted into heat. The heat is then trapped inside the greenhouse so that the temperature of the air inside increases. The greenhouse effect works in the same way. Heat is constantly given off by the processes that occur on the Earth, and some of this **heat escapes (J)** through the cloud cover of carbon dioxide. However, much of the **heat is reflected (K)** from this increased carbon dioxide cloud cover and returns to the Earth.

Many scientists believe that the greenhouse effect may cause a rise in the average global temperature as the decades pass. They assert that the Arctic ice cap and glaciers may eventually melt, which would cause sea levels to rise and coastal wetlands to flood. The result of this decrease in size of the continents and the cooling of the climate would be disastrous for many life-forms.

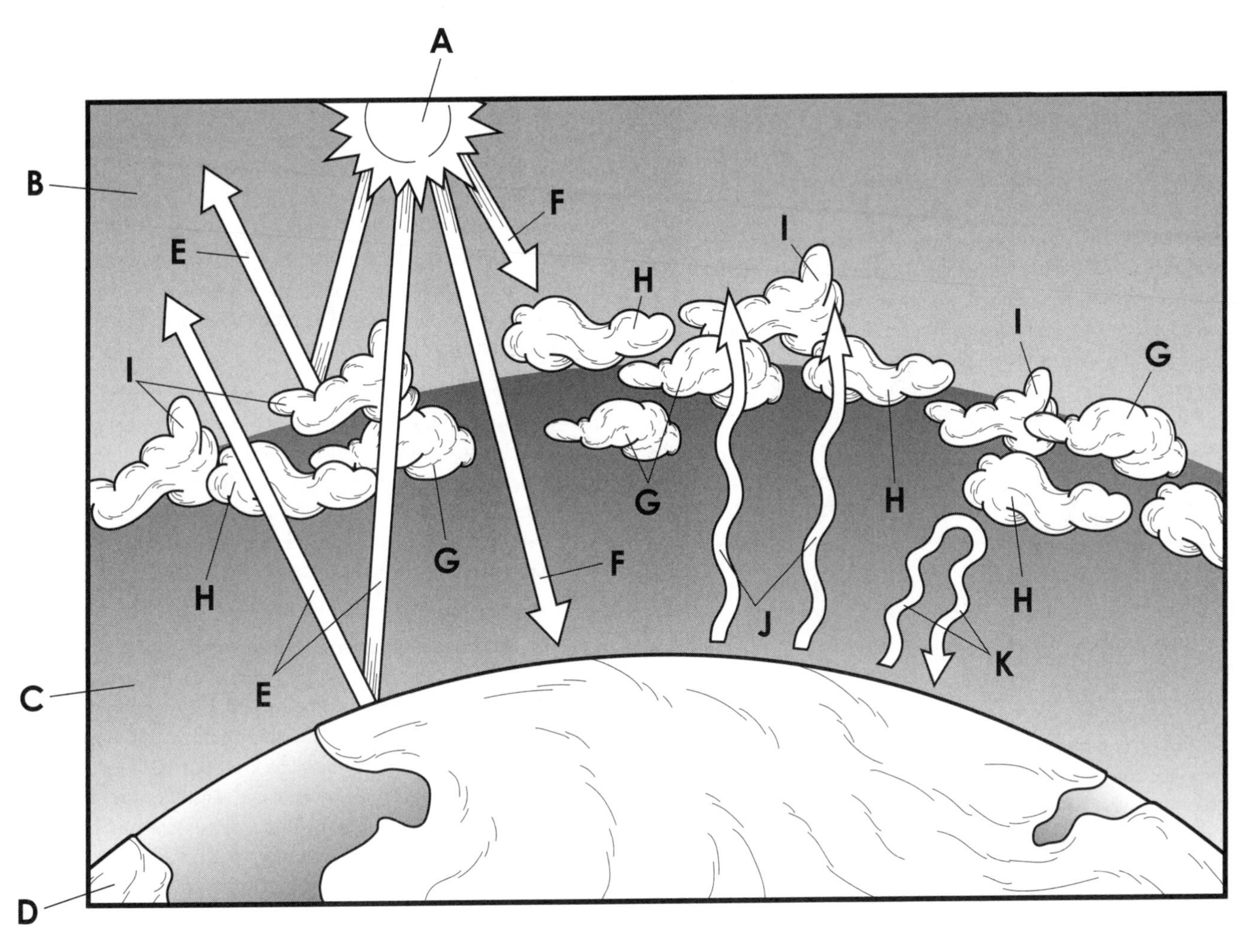

Sun	A	❍	Sun's Reflected Heat	E	❍	Chlorofluorocarbons	I	❍
Outer Space	B	❍	Sun's Absorbed Heat	F	❍	Earth's Heat Escaping	J	❍
Earth's Atmosphere	C	❍	Carbon Dioxide	G	❍	Earth's Reflected Heat	K	❍
Earth	D	❍	Methane	H	❍			

HUMAN POPULATION ECOLOGY

An important aspect of ecology is the study of the dynamics of the human population. The human population consists of individuals of species living together in an environment. To understand the dynamics of the human population, ecologists study its density, the distribution of its individuals, and its growth rate. The density of the population is influenced by factors such as natality (birth into a population), immigration, mortality (the death of individuals), and emigration (movement out of an area).

Population biologists study the age-sex structure of a population to help them predict future changes in it. They construct bar graphs using five-year age categories and sexes, as shown in this plate. The result is a population pyramid that can be broad-based, inverted, or thin at the margins. We will use the population pyramid to study three populations and their future potentials.

This plate displays the population pyramids associated with three countries in 1980: Mexico, the United States, and Sweden. We will see how the age-sex distributions of the three countries have an effect on their structures.

Five-year categories are displayed in the three population pyramids shown in this plate. Ecologists place males to the left of the central, vertical axis, and females to the right (shaded region). The rate of population growth is influenced by the numbers of individuals that are of reproductive age.

Begin your work by focusing on the pyramid to the left, which represents **Mexico (A).** A darker color can be used to color the box at the top, and variations of that color can be used to denote **Mexican males (A_1)** to the left of the axis, and **Mexican females (A_2)** to the right of the axis.

The population pyramid for Mexico is broad at the base and narrow at the top, which is typical of a developing country. Note that the majority of individuals are under the age of five. Ecologists predict that in the future, the population of Mexico will increase dramatically, because the country has more people yet to reproduce than people who have already reproduced or are presently reproducing. The population of Mexico is potentially explosive and will probably grow faster than that of other countries. Here the number of children that are less than 15 years old exceeds 40% of the total population.

We now move to the second pyramid, which represents the United States of America. Continue your coloring as you read about this country in the paragraphs below. Three new colors should be used for this pyramid.

In the second pyramid, we see the age-sex structure for the **United States of America (B)** as of 1980. **American males (B_1)** are shown to the left of the axis, and **American females (B_2)** to the right. This is an example of a relatively stable population, the number of prereproductive individuals nearly balances the number of postreproductive individuals.

Several items in the United States 1980 bar graph are notable. Note that in the age group 45–49, the number of individuals is much smaller than in the group of 20–24. The former represents the small number of children born in the Depression era, while the latter represents the large number of children born during the post–World War II "baby boom." The forecast is that our population will not expand rapidly in the decades ahead, because similar numbers of reproductive and prereproductive individuals exist.

Having contrasted the populations of developing and developed countries, we now examine the age-sex structure of a population that is declining. This is the population of Sweden. Continue your coloring as before, using three new colors to represent the third example of population ecology.

A declining population is represented by the third population pyramid. This bar graph represents the population of **Sweden (C). Swedish males (C_1)** are at the left, and **Swedish females (C_2)** are at the right. In declining populations, the pyramids are tapered at the bottom. They show that more individuals who have reproduced exist than do individuals of prereproductive age. Recent population increases have been comparatively slow in Sweden, so that the base of the bar graph is narrow. People tend to live longer in the developed countries, so the bar graph tends to be wider toward the top of the pyramid. Predictors of population fluctuations forecast a continued decline in the population of this country.

It should be noted that in recent years, the birth rate in Mexico has declined. However, a large number of individuals have entered their reproductive years, so that a great population increase is anticipated. By contrast, the birth rates of the United States and Sweden remain low, and fewer individuals are entering their reproductive years.

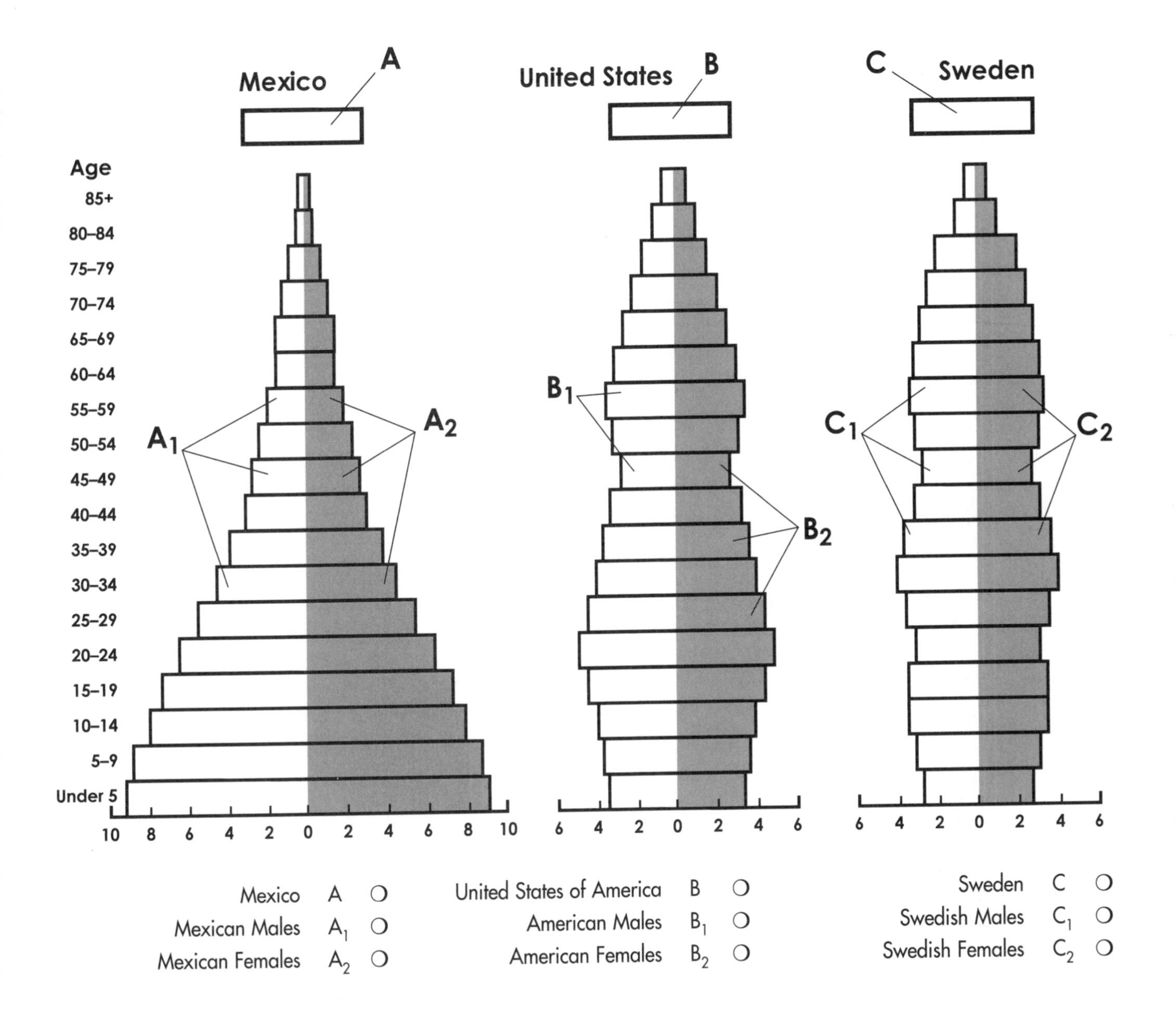

Mexico A ❍
Mexican Males A_1 ❍
Mexican Females A_2 ❍

United States of America B ❍
American Males B_1 ❍
American Females B_2 ❍

Sweden C ❍
Swedish Males C_1 ❍
Swedish Females C_2 ❍

INDEX

Symbols

A

B

C

D

E

F

G

H

I

J

K

L

M

N

O

P

Q

R

S

T

U

V

W

X

Y

Z

NOTES

1 tomatoes

2 Corn

3 potatoes

4 Pork

5 apples

6 beef

7

NOTES

NOTES

NOTES

NOTES

NOTES

NOTES

NOTES

International Offices Listing

China (Beijing)
1501 Building A,
Disanji Creative Zone,
No.66 West Section of North 4th Ring Road Beijing
Tel: +86-10-62684481/2/3
Email: tprkor01@chol.com
Website: www.tprbeijing.com

China (Shanghai)
1010 Kaixuan Road
Building B, 5/F
Changning District, Shanghai, China 200052
Sara Beattie, Owner: Email: sbeattie@sarabeattie.com
Tel: +86-21-5108-2798
Fax: +86-21-6386-1039
Website: www.princetonreviewshanghai.com

Hong Kong
5th Floor, Yardley Commercial Building
[illegible] Connaught Road West, Sheung Wan, Hong Kong
(MTR Exit C)
Sara Beattie, Owner: Email: sbeattie@sarabeattie.com
Tel: +852-2507-9380
Fax: +852-2827-4630
Website: www.princetonreviewhk.com

India (Mumbai)
Score Plus Academy
Office No.15, Fifth Floor
Manek Mahal 90
Veer Nariman Road
Next to Hotel Ambassador
Churchgate, Mumbai 400020
Maharashtra, India
Ritu Kalwani: Email: director@score-plus.com
Tel: + 91 22 22846801 / 39 / 41
Website: www.score-plus.com

India (New Delhi)
South Extension
K-16, Upper Ground Floor
South Extension Part–1,
New Delhi-110049
Aradhana Mahna: aradhana@manyagroup.com
Monisha Banerjee: monisha@manyagroup.com
Ruchi Tomar: ruchi.tomar@manyagroup.com
Rishi Josan: Rishi.josan@manyagroup.com
Vishal Goswamy: vishal.goswamy@manyagroup.com
Tel: +91-11-64501603/ 4, +91-11-65028379
Website: www.manyagroup.com

Lebanon
463 Bliss Street
AlFarra Building - 2nd floor
Ras Beirut
Beirut, Lebanon
Hassan Coudsi: Email: hassan.coudsi@review.com
Tel: +961-1-367-688
Website: www.princetonreviewlebanon.com

Korea
945-25 Young Shin Building
25 Daechi-Dong, Kangnam-gu
Seoul, Korea 135-280
Yong-Hoon Lee: Email: TPRKor01@chollian.net
In-Woo Kim: Email: iwkim@tpr.co.kr
Tel: + 82-2-554-7762
Fax: +82-2-453-9466
Website: www.tpr.co.kr

Kuwait
ScorePlus Learning Center
Salmiyah Block 3, Street 2 Building 14
Post Box: 559, Zip 1306, Safat, Kuwait
Email: infokuwait@score-plus.com
Tel: +965-25-75-48-02 / 8
Fax: +965-25-75-46-02
Website: www.scorepluseducation.com

Malaysia
Sara Beattie MDC Sdn Bhd
Suites 18E & 18F
18th Floor
Gurney Tower, Persiaran Gurney
Penang, Malaysia
Email: tprkl.my@sarabeattie.com
Sara Beattie, Owner: Email: sbeattie@sarabeattie.com
Tel: +604-2104 333
Fax: +604-2104 330
Website: www.princetonreviewKL.com

Mexico
TPR México
Guanajuato No. 242 Piso 1 Interior 1
Col. Roma Norte
México D.F., C.P.06700
registro@princetonreviewmexico.com
Tel: +52-55-5255-4495
+52-55-5255-4440
+52-55-5255-4442
Website: www.princetonreviewmexico.com

Qatar
Score Plus
Office No: 1A, Al Kuwari (Damas)
Building near Merweb Hotel, Al Saad
Post Box: 2408, Doha, Qatar
Email: infoqatar@score-plus.com
Tel: +974 44 36 8580, +974 526 5032
Fax: +974 44 13 1995
Website: www.scorepluseducation.com

Taiwan
The Princeton Review Taiwan
2F, 169 Zhong Xiao East Road, Section 4
Taipei, Taiwan 10690
Lisa Bartle (Owner): lbartle@princetonreview.com.tw
Tel: +886-2-2751-1293
Fax: +886-2-2776-3201
Website: www.PrincetonReview.com.tw

Thailand
The Princeton Review Thailand
Sathorn Nakorn Tower, 28th floor
100 North Sathorn Road
Bangkok, Thailand 10500
Thavida Bijayendrayodhin (Chairman)
Email: thavida@princetonreviewthailand.com
Mitsara Bijayendrayodhin (Managing Director)
Email: mitsara@princetonreviewthailand.com
Tel: +662-636-6770
Fax: +662-636-6776
Website: www.princetonreviewthailand.com

Turkey
Yeni Sülün Sokak No. 28
Levent, Istanbul, 34330, Turkey
Nuri Ozgur: nuri@tprturkey.com
Rona Ozgur: rona@tprturkey.com
Iren Ozgur: iren@tprturkey.com
Tel: +90-212-324-4747
Fax: +90-212-324-3347
Website: www.tprturkey.com

UAE
Emirates Score Plus
Office No: 506, Fifth Floor
Sultan Business Center
Near Lamcy Plaza, 21 Oud Metha Road
Post Box: 44098, Dubai
United Arab Emirates
Hukumat Kalwani: skoreplus@gmail.com
Ritu Kalwani: director@score-plus.com
Email: info@score-plus.com
Tel: +971-4-334-0004
Fax: +971-4-334-0222
Website: www.princetonreviewuae.com

Our International Partners

The Princeton Review also runs courses with a variety of partners in Africa, Asia, Europe, and South America.

Georgia
LEAF American-Georgian Education Center
www.leaf.ge

Mongolia
English Academy of Mongolia
www.nyescm.org

Nigeria
The Know Place
www.knowplace.com.ng

Panama
Academia Interamericana de Panama
http://aip.edu.pa/

Switzerland
Institut Le Rosey
http://www.rosey.ch/

All other inquiries, please email us at
internationalsupport@review.com